Introduction to Classical and Modern Optics

Second Edition

Introduction to
Classical and Modern
Optics

Jurgen R. Meyer-Arendt, M.D.

Professor of Physics, Pacific University

Prentice-Hall, Inc., Englewood Cliffs, New Jersey 07632

Library of Congress Cataloging in Publication Data

Meyer-Arendt, Jurgen R.
    Introduction to classical and modern optics.

    Bibliography: p.
    Includes index.
    1. Optics. I. Title.
QC355.2.M49   1984   535   83–10902
ISBN 0–13–479303–X

Editorial/production supervision
        and interior design: *Kathleen M. Lafferty*
Manufacturing buyer: *John B. Hall*
Cover design: *Diane Saxe*
Cover illustration: *Moiré pattern showing surface contour of a lacrosse ball, photographed using the arrangement illustrated in Figure 2.4-10. Courtesy of Gerald S. Komarnicky.*

Printed in the United States of America

10 9 8 7 6 5 4 3 2 1

ISBN 0-13-479303-X

Prentice-Hall International, Inc., *London*
Prentice-Hall of Australia Pty. Limited, *Sydney*
Editora Prentice-Hall do Brasil, Ltda., *Rio de Janeiro*
Prentice-Hall Canada Inc., *Toronto*
Prentice-Hall of India Private Limited, *New Delhi*
Prentice-Hall of Japan, Inc., *Tokyo*
Prentice-Hall of Southeast Asia Pte. Ltd., *Singapore*
Whitehall Books Limited, *Wellington, New Zealand*

Contents

Preface

THE PURPOSE AND EMPHASIS of this completely revised second edition remain the same: to provide a clear and readable *Introduction to Classical and Modern Optics*. Its level is intermediate, for advanced undergraduates and for a course spanning two semesters or the equivalent.

I have made the text as self-contained as practical, set out the motivation for each step, and avoided shortcuts of the it-can-be-shown-that type. Also, I have used extensive cross-references within the text to make it useful as a reference tool, after formal classwork has ended.

Compared with the first edition, there are numerous major changes. I have adopted the rational, *Cartesian sign convention*. This convention, long used in ophthalmic optics but until recently somewhat slighted in physics, is essential for the concept of *vergence* and is mandatory in any *lens design*. This dual connection identifies the two groups of readers to which this *Introduction to Optics* is addressed in particular: those interested in the *scientific and engineering applications of optics* and those preparing for the *ophthalmic professions*.

Another useful innovation is that virtually every chapter on *geometric optics* opens with the same triplet combination of lenses. Each time, the light progresses a little further, showing the logic in the sequence of topics:

from the propagation of light, to refraction, to aberrations, to the design of complex systems.

Likewise, every chapter on *physical optics* has been rewritten, most extensively the chapters on interference, diffraction, diffraction gratings, radiometry, absorption, and lasers. The theory of the Fabry–Perot interferometer, light scattering, and atomic spectra are derived step by step. Clerk Maxwell's equations and Fresnel's equations, discussed in detail, are now placed together in one chapter. Electron optics has been deleted.

As before, ample space has been allocated to classical topics such as thin lenses, aberrations, lens systems, and polarization. But modern subjects, such as the use of matrix methods in lens design, catadioptric systems, gradient-index lenses, fiber waveguides, integrated optics, multilayer antireflection coatings, Fourier transform spectroscopy, transfer functions, optical data processing, holography, and of course lasers and laser safety, are also presented in reasonable detail. Radiometry and photometry are discussed in the most recently adopted terminology. Relativistic optics, likely to play an important part in celestial navigation, is fun to read, aside from its utilitarian aspect.

As a *prerequisite,* all that is needed is a good background in general physics and a working knowledge of how to use a hand-held calculator (preferably including inverse trigonometric and exponential functions). A concurrent course in calculus is desirable though not essential.

Numerous worked-out *examples,* from penumbras to relativistic reflection, are interspersed throughout the text. *Problems,* ranging from easy to difficult, are found at the end of each chapter, following the *Suggestions for Further Reading.* Most of the examples and problems are new. All have been use-tested and modified where needed. (Answers to the odd-numbered problems are found in the back of the book.) None of the problems is intended as "busy work"; all are realistic and apply to practical situations; some, in fact, were drawn from consulting work I do from time to time. *Lecture demonstrations* are referred to on occasion. A great many *historical footnotes* have been included to reveal the human side of optics' great masters, adding color to the description of their accomplishments. (For the statistical-minded, there are 28 chapters, 368 illustrations, 197 examples, 510 problems, 134 historical footnotes, and 148 Suggestions for Further Reading, not counting the references.)

I have enjoyed writing this new edition. It took hundreds of hours of lecturing. It also took the patience of a great many individuals, questioning, challenging, attentive, and at times not so attentive, for me to discover —often by trial and error—how to get a concept across. How do I best present the idea of principal planes, the use of Cornu's spiral, the Poincaré sphere, the theory of stimulated emission? How do I present the fantastically wide field of optics at a reasonable level and within a reasonable length of time? There is no final answer, just steps of successive improvement.

Acknowledgments

Many of my colleagues have helped, in ways large and small, with the preparation and revision of the manuscript. Much of my appreciation goes to my students. But comments, both laudatory and critical, have come also from readers I have never met. Helpful suggestions were made in particular by R. Barer, Denis R. Holmes, Ernest V. Loewenstein, D. J. Lovell, Robin G. Simpson, and C. Michael Smith. Other help came from librarians Nancy Blase and Laurel Gregory, from physicians Henry B. Garrison and Albert Starr, from Susan R. Palmer and R. Michael McClung who once more checked the typescript, and from Patricia R. Wakeling, Managing Editor of *Applied Optics*. I also wish to thank Doug Humphrey, Kathleen Lafferty, and the editorial and production staff of Prentice-Hall, Inc. Most of all, my thanks and appreciation go to Thomas B. Greenslade, Jr., of Kenyon College, Gambier, Ohio, one of the reviewers of the manuscript. Without his painstaking attention to detail, and his sometimes caustic remarks, my manuscript would never have become the book it now is.

Jurgen R. Meyer-Arendt

Introduction

I BEGIN THIS INTRODUCTION TO OPTICS by presenting right away a complex practical problem. Look at Figure 0-1. This is a combination of lenses, a lens *system*, containing a positive lens in front (on the left), a negative lens behind it, and another positive lens in the back (on the right). These lenses have certain surface characteristics, they have certain thicknesses, and they are certain distances apart. Probably, they are made out of different types of glass. By choosing these parameters correctly, we can obtain a system of superior performance. In fact, the system shown is the basis for some of the best, and best known, photographic camera lenses.

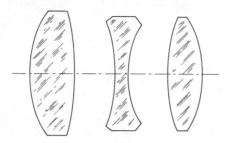

Figure 0-1 Example of optical system used for introducing various aspects of geometric optics.

How does the system work? Why are the lenses of the type shown preferable to other lenses? Why is this system superior to other systems? These are questions that we are not yet ready to answer. We will use this system as an example and a guide to introduce many of the concepts of optics.

More specifically, we will use this model to introduce *geometric optics*. Geometric optics is that part of optics where the wave nature of light can be neglected. *Physical optics* is more inclusive. In fact, image formation can be fully understood only by considering *wave optics* and by considering optical *transformations*. This is a very modern concept, fundamental to optical data processing, pattern recognition, and holography. In *quantum optics* we discuss spectra, absorption, and the ubiquitous laser. *Relativistic optics* points to the future.

Some phenomena in optics are easy to see. Others are very subtle. Tonight, look at a street lamp through the fabric of an open umbrella. You

Figure 0-2 Artist's conception of a star. Stained-glass window in St. John's Church, Herford, Germany, fourteenth centruy.

will see light *fans*, extending in various directions. These are due to *diffraction*.

Or look at a fairly bright star (without the umbrella). You may see faint light fans that seem to come right out of the star. In reality, they do not; the effect is due to your eye. Ordinarily, the pupil of the eye is perfectly round and no such fans are seen. But under certain conditions, the circumference of the pupil has a few straight segments and these cause the fans. And so, in the art of all ages, stars have traditionally been represented as objects with a multitude of points. However, a star with an odd number of points, as shown in Figure 0-2, is wave-optically impossible because the light fans must, by necessity, always occur in pairs.

This brief introduction to some areas of optics may have shown you that optics is a field of science that is lucid, logical, challenging, and beautiful. Most of our appreciation of the outside world—nature, art—comes to us through light. We will now proceed to discuss its many aspects in the way I have outlined.

A BIBLIOGRAPHY ON OPTICS

E. Hecht and A. Zajac, *Optics* (Reading, MA: Addison-Wesley Publishing Company, Inc., 1975).

F. A. Jenkins and H. E. White, *Fundamentals of Optics,* 4th edition (New York: McGraw-Hill Book Company, 1976).

C. S. Williams and O. A. Becklund, *Optics: A Short Course for Engineers & Scientists* (New York: Wiley-Interscience, 1972).

A. Nussbaum and R. A. Phillips, *Contemporary Optics for Scientists and Engineers* (Englewood Cliffs, NJ: Prentice-Hall, Inc., 1976).

W. H. A. Fincham and M. H. Freeman, *Optics,* 9th edition (Woburn, MA: Butterworth Publishers, Inc., 1980).

D. D. Michaels, *Visual Optics and Refraction,* 2nd edition (St. Louis: The C. V. Mosby Company, 1980).

If you want to work on some research project in optics, you should consult, in addition,

M. Born and E. Wolf, *Principles of Optics: Electromagnetic Theory of Propagation, Interference, and Diffraction of Light,* 6th edition (Elmsford, NY: Pergamon Press, Inc., 1980).

R. S. Longhurst, *Geometrical and Physical Optics,* 3rd edition (New York: Longman, Inc., 1974).

W. G. Driscoll and W. Vaughan, editors, *Handbook of Optics* (New York: McGraw-Hill Book Company, 1978).

R. G. Lerner and G. L. Trigg, editors, *Encyclopedia of Physics* (Reading, MA: Addison-Wesley Publishing Company, Inc., 1981).

You should also read, at least, the following journals:

Applied Optics
Journal of the Optical Society of America
Scientific American

Part 1

Elementary Treatment of Image Formation

As an introduction to geometric optics, and in order to see how images are formed by a lens, a mirror, or a telescope, we need to know certain general rules and mathematical formulations, and have an awareness of design parameters, aberrations, and bundle limitations. This approach is taken in the following *elementary treatment of image formation*.

1.1

The Propagation of Light

RETURNING TO OUR EXAMPLE of a complex optical system, consider first the light itself, before it enters the system. If the light comes from a point source, it spreads out in the form of divergent *rays*, shown in Figure 1.1-1. Rays are lines drawn in the direction of propagation. Rays as such do not exist, nor can they be isolated experimentally; they exist only in theory and on the chalkboard. All that we can isolate are *pencils*, narrow bundles of light. Hypothetical surfaces that are perpendicular to rays are called *wavefronts*. *Wavefront normals* are essentially the same as rays.

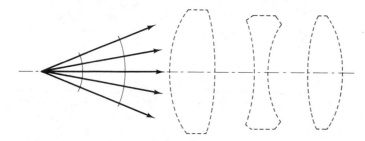

Figure 1.1-1 Divergent light before entering system. Note the *wavefronts* (curved solid lines on left).

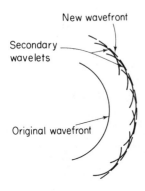

New wavefront

Secondary
wavelets

Original wavefront

Figure 1.1-2 Huygens' construction. In three-dimensional space, the wavefronts, as well as the wavelets, are spherical surfaces.

Rectilinear Propagation

Light as it advances through free space or through homogeneous, isotropic matter follows the *law of rectilinear propagation*. Look at sunlight on a misty morning as it breaks through the dense foliage of a tree. The light, made visible by the mist in the air, forms beautifully straight lines. The reason for such propagation derives from *Huygens' principle.* *

Huygens' principle states that each point on a wavefront may be considered the origin of new, secondary wavelets. The envelope over all these wavelets forms a new wavefront. Points on that wavefront again are the source of secondary wavelets, these form another wavefront, and so on. In this way the wave moves forward (Figure 1.1-2).

Shadows.   One of the consequences of the rectilinear propagation of light is the formation of *shadows*. With a point source, the boundary of a shadow is sharp (neglecting diffraction). With an extended source, a partial shadow, or *penumbra*, will form surrounding the full, inner shadow, or *umbra* (Figure 1.1-3). The size and shape of these shadows are found by similar triangles, as we see from the following example.

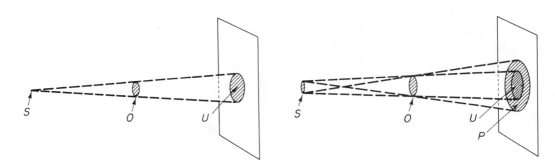

Figure 1.1-3   Shadows as formed by a point source (*left*) and by an extended source (*right*). S, light source; O, obstacle; U, shadow; P, penumbra.

---

*Christiaan Huygens (1629–1695), Dutch mathematician, physicist, and astronomer. Huygens made noteworthy contributions to many fields, invented the pendulum clock, formulated the laws governing centrifugal motion, discovered Saturn's sixth moon, Titan, and built a planetarium that still stands today in Leiden. In 1678 he presented the concept of what is now known as *Huygens' principle* to the French Academy of Sciences: C. Huygens, *Traité de la lumière* (Leiden: Van der Aa, 1690).

Example. *A diffuse, extended light source,*
shadow of a circular obstacle, 20 cm in diameter,
the light source the umbra is 40 cm in diameter, find:
(a) The distance between source and obstacle.
(b) The width of the penumbra.

9

Solution. Draw two parallel lines from the edges of the source to the screen (the dashed lines in Figure 1.1-4).

(a) Then from the construction it follows that

$$\frac{(O - S)/2}{x} = \frac{(U - S)/2}{300}$$

and, inserting the actual figures, we get

$$\frac{(20 - 10)/2}{x} = \frac{(40 - 10)/2}{300}$$

Cross-multiplication and solving for x gives

$$x = \frac{(5)(300)}{15} = \boxed{100 \text{ cm}}$$

which is the distance between source and obstacle.

(b) Since, in this example, the ratio of the distances source–obstacle and obstacle–screen is $1 : 2$ (no matter what the obliquity of the light), the width of the penumbra is

$$P = \frac{S}{x} (300 - x) = \frac{10}{100} (300 - 100) = \boxed{20 \text{ cm}}$$

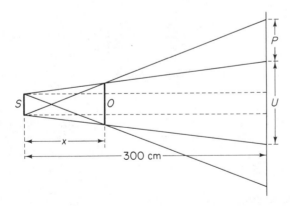

Figure 1.1-4

Path Length, Vergence, Sign Convention

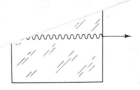

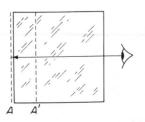

A A'

Figure 1.1-5 Whereas light passing through a block of glass encounters a *longer* path (*top*), to an observer it appears as if A has moved to A' and hence is closer (*bottom*).

Path length. Imagine that two points, A and B, are a certain distance, L, apart from each other. The distance as such has nothing to do with the medium that may be present between A and B. For example, if the two points were submerged in water, the distance between them would still be L.

Now consider that light is propagating from A to B. In air it takes the light a certain length of time to travel from A to B. But if A and B, and the path between them, are immersed in water, it will take the light *longer*. In fact, it will take the light n times as long to travel from A to B. The factor n is called the "refractive index" (of the water), an important term that we will discuss in more detail in the next chapter. Therefore, another "length" (in place of L) is needed to account for the time delay. This length is the *optical path length, S,* the product of length (distance) times refractive index:

$$S = Ln \qquad [1.1\text{-}1]$$

But then consider the situation from the opposite point of view (Figure 1.1-5). An observer may be looking at an object A through a medium of index n. Whereas in reality the object is in plane A, to the observer the object *appears* to be in plane A'. Both distances, the path length in air, Ln_0, and the path length in matter, $L'n$, are the same:

$$Ln_0 = L'n$$

and therefore, since the index of air $n_0 \approx 1$,

$$L' = \frac{L}{n} \qquad [1.1\text{-}2]$$

where L' is called the *reduced distance*.

Units. Since refractive index is a dimensionless quantity, all distances, actual distance, optical path length, and reduced distance are measured in the same units as length in general, preferably in meters or millimeters.

Example. *If two points, A and B, are 10 cm apart, their* actual distance *is $L = 10$ cm. But if the two points and the path between them are immersed in a liquid of refractive index 1.6:*
(a) What is the optical path length?

(b) What is the reduced distance *at which A will appear to an observer at B?*

Solution. (a) Whereas the actual distance from A to B is 10 cm, with a medium of index 1.6 present, it will take the light 1.6 times as long to go from A to B, and therefore

$$S = (10)(1.6) = \boxed{16 \text{ cm}}$$

(b) To an observer at B, plane A will appear at a distance of

$$L' = \frac{10}{1.6} = \boxed{6.25 \text{ cm}}$$

Sometimes it may seem difficult to decide when to use optical path length and when reduced distance. From the "point of view" of the light, use optical path length. From the point of view of the observer, use reduced distance.

Vergence.   Vergence is a concept whose *significance to geometric optics cannot be overemphasized*. Light that spreads out, as in Figure 1.1-6, left, is called *divergent*. Note that the wavefronts are concave toward left. Light that comes together is called *convergent*; its wavefronts are concave toward the right.

With different distances from the point of origin, or from the point of convergence, the curvature of the wavefronts changes. More precisely, the *curvature, C* (of any surface), is inversely proportional to the length of the *radius of curvature, R*:

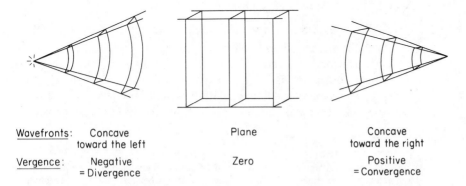

| Wavefronts: | Concave toward the left | Plane | Concave toward the right |
| Vergence: | Negative = Divergence | Zero | Positive = Convergence |

Figure 1.1-6   Wavefronts and vergence of light coming from a point source (*left*) and traveling toward a point image (*right*).

$$C \equiv \frac{1}{R} \qquad \qquad \text{[1.1-3]}$$

If the radius is longer, the curvature is less, and the surface is flatter.

The term *vergence* is essentially the same as "curvature." It is the reciprocal of a distance. However, vergence refers to wavefronts; that is, it refers to *light*. But since the light may propagate in a medium other than air, again the refractive index of that medium must be included:

$$\boxed{\mathbf{V} \equiv \frac{n}{L}} \qquad \qquad \text{[1.1-4]}$$

where $\mathbf{V}$ is called the *reduced vergence*.

Note: Whenever we use vergences, all distances must be given *in meters*. Therefore, the unit of vergence is the *reciprocal meter,* m^{-1}, often called a *diopter*.

Sign convention.
How do we make clear whether the wavefronts are concave to the left or concave to the right? To avoid ambiguity, there are certain rules, known as a *sign convention*. The two major sign conventions in optics are the *rational system,* based on Cartesian coordinates, and the *empirical system*. In principle we are free to choose whichever convention we prefer. But after we have made a choice, we must be consistent and adhere to the convention and use only equations that conform to it.*

In this Introduction to Optics we use the *rational* or *Cartesian sign convention*. The Cartesian sign convention has the advantage that it connects easily with the concept of vergence. It is also used generally in more advanced applications of optics, from lens design to polarization. Specifically, in the Cartesian system:

1. All figures are drawn with the light traveling from left to right.
2. All distances are measured *from* a certain reference surface, such as a wavefront or a refracting surface, located at (0, 0). The x axis, therefore, as in any x–y plot, extends from $x = -\infty$ to $x = +\infty$. Distances measured to the *left* of the reference surface, in a direction opposite to the direction of propagation of the light, are considered

---

*Convincing arguments in favor of the Cartesian sign convention are found in the Report of the Committee [of] *The Physical Society of London* [on] "The Teaching of Geometrical Optics" (Cambridge: Cambridge University Press, 1934).

negative. Distances measured to the *right*, in the direction of propagation of the light, are considered *positive*.

Accordingly, the radii of curvature of the wavefronts in Figure 1.1-6 are measured *from* the wavefronts, *toward* the center of curvature. If these radii extend to the *left*, the radii are *negative* and the light is *divergent* and the vergence *negative*. If the radii extend to the *right*, the radii are *positive* and the light is *convergent* and the vergence *positive*. If the wavefronts are plane, the rays are parallel and the light has zero vergence.

Example. *What is the vergence of light 2 m from a point source?*

Solution. Evidently, the light is *di*vergent and the radius of curvature of the wavefronts and the vergence are negative. Therefore,

$$\mathbf{V} = \frac{1}{-2 \text{ m}} = \boxed{-0.5 \text{ m}^{-1}}$$

If the light were *con*vergent, the radii and the vergences would be positive. For example, 8 cm *before* the light comes to a focus, its vergence is

$$\mathbf{V} = \frac{1}{+0.08 \text{ m}} = \boxed{+12.5 \text{ m}^{-1}}$$

Note: The more curved a wavefront, the higher the (absolute value of the) *vergence*.

How does the vergence change if we proceed, as in Figure 1.1-7, from one plane to another? In plane 1 the vergence is

$$\mathbf{V}_1 = \frac{n}{L}$$

In plane 2 it is

$$\mathbf{V}_2 = \frac{n}{L - d}$$

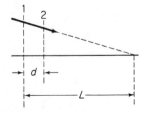

Figure 1.1-7 Deriving change of vergence.

Solving for L in the first equation, substituting in the second equation, and multiplying both numerator and denominator by $\mathbf{V}_1/n$ gives

$$\mathbf{V}_2 = \frac{n}{(n/\mathbf{V}_1) - d} = \frac{n(\mathbf{V}_1/n)}{(n/\mathbf{V}_1 - d)(\mathbf{V}_1/n)}$$

from where

$$V_2 = \frac{V_1}{1 - V_1(d/n)}$$

[1.1-5]

This is an important equation that we will use often as we go on (for example, on pages 51 and 74).

Some characteristics of light.

Under certain conditions, light can be considered to consist of *waves*. Therefore, we will examine some characteristics of waves. If we set a rope in oscillatory motion, waves will propagate along the rope. Actually, a given point on the rope moves only up and down. It does not move forward. What moves forward is the *wave configuration*. Wave configurations, in general, can move along in one, two, or three dimensions. In a rope, the wave moves along in one dimension. Surface ripples on a pond expand in two dimensions. Sound and light propagate in three dimensions. Some terms that we need to know are:

Amplitude, **A**, is the *height* of a wave above the average position. The amplitude may vary between a maximum of $+A$ and a maximum of $-A$ and assume any value in between (Figure 1.1-8).

Wavelength, λ, is a measure of *length*. It is the length of one full wave, from peak to peak, or from one point on a given wave to the equivalent point on the next wave. Wavelength is measured in the same units as length in general. The basic unit of length in the SI (Système International des Unités, International System of Units) is the meter, m. But since the wavelength of light is exceedingly short, fractional units of a meter are needed. These are:

1 millimeter, mm, 10^{-3} m.

1 mi'crometer, μm, 10^{-6} m. The prefix μ alone must not be used. (Do not confuse with micro'meter, accent on the o, which is an instrument for measuring small thicknesses.)

1 nanometer, nm, 10^{-9} m. This is the *preferred unit of wavelength of light* (in the visible part of the spectrum).

1 Ångström, Å, 10^{-10} m.*

---

*Named for Anders Jonas Ångström (1814–1874), Swedish spectroscopist and professor of physics and astronomy at the University of Uppsala. Ångström found that a cold gas absorbs the same wavelengths of light that it emits when hot, discovered hydrogen in the spectrum of the sun, carefully determined the wavelengths of all other lines he could see, and published the results in *Recherches sur le spectra solaire* (Uppsala, 1869).

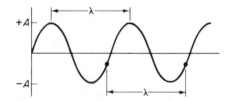

Figure 1.1-8 Amplitude and wavelength of a sinusoidal wave.

Frequency, ν, is the number of oscillations, or cycles, *per second.* The unit of frequency is the hertz,[*] Hz. For example, if an object makes 12 complete (back and forth) oscillations per minute, its frequency is

$$\nu = \frac{12 \text{ osc}}{1 \text{ min}} = \frac{12 \text{ osc}}{60 \text{ s}} = 0.2 \text{ Hz}$$

The *period, T,* of a wave is a measure of *time.* It is the time it takes the object to make one oscillation. Period and frequency, therefore, are reciprocal to one another:

$$T = \frac{1}{\nu} \qquad\qquad [1.1\text{-}6]$$

Velocity, v, in general, is the ratio of distance to time:

$$v = \frac{s}{t}$$

If $s = 1 \lambda$ and $t = T$, then

$$v = \frac{\lambda}{T}$$

But $1/T = \nu$ and hence

$$\boxed{v = \lambda\nu} \qquad\qquad [1.1\text{-}7]$$

This again is an important equation; it holds for any wave. The velocity of *light,* for example (which we will discuss in detail later, Chapter 3.2), is approximately 3×10^8 m s^{-1}.

---

[*]Named after Heinrich Rudolf Hertz (1857–1894), German physicist and professor of physics in Karlsruhe, later at the University of Bonn. Hertz found experimentally the electromagnetic waves that had been predicted by James Clerk Maxwell. When in 1924 the German Physical Society proposed the name "hertz" for the unit cycles per second, Walter Hermann Nernst (1864–1941), German physical chemist, objected, saying: "I do not see the need for introducing a new name; by the same reasoning one could as well call one liter per second a 'falstaff'."

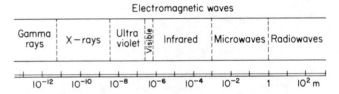

Figure 1.1-9 Electromagnetic specturm. The spectrum extends from gamma and X rays (*left*) to radio waves (*right*). Note how narrow the visible part is.

The electromagnetic spectrum.   Light is only part of the electromagnetic spectrum. The wavelength of visible light extends from about 380 nm for violet-blue to about 750 nm for deep red. "Pure" blue has a peak wavelength of about 475 nm, green of about 520 nm, yellow 575 nm, and red 630 nm. Below 380 nm there is the *ultraviolet,* UV, above 750 nm the *infrared,* IR (Figure 1.1-9).

X *rays* have wavelengths from as short as 10^{-12} m to about 50 nm. Ultraviolet of less than 200 nm is absorbed by air and therefore is called *vacuum ultraviolet. Microwaves* extend from 1 mm to 30 cm and *radio waves* from there up to perhaps 30 km wavelength.

All these limits are didactic and for tabulation only. They do not represent division lines. The properties of one category merge with those of the next, and the methods of production and detection overlap.

SUGGESTION FOR FURTHER READING

M. Young, "Pinhole Imagery," *Am. J. Physics* **40** (1972), 715–20.

PROBLEMS

1.1-1. What is the height of a tower that casts a shadow 48 m in length on level ground if the shadow of the observer, who is 1.80 m tall, is 3.2 m long?

1.1-2. If in a *pinhole camera* (Figure 1.1-10) the distance between an object, left, and its image, right, is 1.5 m, how long must the cam-

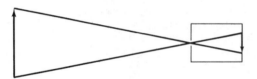

Figure 1.1-10

era be in order to make the image one-fifth the size of the object?

1.1-3. A small hole is made in the shutter of a dark room, and a screen is placed at a distance of 2 m from the shutter. If a tree outside is 40 m away from the shutter and if the tree casts on the screen an image 40 cm high, how tall is the tree?

1.1-4. A pinhole camera produces a 5-cm-high image of a telephone pole. Moving the camera 5 m farther away from the pole reduces the image size to 4 cm. To make the image 5 cm high again, the camera has to be made 4 cm longer. How tall is the telephone pole?

1.1-5. A light source is 10 cm in diameter and located 3 m to the left of a screen. A solid round object, 15 cm in diameter, is placed halfway between the light source and the screen.
(a) What is the diameter of the (central) shadow?
(b) What is the width of the penumbra?

1.1-6. A solid round body 20 cm in diameter is suspended from the ceiling of a room. When 80 cm above the floor, the body casts a (central) shadow 10 cm in diameter on the floor. The penumbra has a (total) diameter of 60 cm. If the light source is round and directly above the body:
(a) How high above the floor is the source?
(b) What is its diameter?

1.1-7. A plane-parallel slab of glass, 4 cm thick and of index 1.6, is laid over a sheet of newsprint. When you are looking down through the glass, by how much will the newsprint appear to be lifted up?

1.1-8. A cylindrical tube is closed at both ends by glass plates 10 mm thick. If the *inside* length of the tube is 7.5 cm and the refractive index of the glass $n = 1.5$, determine:
(a) The *outside* length of the tube.
(b) The optical path length between the outer surfaces if the tube is empty.
(c) The optical path length between the outer surfaces if the tube is filled with water ($n = \frac{4}{3}$).

1.1-9. A microscope is focused on a test target. When a sheet of glass 6 mm thick is placed over the target, the tube of the microscope must be raised 2.25 mm to bring the target back into focus. What is the refractive index of the glass?

1.1-10. Two beams of light pass through a distance of 8.5 cm. In the path of one beam is placed a 1.5-cm-thick plate of a transparent material. If the *difference* between the optical path lengths of the two beams is 7.2 mm, what is the refractive index of the material?

1.1-11. A spherical metal surface of -16-cm radius of curvature is immersed in a liquid of refractive index 1.76. What is the *curvature* of the surface?

1.1-12. Expanding spherical wavefronts of the same (-16 cm) radius of curvature pass through the same liquid ($n = 1.76$) as in Problem 1.1-11. What is the reduced *vergence* of the light?

1.1-13. Light originates at a point light source. What is the vergence of the light at distances of:
(a) 12.5 mm?
(b) 20 cm?
(c) 1.6 m?

1.1-14. Light from a source 10 cm in diameter falls on a flat circular screen, also 10 cm in diameter. If the distance from the source to the screen is 8 cm and if the refractive index of air can be neglected, what is the vergence of the light when reaching the screen?

1.1-15. A 4-m-long rope is set in oscillatory motion so that $2\frac{1}{2}$ waves are present on the rope at any one time. If the waves travel at a velocity of 10 m/s, what is the frequency of oscillation?

1.1-16. Ocean waves follow each other at distances of 40 m, a wave crest passing a given point every 5 s.
(a) What is the velocity of the waves?
(b) What is their frequency?

1.1-17. Yellow sodium light has an average wavelength of 589.3 nm.
(a) How many waves are there in 1 mm?
(b) What is the frequency of the light?

1.1-18. Find the frequency and the period of light of 625 nm wavelength.

1.2

The Refraction
of Light

FOLLOWING THE PATH OF THE LIGHT, we now come to the first surface of our system. Assume that the first surface is *plane* (Figure 1.2-1).

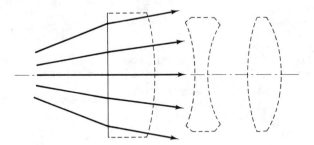

Figure 1.2-1 Light being refracted at a plane surface.

Introduction

If the medium behind the surface is clear, for example if it is glass or water, the light will go right through without much loss. A medium of this

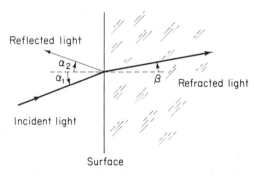

Reflected light

α_2
α_1

Incident light

β Refracted light

Surface

Figure 1.2-2   Reflection and refraction at the surface between two media. (In this example, air is assumed to be on the left of the surface, glass on the right.)

kind is called *transparent.* * If the light energy passes through but news-print could not be read through it, the medium is called *translucent.* An example is frosted or milky glass. If no light will go through, the medium is *opaque.*

Usually, part of the incident light is *reflected,* and part is *refracted.* Reflection means that the light is *returned* to the first medium from which it came. Refraction means that on entering the second medium the light *deviates* from its initial direction.

The point where a ray of light intersects the surface is called the *point of incidence.* A line constructed at this point, perpendicular or *normal* to the surface, is the *surface normal.* The angle subtended by the surface normal and the incident ray is called the *angle of incidence,* α_1 (Figure 1.2-2). The angle subtended by the surface normal and the reflected ray is the *angle of reflection,* α_2, and the angle subtended by the normal and the refracted ray is the *angle of refraction,* β.

Fermat's principle.   According to Fermat's principle, *light takes the path that requires the least time.*[†] This is an important theorem from which, in fact, all laws of reflection and refraction can be derived.

---

*We should realize, however, that the light does not simply "go right through." Light interacts with matter; it is being absorbed and almost instantly *reradiated,* a process that will occupy us further when we come to the scattering of light (Chapter 3.3) and to Fizeau's experiment (page 537).

[†]This theorem was first formulated, in 1657, by Pierre de Fermat (1601–1665), French jurist and mathematician, counsel to the Toulouse parliament. Fermat used to write his thoughts about analytic geometry, calculus, probability, and number theory into the margins of other books, pursuing his avocation mostly for his own enjoyment. In this he obviously succeeded. Five years after his death his notes were published by his son.

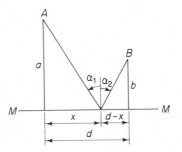

Figure 1.2-3 Fermat's principle for reflection.

Consider first the case of *reflection*. Light coming from a point A is reflected at a mirror MM toward a point B (Figure 1.2-3). Angles α_1 and α_2 are the angles of incidence and reflection, respectively. The distances of A and B from mirror MM are called a and b. From the construction in Figure 1.2-3 and from Pythagoras' theorem we find that the total path length, L, from A to MM to B, is

$$L = \sqrt{a^2 + x^2} + \sqrt{b^2 + (d - x)^2} \qquad [1.2\text{-}1]$$

For Fermat's principle to hold, the derivative of L with respect to x must be zero:

$$\frac{dL}{dx} = 0$$

Differentiating Equation [1.2-1] gives

$$\frac{dL}{dx} = \frac{1}{2}\frac{1}{\sqrt{a^2 + x^2}}2x + \frac{1}{2}\frac{1}{\sqrt{b^2 + (d - x)^2}}2(d - x)(-1) = 0$$

which is equivalent to

$$\frac{x}{\sqrt{a^2 + x^2}} = \frac{d - x}{\sqrt{b^2 + (d - x)^2}} \qquad [1.2\text{-}2]$$

Construction shows that this is equal to

$$\sin \alpha_1 = \sin \alpha_2$$

Fermat justified his principle of least time on the grounds that nature is "economical." Indeed, the time spent by light traveling from one point to another is most often a *minimum*. However, sometimes the path length is a *maximum*, or at least it is *stationary*. The essential point is that light follows a path whose length does not appreciably change as the path is changed.

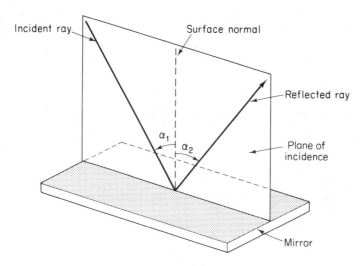

Figure 1.2-4 Law of reflection.

from which it follows that *the angle of incidence,* α_1, *is equal to the angle of reflection,* α_2,

$$\boxed{\alpha_1 = \alpha_2} \qquad [1.2\text{-}3]$$

which is (part of) the *law of reflection.*

In addition, the incident ray, the surface normal, and the reflected ray all lie in the same plane, called the *plane of incidence* (Figure 1.2-4).

Next we turn to *Fermat's principle for refraction* (Figure 1.2-5). Now we have a boundary, S–S, between two media but again the light goes from A to B. If the refractive index on both sides of the boundary were the same, no matter what its magnitude, the path from A to B would be a straight line. But if the two media have different indices, the path is no longer straight. Instead, the light is *refracted.*

We consider the velocities of the light, v_1 and v_2, on both sides of the boundary. Since $v = L/t$, the time it takes the light to traverse the two paths, from A to the point of incidence and from there to B, is

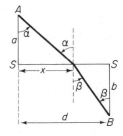

Figure 1.2-5 Fermat's principle for refraction.

$$t = \frac{\sqrt{a^2 + x^2}}{v_1} + \frac{\sqrt{b^2 + (d - x)^2}}{v_2} \qquad [1.2\text{-}4]$$

Again we differentiate Equation [1.2-4] with respect to x and set it equal to zero:

$$\frac{dt}{dx} = \frac{x}{v_1\sqrt{a^2 + x^2}} - \frac{d - x}{v_2\sqrt{b^2 + (d - x)^2}} = 0$$

Construction shows that this is equal to

$$\frac{\sin \alpha}{v_1} - \frac{\sin \beta}{v_2} = 0 \qquad\qquad [1.2\text{-}5]$$

The ratio of the velocities in the two media is called the *relative refractive index* (of medium 2 with respect to 1):

$$\frac{v_1}{v_2} = n_{21}$$

The term *absolute refractive index* refers to the same ratio but now v_1 is replaced by the velocity of light in a vacuum, c, and the subscripts denote the order in which the two media are traversed by the light:

$$\frac{c}{v_1} = n_1 \qquad \text{and} \qquad \frac{c}{v_2} = n_2 \qquad\qquad [1.2\text{-}6]$$

Rearranging Equation [1.2-5] and substituting Equations [1.2-6] solved for v yields

$$\frac{\sin \alpha}{\sin \beta} = \frac{v_1}{v_2} = \frac{c/n_1}{c/n_2} = \frac{n_2}{n_1}$$

The two terms

$$\boxed{\frac{\sin \alpha}{\sin \beta} = \frac{n_2}{n_1}} \qquad\qquad [1.2\text{-}7]$$

are known as *Snell's law of refraction.* * Snell's law shows that *light, on passing from a rarer medium* (of lower index) *into a denser medium* (of higher index) *is bent toward the surface normal*. If the light goes the opposite way, from a denser into a rarer medium, it is bent away from the normal: The direction of light traveling through a given path is *reversible*. Again, as in reflection, the incident ray, the surface normal, and the refracted ray all lie in the same *plane of incidence*.

Table 1.2-1 is a list of some representative refractive indices.

---

*Named after Willebrord Snel van Royen (1591–1626), Dutch astronomer and mathematician. At age 21, Snell succeeded his father as professor of mathematics at the University of Leiden. At 26, he determined the size of the Earth from measurements of its curvature between Alkmaar and Bergen-op-Zoom in Holland. Snell's original statement, based on empirical observation, was that the ratio of the cosecants of the angles of incidence and refraction is constant. The law in its present form is due to René Descartes (Renatus Cartesius) (1596–1650), French philosopher and mathematician, who published it in his *La Dioptrique*, 1637.

Table 1.2-1 INDICES OF REFRACTION (FOR SODIUM D LIGHT, 589 nm)

Vacuum	1.000000
Air	1.0003
Water	1.333
Spectacle crown, C-1	1.5230
Carbon disulfide	1.63
Extra-dense flint, EDF-3	1.7200
Methylene iodide	1.74
Diamond	2.42

Graphical construction. The incident ray, line 1 in Figure 1.2-6, intersects the surface at a point A. Erect a surface normal at A (line 2). Draw two arcs around A, making the lengths of their radii proportional to the two refractive indices, n_1 and n_2. Continue ray 1 until it intersects the n_1 arc. This gives point P. Draw line 3 through P and parallel to line 2 until it intersects the n_2 arc. This gives point B. Connect A and B and continue. This is the refracted ray (4).

Proof

$$\frac{PQ}{AP} = \sin \alpha, \qquad \frac{BC}{AB} = \sin \beta$$

$$PQ = BC$$

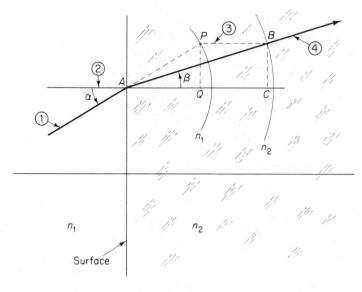

Figure 1.2-6 Snell's method of ray tracing through a plane surface. $n_1 < n_2$.

Therefore,

$$\frac{\sin \alpha}{\sin \beta} = \frac{PQ/AP}{BC/AB} = \frac{AB}{AP} = \frac{n_2}{n_1}$$

The molecular basis of refractivity. We have seen that light travels more slowly in a denser medium. A rigorous interpretation of this observation requires quantum mechanics. Perhaps it is sufficient to say that in free space light encounters no atoms, and therefore its velocity is highest. In matter, the velocity is still the same *between* atoms, but *near* atoms the light interacts with orbiting electrons and hence its velocity is *less*. The resultant velocity, then, is inversely proportional to the density of the atoms along the path. Since for free space $n \approx 1$, the density, ρ, is proportional to $n - 1$:

$$n - 1 = k\rho \qquad [1.2\text{-}8]$$

a relationship known as the *Dale–Gladstone law,** where $k = 2.26 \times 10^{-4}$ m^3 kg^{-1}. A similar equation was deduced later, independently and at the same time, by two men whose names were almost identical, the *Lorentz–Lorenz formula,*[†]

$$\left(\frac{n^2 - 1}{n^2 + 2}\right)M = K\rho \qquad [1.2\text{-}9]$$

where M is the molecular weight and K is another constant, called the *molecular refractivity*.

Critical angle. Assume that light is incident on the surface of a transparent medium at an angle of incidence $\alpha = 0$. The light will go through without deviation. As α is made increasingly larger, the angle of refraction, β, becomes larger too, although not as rapidly as α. As α reaches 90°, its sine becomes unity, and Snell's law reduces to

$$\sin \beta = \frac{n_1}{n_2} \qquad [1.2\text{-}10]$$

which describes the *critical angle of refraction* (Figure 1.2-7, left).

---

*T. P. Dale and J. H. Gladstone, "On the Influence of Temperature on the Refraction of Light," *Phil. Trans. Roy. Soc. London* **148** (1858), 887–94.

[†]H. A. Lorentz, "Ueber die Beziehung zwischen der Fortpflanzungsgeschwindigkeit des Lichtes und der Körperdichte," *Ann. Physik* (Neue Folge) **9** (1880), 641–65. L. Lorenz, "Ueber die Refractionsconstante," *Ann. Physik* (Neue Folge) **11** (1880), 70–103.

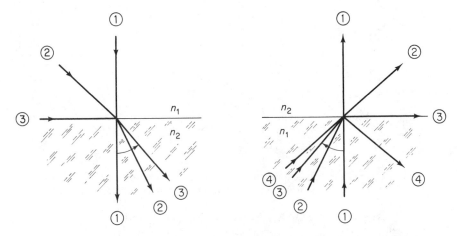

Figure 1.2-7 Critical angle of refraction (*left*) and total internal reflection (*right*). Note the limiting angles.

Now let the light come from the side of the denser medium. The angle of incidence is then *below* the surface, and the light is bent *away* from the normal. As α is made larger, it will reach a value such that β becomes 90° and

$$\sin \alpha = \frac{n_2}{n_1} \qquad [1.2\text{-}11]$$

which is the *minimum angle of total internal reflection* (right). At angles of incidence larger than this limiting angle, the light is reflected back into the first medium. Clearly, the critical angle (of refraction) and the minimum angle (of total internal reflection) are numerically equal; they both lie in the *denser* medium.

Prisms

Reflecting prisms. We distinguish two major groups of prisms, *reflecting prisms* and *dispersing prisms*. Reflecting prisms, unless they have a reflective coating, make use of total internal reflection. In Figure 1.2-8 we show several examples. In a *right-angle prism* the light is reflected at the hypotenuse. If the same prism is oriented so that the light is passing (twice) through the hypotenuse, we have a *roof* or *Porro prism*. Still the same prism, with or without its apex cut off, is called a *Dove* or *erecting prism*. As the Dove prism is rotated about the line of sight, the image rotates through twice the angle. A *pentagonal prism* deflects the light through a constant 90°, without changing the "handedness" of the image.

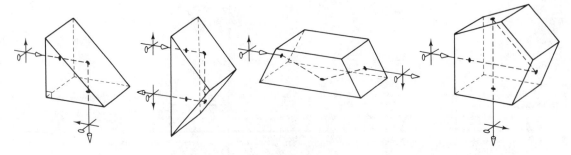

Figure 1.2-8 Some reflecting prisms, based on total internal reflection, are *(from left to right)* right-angle prism, roof prism, Dove prism, and penta prism.

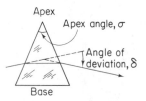

Figure 1.2-9 Elements of a prism.

Dispersing prisms. If the light contains more than one wavelength, a refracting or *dispersing prism* will disperse the light; that is, it will separate the light according to wavelengths.

Generally, a prism has two plane surfaces that subtend an angle, σ, called the *apex-angle* (Figure 1.2-9). The face opposite the apex is known as the *base*. The total angle by which the light changes direction is the *angle of deviation, δ.*

From the construction in Figure 1.2-10 we find that

$$\alpha_1 = \beta_2, \qquad \beta_1 = \alpha_2 = \frac{\sigma}{2}$$

$$\delta_1 = \delta_2 = \frac{\delta}{2} = \beta_2 - \alpha_2 = \beta_2 - \frac{\sigma}{2}$$

$$\beta_2 = \frac{\sigma}{2} + \frac{\delta}{2}$$

From Snell's law,

$$\sin \beta_2 = \frac{n_{prism}}{n_0} \sin \alpha_2$$

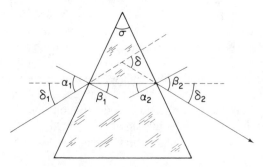

Figure 1.2-10 Deriving the prism equation for minimum deviation.

where n_{prism} is the refractive index of the material of the prism and n_0 is the index outside the prism. Replacing the two angles by the expressions just derived gives

$$\sin\left(\frac{\sigma + \delta}{2}\right) = \frac{n_{prism}}{n_0} \sin\left(\frac{\sigma}{2}\right)$$

from which

$$\boxed{\frac{n_{prism}}{n_0} = \frac{\sin \frac{1}{2}(\sigma + \delta)}{\sin \frac{1}{2}\sigma}} \qquad [1.2\text{-}12]$$

This is the *prism equation,* which holds, however, as we see from the geometry in Figure 1.2-10, only for light deflected through the *angle of minimum deviation*. At other angles the deviation is larger than minimum (a fact that can be verified both by experimental observation and by applying Snell's law successively to both surfaces of the prism). At the angle of minimum deviation the light passes through the prism symmetrically and the angles of incidence and emergence are equal.

Thin prisms.   Ophthalmic or *thin prisms* have apex angles and angles of deviation considerably smaller than those of other prisms. The sines of these angles can then be set equal to the angles themselves, measured in radians, and if the prism is in air ($n_0 \approx 1$), Equation [1.2-12] reduces to

$$\boxed{\delta = \sigma(n_{prism} - 1)} \qquad [1.2\text{-}13]$$

As a thin prism is rotated about the line of sight, the magnitude of displacement changes with the angle of rotation, as illustrated in Figure 1.2-11.

The *refractive power* of a prism is often measured in terms of *prism diopters,* $^{\triangle}$. A prism diopter is defined as causing a deflection of light by 1 cm on a straight line 1 m away from the prism. But I prefer the unit *centrad,* $^{\triangledown}$. A centrad is defined as causing a deflection through 1 cm on the circumference of a circle of 1 m radius; it is *1/100 of a radian.* For thin prisms, prism diopters and centrads are almost the same. But centrads hold also for thick prisms, and using centrads avoids the ambiguity of having the same term "diopter" for both lens powers

Figure 1.2-11   Displacement of a line as seen through a prism.

(where it means $1/m$) and prism powers (where it means cm/m). The *orientation* of a prism is always given with respect to the *base*.

Example. *A prism is 1 mm thick at the apex and 6 mm at the base. The distance between apex and base is 50 mm and the refractive index 1.61.*
(a) Determine the apex angle.
(b) Determine the angle of deviation and the power of the prism.
(c) What is the power of the prism if it is immersed in water ($n = \frac{4}{3}$)?

Solution. (a) A thickness difference of 1 mm versus 6 mm is the same as 0 versus 5 mm; thus

$$\sigma = (2)\ \tan^{-1}\frac{2.5}{50} = \boxed{5.7°}$$

(b) Then, from Equation [1.2-13],

$$\delta = 5.7°(1.61 - 1) = \boxed{3.5°} = 0.06\ \text{rad} = \boxed{6^\triangledown}$$

(c) Because the prism is "thin," we may simplify Equation [1.2-12] and write

$$\frac{n_{\text{prism}}}{n_0} = \frac{\sigma + \delta}{\sigma} = 1 + \frac{\delta}{\sigma}$$

Thus

$$\delta = \sigma\left(\frac{n_{\text{prism}}}{n_0} - 1\right) = 5.7°\left(\frac{1.61}{4/3} - 1\right) = 0.02\ \text{rad} = \boxed{2^\triangledown}$$

Dispersion

I have often referred to "the" refractive index of a material and never mentioned that the refractive index might change as a function of wavelength. In fact, there is no material known whose refractive index is constant throughout the spectrum. Generally, the index *decreases* as the wavelength *increases*. Consequently, light passing through a material medium will be separated according to wavelengths, an effect known as *chromatic dispersion*.

With the light at normal incidence, blue light (for which the index is higher and the velocity less) would merely lag behind red light (Figure 1.2-12, left). With the light incident obliquely, the colors become separated in space (center), both surfaces refracting blue more than red. With

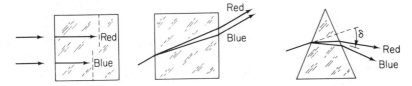

Figure 1.2-12 Dispersion in a block of glass at normal incidence (*left*) and at oblique incidence (*center*); dispersion caused by a prism (*right*).

a prism, the separation is even more pronounced (right). The angle subtended by the directions of the two colors is called the *angular dispersion*.

Example. *A hollow 60° prism is filled with carbon disulfide, whose index of refraction for blue light is 1.652, for red light 1.618. What is the angular dispersion?*

Solution. Using Equation [1.2-12], we determine first the deviation for blue:

$$\frac{1.652}{1.000} = \frac{\sin \frac{1}{2}(60° + \delta)}{\sin \frac{1}{2}(60°)}$$

$$0.826 = \sin \tfrac{1}{2}(60° + \delta)$$

$$111.38° = 60° + \delta$$

$$\delta = 51.38°$$

For red,

$$0.809 = \sin \tfrac{1}{2}(60° + \delta')$$

$$108° = 60° + \delta'$$

$$\delta' = 48.0°$$

The difference between the two angles, that is, the *angular dispersion* between blue and red, is

$$51.38° - 48.00° = \boxed{3.38°}$$

Perhaps we are led to think that materials of high refractive index also have high dispersion and, hence, that dispersion is proportional to refraction. But refraction and dispersion bear no simple relationship to one another. Some glasses have a high index of refraction and little dispersion, others have just the opposite (Figure 1.2-13).

Figure 1.2-13 (*Left*) Prism of high refraction and low dispersion (note how much the light is deviated but how close the two colors are). (*Right*) Prism of low refraction and high dispersion (little deviation but the two colors are spread apart much farther).

Clearly, we need more than one refractive index. In practice, *three* indices are chosen, one for each of three colors identified by *Fraunhofer lines*. These are the blue F line produced by hydrogen, wavelength 486.1 nm, defining the refractive index n_F; the yellow sodium D lines, 589.3 nm, defining n_D; and the red hydrogen C line, 656.3 nm, defining n_C.*

The numerical difference between the two refractive indices n_F and n_C is called the *mean dispersion*. The ratio $(n_F - n_C)/(n_D - 1)$ is the *dispersive power*. Its inverse is *Abbe's number,* also called the ν *value* or *V number:*

$$\nu = \frac{n_D - 1}{n_F - n_C}$$ [1.2-14]

Glasses where $\nu \geqq 55$ (low dispersion) are called *crowns*. Glasses where $\nu \leqq 50$ (high dispersion) are called *flints*. Some typical values are listed in Table 1.2-2.

Table 1.2-2 REFRACTIVE INDICES OF CROWN AND FLINT
AT DIFFERENT WAVELENGTHS

Fraunhofer line	Color	Wavelength (nm)	Spectacle crown	Extra-dense flint
			Refractive index	
F	Blue	486.1	1.5293	1.7378
D	Yellow	589.3	1.5230	1.7200
C	Red	656.3	1.5204	1.7130
			ν value	
			59	29

Glass. Glasses are solutions, rather than chemical compounds. In a way, glass is a *state of matter*. In principle, the arrangement of the ions in a glass is similar to that in a crystal. In a quartz, or silica, *crystal,* SiO_2, for example, each silicon cation is surrounded by four oxygen anions, forming a tetrahedron with the ions at the corners. Each oxygen is bound to the silicon by only one valence, leaving the other valence to connect to the next tetrahedron. Thus every oxygen ion belongs to two tetrahedra, and the ratio between silicon and oxygen is 1:2, in accordance with the formula SiO_2. In silica *glass,* on the other hand, the individual tetrahedra are there but they are distorted and linked together at random.

---

*Fraunhofer lines will be discussed in more detail in Chapter 5.3. In Europe, the green mercury line, 546.1 nm, is used instead of sodium D.

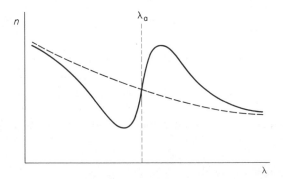

Figure 1.2-14 Normal dispersion (dashed curve) and anomalous dispersion (solid curve). λ_a, center wavelength of absorption band.

About 95 percent of all glasses are of the "soda–lime" type. They contain silica, SiO_2; soda, Na_2O; and lime, CaO. Besides SiO_2, there are other "glass formers," most notably boron oxide, B_2O_3; phosphorus pentoxide, P_2O_5; and germanium oxide, GeO_2. A great many other ingredients may be added, changing the refractive index, dispersion, thermal expansion, electric conductivity, and other properties of the glass. *Crown* glass is a soda–lime–silica composite. *Flint* contains 45 to 65 percent lead oxide; its specific gravity, refractive index, and dispersion all are high, its softening temperature low. Barium glass has barium rather than lead oxide; its index is about as high as that of most flints, but its dispersion is not. The refractive index of crown ranges from 1.42 to 1.53, of flint from 1.5 to 1.76, rare earth flint 1.7 to 1.84, lanthanum flint 1.82 to 1.98; arsenic trisulfide glass has 2.04.

Plastics have refractive indices generally lower than those of glasses. "Crown"-type plastics, usually polymethylmethacrylate, Plexiglas, have about $n = 1.49$; "flint"-type plastics, acrylonitrile–styrene copolymers, have 1.57. Allyl diglycol carbonate, CR-39, has 1.498; it is particularly scratch resistant.

Virtually all transparent materials have refractive indices that are higher at shorter wavelengths and lower at longer wavelengths. A plot of n versus λ, therefore, follows the *dashed curve* in Figure 1.2-14.

A curve of this shape can be represented by *Cauchy's formula.**

$$n = A + \frac{B}{\lambda^2} + \frac{C}{\lambda^4} + \cdots \qquad [1.2\text{-}15]$$

where n is the refractive index, λ is the wavelength, and A, B, and C are empirically determined material constants. The C term is often so small that a plot of n versus $1/\lambda^2$ gives a straight line from blue to red. By differentiation, we find that

---

*Augustin Louis Cauchy (1789–1857), French mathematician. His formula appears in *Mémoire sur la Dispersion de la Lumière* (Prague: J. G. Calve, 1836).

$$\frac{dn}{d\lambda} = -\frac{2B}{\lambda^3} \qquad [1.2\text{-}16]$$

which shows that the dispersion varies as the inverse third power of the wavelength. At 400 nm it is about eight times as high as at 800 nm.

Cauchy's formula holds for what in earlier times has been called *normal* dispersion. Such dispersion exists in colorless, transparent matter such as water or glass. But if there is any absorption, at wavelength λ_a, the dispersion follows the solid line in Figure 1.2-14, and is called *anomalous*. Today the terms "normal dispersion" and "anomalous dispersion" are of historic interest only. If the UV and the IR are included, most any substance will show some absorption—and hence "anomalous" dispersion—somewhere throughout the spectrum.

Applications of dispersing prisms.
The instrument most often used with a dispersing prism is the *prism spectroscope*. A spectro*scope* is used for viewing a spectrum. A spectro*meter* is equipped for measuring. A spectro*graph* is built for photography (Figure 1.2-15). In a spectro*photometer* a photocell takes the place of the photographic film. A spectro*gram* is the result of spectrography or spectrophotometry.

The *resolvance* ("chromatic resolving power") of a prism is a measure of its ability to separate closely adjacent spectrum lines. Resolvance is defined as the ratio of wavelength to the difference of wavelengths (of two such lines), $\lambda/\Delta\lambda$. We will discuss this term in detail later (on pages 251, 400).

An *achromatic prism* deviates the light but gives no dispersion. This is possible by using two prisms of equal dispersion, one made of crown and the other of flint, and combining them so that the base of one prism rests against the apex of the other. A *direct-vision prism,* in contrast, gives dispersion but no deviation (for one wavelength). A *monochromator* is an instrument for selecting light of different wavelengths. For use in the

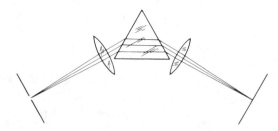

Figure 1.2-15   Diagram of a prism spectrograph. The two lenses (collimating lens left, focusing lens right) allow parallel light to pass through the prism, oriented for minimum deviation.

visible spectrum, prisms and lenses are made out of glass. Outside the visible spectrum, glass shows absorption, and thus must be replaced, by quartz or fluorite in the UV, and rock salt (NaCl, KCl) or sapphire (Al_2O_3) in the IR.

Example. *Assume that 14° is the apex angle of a crown glass prism. What should be the apex angle of a flint prism:*
(a) If the combination of both is to be achromatic *for blue and red?*
(b) If the prism is to have no deviation *for yellow?*

Solution. (a) The mean dispersion of a 14° prism, from Table 1.2-2, is

$$\delta_F - \delta_c = (\sigma_1)(n_F - n_C) = (14)(1.5293 - 1.5204) = 0.1246$$

For flint,

$$(\sigma_2)(n_F - n_C) = (\sigma_2)(1.7378 - 1.7130) = 0.0248\,\sigma_2$$

For the combination to be achromatic,

$$(\sigma_1)(n_{1F} - n_{1C}) + (\sigma_2)(n_{2F} - n_{2C}) = 0$$

Thus

$$\sigma_2 = -\,\frac{0.1246}{0.0248} = \boxed{-5°}$$

the minus sign indicating that the apex of one prism adjoins the base of the other.
(b) For the direct-vision prism,

$$\frac{\sigma_1}{\sigma_2} = \frac{n_{2D} - 1}{n_{1D} - 1}$$

Hence,

$$\sigma_2 = (14)\left(\frac{1.5230 - 1}{1.7200 - 1}\right) = \boxed{10.2°}$$

SUGGESTIONS FOR FURTHER READING

R. P. Feynman, R. B. Leighton, and M. Sands, *The Feynman Lectures on Physics,* Vol. 2, Sec. 26 (Reading, MA: Addison-Wesley Publishing Company, Inc., 1964).

W. J. Smith, *Modern Optical Engineering* (New York: McGraw-Hill Book Company, 1966).

R. H. Doremus, *Glass Science* (New York: John Wiley & Sons, Inc., 1973).

W. L. Wolfe, "Properties of Optical Materials," in W. G. Driscoll and W. Vaughan, editors, *Handbook of Optics,* Sec. 7, pp. 1–157 (New York: McGraw-Hill Book Company, 1978).

PROBLEMS

1.2-1. A ray of light in water ($n = \frac{4}{3}$) is incident on a plate of crown glass at an angle of 65°. What is the angle of refraction in the glass?

1.2-2. A layer of oil of unknown refractive index is floating on top of a layer of carbon disulfide. If the angle of incidence at the air–oil boundary is 60°, what is the angle of refraction in the carbon disulfide?

1.2-3. What is the maximum apparent zenith distance (that is, the angular distance from true vertical) of a star to an eye under water ($n = \frac{4}{3}$)?

1.2-4. A point light source is located 2 m below the surface of a lake ($n = \frac{4}{3}$). Determine the area of the surface that transmits light coming from the source.

1.2-5. A prism of 72° apex angle and 1.66 refractive index is immersed in water ($n = \frac{4}{3}$). Find the angle of minimum deviation.

1.2-6. A 30° prism made out of BK7 glass ($n = 1.515$) is oriented first so that the light is incident at right angles. The prism is then turned to the position of minimum deviation. By how much do the two angles of deviation differ?

1.2-7. When looking through a thin prism of $n = 1.625$ at a meter stick 2 m away, the marks on the stick are seen to be displaced by 3.5 cm. Find:
(a) The deviation of the prism.
(b) The power of the prism.
(c) The apex angle of the prism.

1.2-8. A prism, made from ophthalmic crown, measures 5 cm from base to apex. If, at the base, the prism is 10 mm thick and at the apex 3.3 mm, what is the angle of deflection?

1.2-9. A plane-parallel slab of ophthalmic crown is 50×50 mm² in size. How much glass must be ground off along one edge to make a prism that deflects light by 3.5 centrads?

1.2-10. An open cylindrical container, resting on its circular base, is 20 cm in diameter and 11.5 cm deep. An observer is looking into the container from such a direction that he can just see the opposite bottom corner (Figure 1.2-16). When the container is filled with a liquid, the observer can see the *center* of the bottom. What is the refractive index of the liquid?

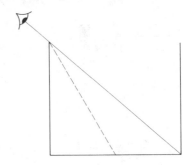

Figure 1.2-16

1.2-11. The light forming a *rainbow* is refracted when it enters each drop of water and refracted again when it leaves. In a *primary* rainbow the light in addition is reflected *once,* at the rear face of the drop (Figure 1.2-17, left). By ray tracing, follow a blue ray and a red ray and indicate whether blue or red is on the outside curvature of the bow.

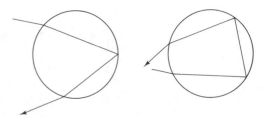

Figure 1.2-17

1.2-12. In a *secondary* rainbow the light is reflected *twice* (Figure 1.2-17, right). By ray tracing, determine how the colors are distributed now.

1.2-13. If white light is incident at an angle of 62° on the polished surface of a slab of extra dense flint, what is the angular dispersion between blue and red within the glass?

1.2-14. A narrow bundle of light falls at 45° on the polished surface of a plane-parallel slab of glass 10 cm thick. If the refractive index of the glass for blue is 1.53 and for red 1.52, by how much will the two colors be separated on leaving the glass?

1.2-15. A 6° crown prism is combined with a flint prism so as to be achromatic for blue and yellow. What should be the apex angle of the flint prism?

1.2-16. What is the angular dispersion for blue and red produced by a direct-vision prism (no deviation for yellow) whose crown component has an apex angle of 11°?

1.3

Thin Lenses

CONTINUING WITH OUR THREE-LENS SYSTEM, now let the light be incident on a *single spherical surface* (Figure 1.3-1).

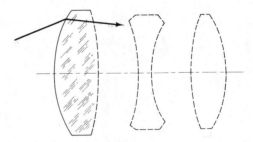

Figure 1.3-1 Ray of light refracted by a single spherical surface.

Single Spherical Surface

Vergence and refractive power. What happens to the light as it passes through the first surface? That depends on the vergences and powers involved. The light as it strikes the surface has a certain

vergence or, more likely, it has a certain *di*vergence (negative vergence). Then, if the *refractive power* of the surface is positive and *higher* than the (absolute) vergence of the light, some excess (positive) vergence will remain and the light after refraction will *converge* (Figure 1.3-2, top). In that case the rays will come together and form a *real image*. A real image contains light energy; the image can be received, and be seen, on a screen.

If the object is moved closer to the surface, the (minus) vergence of the incident light will become higher, and may be counterbalanced by the (plus) power of the surface. The resultant vergence then is zero, which means that the emergent light is parallel and the image is projected out into infinity (Figure 1.3-2, center).

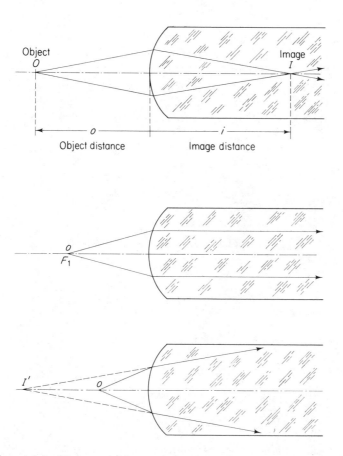

Figure 1.3-2   When a real object, *O*, is brought successively closer to a (positive) refracting surface, the image is first real, *I*, (*top*); projected into infinity (*center*); and then virtual, *I'*, (*bottom*).

If the object is brought even closer, the vergence of the incident light becomes still more negative, the surface does not have enough power to compensate for it, and the emergent light *remains divergent* (Figure 1.3-2, bottom). Now the rays will *not* come together on the right-hand side. Instead, the light seems to come from a point, I', to the left of the surface. Such a point is part of a *virtual image*.*

The vergence of the light when it reaches the surface is called the *entrance vergence*, **V**. The vergence of the light when it leaves the surface is called the *exit vergence*, **V'**. Entrance vergence, **V**, surface power, $\mathfrak{P}$, and exit vergence, **V'**, as we will see in more detail on page 43, are connected as

$$\mathbf{V} + \mathfrak{P} = \mathbf{V'} \qquad [1.3\text{-}1]$$

The units of all three terms are the same: reciprocal meter, m^{-1}, or *diopter*.

If *parallel light is incident* on a single surface, as in Figure 1.3-3, top left, *it forms the second focal point, F_2*. If the *light is made parallel* by the surface, the light *comes from the first focal point, F_1* (top right). This holds as well for a diverging, negative surface (bottom examples). Planes constructed *at* the focal points, normal to the optical axis, are called *focal planes*.

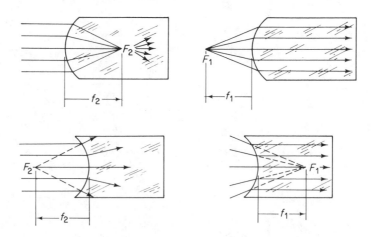

Figure 1.3-3   Focal points, F, and focal lengths, f.

---

* A good example of a virtual image is what you see in a mirror. The object and the observer are both in front of the mirror, but the image is behind it. No light is actually going to where the image *appears to be;* the image is *virtual*. In a way, the terms "real" and "virtual" are misleading. The word "real" seems to imply that "virtual" has something "unreal" to it. That is not so; a virtual image can be seen, and be photographed, just as well as a real image; the only difference is that it cannot be received on a screen.

The point where the optic axis intersects the surface is called the *vertex*. The distance from the vertex to the object is the *object distance, o.* The distance from the vertex to the image is the *image distance, i.* For every position of the object there is a corresponding position of the image. Since the image may be either real or virtual, and may lie on either side of the surface, the *image space* extends from infinity on one side to infinity on the other. The same holds for the *object space;* both completely overlap. Whether a given point is in the object space or the image space depends on whether it is part of a ray *before* or *after* refraction.

Corresponding points that relate to each other as object and image are called *conjugate points*. Planes on which they lie are *conjugate planes*, and the distances from the vertex to these planes are *conjugate distances.*

Example. *A point light source is placed 0.5 m to the left of a* convex *surface of +6.00 diopters power, ground on the end of a solid cylinder of glass of n = 1.6. Find the vergence of the light at various distances from the object and determine the image distance.*

Solution. The light source, as shown in Figure 1.3-4, is at *A*. Since vergence is the reciprocal of the distance, in *meters,* the vergence at *B*, 0.2 m away from the source, is

$$V = \frac{n}{L} = \frac{1.0}{-0.2 \text{ m}} = -5 \text{ m}^{-1}$$

The minus sign indicates *di*vergence.

The *entrance* vergence, at *C*, is

$$V = \frac{1}{-0.5 \text{ m}} = -2 \text{ m}^{-1}$$

The *exit* vergence, according to Equation [1.3-1], is

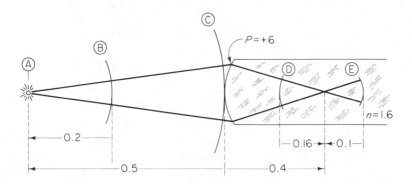

Figure 1.3-4 Vergence of light before and after refraction at a single spherical surface. All distances in meters.

$$\mathbf{V'} = \mathbf{V} + \mathbf{\mathscr{P}} = (-2) + (+6) = +4 \text{ m}^{-1}$$

Note the reversal of the signs on the left and right of the surface. If **V'** is positive, as here, the image is inverted and real.

The light now proceeds in the medium of higher index, $n_2 = 1.6$. Therefore, the *image distance* is

$$i = \frac{n_2}{\mathbf{V'}} = \frac{1.6}{+4} = \boxed{+0.4 \text{ m}}$$

At D, 0.16 m *before* the light reaches the image, the light is still *con*vergent:

$$\mathbf{V} = \frac{1.6}{+0.16} = +10 \text{ m}^{-1}$$

but at E, *to the right* of the image, it is *di*vergent:

$$\mathbf{V} = \frac{1.6}{-0.1} = -16 \text{ m}^{-1}$$

If the surface had only $+1.50$ diopters power, rather than $+6.00$, how would the image change? In that case

$$\mathbf{V'} = (-2) + (+1.5) = -0.5 \text{ m}^{-1}$$

which means that the light remains divergent, and the image is virtual and *to the left* of the surface (as in Figure 1.3-2, bottom), at a distance of

$$i = \frac{1.6}{-0.5} = \boxed{-3.2 \text{ m}}$$

This is an important case. While the medium to the left of the surface is air ($n \approx 1.0$), the *index related to the image space is 1.6*. In other words, with any image, real or virtual, *always use the image space index*. Do not use the index of the medium where the image merely *seems* to be.

Finally, if we had a *concave,* negative surface, the image, no matter what the distances, would always be *virtual*.

From this example we see that our earlier sign convention (page 12) needs some expansion. In particular:

3. The refractive *power* of a surface that makes light rays more convergent, or less divergent, is positive. Such a surface is converging; its (second) *focal length* is positive also. The power and the (second) focal length of a diverging surface are negative.

4. Since all distances are measured in Cartesian coordinates, *the distance of a real object is negative*.

5. For the same reason, the *distance of a real image is positive*. The distance of a virtual image is negative.

6. *Heights* above the optic axis are positive; distances below the axis are negative.

7. *Angles* measured clockwise, beginning at the optic axis or at the normal erected on a surface at a point of incidence, are positive. Angles measured counterclockwise are negative.

Note that there is a conceptual chance for error. Although the object distance, o, is measured to the left and, following the Cartesian sign convention, is considered negative, the letter symbol "o" as such is positive. Only when we substitute a numerical value for "o" that is negative, then o becomes negative. The same rule applies to angles, heights above the axis, and so on.

Gauss' formula and the surface power equation. We
are now ready to derive the numerical relationship between the shape of a surface and its power. Consider, for example, a convex surface, ground on a block of glass. The radius of curvature of the surface, according to our sign convention, is positive. A ray (1) may originate at an axial point object, O, and subtend an angle γ with the axis. At the image, I, the ray subtends angle γ'. From the construction in Figure 1.3-5 it follows that

$$\alpha = \gamma - \phi \quad \text{and} \quad \beta = -(\phi - \gamma') = \gamma' - \phi$$

Assume that the rays under consideration are close to the optic axis (*paraxial rays*).* Then the angles involved are small and their sines and tangents are nearly equal to the angles themselves, measured in radians. With this approximation, Snell's law, $n_1 \sin \alpha = n_2 \sin \beta$, can be written $n_1 \alpha = n_2 \beta$, from which

$$n_1 \gamma - n_1 \phi = n_2 \gamma' - n_2 \phi$$

Furthermore, for paraxial rays we may neglect d, as compared to o or i; thus

$$\gamma = \frac{h}{o} \qquad \gamma' = \frac{h}{i} \quad \text{and} \quad \phi = \frac{h}{R} \qquad [1.3\text{-}2]$$

Substituting these terms in the preceding equation gives

$$\frac{n_1 h}{o} - \frac{n_1 h}{R} = \frac{n_2 h}{i} - \frac{n_2 h}{R}$$

and canceling h and condensing yields

$$\boxed{\frac{n_1}{o} + \frac{n_2 - n_1}{R} = \frac{n_2}{i}} \qquad [1.3\text{-}3]$$

---

*From the Greek παρὰ = nearby, closely adjacent; that is, paraxial = close to the optic axis.

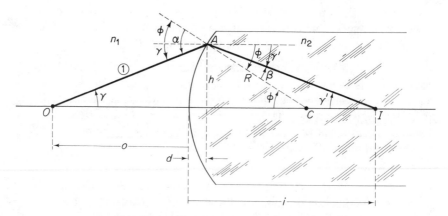

Figure 1.3-5   Refraction at a single spherical surface.

which is *Gauss' formula for refraction at a single surface,* * written in a form that is easy to remember: just think of the light propagating from object (o) to surface (R) to image (i). Although I have derived Gauss' formula for a convex surface, and for a real object and image, the formula holds as well for all other conditions.

If the object is moved to infinity, o becomes $-\infty$, the first term in Equation [1.3-3] drops out, and the image moves to the second focal point, $i \rightarrow \mathfrak{f}_2$; thus

$$\frac{n_2}{\mathfrak{f}_2} = \frac{n_2 - n_1}{R} \qquad\qquad [1.3\text{-}4]$$

If we write the first term in the form $1/(\mathfrak{f}_2/n_2)$, we see that this is the inverse of the *reduced* focal length and, by definition, the refractive power, $\mathfrak{P}$, of the surface. Thus

$$\boxed{\;\mathfrak{P} = \frac{n_2 - n_1}{R}\;} \qquad\qquad [1.3\text{-}5]$$

which is the *surface power equation.*

---

* Named after Karl Friedrich Gauss (1777–1855), German mathematician and astronomer. A child prodigy even before entering elementary school, Gauss discovered the binomial theorem, the prime number theorem, and the method of least squares (used in statistics). As an astronomer he found a way to calculate the orbits of asteroids and designed the eyepiece named after him. Other accomplishments that bear his name are the derivation of the surface power equation, the thin lens equation, the law of electrostatics, and the divergence theorem, the latter two leading up to Maxwell's equations. His book, *Dioptrische Untersuchungen,* was published in 1843. Gauss was the first to use the paraxial-ray approximation; this is known as *first-order, paraxial,* or *Gaussian optics.*

Consider again Gauss' formula, Equation [1.3-3], and recall from Equation [1.1-4] that (reduced) vergence is the reciprocal of a distance, $V = n/L$. Therefore, the first term in Equation [1.3-3] is simply the entrance vergence and the last term is the exit vergence. Substituting for the second term the surface power equation, we find that

$$\boxed{V + \mathfrak{P} = V'} \qquad\qquad [1.3\text{-}1]$$

This is an important relationship that we will use frequently.

Example 1. *A small object is located 0.1 m in front of a convex surface, ground with a 0.03-m radius on a block of glass of index 1.66. Find the position of the image:*
(a) If the object and glass are in air.
(b) If both are immersed in water (n = $\frac{4}{3}$).

Solution. (a) From Gauss' formula we find that

$$\frac{1.66}{i} = \frac{1.00}{-0.1} + \frac{1.66 - 1.00}{+0.03}$$

$$i = \frac{1.66}{12} = \boxed{+0.138 \text{ m}}$$

which means that the image is real, inside the glass, and 13.8 cm to the right of the surface.

(b) If the object and glass are in water,

$$\frac{1.66}{i} = \frac{4/3}{-0.1} + \frac{1.66 - 4/3}{+0.03}$$

$$i = \frac{1.66}{-2.444} = \boxed{-0.679}$$

which means that now the image is virtual, inside the water, and 67.9 cm to the left of the surface.

Example 2. *A reduced eye is a much simplified model of the human eye; it has only one refracting surface, assumed to have a radius of 5.7 mm (Figure 1.3-6). The refractive index to the left of the surface is 1.00, the refractive index to the right is 1.34. Find the focal length(s) and power(s) of the surface.*

Solution. The focal length inside the eye, to the right of the surface, is found from Equation [1.3-4]:

$$f_2 = \frac{n_2 R}{n_2 - n_1} = \frac{(1.34)(0.0057)}{1.34 - 1.00} = \boxed{+22.5 \text{ mm}}$$

In order to find f_1, to the left of the surface, we let the image distance in

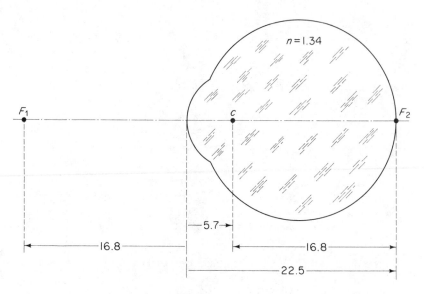

Figure 1.3-6 The reduced eye. Dimensions in millimeters.

Equation [1.3-3] become infinity, $i \to \infty$. Then $o \to f_1$, and

$$f_1 = \frac{n_1 R}{n_1 - n_2} = \frac{(1)(0.0057)}{1.00 - 1.34} = \boxed{-16.8 \text{ mm}}$$

We have *two focal lengths,* their ratio, neglecting the sign, being equal to the ratio of the refractive indices,

$$\frac{22.5}{16.8} = \frac{1.34}{1.00}$$

But we have only *one power,*

$$\mathfrak{P} = \frac{n_2 - n_1}{R} = \frac{1.34 - 1.00}{0.0057} = \boxed{+60 \text{ diopters}}$$

Thin Lenses

We now turn to a combination of two surfaces that enclose a medium of a refractive index different from that outside the surfaces. Such a combination is known as a *lens*. A lens is considered *thin* if its thickness is much less than the radii of curvature of its two surfaces; the *refractive power of a thin lens, therefore, is the algebraic sum of the powers of the two surfaces.*

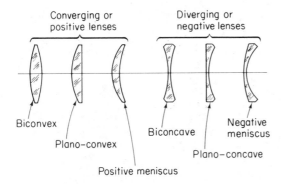

Figure 1.3-7 Types of lenses.

There is no numerical limit where a thin lens ends and a thick lens begins. That is entirely a matter of precision required for solving a given problem; the same lens may well be considered "thin" for a preliminary, and "thick" for a rigorous solution.

There are two general classes of lenses, depending on whether they make a parallel bundle of light convergent or divergent (Figure 1.3-7). A converging, or positive, lens is thicker in the center than at the periphery (provided that its refractive index is higher than the index outside); a diverging, or negative, lens is thinner in the center.

A lens has two centers of curvature and two radii of curvature, one for each surface. The line connecting the two centers is the *optic axis of the lens*. The points where this axis intersects the surfaces of the lens are the *front vertex* and the *back vertex,* respectively. If parallel light is incident on the lens, it forms the *second* focal point (Figure 1.3-8). Conversely, if the light is made parallel by the lens, it comes from the *first* focal point. As the object is brought closer to the lens, or moved away from it, the image distance changes accordingly (Figure 1.3-9).

The power of a converging lens is positive, that of a diverging lens negative. But whereas the second focal length of a converging lens is positive, the first, according to the Cartesian sign convention, is negative. With a diverging lens, the reverse is true. The two focal lengths of a lens, therefore, are numerically equal but of opposite sign. Customarily, whenever reference is made to "*the* focal length" of a lens, that always means the second focal length.

Lens equations: The lens-makers formula. I mentioned earlier, and will show in detail on page 72, that the power of a thin lens is the algebraic sum of the powers of its two surfaces:

$$P_{\text{lens}} = \mathfrak{P}_1 + \mathfrak{P}_2 \qquad\qquad [1.3\text{-}6]$$

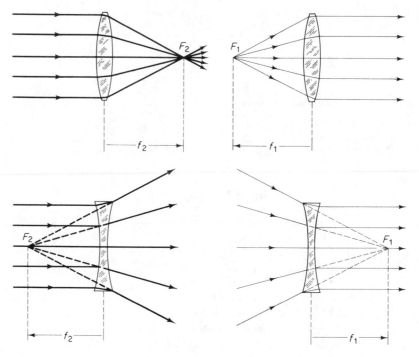

Figure 1.3-8   Focal points, F, and focal lengths, f, of a lens.

Then, if we replace the P's by the surface power equation,

$$\mathfrak{P}_1 = \frac{n_{\text{lens}} - n_0}{R_1} \qquad \text{and} \qquad \mathfrak{P}_2 = \frac{n_0 - n_{\text{lens}}}{R_2} = \frac{n_{\text{lens}} - n_0}{-R_2}$$

and substitute these terms in Equation [1.3-6], we find that

$$\boxed{P_{\text{lens}} = (n_{\text{lens}} - n_0)\left(\frac{1}{R_1} - \frac{1}{R_2}\right)} \qquad [1.3\text{-}7]$$

which is the *lens-makers formula*.

Example. *A biconvex lens is made out of glass of $n = 1.52$. If one surface has twice the radius of curvature of the other, and if the focal length is 5 cm, what are the two radii?*

Solution. First we convert the focal length into power:

$$P = \frac{n}{f} = \frac{1.00}{0.05 \text{ m}} = 20 \text{ m}^{-1}$$

Since R_1 is to be equal to twice the (absolute) value of R_2, or

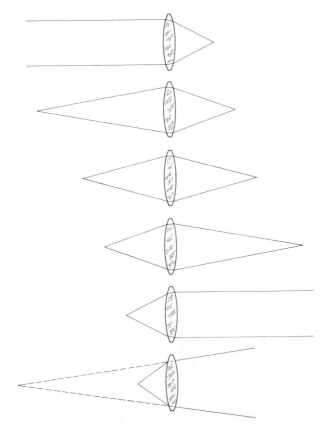

Figure 1.3-9 As an object at infinity (*top*) is brought closer to the lens, the image moves away from it. (*Bottom*) The image is *virtual* and to the left of the lens.

$$R_2 = -\tfrac{1}{2}R_1$$

it follows from the lens-makers formula that

$$20 = (1.52 - 1.00)\left(\frac{1}{R_1} - \frac{1}{-\tfrac{1}{2}R_1}\right) = \frac{1.56}{R_1}$$

Thus

$$R_1 = \frac{1.56}{20} = \boxed{7.8 \text{ cm}}$$

and

$$R_2 = \boxed{-3.9 \text{ cm}}$$

Newton's lens equation. The lens-makers formula tells the lens-maker what is needed to know to grind a lens of a given power: the refractive index of the glass and the two radii of curvature. Now we derive two *"lens-users equations"* for the practical application: the relationship between focal length and object and image distances.

The *height* of the object, or *object size,* in Figure 1.3-10 is called h and the height of the image, or *image size,* h'. From the similar triangles to the left of the lens we find that

$$\frac{h}{N_o} = \frac{h'}{f_1}$$

To the right of the lens we have

$$\frac{h}{f_2} = \frac{h'}{N_i}$$

Rearranging both equations and combining them by eliminating h/h' gives

$$\frac{N_o}{f_1} = \frac{f_2}{N_i}$$

and

$$\boxed{N_o N_i = f_1 f_2} \qquad\qquad [1.3\text{-}8]$$

which is *Newton's lens equation.* The terms N_o and N_i are Newton's *extra-focal* object and image distances, respectively; they are measured from the foci, rather than from the lens. Consequently, Newton's equation can be used with both "thin" and "thick" lenses.

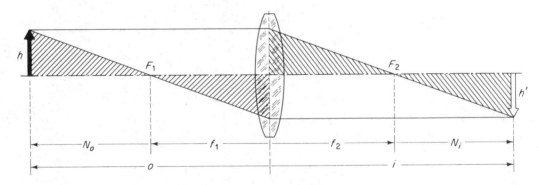

Figure 1.3-10 Deriving Newton's lens equation.

Gauss' thin-lens equation. We begin with Gauss' formula,

$$\frac{n_1}{o} + \frac{n_2 - n_1}{R} = \frac{n_2'}{i}$$
[1.3-3]

and substitute for the second term the surface power equation, $\mathfrak{P} = (n_2 - n_1)/R$, Equation [1.3-5]. But P is the reciprocal of the (reduced) focal length, f/n, and therefore,

$$\boxed{\frac{n_1}{o} + \frac{n_2}{f_2} = \frac{n_2}{i}}$$
[1.3-9]

where n_1 refers to the index of the medium to the left of the lens and n_2 to the index of the medium to the right of it. (The index of the material of the lens is *not* part of this equation; it occurs only in the lens-makers formula.) If the lens is in air, where $n \approx 1$,

$$\boxed{\frac{1}{o} + \frac{1}{f_2} = \frac{1}{i}}$$
[1.3-10]

This is *Gauss' thin-lens equation*, written in a form similar to that of Gauss' formula for a single surface, proceeding from object (o) to lens (f) to image (i).

Example 1. *An axial point object is located 25 cm to the left of a +10.00-diopter lens. A ray of light entering the lens subtends an angle $\gamma = -5°$ with the axis. What is the angle, γ', after refraction?*

Solution. First we determine the entrance *vergence*,

$$\mathbf{V} = \frac{n}{o} = \frac{1.00}{-0.25} = -4 \text{ diopters}$$

Then we find the exit vergence,

$$\mathbf{V'} = \mathbf{V} + P = (-4) + (+10) = +6 \text{ diopters}$$

The two angles are proportional to the vergences,

$$\frac{\gamma}{\gamma'} = \frac{\mathbf{V}}{\mathbf{V'}}$$

Thus

$$\gamma' = (-5°)\frac{+6}{-4} = \boxed{+7.5°}$$

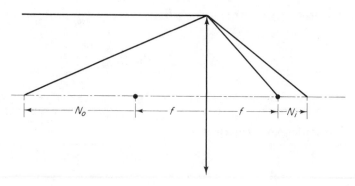

Figure 1.3-11

Example 2. *An object is placed first at infinity and then 20 cm from the object-side focal plane of a converging lens. The two images thus formed are 8 mm apart from each other. Determine the focal length of the lens.*

Solution. For the second of the two positions, and from Figure 1.3-11 and *Newton's equation,*

$$N_o N_i = f_1 f_2 \qquad \text{or} \qquad -N_o N_i = f_2^2$$

$$-(-20 \text{ cm})(0.8 \text{ cm}) = 16 \text{ cm}^2 = f_2^2$$

$$f_2 = \boxed{+4 \text{ cm}}$$

Example 3. *When an object is placed 6 cm in front of a converging lens, its image is three times as far away from the lens as if the object were at infinity. Find the focal length of the lens.*

Solution. Whereas for the object at infinity $i = f$, for the close-by object,

$$i = 3f$$

Then from *Gauss' thin-lens equation,* solved for f,

$$f = \frac{oi}{o - i} = \frac{(o)(3f)}{(o) - (3f)} = \frac{(-6)(3f)}{(-6) - (3f)}$$

$$f(-6 - 3f) = (-6)(3f)$$

$$-6f - 3f^2 = -18f$$

$$3f^2 = 12f$$

$$f = \boxed{4 \text{ cm}}$$

Example 4. *Three lenses are lined up coaxially, at distances of 6 cm from each other. The first lens has a focal length of +20 cm, the second lens −6.25 cm, and the third lens +10 cm. If the object is at infinity, how far from the last lens is the image?*

Solution. Problems of this kind can be solved in different ways.
(a) For example, we could use *Gauss' equation*. For the first lens, $o_1 = -\infty$, and therefore,

$$i = +20 \text{ cm}$$

This image then becomes the object for the second lens but, since the two lenses are 6 cm apart, the object distance (for the second lens) becomes 6 cm *less* than the image distance (for the first lens):

$$o_2 = (+20) - (+6) = +14 \text{ cm}$$

Again from Gauss' equation,

$$i_2 = \frac{of}{o + f} = \frac{(14)(-6.25)}{14 + (-6.25)} = -11.29 \text{ cm}$$

For the third lens,

$$o_3 = (-11.29) + (-6) = -17.29 \text{ cm}$$

$$i_3 = \frac{(-17.29)(10)}{(-17.29) + (10)} = \boxed{+23.72 \text{ cm}}$$

(b) Or we could use the *change-of-vergence equation*, Equation [1.1-5]. Converting first the focal lengths into powers, we obtain

$$P_1 = \frac{1}{+0.2} = +5, \quad P_2 = -16, \quad P_3 = +10 \text{ diopters}$$

Then for the "translation" from lens 1 to 2,

$$V_2 = \frac{5}{1 - (5)(0.06)} = 7.1429 \text{ m}^{-1}$$

On leaving lens 2, the vergence is

$$V' = V + P = (+7.1429) + (-16) = -8.8571 \text{ m}^{-1}$$

For the translation from lens 2 to 3,

$$V' = \frac{-8.8571}{1 - (-8.8571)(0.06)} = -5.7836 \text{ m}^{-1}$$

On leaving lens 3,

$$V'' = (-5.7836) + (+10) = +4.2164 \text{ m}^{-1}$$

The inverse of that is the image distance,

$$i_3 = \frac{1}{+4.2164} = \boxed{+23.72 \text{ cm}}$$

the same result as in (a).

Graphical construction.   In recent years the art of graphical ray tracing has lost some of its earlier prominence. This is probably due to the growing sophistication of numerical ray tracing, much of it based on matrix methods which I will present in the next chapter. For this reason I show here only a few examples of graphical ray tracing.

There are two methods. In the *parallel-ray method* we proceed in a way similar to that used with a plane surface (page 23). The incident ray, 1 in Figure 1.3-12, intersects the surface at point A. Line 2 is *normal to the surface;* that is, it is a *radius,* drawn from A to the center of curvature, C. Two arcs are drawn around A; they represent n_1 and n_2. Continue line 1 until it intersects the n_1 arc. Make line 3 parallel to 2; 3 will intersect the n_2 arc at B. Connect A and B and continue. This is the refracted ray (4).

The *oblique-ray method* is illustrated in Figure 1.3-13. The incident ray (1) intersects a lens at A. Draw the (second) focal plane of the lens (2).

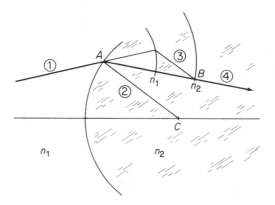

Figure 1.3-12 Parallel-ray method of ray tracing through a convex surface. $n_1 < n_2$.

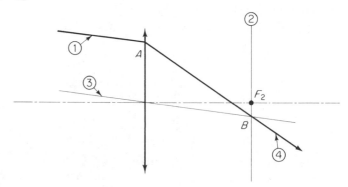

Figure 1.3-13 Ray tracing through lens using the oblique-ray method. The converging lens is identified by the double arrow, pointing outward. (A diverging lens would be shown as a double arrow pointing inward.)

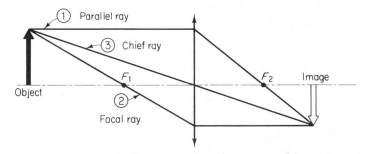

Figure 1.3-14 Designation of rays used for image construction. The object is *outside* the focal length of a converging lens.

Construct a ray (3) parallel to ray 1, through the center of the lens. Such a ray is called a *chief ray*. The chief ray intersects ray 2 at *B*. Connect *A* and *B* and continue. This is the refracted ray (4).

Assume next that instead of a given *ray*, we have an *object* and wish to find the conjugate image. To do so, we need at least two out of three rays shown in Figure 1.3-14.

1. The *parallel ray* is first parallel to the axis and then, after refraction, passes through F_2.
2. The *focal ray* passes through F_1 and then is rendered parallel to the axis.
3. The *chief ray* goes through the center of the lens, without deviation and also without displacement (because the lens is assumed to be infinitesimally *thin*).

If the object lies *within* the focal length, we follow the same rules as in Figure 1.3-14. However, now rays 1–2–3, after refraction, are still divergent and do not intersect on the right (Figure 1.3-15). Instead, they

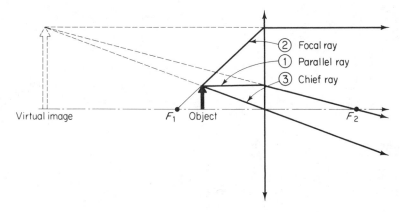

Figure 1.3-15 Converging lens with object *inside* the focal length.

seem to come from a point on the left. The image, therefore, is virtual, upright, and larger than the object.

If we have a *virtual object* (which usually is a virtual image formed by a preceding lens), the rays seem to head toward a point in the object, but they do not actually reach it. The image is real, upright, and reduced in size. With a *diverging lens*, the image, no matter where the object is placed, is always virtual, upright, and reduced in size.

Magnification

The term "magnification" means many things. To be precise we distinguish the following three types.

1. *Transverse*, or *lateral, magnification, M_t, is defined as the ratio of image size, h', to object size, h:*

$$\boxed{M_t = \frac{h'}{h}} \qquad [1.3\text{-}11]$$

If the refractive indices on either side of the lens are the same, then from the similar triangles in Figure 1.3-16, $h/o = h'/i$ and thus

$$M_t = \frac{i}{o} \qquad [1.3\text{-}12]$$

If the refractive indices are different, as they must be on different sides of a *single surface*, then from Equation [1.3-2], $h/o = \gamma$ and $h/i = \gamma'$. Dividing the first of these equations by the second and eliminating h gives

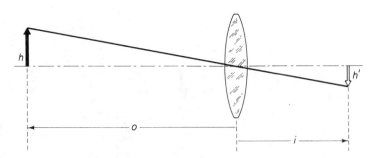

Figure 1.3-16   Deriving transverse magnification.

$$\frac{i}{o} = \frac{\gamma}{\gamma'} \qquad\qquad [1.3\text{-}13]$$

If the refractive indices on both sides of a *lens* are different, too, then o or i, or both, must be replaced by the *reduced* distances, $o \rightarrow o/n_1$ and $i \rightarrow i/n_2$. We divide both sides in Equation [1.3-13] by n_2/n_1:

$$\frac{i/n_2}{o/n_1} = \frac{\gamma/n_2}{\gamma'/n_1}$$

and thus, instead of Equation [1.3-12],

$$M_t = \frac{n_1\, i}{n_2\, o} = \frac{n_1\, \gamma}{n_2\, \gamma'} \qquad\qquad [1.3\text{-}14]$$

But $n\gamma$ is the reduced *vergence* and hence

$$\boxed{M_t = \frac{\mathbf{V}}{\mathbf{V'}}} \qquad\qquad [1.3\text{-}15]$$

Finally, combining Equations [1.3-14] and [1.3-11] by eliminating M_t leads to

$$\boxed{n_1\, h\, \gamma = n_2\, h'\, \gamma'} \qquad\qquad [1.3\text{-}16]$$

which is the *Smith–Helmholtz relationship,* also known as *Lagrange's theorem.* The product $nh\gamma$ is called *Lagrange's invariant.* Return for a moment to Figure 1.3-9. As the object is far away, peripheral rays subtend at the image a *large* angle but the image is *small.* If the object is close (as shown in the fourth example), the rays subtend a small angle and the image is large, which suggests the constancy of the product $h\gamma$.*

If the object is farther away from the lens than twice the focal length, the image is reduced in size. If it is closer, the image is magnified. If both

---

*Robert Smith (1689–1768), British physicist, Plumian professor of physics at Trinity College. Smith seems to have been the first to find this relationship. He wrote two books, *A Compleat System of Opticks in Four Books, viz. A Popular, a Mathematical, a Mechanical, and a Philosophical Treatise* (Cambridge, 1738), a textbook widely used in the eighteenth century, and *Harmonics, or the Philosophy of Musical Sounds* (Cambridge, 1749).

The modern form, Equation [1.3-13], is due to Hermann Ludwig Ferdinand von Helmholtz (1821–1894), German physician and physicist. While still in medical school, von Helmholtz investigated heat production in animals and found the law of conservation of energy. After working for six years as an army surgeon, he became an instructor of physiology in Königsberg, professor of anatomy and physiology in Bonn, then professor

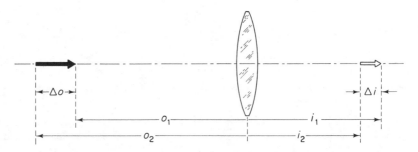

Figure 1.3-17 Deriving axial magnification.

object distance and image distance are equal to twice the focal length, object and image are of the same size. According to the Cartesian sign convention, if an image is inverted, its size, and therefore M_t, are negative. If the image is upright, its size, and M_t, are positive.

2. *Axial magnification.* So far we have always assumed that object and image are both two-dimensional and that they lie in planes normal to the optic axis. But many objects are three-dimensional and thus we should consider the *axial*, or *longitudinal*, magnification. Axial magnification, M_a, is defined as the ratio of a short length, or depth, in the image, measured *along the axis*, to the conjugate length in the object. Using the notation in Figure 1.3-17, we have

$$M_a = \frac{\Delta i}{\Delta o} \qquad [1.3\text{-}17]$$

In order to connect axial and transverse magnification, we take Gauss' lens equation, solve it for f, and apply it to the *head* of the arrow and then to its *foot*:

$$f = \frac{o_1 i_1}{o_1 - i_1} = \frac{o_2 i_2}{o_2 - i_2}$$

$$o_1 i_1 (o_2 - i_2) = o_2 i_2 (o_1 - i_1)$$

Multiplying and rearranging gives

of physics in Berlin. Von Helmholtz wrote two books, covering similar ground as Smith before him, *Handbuch der physiologischen Optik* (Leipzig: L. Voss, 1867) and *Die Lehre von den Tonempfindungen als physiologische Grundlage für die Theorie der Musik* (Braunschweig: F. Vieweg & Sohn, 1863). He invented the ophthalmoscope and the keratometer; extended Young's theory of color vision; contributed to diffraction theory, thermodynamics, fluid dynamics, and meteorology; and showed that the character of a musical tone depends on harmonics present in addition to the fundamental.

Joseph Louis Lagrange (1736–1813), French mathematician, introduced the concept of conjugate elements. As a member of the Committee on Weights and Measures, together with Pierre Simon Laplace and Adrien Marie Legendre, he devised the metric system as we use it today.

$$o_1 o_2 i_2 - o_1 i_1 o_2 = i_1 o_2 i_2 - o_1 i_1 i_2$$

$$(i_2 - i_1)(o_1 o_2) = (o_2 - o_1)(i_1 i_2)$$

and, since $i_2 - i_1 = \Delta i$ and $o_2 - o_1 = \Delta o$,

$$\frac{i_2 - i_1}{o_2 - o_1} = M_a = \frac{i_1 i_2}{o_1 o_2} = M_{t1} M_{t2}$$

$$M_a = M_t^2 \qquad\qquad [1.3\text{-}18]$$

which shows that the axial magnification is equal to the *square* of the transverse magnification.

Example. *The object is a transparent cube, 4 mm across, placed 60 cm in front of a lens of +20 cm focal length. Compare the transverse and axial magnifications.*

Solution. From Gauss' equation, solved for i, we find for the rear surface of the cube (the face closer to the lens) that

$$i = \frac{of}{o + f} = \frac{(-60)(20)}{-60 + 20} = +30 \ cm$$

For the front surface (the face farther away from the lens),

$$i' = \frac{(-60.4)(20)}{-60.4 + 20} = +29.9 \ cm$$

The *transverse* magnification for the rear surface is

$$M_t = \frac{+30}{-60} = \boxed{-0.5 \times}$$

But the *axial* magnification is

$$M_a = \frac{i - i'}{o - o'} = \frac{30 - 29.9}{-60 - (-60.4)} = \boxed{+0.25 \times}$$

which is the square of the transverse magnification.

3. *Angular magnification*, sometimes called "magnifying power," is probably the most important of all. The actual size of an object has little to do with how large it appears to an observer, because this perception depends also on how far away the object is; it is the *angular* size that counts.

Consider first an object held at arm's length away from the eye. As the object is brought closer, the image formed on the retina grows larger. But there is a limit to how close the object can be brought and still remain in focus. Although this limit is different for different persons, the *distance of most distinct vision* is taken to be *25 cm*. At this distance, as we see from

Figure 1.3-18, top, the object subtends at the eye an angle

$$\theta = \tan^{-1} \frac{h}{-25} \qquad [1.3\text{-}19]$$

But there is a way to move the object even closer (and hence see it larger) and still *keep it in focus:* Place a converging lens in front of the eye and move the object into the left-hand focal plane of the lens. A lens used this way is called a *magnifier* (Figure 1.3-18, center). In this case the angle subtended at the eye is

$$\theta' = \tan^{-1} \frac{h}{-f} \qquad [1.3\text{-}20]$$

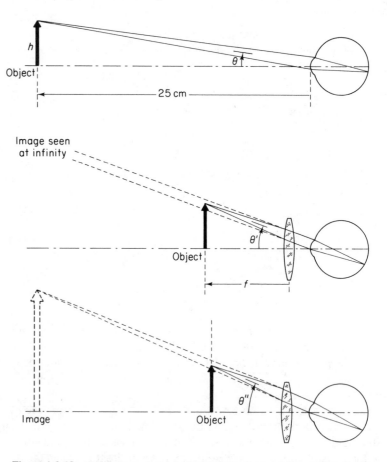

Figure 1.3-18 (*From top to bottom*) Object at the distance of most distinct vision; object as viewed through magnifier and seen at infinity; object as viewed through magnifier and seen again at 25 cm distance.

The angular *magnification,* M_θ, is the ratio of these two angles,

$$M_\theta = \frac{\theta'}{\theta} \qquad\qquad [1.3\text{-}21]$$

If we then insert Equations [1.3-20] and [1.3-19] in [1.3-21], we find that

$$M_\theta = \tan^{-1}\left(\frac{h/-f}{h/-25}\right)$$

For small angles, the tangents can be replaced by the angles themselves so that

$$M_\theta = \frac{25}{f} \qquad\qquad [1.3\text{-}22]$$

where f is given in centimeters. Note that this equation holds only if (a) the eye is focused for infinity *and* (b) the magnifier is placed close to the eye.

   If the eye is *not* focused for infinity, that is, if the observer *accommodates,* the situation becomes slightly more complicated (Figure 1.3-18, bottom). The light entering the eye is then divergent, the same as when reading, and the object can be moved even closer, to within the focal length of the magnifier. This makes the object distance

$$o = \frac{if}{f - i} = \frac{(-25)(f)}{f - (-25)}$$

and θ', instead of being equal to $h/-f$, becomes

$$\theta'' = \frac{(h)(f + 25)}{-25f}$$

Inserting *this* term in Equation [1.3-21] yields

$$M_\theta = \frac{h(f + 25)/-25f}{h/-25} = \frac{25}{f} + 1 \qquad\qquad [1.3\text{-}23]$$

which shows that when focusing at the distance of most distinct vision, at the *near point,* the angular magnification is a little higher than when focusing at infinity.

   Example. *So far, the magnifying lens was thought to be in contact with the eye. What if it is not? Assume that the object is 4 cm in front of a +20.00-diopter lens and the lens is 5 cm in front of the eye. What is the angular magnification?*

   Solution. First consider the lens alone (Figure 1.3-19). Since $o = -4$ cm and $f = 1/+20$ m^{-1} = +5 cm, the image distance (of the virtual image formed by the lens) is

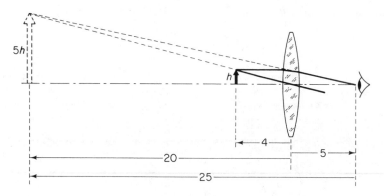

Figure 1.3-19

$$i = \frac{of}{o + f} = \frac{(-4)(5)}{(-4) + 5} = -20 \; cm$$

The *transverse* magnification of this image is

$$M_t = \frac{i}{o} = \frac{-20}{-4} = +5 \times$$

The virtual image now serves as the object. Its distance from the eye is

$$o' = (-20) + (-5) = -25 \; cm$$

If the original object has a height h, then with $5\times$ transverse magnification the virtual image has a height $5h$ and the *angular* magnification, again from Figure 1.3-19, is

$$M_\theta = \frac{\theta'}{\theta} = \frac{5h/-25}{h/-(4 + 5)} = \frac{(5)(-9)}{-25} = \boxed{+1.8 \times}$$

Since angular magnification is given as θ'/θ, and since vergences are the reciprocals of distances, we can also write

$$M_\theta = \frac{\theta'}{\theta} = \frac{h/i}{h/o} = \frac{\mathbf{V'}}{\mathbf{V}} \qquad [1.3\text{-}24]$$

But this is the inverse of Equation [1.3-15], $M_t = \mathbf{V}/\mathbf{V'}$. Therefore, angular magnification and transverse magnification are reciprocal to each other,

$$M_\theta = \frac{1}{M_t} \qquad [1.3\text{-}25]$$

The *total* magnification, M, of a combination of lenses is the *product* of the magnifications of the individual lenses—where we are free to multiply transverse and angular magnifications with each other without hesitation.

Determining Focal Length.   There are several ways of determining the focal length of a lens.

1. Measure the object distance and the image distance, and calculate the focal length using the *thin-lens equation*, $f = oi/(o - i)$.

2. *Bessel's method.* Make the distance between object and image longer than four times the focal length of the lens. Slowly move the lens back and forth along the axis; you will find *two* positions where the image is in focus. In one position the image is magnified and in the other it is reduced in size. Using the notation in Figure 1.3-20, determine the focal length from

$$f = \frac{L^2 - d^2}{4L}$$ [1.3-26]

The advantage of this method is that we need not measure the object distance or image distance.

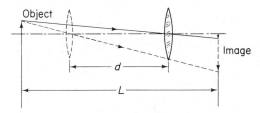

Figure 1.3-20   Bessel's method.

3. *Autocollimation.* (Collimation means placing a source or target in the left-hand focal plane of a converging lens so as to obtain parallel light. Autocollimation means that the light, in addition, is reflected back along the same path and comes to a focus in the plane of the source. In a way, collimation is one-half of autocollimation.)

Mount an illuminated target + image screen at one end of the bench and at some distance from it the lens to be measured and a plane mirror behind it (Figure 1.3-21). Slide the lens + mirror back and forth along the axis until the image of the target is in focus. The distance between the object–image screen and the lens is the focal length. This method is well suited for the student laboratory.

4. *Abbe's method.* Form a real image on the right. Record the position of the object. Measure object size and image size. Determine the transverse

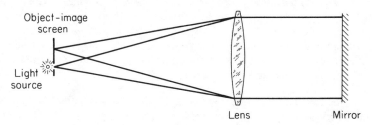

Figure 1.3-21   Autocollimation.

magnification, M_1. Then, leaving the lens in place, move the object a short distance $\mathfrak{A}$ to the right (which makes $\mathfrak{A}$ positive) and focus again, moving the image screen only. Determine the new magnification, M_2. The focal length is then found from Abbe's equation:

$$f = \frac{\mathfrak{A} M_1 M_2}{M_1 - M_2} \qquad [1.3\text{-}27]$$

This method is preferred in industrial laboratories; it requires a microscope for the precise reading of the image size.

  5. The focal length of a *diverging lens* cannot be determined directly as it can with a converging lens. Therefore, another approach is needed. In the *virtual-object method*, for example, we combine the minus lens with a (stronger) plus lens but the two lenses need not be in contact. First form an image using the plus lens only. Note where the image is in focus; this is screen position 1 (Figure 1.3-22).

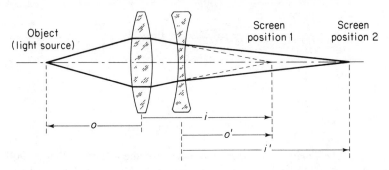

Figure 1.3-22 Determining the power of a diverging lens.

  Then insert the minus lens. Move the screen to the right to bring the image back into focus (screen position 2). The earlier image (position 1) now acts as a virtual object for the minus lens. (The unprimed distances in Figure 1.3-22 refer to the plus lens, the primed distances to the minus lens.) The image distance i, *less* the separation of the two lenses, therefore becomes the object distance o'. The power of the minus lens then follows from $P' = (o' - i')/(o' \, i')$.

  (If the plus lens is too weak, or the minus lens too strong, the combination may not have enough power to form a real image at a reasonable distance or to form an image at all.)

  6. A *lens clock*, also called "lens measure" or "lens gauge," actually measures surface *curvature* but, for a given refractive index, can be calibrated in terms of surface *power*. The clock consists of two outer points and a movable inner point. It is set on the surface so that all three points touch. On a plane surface the points form a straight line; on a curved surface the center point either protrudes, or is pushed in, by a certain distance (Figure 1.3-23).

  If the outer points are a distance D apart, then the center point (on a concave surface) protrudes by the *vertex depth* or *sagitta, s:*

$$s = R - \sqrt{R^2 - \left(\frac{D}{2}\right)^2}$$

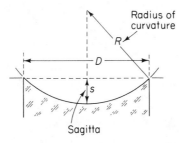

Figure 1.3-23 Deriving the vertex depth formula.

Rearranging, squaring, and, because $s \ll R$, neglecting s^2 gives

$$2Rs \approx \left(\frac{D}{2}\right)^2 \qquad\qquad [1.3\text{-}28]$$

Solving for R and substituting in the surface power equation then leads to

$$\mathfrak{P} \approx \frac{2(n_2 - n_1)\,s}{(D/2)^2}$$

which shows that, D being constant, P is directly proportional to s.

7. A *lensometer* is used to determine the power, and other characteristics, of a lens. A lensometer consists of a light source, a target that can be moved back and forth along the axis, a stationary "standard" lens, and a telescope focused for both object and image at infinity. If there is no other lens present in the path, the target is seen in focus (Figure 1.3-24, top; see next page).

However, if another lens is placed between standard lens and telescope (usually 50 mm from the standard lens), the light entering the telescope is not collimated and the target will not be seen in focus (center). To make the light parallel again, the light to the left of a plus test lens must be divergent, and the light to the left of the standard lens must be more divergent than it was before; hence, the target must be moved to the *right* (bottom). The opposite holds true for a minus lens.

8. In recent years, fully *automated lensometers* have become available. The lens is simply inserted into the instrument and within seconds the built-in optical system and subsequent microprocessor produce the result, often in the form of both a numerical display and a printout.

SUGGESTIONS FOR FURTHER READING

R. S. Longhurst, *Geometrical and Physical Optics,* 3rd edition, pp. 7–15, 30–41, 61–71 (New York: Longman, Inc., 1974).

W. H. A. Fincham and M. H. Freeman, *Optics,* 9th edition, pp. 61–109 (Woburn, MA: Butterworth Publishers, Inc., 1980).

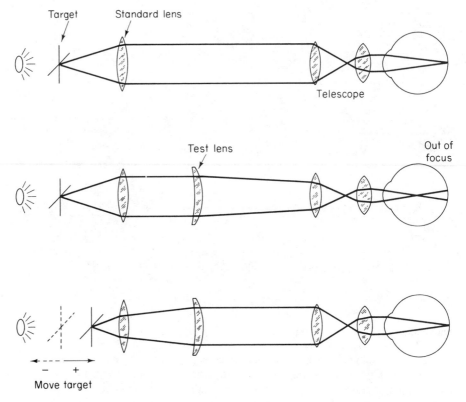

Figure 1.3-24 Ray diagram of a lensometer. (*Top*) Standard lens produces collimated light. Telescope is focused for infinity. Target is seen in focus. (*Center*) Adding unknown lens (plus lens in the example shown) will defocus image. (*Bottom*) Moving target brings image back into focus.

D. D. Michaels, *Visual Optics and Refraction,* 2nd edition, pp. 30–53, 66–72 (St. Louis: The C. V. Mosby Company, 1980).

M. L. Rubin, *Optics for Clinicians,* 2nd edition, pp. 3–115 (Gainsville, FL: Triad Scientific Publishers, 1974).

PROBLEMS

1.3-1. An object is located 2 cm to the left of the convex end of a glass rod which has a radius of curvature of 1 cm. The index of refraction of the glass is $n = 1.5$. Find the image distance.

1.3-2. A large water tank has on its left-hand side a protruding window (of negligible thickness) that is part of a sphere of 4 cm radius. A light source is placed 30 cm to the left of the window, outside the tank. Find the image distance inside the water ($n = \frac{4}{3}$).

1.3-3. A -10.00-diopter biconcave lens has 12 cm radii of curvature for either surface. What is the refractive index of the glass?

1.3-4. A thin lens of refractive index 1.5 has a

focal length of 20 cm. If the radius of curvature of the back surface is -8 cm, what is the radius of the front surface?

1.3-5. A thin lens made of glass of index 1.50 has a front surface of $+11.00$ diopters power and a back surface of -6.00 diopters. What is the power of the lens when submerged in a liquid of index 1.60?

1.3-6. Indicate whether the various surfaces of the lenses in Figure 1.3-25, and the lenses themselves, have positive or negative powers. Also show how parallel light incident from the left will be refracted and how it will emerge from the lenses.

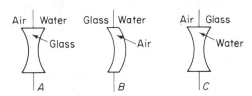

Figure 1.3-25

1.3-7. A thin negative meniscus has radii of curvature of $+50$ cm and $+25$ cm, respectively. If the refractive index of the glass is 1.5, and if air is to the left of the lens and oil ($n = 1.8$) to the right, what is the power of the meniscus?

1.3-8. A biconvex lens, with equal radii of curvature and of index 1.6, separates air to the left of it from water ($n = \frac{4}{3}$) to the right. If the lens is to have $+5.75$ diopters power, what radius of curvature is needed?

1.3-9. An object is placed 30 cm from a lens having a focal length of $+10$ cm. Determine the image distance:
(a) By the Gaussian form of the lens equation.
(b) By the Newtonian form of the lens equation.

1.3-10. Light is made to converge toward a certain point. If then a -6.00-diopter lens is placed 25 cm ahead of this point, where will the light converge, or appear to come from?

1.3-11. When an object is placed 15 cm in front of a thin lens, a virtual image is formed 5 cm away from the lens. What is the focal length of the lens?

1.3-12. A $+5.00$-diopter lens forms a real image on a screen placed 90 cm away from the object. Find the two object distances possible. Solve by the vergence method.

1.3-13. An axial point object is viewed through a diverging lens located $1\frac{1}{2}$ times its (absolute) focal length from the object. Find the conjugate image by ray tracing.

1.3-14. Find the image conjugate to point object O using the combination of a converging and a diverging lens illustrated by Figure 1.3-26. The primed focus refers to the right-hand lens.

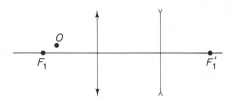

Figure 1.3-26

1.3-15. Light from a point object, located 2 cm off the optic axis, falls on a thin lens of $+10.00$ diopters power. After refraction the light is still divergent, in such a way that the ray, which emerges from the object parallel to the optic axis, after refraction subtends an angle of $7°$ with the chief ray. What is the image distance?

1.3-16. A great many converging lenses, all of the same focal length f, are placed along the axis at distances, between each other, of $d = 3f$, $d = 2f$, or $d = f$. Assuming that the incident light is parallel, trace a ray through each of these combinations.

1.3-17. An object is located 1.25 m in front of a screen. Determine the focal length of a lens that forms on the screen a real, inverted image magnified four times.

1.3-18. If an object is placed 50 cm in front of a -3.00-diopter lens, what is the transverse magnification of the image?

1.3-19. If a lens of 20 mm focal length forms of a given object a real image 10 times magnified, what is:
(a) The object distance?
(b) The image distance?

1.3-20. If the real image formed by a lens is twice as large as the object, and if the distance between object and image is 90 cm, what is the power of the lens?

1.3-21. The least distance between a real object and its *real* image for a certain lens is 40 cm. If a *virtual* image is then formed that is magnified two times, how far from the lens must the object be placed?

1.3-22. If a +8.50-diopter magnifying glass is held close to the eye and the image is seen at the distance of most distinct vision, 25 cm, how far away from the lens is the object?

1.3-23. An object is located 12 cm to the left of a lens of −20 cm focal length. A lens of +4 cm focal length is placed 2 cm to the right of the first lens. Find the image distance, as measured from the second lens.

1.3-24. An object is located 30 cm to the left of a +5.00-diopter lens. A second lens of +10.00 diopters is placed 10 cm to the right of the first lens. Determine the image distance as measured from the second lens and the total magnification.

1.3-25. An object 12 mm high is located 80 cm to the left of a single surface of +2.50 diopters power ground on a glass cylinder of $n = 1.5$. Find:
(a) The image *distance* from Gauss' formula.
(b) The image *size* by Lagrange's theorem.

1.3-26. The image formed on a screen 60 cm from an object is 2 cm in size. If the lens forming the image is moved a certain distance, with-out moving either object or screen, another image comes into focus that is 8 cm in size.
(a) How large is the object?
(b) What is the power of the lens?

1.3-27. The distance between object and screen is 100 cm. If a lens produces an image on the screen when placed at either of two positions 38 cm apart, what is the power of the lens?

1.3-28. Using Abbe's method, find the focal length of a lens from the following data: initial position of the target on the bench 12.0 cm; subsequent position 17.0 cm; target size 12 mm; image size first 6 mm, then 8 mm.

1.3-29. The outer points of a lens clock are 20 mm apart. If the center point is pushed in by 0.8 mm, a reading of +9.25 diopters results. What is the refractive index the lens clock is calibrated for?

1.3-30. If the standard lens in a lensometer has a focal length of 50 mm, how much, and in which direction, must the target be moved when inserting:
(a) a +4.00-diopter lens
(b) a −10.00-diopter lens
and focusing again on the target? *Hint:* The light leaving the test lens must have zero vergence.

1.4

Lens Design

WE ARE NOW READY TO CONSIDER REAL LENSES, often called *thick* lenses. We will also study combinations of such lenses; in fact, the term *lens design* does not necessarily refer to the design of a single lens. A "wide-angle lens," for example, is actually a composite of several lenses.

Any ray that passes through a thick lens, or through a combination of lenses, can be described by two parameters: by its *direction*, with respect to the axis, and by its *height*, above the axis. As we see in Figure 1.4-1, the ray changes direction *at* the various surfaces, and it changes height *between* the surfaces.

There are two ways of following such changes: We can *assume* that the rays change direction at certain hypothetical planes called *principal*

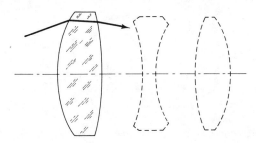

Figure 1.4-1 Changes of height and direction of a ray passing through a thick lens.

planes. If we make this assumption, we can continue using convenient, simple thin-lens equations. Or, we can take the more rigorous, and more modern, approach of using *matrix algebra*. In this chapter we will do both.

Principal Planes

The focal length of a thin lens is measured from the center of the lens. But for a thick lens, or a combination of lenses, we measure the focal length, and object distances and image distances as well, from certain hypothetical surfaces called *principal planes*.

Principal planes are defined as the loci where refraction is assumed to occur without reference as to where it actually does occur. Principal planes do not really exist within a lens; they are conceptual planes. The points where the optic axis intersects these planes (actually they are slightly curved surfaces) are called *principal points*. In a symmetric lens, the two principal planes, H_1 and H_2, straddle the center of the lens, (as in Figures 1.4-2, 1.4-3, and 1.4-6). In a plano-convex lens, and even more so in a meniscus (as in Figure 1.4-4), the two H planes move to one side, and may even move outside of the lens, toward the side of the steeper curvature. And important to note, even a complex lens system has only two principal planes.

Assume that we have a "black box" which, as we find out experimentally, forms an image (Figure 1.4-2). Without considering what might be inside the box, we draw three rays, the parallel ray (1), the focal ray (2), and the chief ray (3), all incident on the box from the left. On the right, these rays emerge from the box and come together, defining the image. If we now continue the incident rays to the right, and back-trace the emergent

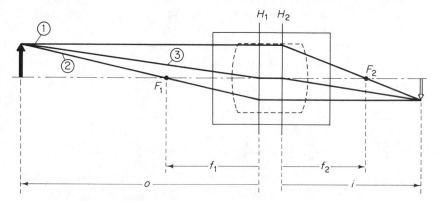

Figure 1.4-2   Principal planes, focal lengths, and object and image distances as they relate to a thick lens.

rays to the left, we can define two planes where *refraction is assumed to take place*.

More specifically, the parallel ray is assumed to be refracted at H_2 and then passes through F_2. The focal ray is refracted at H_1 and thus made parallel (to the axis). The chief ray goes to the first principal point (intersection of H_1 plane and axis), is translated (shifted) to the second principal point, and continues from there, parallel to its initial direction. Between H_1 and H_2 all rays are parallel to the axis.

We see, then, that *the first focal length, f_1, and the object distance, o, are measured from H_1*, and that *the second focal length, f_2, and the image distance, i, are measured from H_2*.

With a negative lens, F_2 lies to the *left* of the lens and the image distance often is negative and hence measured to the left also. But, as shown, f_1 and o are measured from H_1, and f_2 and i are measured from H_2.

The location of the principal planes is invariant for a given lens or lens system. Their positions do not change with the object distance and image distance actually used in a particular experiment.

The concept of equivalent power. How do we find the principal planes? This can be done in two ways: analytically (as I show next) and experimentally, by the nodal slide method (as I will show later). In Figure 1.4-3 we have a system composed of two thick lenses. Each of

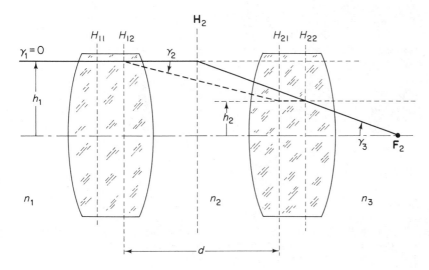

Figure 1.4-3 Deriving the equivalent power equation.

these lenses has two principal planes, H_{11} and H_{12} for the left-hand lens, and H_{21} and H_{22} for the right-hand lens. The *system* as a whole also has two principal planes, $\mathbf{H}_1$ and $\mathbf{H}_2$, but only $\mathbf{H}_2$ is shown.* The second focal point of the system is $\mathbf{F}_2$. We will now derive the focal length as it pertains *to the system*.

In order to do so, we apply Gauss' thin-lens equation to the first lens:

$$\frac{n_2}{i} - \frac{n_1}{o} = \frac{n_2}{f}$$

Replacing n_2/f by power, P, and multiplying both sides by h_1 gives

$$\frac{n_2 h_1}{i} - \frac{n_1 h_1}{o} = P h_1$$

By definition, and following Equation [1.3-2], the slopes before and after refraction are

$$\gamma_1 = \frac{h_1}{o} \quad \text{and} \quad \gamma_2 = \frac{h_1}{i}$$

If $\gamma_1 = 0$, as in this case, then

$$n_2 \gamma_2 = P_1 h_1 \qquad [1.4\text{-}1]$$

We repeat the process for the second lens:

$$n_3 \gamma_3 - n_2 \gamma_2 = P_2 h_2 \qquad [1.4\text{-}2]$$

Adding Equations [1.4-1] and [1.4-2] gives

$$n_3 \gamma_3 = P_1 h_1 + P_2 h_2 \qquad [1.4\text{-}3]$$

Now consider both lenses together. The focal length, **f,** *of the system* is measured from $\mathbf{H}_2$. Since the height of the ray when incident on $\mathbf{H}_2$ is h_1,

$$n_3 \gamma_3 = \mathbf{P} h_1 \qquad [1.4\text{-}4]$$

where $\mathbf{P}$ is the power of the system. Substituting Equation [1.4-4] into [1.4-3] gives

$$\mathbf{P} h_1 = P_1 h_1 + P_2 h_2 \qquad [1.4\text{-}5]$$

From Figure 1.4-3 we also see that

$$\gamma_2 = \frac{h_1 - h_2}{d}$$

---

*Interestingly, with a combination of two lenses the sequence of the **H** planes is *reversed:* $\mathbf{H}_2$ is on the left and $\mathbf{H}_1$ is on the right (see, for example, Figures 1.4-5 and 1.4-8), just the opposite as with a single thick lens.

Thus

$$h_2 = h_1 - \gamma_2 d$$

Substituting Equation [1.4-1] gives

$$h_2 = h_1 - \frac{P_1 h_1 d}{n_2} \qquad [1.4\text{-}6]$$

and substituting this in Equation [1.4-5] yields

$$\mathbf{P}h_1 = P_1 h_1 + P_2 h_1 - \frac{P_1 P_2 h_1 d}{n_2}$$

Dividing by h then leads to

$$\boxed{\mathbf{P} = P_1 + P_2 - P_1 P_2 \frac{d}{n_2}} \qquad [1.4\text{-}7]$$

which is *Gullstrand's equation.** This is an important equation which *describes the equivalent power of the system.*

Converting the powers in Equation [1.4-7] into focal lengths, $P \rightarrow 1/f$, and multiplying the right-hand side by $f_1 f_2 / f_1 f_2$ yields

$$\boxed{\mathbf{f} = \frac{f_1 f_2}{f_1 + f_2 - d/n_2}} \qquad [1.4\text{-}8]$$

which is the *equivalent focal length of the system.* The term n_2 in both Equations [1.4-7] and [1.4-8] is the refractive index of the medium *between* the two lenses; it is *not* the refractive index of the glass. But if the medium between the lenses *were* glass, we would in effect have a *thick lens.*

If the two lenses are in air, then $n_2 \approx 1$, and

$$\mathbf{P} = P_1 + P_2 - P_1 P_2 d \qquad [1.4\text{-}9]$$

---

*Allvar Gullstrand (1862–1930), Swedish ophthalmologist and professor of physiological and physical optics at the University of Uppsala. Gullstrand is best known for the *schematic eye* and for his development of the slit lamp and the ophthalmoscope. He also contributed to the tracing of skew rays, caustic surfaces (in aberrations), accommodation, aspheric lenses for aphakics, and to the propagation of light through gradient-index media (such as the crystalline lens of the eye). In 1911 he received the Nobel prize in physiology for his "work on the diffraction of light by lenses applied to the eye" and, a few years later, published a detailed exposition of geometric-optical image formation, replete with numerous equations, "Das allgemeine optische Abbildungssystem," *Kungliga Svenska vetenskapsakademiens handlingar* (4) **55** (1915–16), 1–139.

Table 1.4-1 Equivalent (and Approximate) Powers and Focal Lengths

Description	Power	Focal length
Two *surfaces* of power $\mathfrak{P}_1$ and $\mathfrak{P}_2$, separated by distance d, and enclosing medium of index n	$\mathbf{P} = \mathfrak{P}_1 + \mathfrak{P}_2 - \mathfrak{P}_1\mathfrak{P}_2\dfrac{d}{n}$	$f = \dfrac{\mathfrak{f}_1\mathfrak{f}_2}{\mathfrak{f}_1 + \mathfrak{f}_2 - \dfrac{d}{n}}$
Two *lenses*, of power P_1 and P_2, separated by distance d	$\mathbf{P} = P_1 + P_2 - P_1P_2d$	$f = \dfrac{f_1f_2}{f_1 + f_2 - d}$
Two elements, of power P_1 and P_2, in contact	$\mathbf{P} = P_1 + P_2$	$f = \dfrac{f_1f_2}{f_1 + f_2}$

and if the two lenses are in contact, $d = 0$ and

$$\mathbf{P} = P_1 + P_2 \qquad\qquad [1.4\text{-}10]$$

which is the *approximate power* of the combination, the same equation, Equation [1.3-6], that we have seen before when considering the power of a thin lens as the sum of the powers of its two surfaces. Table 1.4-1 summarizes these findings.

From Equation [1.4-7] we derive another useful equation. Recall that the surface powers of a lens, in air, are

$$\mathfrak{P}_1 = \frac{n-1}{R_1} \quad \text{and} \quad \mathfrak{P}_2 = \frac{1-n}{R_2} = \frac{n-1}{-R_2}$$

Substituting these terms in Equation [1.4-7] gives

$$\mathbf{P} = \frac{n-1}{R_1} - \frac{n-1}{R_2} + \frac{n-1}{R_1}\left(\frac{n-1}{R_2}\right)\frac{d}{n}$$

and thus

$$\boxed{\mathbf{P} = (n-1)\left(\frac{1}{R_1} - \frac{1}{R_2}\right) + \frac{(n-1)^2 d}{nR_1R_2}} \qquad\qquad [1.4\text{-}11]$$

which is the *lens-makers formula for a thick lens*. Note that the first part of this equation is simply the familiar lens-makers formula for a thin lens. If we imagine that the two surfaces of the lens are being pulled apart through distance d, the (equivalent) power of the lens changes by the last term.

Front vertex power and back vertex power.   Equivalent power is measured *at* the respective principal plane. This statement makes more sense when we realize that the reciprocal of equivalent power is the equivalent focal length, and this focal length is measured *from* the respective principal plane. Unfortunately, we often do not know where the principal planes are, and when we do know, they often are inaccessible within the lens, or lens system. What we need, then, are clearly defined and easily accessible reference points from which to measure. These reference points are the *vertices* of the lens. In Figure 1.4-4, for example, a_1 is the *front vertex focal length* and a_2 is the *back vertex focal length*. The reciprocals of these focal lengths are the *front vertex power*, P_1, and the *back vertex power, P_2*.

Be sure to distinguish between back *surface* power and back *vertex* power. Back surface power is simply the power of the back surface. But back vertex power refers to the power of the whole lens (front surface plus back surface), *as measured at the back vertex.*

The two focal lengths a_1 and a_2 can be determined, very conveniently, by autocollimation. Use an illuminated slit or other target in a screen to the left of the lens and a plane mirror to the right. Move the lens back and forth and align the mirror until the image of the slit appears in focus on the screen, in the same plane as the slit. Then the screen is at F_1 and it is easy to measure a_1. Next turn the lens around so that its right-hand side faces the target, giving a_2.

The *front vertex power* is also called "neutralizing power" (because lenses, of the same power but opposite sign, are placed in contact with the front vertex of a spectacle lens for neutralization). With a lens of appreciable thickness, or with two lenses separated by an appreciable distance, we cannot use the approximate power equation, $\mathbf{P} = P_1 + P_2$, Equation

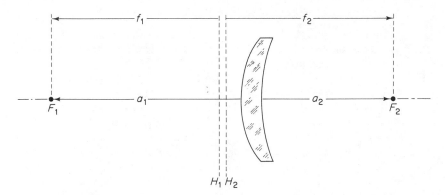

Figure 1.4-4   Conventional focal lengths, f_1 and f_2, and front vertex and back vertex focal length, a_1 and a_2, of a thick positive meniscus.

[1.4-10]. At the front vertex of a thick lens, or at the first of two thin lenses, the first power term, P_1, would be measured correctly but because of the separation, the second power term, P_2, would not be. Therefore, P_2 in Equation [1.4-10] has to change. This is done by replacing the vergences in Equation [1.1-5] by powers and thus

$$\mathbf{P}_1 = P_1 + \frac{P_2}{1 - P_2(d/n)} \qquad [1.4\text{-}12]$$

which is approximately equal to

$$\mathbf{P}_1 \approx P_1 + P_2 + P_2^2\frac{d}{n} \qquad [1.4\text{-}13]$$

Back vertex power, or, as it is sometimes called, "effective power," is the most important of all. Virtually all ophthalmic procedures, from determining visual acuity to writing lens prescriptions to using a lensometer, refer to the back vertex power. In analogy with Equation [1.4-12] we now have

$$\mathbf{P}_2 = \frac{P_1}{1 - P_1(d/n)} + P_2 \qquad [1.4\text{-}14]$$

or, approximately,

$$\mathbf{P}_2 \approx P_1 + P_2 + P_1^2\frac{d}{n} \qquad [1.4\text{-}15]$$

Example. *What happens if two positive lenses of equal power are first in contact and then are gradually pulled apart from each other?*

Solution. When the lenses are in contact, the power of the combination is highest, the focal length is least, and the two **H** planes are close together inside the system (Figure 1.4-5a).

Now, as we separate the lenses, the **H** planes separate too. For example, if the distance between the lenses is equal to their focal length, the **H** planes coincide with the lenses (b). When the distance between the lenses is less than twice their focal length, the **H** planes have moved outside the lenses (c); and when the distance is equal to $2f$ and the system becomes *afocal,* $\mathbf{H}_2$ and $\mathbf{H}_1$ have moved out to infinity (d). Predictably, as we separate the lenses even further, the power of the system becomes negative: Note that the ray emerging on the right-hand side has continued to turn counterclockwise and now, in (e), is pointing *upward,* apparently refracted at $\mathbf{H}_2$, as it should.

The question of whether a system has positive or negative power is easy to answer: If the emergent ray, after refraction at $\mathbf{H}_2$, points toward the

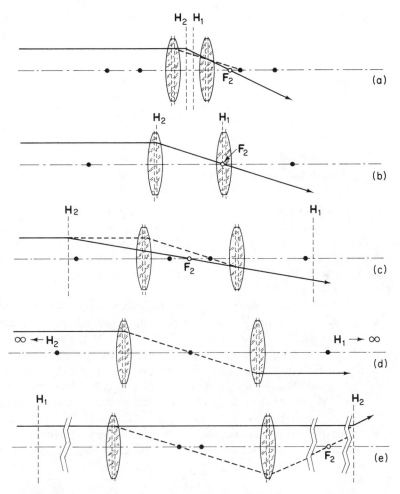

H₂ H₁

F₂

(a)

H₂   H₁

F₂

(b)

H₂   H₁

F₂

(c)

∞ ← H₂   H₁ → ∞

(d)

H₁   H₂

F₂

(e)

Figure 1.4-5 Location of principal planes of a system of two lenses as a function of distance between the lenses. Focal points of individual lenses are identified by dots, second focal points of system, **F₂**, by hollow circles.

axis and crosses it at **F₂** [(a), (b), and (c)], the system is positive. If the ray turns away from the axis, without crossing it (e), the system is negative.

Assume, for example, that the two lenses have +8.00 diopters power each and that they are separated by a distance d. At which distance does the power of the combination change from positive to negative?

Numerical solution. We take Gullstrand's equation, Equation [1.4-7], and set it equal to zero:

$$(+8.00) + (+8.00) - (+8.00)(+8.00)(d) = 0$$

Then

$$16 = 64d$$

and

$$d = \frac{16}{64} = 0.25 \ m = \boxed{25 \ cm}$$

Since $1/8.00 = 12.5$ cm, this is twice the focal length of either lens; at that distance the system is *afocal*.

Nodal points.   We have seen that the principal planes are where all refraction is assumed to occur. In contrast, the nodal points are where *no refraction occurs*. In a thin lens the nodal point is at the center of the lens: light passing through the center does so without refraction. In a thick lens, as with the principal points, this center separates, producing *two nodal points*.

Whenever the refractive indices on either side of the lens are the same, the nodal point N_1 coincides with the principal point H_1, and N_2 coincides with H_2. This is shown in Figure 1.4-6. Note that the chief ray, coming from the tip of the object, is headed toward N_1. Between N_1 and N_2 the chief ray is merely displaced, with no change in direction; it emerges from N_2 parallel to its initial direction. (If the refractive indices on the two sides of the lens were different, the N points would move away from the H planes, *toward* the side of *the higher index*.) Now, if we knew the locations of the N points, we would also know the locations of the H planes (which are important because from there the focal lengths and the object and image distances are measured). We find the N points using the *nodal slide method*.

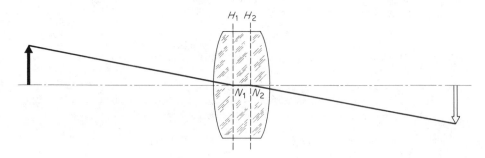

Figure 1.4-6   Nodal points. Note the displacement of the chief ray.

A nodal slide is a particular kind of a lens support that can be swiveled back and forth about a vertical axis. The image formed by the lens or lens system is watched for any sideways motion. To operate, cautiously move the lens support *on top of* the nodal slide in one direction and move the nodal slide *as a whole* an equal distance in the opposite direction. *Keep the image in focus all the time*! At the position where, on rotating the nodal slide, there is *no sideways image motion,* the axis of rotation of the nodal slide goes through the (second) nodal point, N_2 in Figure 1.4-7. Next, using autocollimation as described on page 73, determine the back vertex focal length, a_2, of the system. The system is then turned around and the procedure repeated, locating N_1, and finding a_1.

Now remove the collimating lens. Move the nodal slide with the lens or lenses on it farther away from the target and form an image of the target on a screen placed some distance to the right. This gives us two conjugate planes, O and I in Figure 1.4-8. The distances of O and I from the respective vertices are called b_1 and b_2. Subtract a_1 from b_1, and a_2 from b_2. This results, by definition, in the two Newton distances, N_o and N_i:

$$b_1 - a_1 = N_o \qquad \text{and} \qquad b_2 - a_2 = N_i$$

The (equivalent) focal length of the system, **f,** is then found from Newton's equation, $N_o N_i = f_1 f_2$ (page 48). Knowing the **f**'s, we can again locate H_1 and H_2, which serves as a check on our earlier determination.

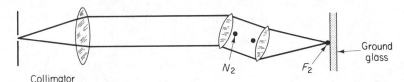

Figure 1.4-7   Principle of the nodal slide method (seen from top). Rotation of lens system about N_2 will cause no image motion at F_2.

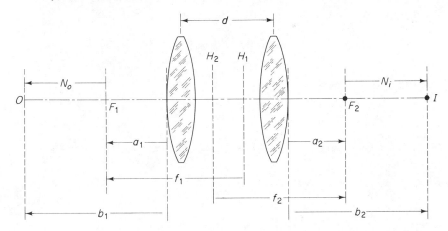

Figure 1.4-8   Notation used with the nodal slide method.

Matrix Methods in Lens Design

From Figure 1.4-1 we are familiar with the changes that occur to a ray of light as it passes through a thick lens or a combination of lenses. At each surface, the direction of the ray changes, but the height does not; this is called *refraction*. Within each interval between surfaces, the height of the ray changes but the direction does not; this is called *translation*. Thus we need two operators, one for the refraction process and the other for the translation process, to fully describe the ray as it progresses through surface after surface. These two operators are the *refraction matrix* and the *translation matrix*.

Matrix algebra is an elegant and powerful tool. It is widely used in more advanced geometric optics such as in any *lens design,* but it plays an important role also in physical optics (for example in polarization, page 344). In particular, matrices connect readily to the concept of vergence on one hand and, since they require little more than multiplication and addition, they can easily be written as computer programs on the other. Matrix methods are the modern way to go.

Refraction matrix.   We start out with Gauss' formula for refraction at a single surface, Equation [1.3-3]:

$$\frac{n_1}{o} + \frac{n_2 - n_1}{R} = \frac{n_2}{i}$$

As before, o is the object distance, i the image distance, R the radius of curvature of the surface, and n_1 and n_2 are the refractive indices of the media in front of, and behind, the surface. Replacing the second term by the power of the surface and rearranging gives

$$\mathfrak{P} + \frac{n_1}{o} = \frac{n_2}{i} \qquad [1.4\text{-}16]$$

From Equation [1.3-2],

$$\gamma = \frac{h}{o} \qquad \text{and} \qquad \gamma' = \frac{h}{i}$$

Solving for o and i, respectively, and substituting in Equation [1.4-16] yields

$$\mathfrak{P} \quad + \quad \frac{n_1 \gamma}{h} \quad = \quad \frac{n_2 \gamma'}{h} \qquad [1.4\text{-}17]$$

refraction   input   output

The term marked *input* refers to the initial coordinates of the ray vector (its direction, the refractive index of the medium in which it lies, and its height above the axis). The term *refraction* refers to refraction at the surface, and *output* refers to the coordinates of the ray as it emerges from the surface, after refraction. Note the sequence in which these terms are written, *refraction* to the *left* of *input*.

Now we convert Equation [1.4-17] into matrix form*:

$$\begin{bmatrix} 1 & \mathfrak{P} \\ 0 & 1 \end{bmatrix} \quad \begin{bmatrix} n_1 \gamma \\ h \end{bmatrix} = \begin{bmatrix} n_2 \gamma' \\ h \end{bmatrix} \qquad [1.4\text{-}18]$$

$$\quad \textit{refraction} \quad \textit{input} \quad \textit{output}$$

The input and output terms are 2×1 matrices (2 rows, 1 column). The left-hand term is a 2×2 matrix (2 rows, 2 columns); it is called the *refraction matrix,* R:

$$\boxed{\; \mathsf{R} = \begin{bmatrix} 1 & \mathfrak{P} \\ 0 & 1 \end{bmatrix} \;} \qquad [1.4\text{-}19]$$

If the surface is plane, the upper right-hand element becomes zero, $\mathfrak{P} = 0$, and we have a *unit matrix.*

In order to use such matrices, and to transform an initial set of coordinates into a new set of coordinates, the matrices are multiplied. Consider, for example, the two matrices

$$\begin{bmatrix} 4 & 3 \\ 6 & 5 \end{bmatrix} \quad \begin{bmatrix} 1 \\ 2 \end{bmatrix}$$

Take the top row in the left-hand matrix, the shaded area containing 4 3, and multiply it with the right-hand matrix. Add the two products. This gives the *upper* element in the resultant matrix:

$$\begin{bmatrix} 4 & 3 \\ \cdot & \cdot \end{bmatrix} \quad \begin{bmatrix} 1 \\ 2 \end{bmatrix} \rightarrow \begin{matrix} 4 \times 1 \\ 3 \times 2 \end{matrix} \rightarrow 4 + 6 \rightarrow \begin{bmatrix} \textcircled{10} \end{bmatrix}$$

---

*The matrices used here are different from those used by other authors. In particular, the power element, $\mathfrak{P}$, in the refraction matrix does not have a minus sign in front of it (but the reduced-distance element, d/n, in the translation matrix does). This facilitates the use of the *Gaussian constants* and of the *vergence matrix*, both to be introduced shortly. Writing matrices this way and deriving the constant k in the vergence matrix are due to Robin G. Simpson, at that time one of my students.

Then take the bottom row. Multiply and add. This gives the *lower* element:

$$\begin{bmatrix} 4 & 3 \\ 6 & 5 \end{bmatrix} \begin{bmatrix} 1 \\ 2 \end{bmatrix} \rightarrow \begin{matrix} 6 \times 1 \\ 5 \times 2 \end{matrix} \rightarrow 6 + 10 \rightarrow \begin{bmatrix} 10 \\ \boxed{16} \end{bmatrix}$$

Example. *A ray of light, emerging from an axial point object, subtends an angle of $\gamma = -6°$ with the optic axis. At a height $h = 2$ cm it intersects a convex spherical surface, ground with a radius of $+5$ cm on glass of index $n = 1.62$. Find the angle the ray subtends with the axis after refraction.*

Solution. First we determine the power of the surface:

$$\mathfrak{P} = \frac{n_2 - n_1}{R} = \frac{1.62 - 1.00}{0.05} = +12.4\,m^{-1}$$

Next we convert the angle into radians, $-6° = -0.1047$ rad. Then from Equation [1.4-18],

$$\begin{bmatrix} 1 & 12.4 \\ 0 & 1 \end{bmatrix} \begin{bmatrix} -0.1047 \\ 0.02 \end{bmatrix} = \begin{bmatrix} -0.1047 + 0.248 \\ 0.02 \end{bmatrix} = \begin{bmatrix} 0.1433 \\ 0.02 \end{bmatrix} = \begin{bmatrix} n_2\gamma' \\ h \end{bmatrix}$$

The angle subtended by the refracted ray is

$$\gamma' = \frac{0.1433}{n_2} = \frac{0.1433}{1.62} = 0.0885 \text{ rad} = \boxed{+5°}$$

Translation matrix.   Next we consider the change of height, Δh, of the ray as it proceeds from one surface to the next. If d is the distance (measured along the axis) between the two surfaces, then

$$\Delta h = h_1 - h_2 = d \tan \gamma' \qquad\qquad [1.4\text{-}20]$$

where γ' is the angle (in radians) which the ray subtends with the axis after refraction. For paraxial rays,

$$h_1 - h_2 = d\gamma'$$

Dividing both sides by $n_2\gamma'$ and rearranging yields

$$-\frac{d}{n_2} + \frac{h_1}{n_2\gamma'} = \frac{h_2}{n_2\gamma'} \qquad\qquad [1.4\text{-}21]$$

$$\textit{translation} \quad \textit{input} \quad \textit{output}$$

which again can be written in matrix form:

$$\begin{bmatrix} 1 & 0 \\ -d/n_2 & 1 \end{bmatrix} \begin{bmatrix} n_2\gamma' \\ h_1 \end{bmatrix} = \begin{bmatrix} n_2\gamma' \\ h_2 \end{bmatrix} \qquad\qquad [1.4\text{-}22]$$

$$\textit{translation} \qquad \textit{input} \qquad \textit{output}$$

The left-hand term is the *translation matrix*, T:

$$T = \begin{bmatrix} 1 & 0 \\ -d/n & 1 \end{bmatrix} \qquad\qquad [1.4\text{-}23]$$

Obviously, when a ray passes through a *lens* (rather than through a single surface), refraction occurs twice, *at* each surface, and translation occurs once, *between* surfaces. Therefore, in order to describe the function of a lens, we need two refraction matrices, R_1 and R_2, and one translation matrix, T. Multiplication of these matrices, written from right to left, leads to the *system matrix,* S:

$$R_2 \quad T \quad R_1 = S$$

and substituting the actual refraction and translation matrices, we have

$$\begin{bmatrix} 1 & \mathscr{P}_2 \\ 0 & 1 \end{bmatrix} \begin{bmatrix} 1 & 0 \\ -d/n & 1 \end{bmatrix} \begin{bmatrix} 1 & \mathscr{P}_1 \\ 0 & 1 \end{bmatrix} = S \qquad [1.4\text{-}24]$$

Multiplication of two 2×2 matrices is carried out as follows:

$$\begin{bmatrix} 6 & 5 \\ 8 & 7 \end{bmatrix} \begin{bmatrix} 2 & 1 \\ 4 & 3 \end{bmatrix}$$

Take the top *row* in the left-hand matrix, the shaded area containing 6   5, and multiply it with the left-hand *column*, the shaded area containing 2   4, in the right-hand matrix. Add the products. This gives the upper left-hand element in the resultant matrix:

$$\begin{bmatrix} 6 & 5 \\ \cdot & \cdot \end{bmatrix} \begin{bmatrix} 2 & \cdot \\ 4 & \cdot \end{bmatrix} \rightarrow \begin{matrix} 6 \times 2 \\ 5 \times 4 \end{matrix} \rightarrow 12 + 20 \rightarrow \begin{bmatrix} \circled{32} & \\ & \end{bmatrix}$$

Continue with the top row and multiply it with the *right*-hand column, 1   3. This gives the upper right-hand element:

$$\begin{bmatrix} 6 & 5 \\ \cdot & \cdot \end{bmatrix} \begin{bmatrix} 2 & 1 \\ 4 & 3 \end{bmatrix} \rightarrow \begin{matrix} 6 \times 1 \\ 5 \times 3 \end{matrix} \rightarrow 6 + 15 \rightarrow \begin{bmatrix} 32 & \circled{21} \\ & \end{bmatrix}$$

Do the same with the *bottom* row of the left-hand matrix and the left-hand column of the right-hand matrix:

$$\begin{bmatrix} 6 & 5 \\ 8 & 7 \end{bmatrix} \begin{bmatrix} 2 & 1 \\ 4 & 3 \end{bmatrix} \rightarrow \begin{matrix} 8 \times 2 \\ 7 \times 4 \end{matrix} \rightarrow 16 + 28 \rightarrow \begin{bmatrix} 32 & 21 \\ \circled{44} & \end{bmatrix}$$

and again with the right-hand column:

$$\begin{bmatrix} 6 & 5 \\ 8 & 7 \end{bmatrix} \begin{bmatrix} 2 & 1 \\ 4 & 3 \end{bmatrix} \rightarrow \begin{matrix} 8 \times 1 \\ 7 \times 3 \end{matrix} \rightarrow 8 + 21 \rightarrow \begin{bmatrix} 32 & 21 \\ 44 & \boxed{29} \end{bmatrix}$$

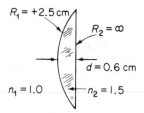

$R_1 = +2.5\,cm$

$R_2 = \infty$

$d = 0.6$ cm

$n_1 = 1.0$ $n_2 = 1.5$

Figure 1.4-9 Plano-convex lens.

Example. *Consider a plano-convex lens of n = 1.5 and center thickness 6 mm. The convex surface has a radius of 2.5 cm and is facing the incident light (Figure 1.4-9). Determine the system matrix.*

Solution. The front surface power is

$$\mathfrak{P}_1 = \frac{n_2 - n_1}{R_1} = \frac{1.5 - 1.0}{0.025} = +20\,m^{-1}$$

and the back surface power

$$\mathfrak{P}_2 = \frac{1.0 - 1.5}{\infty} = 0$$

Thus the system matrix is

$$\begin{bmatrix} 1 & 0 \\ 0 & 1 \end{bmatrix} \begin{bmatrix} 1 & 0 \\ -0.006/1.5 & 1 \end{bmatrix} \begin{bmatrix} 1 & 20 \\ 0 & 1 \end{bmatrix}$$

Multiplying the two right-hand matrices first, we obtain

$$\begin{bmatrix} 1 & 20 \\ -0.004 & 0.92 \end{bmatrix}$$

Further multiplication (with the left-hand matrix) does not change the result because R_2 is a unit matrix, and hence

$$\mathsf{S} = \begin{bmatrix} 1 & 20 \\ -0.004 & 0.92 \end{bmatrix}$$

To check the result, calculate the *determinant,* multiplying the elements on the rightward arrow and subtracting the product of the elements on the leftward arrow:

$$\begin{matrix} 1 & 20 \\ -0.004 & 0.92 \end{matrix}$$

$$(1 \times 0.92) - (20 \times -0.004) = 0.92 + 0.08 = 1.00$$

The result should be *unity*, as indeed it is.

System matrix.

The system matrix of a realistic ("thick") lens contains, as we have seen from Equation [1.4-24], two refraction matrices and one translation matrix:

$$\mathsf{R}_2\mathsf{T}\mathsf{R}_1 = \mathsf{S} \qquad\qquad [1.4\text{-}25]$$

If we have two such (realistic) lenses, separated by a certain distance from each other,

$$R_4 T_3 R_3 T_2 R_2 T_1 R_1 = S \qquad [1.4\text{-}26]$$

where R_1, T_1, and R_2 refer to the first lens, T_2 to the translation between the two lenses, and R_3, T_3, and R_4 to the second lens. The important point is that any optical system, no matter how complex, can be fully described by *one* system matrix, just as such a system can be represented by *one* set of principal planes.

A system matrix contains four elements, a, b, c, d, known as the *Gaussian constants:*

$$S = \begin{bmatrix} b & a \\ d & c \end{bmatrix} \qquad [1.4\text{-}27]$$

These Gaussian constants represent various parameters of the system. Constant a, for example, is the *equivalent power*. Its reciprocal is the *equivalent focal length,*

$$f_1 = -\frac{n_1}{a} \quad \text{and} \quad f_2 = \frac{n_3}{a} \qquad [1.4\text{-}28]$$

Constant b is the *angular magnification* of the system. Constants a, b, and c define the *front vertex focal length*, a_1 (see Figure 1.4-4):

$$a_1 = -\frac{bn_1}{a} \qquad [1.4\text{-}29]$$

and the *back vertex focal length*, a_2:

$$a_2 = \frac{cn_3}{a} \qquad [1.4\text{-}30]$$

These same constants also describe the distances from the vertices to the principal planes. For example, the distance from the first, left-hand vertex, V_1, of a lens to the first principal plane, H_1, is

$$V_1 H_1 = n_1 \left(\frac{1-b}{a} \right) \qquad [1.4\text{-}31]$$

and the distance from the second, right-hand vertex, V_2, to the H_2 plane is

$$V_2 H_2 = n_3 \left(\frac{c-1}{a} \right) \qquad [1.4\text{-}32]$$

Constant d is the (negative, reduced) *thickness* of the lens, $-d/n_2$. In the case of two thin lenses (in air), d is the *separation* of the lenses and n_2 drops out. The system matrix of a realistic, thick lens, therefore, can be written in detail as

$$S = \begin{bmatrix} b & a \\ d & c \end{bmatrix} = \begin{bmatrix} 1 - \mathscr{P}_2\dfrac{d}{n_2} & \mathscr{P}_1 + \mathscr{P}_2 - \mathscr{P}_1\mathscr{P}_2\dfrac{d}{n_2} \\ \dfrac{-d}{n_2} & 1 - \mathscr{P}_1\dfrac{d}{n_2} \end{bmatrix} \qquad [1.4\text{-}33]$$

For a thin lens, where $d \to 0$, this reduces to

$$S = \begin{bmatrix} 1 & \mathscr{P}_1 + \mathscr{P}_2 \\ 0 & 1 \end{bmatrix} \qquad [1.4\text{-}34]$$

Note that **a**, for a thin lens, is the *approximate* power, whereas for a thick lens it is the *equivalent* power. As for **b**, we realize that a thin lens cannot provide angular magnification. Angular magnification requires two lenses, or at least it requires two surfaces separated by an appreciable distance; it requires a *very thick* lens. We will discuss angular magnification in more detail when we come to telescopes (page 148).

Example. *A single thick lens has radii of curvature $R_1 = +8$ cm, $R_2 = -5$ cm, a center thickness $d = 1$ cm, and a refractive index $n = 1.6$. Determine:*
(a) The system matrix.
(b) The focal length.
(c) The positions of the principal planes.

Solution. (a) The surface powers are

$$\mathscr{P}_1 = \frac{n_2 - n_1}{R} = \frac{1.6 - 1.0}{+0.08\ m} = +7.5\ m^{-1}$$

and

$$\mathscr{P}_2 = \frac{1.0 - 1.6}{-0.05\ m} = +12\ m^{-1}$$

Thus the system matrix is

$$S = \begin{bmatrix} 1 & 12 \\ 0 & 1 \end{bmatrix} \begin{bmatrix} 1 & 0 \\ \dfrac{-0.01}{1.6} & 1 \end{bmatrix} \begin{bmatrix} 1 & 7.5 \\ 0 & 1 \end{bmatrix} = \begin{bmatrix} 1 & 12 \\ 0 & 1 \end{bmatrix} \begin{bmatrix} 1 & 7.5 \\ -0.00625 & 0.953 \end{bmatrix}$$

$$= \begin{bmatrix} 0.925 & 18.936 \\ -0.00625 & 0.953 \end{bmatrix}$$

As a check,

$$(0.925)(0.953) - (18.936)(-0.00625) = 1$$

(b) The focal length or, more precisely, the second, right-hand, equivalent focal length is simply the reciprocal of **a**:

$$\mathbf{f_2} = \frac{1}{18.936} = 0.0528 = \boxed{+5.28 \text{ cm}}$$

(c) The principal planes, with respect to the vertices, are located at distances of

$$V_1 H_1 = \frac{1 - b}{a} = \frac{1 - 0.925}{18.936} = 0.00396 \approx \boxed{4 \text{ mm}}$$

and

$$V_2 H_2 = \frac{c - 1}{a} = \frac{0.953 - 1}{18.936} = -0.00248 \approx \boxed{-2.5 \text{ mm}}$$

We may use Equations [1.4-29] and [1.4-30] as a check because the "full" focal lengths are the sums of $a_1 + V_1 H_1$ and $H_2 V_2 + a_2$. Indeed,

$$\mathbf{f_1} = -\left(\frac{b}{a} + V_1 H_1\right) = -\left(\frac{0.925}{18.936} + 0.004\right) = -0.0528 = \boxed{-5.28 \text{ cm}}$$

and

$$\mathbf{f_2} = H_2 V_2 + \frac{c}{a} = +0.0025 + \frac{0.953}{18.936} = 0.0528 = \boxed{+5.28 \text{ cm}}$$

the same result as in (b).

The matrix description of image formation. The system matrix relates to the properties of a lens or combination of lenses in a way similar to the lens-makers formula. Now we turn again to lens-users equations and consider image formation. Consider the "typical" case of an axial point object, a converging lens, and a real image. Clearly, there is first a translation, from the object to the lens, then the system matrix, and then another translation, from the lens to the image.

The distances in these two translations are the object distance, o, and the image distance, i, or if the system is in a medium other than air, they are the *reduced* object and image distances, o/n_1 and i/n_2. The whole sequence, from object to image, can then be written in the form of an *object–image matrix*:

$$\underbrace{\begin{bmatrix} 1 & 0 \\ -i/n_2 & 1 \end{bmatrix} \begin{bmatrix} b & a \\ d & c \end{bmatrix} \begin{bmatrix} 1 & 0 \\ -o/n_1 & 1 \end{bmatrix}}_{system} \underbrace{\begin{bmatrix} n_1 \gamma \\ h \end{bmatrix}}_{input} = \underbrace{\begin{bmatrix} n_2 \gamma' \\ h \end{bmatrix}}_{output} \qquad [1.4\text{-}35]$$

Using this matrix, we can transform the coordinates of a ray emerging from the object into the coordinates of the same ray reaching the image.

But since I have presented geometric optics in terms of vergence, I now introduce the *vergence matrix*, V. That is simply a 2×1 matrix which can be derived from Equation [1.4-17]. By definition, $n_1\gamma/h$ in that equation is the entrance vergence, $\mathbf{V}$, and $n_2\gamma'/h$ is the exit vergence, $\mathbf{V}'$. Multiplying by $1/h$ and substituting in Equation [1.4-18] gives

$$\begin{bmatrix} 1 & \mathfrak{P} \\ 0 & 1 \end{bmatrix} \begin{bmatrix} \mathbf{V} \\ 1 \end{bmatrix} = \begin{bmatrix} \mathbf{V}' \\ 1 \end{bmatrix} \qquad [1.4\text{-}36]$$

In simple cases the vergence matrix takes the form shown here, with $\mathbf{V}$ being the upper element and 1 the lower. However, when used with a system matrix, the lower element of the resultant is rarely 1. Multiplying $\mathsf{S} \times \mathsf{V}$, for example, gives

$$\begin{bmatrix} b & a \\ d & c \end{bmatrix} \begin{bmatrix} \mathbf{V} \\ 1 \end{bmatrix} = \begin{bmatrix} b\mathbf{V} + a \\ d\mathbf{V} + c \end{bmatrix} \qquad [1.4\text{-}37]$$

Using the determinant of the system matrix, $\mathsf{bc} - \mathsf{ad} = 1$, and combining Equations [1.4-36] and [1.4-37], we find that

$$\frac{b\mathbf{V} + a}{d\mathbf{V} + c} = \mathbf{V} + \mathfrak{P} = \mathbf{V}' \qquad [1.4\text{-}38]$$

Then, setting $d\mathbf{V} + c$ equal to k and substituting this and Equation [1.4-38] into the right-hand matrix of Equation [1.4-37] yields

$$\mathsf{S} \begin{bmatrix} \mathbf{V} \\ 1 \end{bmatrix} = \begin{bmatrix} k\mathbf{V}' \\ k \end{bmatrix} \qquad [1.4\text{-}39]$$

Eliminating k is easy:

$$\frac{1}{k} \begin{bmatrix} k\mathbf{V}' \\ k \end{bmatrix} = \begin{bmatrix} \mathbf{V}' \\ 1 \end{bmatrix} \qquad [1.4\text{-}40]$$

where, as desired, the lower element in the right-hand matrix has become *unity*.

The use of the vergence matrix, as with vergences in general, is limited to *on-axis* points. (The vergence matrix is no substitute for exact ray tracing.) Nevertheless, it is a useful approach because vergences, in the paraxial approximation, are independent of the height above the axis. A rigorous solution, of course, that also includes off-axis points, and even *skew rays,* requires taking into account, at each surface, the ray's new height and direction, a complex procedure that I will briefly outline, following the examples.

Example 1. *The light may come from an axial point object 40 cm to the left of a convex spherical surface, ground with a 6-cm radius of curvature on a solid glass cylinder of n = 1.63. Find the image distance, using the matrix method.*

Solution. The light, on entering the lens, has a vergence of $V = 1/-0.4$ m $= -2.5$ m^{-1}. Therefore, the vergence matrix is

$$V = \begin{bmatrix} V \\ 1 \end{bmatrix} = \begin{bmatrix} -2.5 \\ 1 \end{bmatrix}$$

Then, inserting the power of the surface,

$$\mathfrak{P} = \frac{n_2 - n_1}{R} = \frac{1.63 - 1.00}{+0.06} = +10.5 \text{ m}^{-1}$$

into the refraction matrix and multiplying it by the vergence matrix gives

$$\begin{bmatrix} 1 & 10.5 \\ 0 & 1 \end{bmatrix} \begin{bmatrix} -2.5 \\ 1 \end{bmatrix} = \begin{bmatrix} 8 \\ 1 \end{bmatrix}$$

The exit vergence, therefore, is

$$V' = +8.00 \text{ m}^{-1}$$

and the image distance

$$i = \frac{n}{V'} = \frac{1.63}{+8:00} = \boxed{+20.4 \text{ cm}}$$

Example 2. *Continue with the same sequence of three lenses as on page 50. The lenses have +5.00, −16.00, and +10.00 diopters power, respectively, and are separated by distances of 6 cm each. Find the image distance as measured from the last lens.*

Solution. Rigorously, each lens has to be represented by a system matrix, but for simplicity we consider the lenses *thin*, and refraction matrices will do:

$$R_1 = \begin{bmatrix} 1 & 5 \\ 0 & 1 \end{bmatrix}, \quad R_2 = \begin{bmatrix} 1 & -16 \\ 0 & 1 \end{bmatrix}, \quad R_3 = \begin{bmatrix} 1 & 10 \\ 0 & 1 \end{bmatrix}$$

The distances between the lenses are represented by translation matrices:

$$T_1 = T_2 = \begin{bmatrix} 1 & 0 \\ -0.06 & 1 \end{bmatrix}$$

If the light entering the system is parallel, then its vergence is zero and the vergence matrix

$$V = \begin{bmatrix} 0 \\ 1 \end{bmatrix}$$

The complete sequence, written from right to left, is

$$\begin{bmatrix} 1 & 10 \\ 0 & 1 \end{bmatrix} \begin{bmatrix} 1 & 0 \\ -0.06 & 1 \end{bmatrix} \begin{bmatrix} 1 & -16 \\ 0 & 1 \end{bmatrix} \begin{bmatrix} 1 & 0 \\ -0.06 & 1 \end{bmatrix} \begin{bmatrix} 1 & 5 \\ 0 & 1 \end{bmatrix} \begin{bmatrix} 0 \\ 1 \end{bmatrix} = \begin{bmatrix} V' \\ 1 \end{bmatrix}$$

Carrying out the multiplication, beginning on the right, gives

$$\begin{bmatrix} 1 & 5 \\ 0 & 1 \end{bmatrix} \begin{bmatrix} 0 \\ 1 \end{bmatrix} = \begin{bmatrix} 5 \\ 1 \end{bmatrix}$$

For the first translation,

$$\begin{bmatrix} 1 & 0 \\ -0.06 & 1 \end{bmatrix} \begin{bmatrix} 5 \\ 1 \end{bmatrix} = \begin{bmatrix} 5 \\ 0.7 \end{bmatrix}$$

For refraction in lens 2,

$$\begin{bmatrix} 1 & -16 \\ 0 & 1 \end{bmatrix} \begin{bmatrix} 5 \\ 0.7 \end{bmatrix} = \begin{bmatrix} -6.2 \\ 0.7 \end{bmatrix}$$

For the second translation,

$$\begin{bmatrix} 1 & 0 \\ -0.06 & 1 \end{bmatrix} \begin{bmatrix} -6.2 \\ 0.7 \end{bmatrix} = \begin{bmatrix} -6.2 \\ 1.072 \end{bmatrix}$$

and for refraction in lens 3,

$$\begin{bmatrix} 1 & 10 \\ 0 & 1 \end{bmatrix} \begin{bmatrix} -6.2 \\ 1.072 \end{bmatrix} = \begin{bmatrix} 4.52 \\ 1.072 \end{bmatrix}$$

Thus the exit vergence on leaving the third lens is

$$\mathbf{V}' = \frac{4.52}{1.072} = +4.2164 \ \text{m}^{-1}$$

The inverse of this is the image distance,

$$i = \frac{1}{+4.2164} = \boxed{+23.72 \ \text{cm}}$$

precisely the same result as before.

Skew rays.

There are two groups of rays that can be traced through a system:

1. *Meridional rays* are those rays that lie in a plane containing the optic axis. Such planes are called *meridional planes*. As a meridional ray passes through the system it will remain in its meridional plane, as long as the system is *centered*, that is, as long as the vertices and centers of curvature of all surfaces lie in a common straight line. Tracing a meridional ray is comparatively easy, because the problem is two-dimensional; it merely requires plane geometry.

2. The term *skew rays* is more general; it includes those rays that pass through the system *outside* of any meridional plane. Tracing a skew ray is much more difficult, because the problem is three-dimensional; it requires solid geometry.

A given skew ray may originate at a surface at the coordinates x_1, y_1, and z_1, z being horizontal and transverse to the axis. As the ray proceeds, it subtends with the x, y, and z axes, respectively, a set of three angles, χ, ψ, and ω. The cosines of these angles are called the *direction cosines*, $\cos \chi$ often abbreviated X; $\cos \psi$, Y; and $\cos \omega$, Z. If both z_1 and Z are zero, the ray is a meridional ray and is equal to the familiar angle γ. The sum of the squares of the three direction cosines is unity,

$$X^2 + Y^2 + Z^2 = 1 \qquad\qquad [1.4\text{-}41]$$

which shows that two of the angles are independent variables and one is dependent.

Knowing the direction cosines, we can determine the coordinates where the ray intersects the next surface, and through a succession of refractions and translations, we can follow the ray through the entire system, a complex procedure that recommends the use of a high-speed computer.

SUGGESTIONS FOR FURTHER READING

R. Kingslake, *Lens Design Fundamentals* (New York: Academic Press, Inc., 1978).

W. J. Smith, *Modern Optical Engineering* (New York: McGraw-Hill Book Company, 1966).

A. Nussbaum and R. A. Phillips, *Contemporary Optics for Scientists and Engineers* (Englewood Cliffs, NJ: Prentice-Hall, Inc., 1976).

Military Standardization Handbook, *Optical Design*, MIL-HDBK-141 (Washington, DC: Defense Supply Agency, 1962).

D. F. Horne, *Optical Production Technology* (New York: Crane, Russak & Company, Inc., 1972).

PROBLEMS

1.4-1. In the optical system shown in Figure 1.4-10, find the object point conjugate to image point I.

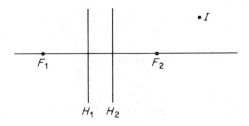

Figure 1.4-10

1.4-2. Two plano-convex lenses are placed a short distance apart, their plane sides facing each other. Trace two rays, coming from an off-axis point object located in the left-hand focal plane of the first lens. Assume refraction at the principal planes only and find the conjugate image. Directly below draw the same *outer* surfaces with the space between the lenses filled with glass, that is, consider a rather thick biconvex lens. Trace two similar rays.

1.4-3. Two lenses of equal power but opposite sign are mounted coaxially a short distance apart from each other.
(a) If parallel light passes first through the pos-

itive lens and then through the negative lens, what will be the equivalent power of the combination?

(b) What will it be if the light passes first through the negative lens?

1.4-4. Continue with Problem 1.4-3 and assume that now the light is *convergent* rather than parallel.

1.4-5. A meniscus, 20 mm thick and of index 1.523, has a back surface of −2.50 diopters power and an (equivalent) focal length of +10 cm. Determine the power of the front surface.

1.4-6. A symmetric, biconvex lens, 20 mm thick and of index 1.523, has an (equivalent) focal length of 10 cm. Find the power of the two surfaces.

1.4-7. What must be the minimum separation of two thin lenses of +6.00 and −4.00 diopters power, respectively, so that the combination of both has an equivalent power of +5.00 diopters?

1.4-8. A light collector is often a combination of two identical plano-convex lenses, their focal length chosen so as to be slightly less than twice the desired equivalent focal length of the system. If the latter is to be 50 mm, then probably $f = 90$ mm is a suitable figure for the former. How far apart from each other should the lenses be placed?

1.4-9. Two thin lenses have a combined (approximate) power of +10.00 diopters. If they become separated by 20 cm, their equivalent power decreases to +6.25 diopters. What are the powers of the two lenses?

1.4-10. Consider two thin lenses, of +5.00 and −10.00 diopters power, respectively. Show in the form of a plot how the equivalent power of the combination changes as a function of the separation of the two lenses.

1.4-11. A thick meniscus has a front surface of +5.00 diopters power, a back surface of −2.00 diopters power, is 3 cm thick, and is made out of glass of $n = 1.5$. Find the differences between the equivalent power on the one hand and the front vertex power and the back vertex power on the other.

1.4-12. Determine both the front vertex and the back vertex focal length of a two-lens system composed of a +8.00-diopter lens 4 cm in front of a −5.00-diopter lens.

1.4-13. Two +6.00-diopter lenses are mounted some distance apart. Find the least distance at which the equivalent power of the combination changes from positive to negative.

1.4-14. Two converging lenses, one with twice the power of the other, form an *afocal* system. If the distance between the two lenses is 50 cm, what are their powers?

1.4-15. A ray of light proceeding parallel to, and 18 mm above, the optic axis is incident on a concave surface, ground with a radius of 7 cm on a solid cylinder of $n = 1.49$. Using matrix algebra, determine the angle of the ray, in degrees, inside the cylinder.

1.4-16. An axial point object is located 25 cm in front of a +10.00-diopter lens. A ray of light entering the lens subtends an angle of −5° with the axis. What is the angle after refraction? (This is our Example 1 on page 49; it should now be solved by matrix algebra.)

1.4-17. Using matrix multiplication, determine the new coordinates x', y' in terms of the initial coordinates x_0, y_0:

$$\begin{bmatrix} 3 & 1 \\ 2 & 0 \end{bmatrix} \begin{bmatrix} 2 & 0 \\ -1 & 3 \end{bmatrix} \begin{bmatrix} x_0 \\ y_0 \end{bmatrix} = \begin{bmatrix} x' \\ y' \end{bmatrix}$$

1.4-18. Find the system matrix of a biconvex lens of $P_1 = P_2 = 2.5$ m^{-1}, thickness 3 cm, and index 1.5. Check by the determinant.

1.4-19. Determine the system matrix of the lens illustrated in Figure 1.4-11. Check by the determinant.

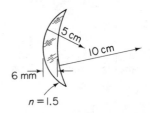

Figure 1.4-11

1.4-20. Find the system matrix of the lens in Figure 1.4-11, but now with its *concave* side facing the light incident from the left. Check by the determinant. (The result shows that matrices are not commutative.)

1.4-21. An object is located 2 cm to the left of the convex surface of a glass rod which has a radius of curvature of 1 cm and an index of refraction of $n = 1.5$. (This is our earlier Problem 1.3-1.) Find the image distance by matrix algebra.

1.4-22. Light comes from an axial point object 1.25 m in front of a lens and converges to an axial point image 40 cm behind the lens. Using matrices, determine the focal length of the lens.

1.4-23. Light emerging from an axial point object 5 cm away is focused by a thin lens of 4 cm focal length. Using vergence matrices, find the image distance.

1.4-24. If a patient who ordinarily wears -5.00-diopter contact lenses wants to wear spectacle lenses instead, and if these lenses are to be placed 15 mm in front of the cornea, what power should they have? Solve by the matrix method.

1.4-25. A lens of -20 cm focal length is placed 2 cm in front of a lens of $+4$ cm focal length. Find the system matrix of the combination using the Gaussian constants and check by the determinant.

1.4-26. Continuing with Problem 1.4-25, assume that an object is located 12 cm in front of the first lens. Determine the image distance, as measured from the second lens, and compare the result with that of Problem 1.3-23.

1.4-27. If an object is placed 50 cm in front of a -3.00-diopter lens, what is the transverse magnification of the image? (This is our earlier Problem 1.3-18; it should now be solved using the Gaussian constants.)

1.4-28. Two lenses are placed 30 cm apart. If the second lens has twice the power of the first lens, and the system is afocal, what are the powers of the two lenses? Use matrices to solve.

1.5

Mirrors

MIRRORS CAN DO THE SAME THINGS that lenses can. In many cases mirrors may be exchanged for lenses. The only difference is that a mirror *returns* the light while a lens allows it to continue.

The Mirror Equation

At first, a mirror may seem difficult to use with our sign convention, because both a real object and a real image lie on the same side, in front of the mirror. But let us continue using Cartesian coordinates and, as with a lens surface, consider the radius of curvature of a mirror concave to the left as negative (Figure 1.5-1). Conversely, the radius of a mirror convex to the left is positive. However, a *concave* mirror has a function equivalent to that of a converging lens and therefore its focal length and power are *positive*. A *convex* mirror is equivalent to a diverging lens and its focal length and power are *negative*.

Now assume that a real object is placed some distance to the left of a concave mirror, as shown. If the object distance chosen is longer than the focal length, the mirror, like a lens, forms a real image. This image lies to the left of the mirror. According to our convention, light, as it travels from the object to the mirror, advances in the $+x$ direction. Therefore, the velocity of the light is positive:

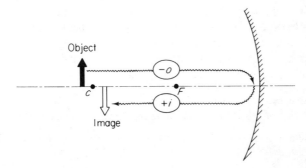

Figure 1.5-1 Concave mirror forming real image. Note negative object distance but positive image distance. Mirror shown acts like a converging lens; it has a positive focal length.

$$\frac{+x}{+t} = +v$$

Light returned by the mirror, however, travels in the $-x$ direction. This means, because time can only go forward ($+t$, rather than $-t$), that the velocity of the light is now negative:

$$\frac{-x}{+t} = -v$$

But since $v_1/v_2 = n_2/n_1$, the index related to the return path becomes negative also:

$$n_2 = -n_1$$

All that is needed then is to substitute $-n_1$ for n_2 in the surface power equation:

$$P = \frac{n_2 - n_1}{R} = \frac{(-n_1) - n_1}{R} = \frac{-2n}{R} \qquad [1.5\text{-}1]$$

where P is the *reflective power* of the mirror, R its radius of curvature, and n the refractive index of the medium in front of the mirror. Then we substitute $P = n/f$ in Equation [1.5-1], cancel n, and combine the result with Equation [1.3-10], solved for $1/f$. This gives

$$\boxed{\frac{1}{f} = \frac{1}{i} - \frac{1}{o} = \frac{-2}{R}} \qquad [1.5\text{-}2]$$

which is the *spherical mirror equation*. Although I have derived this equation for a concave mirror and for a real object and real image, it holds as well for plane and convex mirrors and for any object and image distance.

In short, if the image is located along the return path, left of the mirror (real image!), the image distance is considered positive, because axial *distances are measured in the direction of propagation of the light.* If the image were located to the right of the mirror (virtual image!), the image distance would be negative.

Example. *An object 16 mm high is placed 13 cm in front of a concave mirror of 26 cm radius of curvature. Find the angular size of the image.*

Solution. Considering only the first and last terms in Equation [1.5-2] and solving for f gives

$$f = \frac{R}{-2} = \frac{-26 \text{ cm}}{-2} = 13 \text{ cm}$$

which means that the object is located *in* the focal plane. Rays emerging from the object and reflected by the mirror, therefore, are projected out to infinity.

Then, from the construction in Figure 1.5-2,

$$\tan \gamma = \frac{16 \text{ mm}}{-130 \text{ mm}}$$

so the angle subtended by the axis and the tip of the arrow projected out to infinity is

$$\gamma = \tan^{-1} \frac{16}{-130} = \boxed{-7°}$$

In contrast to a lens, the refractive index of the medium in front of a mirror is (almost) irrelevant. A mirror in air has the same focal length as the same mirror under water. However, the *power* of the mirror changes, as seen from the first and last terms of Equation [1.5-1].

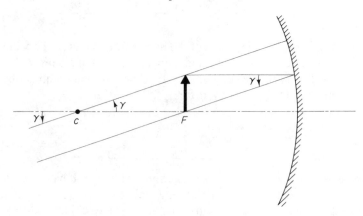

Figure 1.5-2

Mirrors and vergence.

Vergences and changes of vergence are handled the same way with a mirror as with a lens.

Example 1. *An object is located 2.5 cm to the left of a concave mirror of 10 cm radius of curvature. Find the image distance using the vergence method.*

Solution. First we determine the entrance vergence,

$$\mathbf{V} = \frac{1}{o} = \frac{1}{-0.025} = -40 \text{ m}^{-1}$$

and the power of the mirror,

$$P = \frac{1}{f} = \frac{-2}{R} = \frac{-2}{-0.1} = +20 \text{ m}^{-1}$$

Adding both gives the exit vergence,

$$\mathbf{V}' = (-40) + (+20) = -20 \text{ m}^{-1}$$

The image, therefore, is virtual and located

$$i = \frac{1}{\mathbf{V}'} = \frac{1}{-20} = -0.05 = \boxed{-5 \text{ cm}}$$

to the *right* of the mirror.

Example 2. *Consider the combination of a lens and a mirror. The object is located 50 cm to the left of a −2.00-diopter lens. A +12.50-diopter mirror is placed 15 cm to the right of the lens. Find the position of the image.*

Solution. In Figure 1.5-3 the vergences of the light coming from the object are shown *above the axis* and the vergences of the light reflected back by the

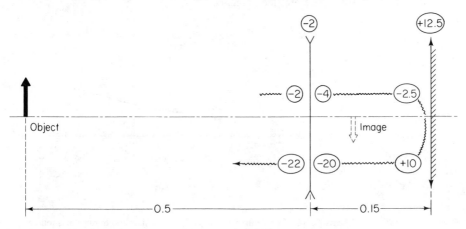

Figure 1.5-3 Schematic representation of light passing through a lens–mirror combination.

mirror *below the axis*. With the object 50 cm away, the entrance vergence at the lens is

$$V = \frac{1}{-0.5} = -2 \text{ m}^{-1}$$

The exit vergence therefore is

$$V' = (-2) + (-2) = -4 \text{ m}^{-1}$$

This means that the center of curvature of the wavefronts leaving the lens is

$$L = \frac{1}{V'} = \frac{1}{-4} = -0.25 \text{ m}$$

to the left of the lens, or 40 cm to the left of the mirror. At the mirror, therefore, the entrance vergence is

$$V = \frac{1}{-0.4} = -2.5 \text{ m}^{-1}$$

and the exit vergence is

$$V' = (-2.5) + (+12.5) = +10 \text{ m}^{-1}$$

At this point a slight complication occurs. From the exit vergence $V' = +10 \text{ m}^{-1}$ we conclude that 10 cm to the left of the mirror the light comes to a focus. But the light does not stop there; it continues, becomes *di*vergent, and passes once more through the lens (Figure 1.5-4).

For the dimensions given, the crossover point (where the light becomes divergent) lies 5 cm before the light reaches the lens. On entering the lens, therefore, the vergence is

$$V = \frac{1}{-0.05} = -20 \text{ m}^{-1}$$

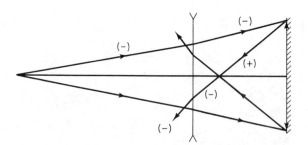

Figure 1.5-4 Lens–mirror combination. The plus and minus signs indicate whether the light is convergent or divergent.

As before, the lens makes the light still more divergent:

$$\mathbf{V}' = (-20) + (-2) = -22 \text{ m}^{-1}$$

The final image, then, is located at a distance

$$i = \frac{1}{-22} = (-)0.045 = \boxed{(-)4.5 \text{ cm}}$$

I have put the minus signs in parentheses. True, the image is located on the "wrong" side considering the direction of the light, but it is to the *right* of the lens. This, and the image being virtual *and* inverted, makes for an odd combination, due to reflection at the mirror.

Reflection matrix.   Since a mirror is so much like a lens, a reflection matrix is very similar to a refraction matrix. We only need to substitute

$$P = \frac{-2n}{R}$$

which is part of Equation [1.5-1], in the refraction matrix. Then we arrive at the *reflection matrix,* M:

$$\mathbf{M} = \begin{bmatrix} 1 & P \\ 0 & 1 \end{bmatrix} = \begin{bmatrix} 1 & -2n/R \\ 0 & 1 \end{bmatrix} \qquad [1.5\text{-}3]$$

As before, n is the refractive index of the medium in front of the mirror and R is the radius of curvature. If the mirror is turned around to face the other way, both the radius of curvature and the index change signs, so the matrix itself remains the same.

Example. *With an object 2 m away, a concave mirror forms a real image 50 cm away. Using matrices, find the radius of curvature of the mirror.*

Solution. Convert the known object and image distances into vergences and these into vergence matrices:

$$\mathbf{V} = \frac{n}{o} = \frac{1}{-2} = -0.5 \text{ m}^{-1}, \qquad \mathbf{V} = \begin{bmatrix} \mathbf{V} \\ 1 \end{bmatrix} = \begin{bmatrix} -0.5 \\ 1 \end{bmatrix}$$

$$\mathbf{V}' = \frac{1}{0.5} = 2 \text{ m}^{-1}, \qquad \mathbf{V}' = \begin{bmatrix} 2 \\ 1 \end{bmatrix}$$

Then, writing the sequence from object to mirror to image,

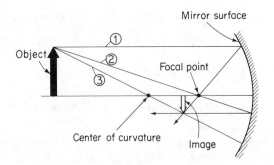

Figure 1.5-5 Image formed by a spherical mirror. Object *outside* the center of curvature: Image is real, inverted, and reduced in size.

$$\begin{matrix} \mathsf{M} & \mathsf{V} & = & \mathsf{V'} \end{matrix}$$

$$\begin{bmatrix} 1 & -2/R \\ 0 & 1 \end{bmatrix} \begin{bmatrix} -0.5 \\ 1 \end{bmatrix} = \begin{bmatrix} 2 \\ 1 \end{bmatrix}$$

$$\begin{bmatrix} (-0.5) + (-2/R) \\ 1 \end{bmatrix} = \begin{bmatrix} 2 \\ 1 \end{bmatrix}$$

$$-0.5 - \frac{2}{R} = 2$$

$$\frac{-2}{R} = 2.5$$

$$R = \frac{-2}{2.5} = -0.8 = \boxed{-80 \text{ cm}}$$

Graphical construction.   In principle very similar to refraction, rays are deflected as follows:

The *parallel ray* (1), after reflection, goes through the focal point.
The *focal ray* (2) is reflected back parallel to the optic axis.
The *chief ray* (3) is returned in its own path.

Note the two examples shown in Figures 1.5-5 and 1.5-6.

Plane Mirrors

A plane mirror has a radius of infinite length, $R = \infty$. Therefore, from Equation [1.5-2],

$$\frac{1}{i} - \frac{1}{o} = \frac{-2}{\infty}$$

so that

$$o = i \qquad\qquad [1.5\text{-}4]$$

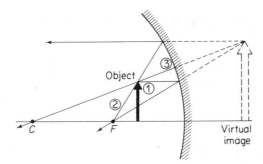

Figure 1.5-6 Object *inside* the focal length: Image is virtual, upright, and magnified.

which means that *the (virtual) image formed by a plane mirror is located at the same distance behind the mirror as the object is in front of it,* no matter how we look at the object or at the mirror. This holds as well for an *extended object*.

Consider in more detail the *orientation* of such an image. The object may be the letter R. If the mirror is standing upright, like a dresser mirror, with its surface parallel to the straight line in the R, the images are upright but right and left are reversed (Figure 1.5-7, left and right). Such images, Я, are called "upright backward" or *reverted*; they are *wrong-reading*.

If the mirror surface is horizontal, like a reflecting pool, the images, Я, are "inverted forward" or simply *inverted*, and also wrong-reading. An image formed by a single plane mirror is always wrong-reading.

If a mirror is tilted *(rotated)* through a given angle, a beam reflected by the mirror will be rotated through twice that angle. This fact plays a role in optical levers, galvanometers, sextants, and similar instruments.

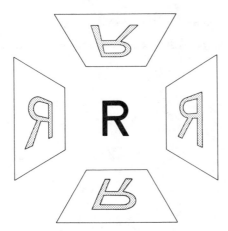

Figure 1.5-7 Looking at an object through a tunnel lined with four plane mirrors.

Multiple plane mirrors.   Two or more plane mirrors, when combined, cause multiple reflections, and therefore multiple virtual images. Consider *two* mirrors, and assume that all rays and the normals to the two mirrors lie in the same plane (Figure 1.5-8). The light may be incident on the first mirror at an angle α. If the two mirrors together subtend an angle γ, light is deviated through δ, and

$$\delta = 2\gamma \qquad\qquad [1.5\text{-}5]$$

which means that the deviation is twice the angle subtended by the two mirrors, *independent of the angle of incidence*. (Solve Problem 1.5-14 to verify this statement.) When $\gamma = 90°$, $\delta = 180°$: In that case the incident and the emergent rays are antiparallel.

Three plane mirrors, combined at right angles, form a *cube-corner reflector*. Such a reflector returns the light even in three-dimensional space (Figure 1.5-9).

Archimedes' mirror which, according to legend, he used to set Roman warships on fire probably were a series of *plane* mirrors. The important point is

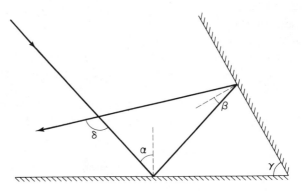

Figure 1.5-8   Reflection on two plane mirrors.

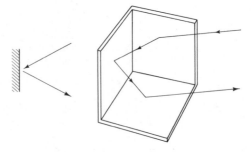

Figure 1.5-9   Conventional *specular* reflection (*left*) and cube corner reflector (*right*).

this: Because of the finite angular size of the sun, the image formed by a concave mirror (of necessarily long focal length) is finite in size also (for $f = 125$ m about 64 cm in diameter). An infinitesimally small plane mirror would give an image of the same size. And, just as the size of the hole in a pinhole camera does not materially affect the size of the image, the image formed by a plane mirror is larger only by the size of the mirror. With enough people standing on the sloping bank of a shoreline, each holding a large plane mirror, enough energy could be conveyed to start a fire. [From K. D. Mielenz, "Eureka!" *Applied Optics* **13** (1974), A14–16.]

Aspheric Mirrors

Parallel light will come to a focus only if reflected in a *paraboloidal* mirror. A paraboloid is the result of rotating a parabola about its axis of symmetry. Such mirrors are used in astronomical telescopes, searchlights, and microwave antennas. In contrast, in an *ellipsoidal* mirror light originating at one focus of the ellipse is projected into the other focus. The two foci thus are conjugate (Figure 1.5-10).

Determining the Focal Length of a Mirror. 1. The power of a mirror can be determined using a lens clock (see page 62). Assume that the clock when held against the mirror gives a reading of x diopters. Then, using the surface power equation, $P = (n_2 - n_1)/R$, setting $n_1 = 1$ and $P = x$ and solving for R gives $R = (n - 1)/x$. Substituting this in the mirror equation yields $P = -2/R = -2x/(n - 1)$ and, if the lens clock is calibrated for $n = 1.5$,

$$P = -4x \qquad\qquad [1.5\text{-}6]$$

which means that the mirror has four times the power, and the opposite sign, of that shown by the clock.

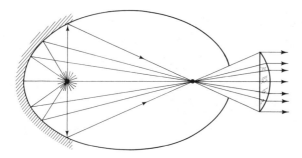

Figure 1.5-10 Ellipsoidal mirror together with collimating lens makes for efficient light collector. Note that the angle subtended by the mirror surface is much larger than that subtended by the lens (with the source then placed in the right-hand focus).

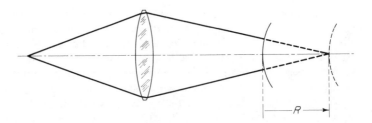

Figure 1.5-11   Measuring the radius of curvature of a mirror.

Optical methods are preferable, in particular if the mirror is small, if it has a long focal length, or if it is a front-surface mirror which could easily be damaged by the points of the clock.

2. Place an object-image screen (illuminated target) in front of a (concave) mirror at such a distance that the image of the target is reflected into the plane of the target. The distance from the target screen to the mirror is then equal to the radius of curvature. One-half of that distance is the focal length.

3. Focus a telescope for infinity and then aim at the (concave) mirror. A small object such as a pin is mounted, slightly off-axis, between mirror and telescope. Moving the object back and forth along the axis, find a position where the image is seen in focus. Light coming from the object and reflected by the mirror must now be parallel and the object located in the mirror's focus.

4. A convex mirror, like a diverging lens, does not form a real image. Proceed as with a minus lens. First form an image using a plus lens only. Focus *on* the mirror surface (dashed lines on the right in Figure 1.5-11). Then move the mirror to the left (solid line) until an image is formed in the plane of the target. The distance between the two positions of the mirror is equal to its radius of curvature. The same procedure can be used with a concave mirror, moved *away* from the lens.

SUGGESTIONS FOR FURTHER READING

I. D. Gluck, *It's All Done with Mirrors* (Garden City, NY: Doubleday & Company, Inc., 1968).

R. E. Hopkins, "Mirror and Prism Systems," in R. Kingslake, editor, *Applied Optics and Optical Engineering*, Vol. III, pp. 269–308 (New York: Academic Press, Inc., 1965).

W. B. Elmer, *The Optical Design of Reflectors*, 2nd edition (New York: John Wiley & Sons, Inc., 1980).

PROBLEMS

1.5-1. An object 1 cm high is placed 4 cm in front of a mirror. What type of a mirror and what radius of curvature is needed to produce an upright image that is 3 cm high?

1.5-2. An object is placed 60 cm in front of a spherical mirror. If the mirror forms a virtual image at a distance of 15 cm, what is the radius of curvature of the mirror?

1.5-3. How far from a concave mirror of 10 cm radius of curvature must a real object be placed so that its image is real and four times the size of the object?

1.5-4. A dentist holds a concave mirror of 4 cm radius of curvature at a distance of 15 mm from a filling in a tooth. What is the magnification?

1.5-5. A concave mirror has a radius of curvature of 40 cm. Find the two distances at which an object may be placed in order to give an image four times as large as the object.

1.5-6. Which two radii of curvature of a mirror will produce an image twice the size of an object 15 cm away from the mirror?

1.5-7. If an object is placed halfway between the focal point and the vertex of a concave mirror, how much will the image be magnified?

1.5-8. If a concave mirror of 60 cm radius of curvature gives an image twice as far away as the object, what is the object distance? Use matrices to solve.

1.5-9. A small object is placed between the center of curvature and the focal plane of a concave mirror. By graphical construction find the orientation, type (real or virtual), and magnification of the image.

1.5-10. Assume that the object is placed in front of a *convex* mirror. As in Problem 1.5-9, determine the character of the image.

1.5-11. A -1.00-diopter lens of a certain diameter is placed in contact with a mirror of $+40$ cm focal length and of the same diameter. What is the focal length of the combination?

1.5-12. Continue with Problem 1.5-11 and assume that the lens and the mirror are separated by a distance of 15 cm. How far from the lens will parallel light come to a focus? Solve by the vergence method.

1.5-13. How long must a (vertical) plane mirror be for a man who stands 1.82 m tall to see his full length?

1.5-14. If two plane mirrors, as in Figure 1.5-8, subtend a certain angle and if light is incident on one mirror, prove that the light reflected from the other mirror is deviated through *twice* that angle.

1.5-15. Two plane mirrors subtend an angle of 35°. At what angle must light be incident on one mirror so that, after reflection at the other mirror, the light exactly retraces its path?

1.5-16. In Figure 1.5-12, A represents the location of one of several images which can be seen in the two mirrors by an observer at E. Find the location of the object which causes these images and label it O. Show the path of the light from O to E.

Figure 1.5-12

1.5-17. Two plane mirrors subtend a certain angle with each other. A ray of light is parallel to one of the mirrors, and after four reflections it exactly retraces its path. What angle do the mirrors subtend?

1.5-18. Two plane front-surface mirrors are set at right angles upright on a table. A postage stamp is held halfway between the mirrors, facing the line joining them. When looking past the stamp toward the mirrors:
(a) How many images of the stamp can be seen?
(b) Are these images *right-reading* or *wrong-reading*?

1.5-19. A $+6.00$-diopter lens is placed 25 cm from a transilluminated target. If then a convex mirror, as in Figure 1.5-11, is set 40 cm from the lens, it returns the light in its original path. What is the power of the mirror?

1.5-20. A lens clock, held against a mirror, gives a reading of -2.50 diopters. If this mirror forms a real image at a distance one-half the object distance, what is the image distance?

1.6

Aberrations

AGAIN, OUR THREE-LENS SYSTEM SERVES as a guide to further exploration. If a few sample rays, as in Figure 1.6-1, are traced *exactly* through the first lens, we find that they do not come together as we would like. The marginal rays intersect the optic axis *closer to the lens* than the paraxial rays. This is an example of an *aberration*.

Introductory remarks. Aberrations are due to inherent shortcomings of a lens, even a lens made of the best glass and free from manufacturing and other defects. Some aberrations occur with monochromatic light; they are called *monochromatic aberrations* (1 through 5

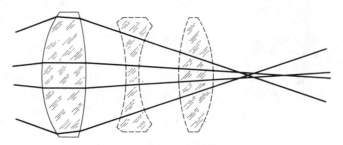

Figure 1.6-1 Spherical aberration.

of the aberrations listed below). Other aberrations occur only with light that contains at least two wavelengths; these are called *chromatic aberrations* (6). We distinguish the following:

1. Spherical aberration
2. Coma
3. Oblique astigmatism
4. Curvature of field
5. Distortion
6. Chromatic aberration

The monochromatic aberrations can be further divided into two groups depending on whether or not they cause the image to deteriorate and become blurred. Image blur occurs in spherical aberration, coma, and oblique astigmatism. Curvature of field and distortion, on the other hand, cause dislocations within the image but no blur. Furthermore, spherical aberration and chromatic aberration both have a longitudinal (axial) and a transverse (lateral) variety, each with its own causes and corrections.

For numerical analysis, aberrations can be handled along two general lines, considering either *rays* or *wavefronts*. In *paraxial-ray tracing,* which we discussed earlier in conjunction with single surfaces and thin lenses, we had assumed that the angles which the rays subtend with the optic axis are small (no larger than a few degrees). If this is so, the sines of these angles can be set equal to the angles themselves (in radians). If the angles are larger, and if we use *exact ray tracing,* this can no longer be done. The sine of an angle must then be represented by a series expansion,

$$\sin \alpha = \alpha - \frac{\alpha^3}{3!} + \frac{\alpha^5}{5!} - \frac{\alpha^7}{7!} + \cdots \qquad [1.6\text{-}1]$$

where the α's are again given in radians.

If only the first term, α, is retained (that is, if we assume that $\sin \alpha = \alpha$), we have an example of paraxial or *first-order theory.* If the next term, $\alpha^3/3!$, is retained also (and we assume that $\sin \alpha = \alpha - \alpha^3/3!$), we have *third-order theory.**

---

*In the expansion of a sine function there are additional, even-numbered terms which have zero coefficients; thus, since only the odd-numbered orders are significant, we go directly from *first*-order to *third*-order to *fifth*-order theory, and so on.

Aberrations based on third-order theory are often called *von Seidel aberrations,* named after Ludwig Philipp von Seidel (1821–1896), German mathematician and astronomer, professor at the University of Munich. After working on divergent and convergent series, von Seidel became interested in stellar photometry, probability theory, and the method of least squares, studied the trigonometry of skew rays, and became the first to establish a rigorous theory of monochromatic third-order aberrations which led to the

If we consider *wavefronts,* we note that a wavefront, in order to converge toward a point, must be spherical. In the case of an image-blur aberration, the light does not converge toward a point and the wavefronts are not truly spherical; they are *distorted.* The degree of departure from true sphericity, that is, the distance between the ideal *reference sphere* and the actual, distorted wavefront, measured along the radius of the reference sphere, is called the *wave aberration.*

Spherical aberration, coma, and astigmatism all have characteristic wavefront distortions. In curvature of field and distortion, the wavefronts are spherical but the spheres are displaced; the wavefront normals come together but at the wrong point. The advantage of using wave aberrations is that the path difference, at a given height above the axis, caused by one surface in a system can be added directly to the path difference caused by the next surface, and so on, to find the total aberration present in the system.

Spherical Aberration

Spherical aberration is the phenomenon wherein rays passing through different zones of a surface, that is, passing through the surface at different heights above the axis, come to different foci. We have already seen that with a single lens rays farther away from the axis are refracted more, and come to a focus closer to the lens, than paraxial rays (Figure 1.6-1). This is called *positive longitudinal spherical aberration,* and a lens acting this way is said to be *undercorrected.* The aberration, therefore, causes every point in the image formed by marginal rays to be surrounded by a diffuse disk of light formed by paraxial rays, and vice versa—which accounts for the blur of the image.

Consider the spherical aberration of a single spherical surface with its center of curvature at C (Figure 1.6-2). Light comes from an axial point object, O, to the left of the surface. A ray intersects the surface at a point P, at some distance above the axis, and is refracted toward I. At O the ray subtends an angle γ with the axis, and at I it subtends an angle γ'.

Look first at triangle OPC. From the law of sines, and assuming initially that all values are positive,

$$\frac{R}{\sin \gamma} = \frac{o + R}{\sin (180° - \alpha)}$$

construction of much improved astronomical telescopes. L. Seidel, "Ueber die Entwicklung der Glieder 3ter Ordnung, welche den Weg eines ausserhalb der Ebene der Axe gelegenen Lichtstrahles durch ein System brechender Medien bestimmen," *Astron. Nachr.* **43** (1856), 289–304, 305–20, 321–32.

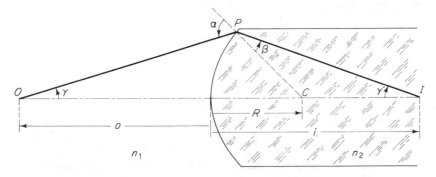

Figure 1.6-2   Spherical aberration of a single refracting surface.

where R is the radius of curvature, o the object distance, and α the angle of incidence on the surface at P.

But following our sign convention, o is measured from the vertex of the surface to the left and, with the actual distance of a real object substituted for o, would become negative. Hence, o must have a minus sign in order to represent the true length of the distance OC. Similarly, and because angle γ is turned counterclockwise, $\sin \gamma$ must have a minus sign too. Thus

$$\frac{R}{-\sin \gamma} = \frac{-o + R}{\sin (180° - \alpha)} \qquad [1.6\text{-}2]$$

Now, since $\sin (180° - \alpha) = \sin \alpha$,

$$\frac{R}{\sin \gamma} = \frac{o - R}{\sin \alpha}$$

so that

$$\sin \alpha = \frac{o - R}{R} \sin \gamma \qquad [1.6\text{-}3]$$

The angle of refraction, β, from Snell's law, is given by

$$\sin \beta = \frac{n_1}{n_2} \sin \alpha \qquad [1.6\text{-}4]$$

In any triangle, for example in triangle OPI, the interior angles add up to 180°:

$$\gamma + (180° - \alpha) + \beta + \gamma' = 180°$$

Thus

$$\alpha + \gamma = \beta + \gamma' \qquad [1.6\text{-}5]$$

Then we apply the law of sines to triangle *PIC:*

$$\frac{R}{\sin \gamma'} = \frac{i - R}{\sin \beta}$$ [1.6-6]

$$R \sin \beta = i \sin \gamma' - R \sin \gamma'$$

and

$$i = \frac{R \sin \beta}{\sin \gamma'} + R$$ [1.6-7]

Ordinarily, *i* refers to the *paraxial* image distance. In Figure 1.6-2 and Equation [1.6-7], however, *i* refers to the image distance relative to a *marginal* ray. The difference between the two distances, therefore, is a measure of the *longitudinal spherical aberration,* LSA, of the surface, at a given height of *P* above the axis:

$$\text{LSA} = i - \left(\frac{R \sin \beta}{\sin \gamma'} + R\right)$$ [1.6-8]

Clearly, with increasing aperture the spherical aberration becomes worse; in fact, the LSA increases with the *square* of the distance of *P* from the axis. Calculating the spherical aberration as a function of height is a laborious undertaking; it virtually requires a ray-tracing computer program.

If we plot the distance at which a given ray intersects the axis as a function of height (of point *P*), we obtain a curve characteristic of the type and magnitude of the spherical aberration. In Figure 1.6-3, the aberration is positive. If the curve were to lean to the right, the aberration would be negative and the lens be called *overcorrected.*

Besides longitudinal spherical aberration, there is a *transverse* (lateral) variety. It is defined as the height above, or below, the axis at which a marginal ray intersects the plane of the paraxial focus (Figure 1.6-4), and it is a more precise measure of image blur. Transverse spherical aberration, TSA, and longitudinal spherical aberration, LSA, are connected as

$$\text{TSA} = \text{LSA} \tan \gamma'$$ [1.6-9]

where γ' refers to the slope of the marginal ray. There is also an *angular spherical aberration:* This is the difference in direction of a marginal ray going through *its* focus rather than through the paraxial focus.

If the screen on which we receive the image is moved back and forth along the axis, we find a position between the marginal focus and the paraxial focus where the cross section of the light will have a minimum diameter. This cross section is called the *circle of least confusion,* a term

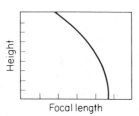

Figure 1.6-3 Spherical aberration of positive surface or lens. Plot of height above axis versus focal length. Arbitrary units.

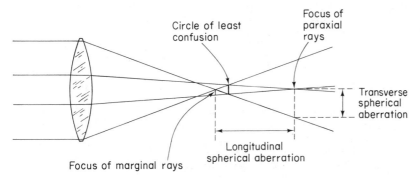

Figure 1.6-4   Longitudinal and transverse spherical aberration.

that we will discuss in more detail later in conjunction with astigmatism and cylinder lenses.

Correction for spherical aberration.

With a single lens spherical aberration cannot be eliminated completely. But it can be minimized. This can be done by making the two surfaces of the lens contribute about equally to the refraction (Figure 1.6-5), similar to setting a prism to the angle of minimum deviation. The lens must have the right shape, or *bending* (Figure 1.6-6). To further reduce spherical aberration, the radii should be as long as practical, which means using *high-index* glass. Also, either or both surfaces could be made *aspheric,* or the lens made with an index of refraction higher in the center than in the periphery (*gradient-index lenses,* page 362). And finally, in a lens *system,* the undercorrection of one lens can be balanced by the overcorrection of another lens.

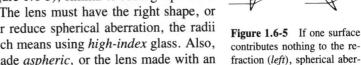

Figure 1.6-5 If one surface contributes nothing to the refraction (*left*), spherical aberration is worse than if both surfaces contribute about equally (*right*).

The shape of a lens at which it has the least amount of spherical aberration is called its *best form.* To find the best form, we need to consider two factors. One is the *Coddington shape factor,** $\mathscr{S}$. It describes the degree of bending as a function of the two radii of curvature, R_1 and R_2:

$$\mathscr{S} = \frac{R_2 + R_1}{R_2 - R_1}$$

[1.6-10]

---

*Henry Coddington (born around 1800, died 1845), English mathematician and cleric. Coddington had many interests, spoke several languages, was a good musician, draftsman, and botanist. He wrote two books on optics, the more important one the two-volume *A System of Optics.* Part I, *A Treatise on the Reflection and Refraction of Light* (Cambridge: Cambridge University Press, 1829), contains a thorough investigation of reflection and refraction. In volume 2, Coddington discusses the eye and the theory and construction of various types of eyepieces, telescopes, and microscopes.

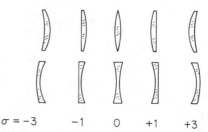

$$\sigma = -3 \qquad -1 \qquad 0 \qquad +1 \qquad +3$$

Figure 1.6-6 Lenses of different bending but (within each row) equal focal length. Coddington shape factor shown below.

For example, when $\mathcal{S} = 0$, the lens is symmetric, either equiconvex or equiconcave. When $\mathcal{S} = -1$, the first surface is plane and the second surface is either convex or concave. When $\mathcal{S} = +1$, the second surface is plane. When $\mathcal{S}$ is less than -1, or larger than $+1$, the lens is of the meniscus type. If the lens is turned around so that the light is incident on the other side, the shape factor retains its absolute value but the sign changes.

Whereas $\mathcal{S}$ is determined by the physical shape of the lens, the *Coddington position factor*, $\mathcal{P}$, depends on the object and image distances actually used:

$$\boxed{\mathcal{P} = \frac{i + o}{i - o}} \qquad\qquad [1.6\text{-}11]$$

When $\mathcal{P} = -1$, the incident light is parallel. When $\mathcal{P} = 0$, object distance and image distance are equal, except for the sign; and when $\mathcal{P} = +1$, the emergent light is parallel. Like the shape factor, $\mathcal{P}$ has no dimension.

Now, to find the best form, we first determine the position factor. Substituting in Equation [1.6-11] the Gaussian thin-lens equation, solved for i, gives

$$\mathcal{P} = \frac{of + o(o + f)}{of - o(o + f)} = -\left(\frac{2f}{o} + 1\right) \qquad\qquad [1.6\text{-}12]$$

and solved for o,

$$\mathcal{P} = \frac{i(f - i) + if}{i(f - i) - if} = 1 - \frac{2f}{i} \qquad\qquad [1.6\text{-}13]$$

Then from the lens-makers formula,

$$\frac{1}{f} = (n - 1)\left(\frac{1}{R_1} - \frac{1}{R_2}\right) = \frac{n - 1}{R_1} - \frac{n - 1}{R_2}$$

$$R_1 = f(n - 1) - \frac{fR_1(n - 1)}{R_2} \qquad [1.6\text{-}14]$$

Next we take the shape factor, Equation [1.6-10], and solve for R_2:

$$R_2 = R_1 \frac{\mathscr{S} + 1}{\mathscr{S} - 1} \qquad [1.6\text{-}15]$$

Substituting Equation [1.6-15] in [1.6-14] gives

$$R_1 = f(n - 1) - \frac{f(n - 1)(\mathscr{S} - 1)}{\mathscr{S} + 1}$$

Multiplying the $f(n - 1)$ term by $(\mathscr{S} + 1)/(\mathscr{S} + 1)$ and simplifying leads to

$$R_1 = \frac{2f(n - 1)}{\mathscr{S} + 1} \qquad [1.6\text{-}16]$$

and similarly,

$$R_2 = \frac{2f(n - 1)}{\mathscr{S} - 1} \qquad [1.6\text{-}17]$$

Differentiation* then shows that *minimum spherical aberration will occur whenever*

$$\boxed{\mathscr{S} = -\frac{2(n^2 - 1)}{n + 2}\,\mathscr{P}} \qquad [1.6\text{-}18]$$

The best form of a single lens, with either o or i infinite, is very nearly plano-convex. If $o = i$, the best form is symmetric.

Example. *Find the radii of curvature for a lens of $f = +10$ cm, $n = 1.5$, which for parallel incident light has minimum spherical aberration.*

Solution. First, determine the position factor:

---

*For details, see J. Morgan, *Introduction to Geometrical and Physical Optics,* pp. 86–94 (New York: McGraw-Hill Book Company, 1953).

$$\mathscr{P} = \frac{i + o}{i - o} = \frac{(+10) + (-\infty)}{(+10) + (-\infty)} = -1$$

Substitute the data given and $\mathscr{P} = -1$ in Equation [1.6-18]:

$$\mathscr{G} = -\frac{(2)(1.5^2 - 1)}{1.5 + 2}(-1) = +0.714$$

Then from Equations [1.6-16] and [1.6-17],

$$R_1 = \frac{2f(n - 1)}{\mathscr{G} + 1} = \frac{(2)(10)(1.5 - 1)}{(+0.714) + 1} = \boxed{+5.83 \text{ cm}}$$

$$R_2 = \frac{2f(n - 1)}{\mathscr{G} - 1} = \frac{(2)(10)(1.5 - 1)}{(+0.714) - 1} = \boxed{-35 \text{ cm}}$$

Thus the lens is biconvex but bent so that the front surface has a radius of +5.83 cm, the back surface a radius of −35 cm.

Coma

Spherical aberration occurs for points that lie *in the optic axis*. For object–image points that are *off-axis* the aberration is called *coma*.*

Consider Figure 1.6-7, top, and compare it with Figure 1.6-1. This time the object is a circular opening of a certain diameter, D. (The image-forming rays, therefore, come from off-axis points.) The lens forms a real image, of diameter D'. But rays 1, which pass through the periphery of the lens, form an image *larger* than rays 2 which pass through the center and, therefore, the image is blurred. [If, as shown on the far right, we consider only a single object point, rays 2 would form a single image point, but rays 1 would give a larger, more diffuse patch. The result is a characteristic *comatic flare*. The diffuse tail of the flare can lie in a direction away from the axis ("outward" coma), as shown, or toward it ("inward" coma), depending on the type of lens used.]

We need to make the different images coincide, as shown in the lower diagram. This time the light comes from an extended source, placed far to the left of the circular opening. Again some of the light proceeds along the axis. Some other light is oblique; it subtends angle γ with the axis. This angle relates to p as

---

*The term *coma* comes from the Greek κόμη, long hair, referring to the long tail of a comet. Coma is easy to see. Hold a magnifying glass in the path of sunlight and tilt it: The image of the sun will elongate into the cometlike shape characteristic of coma.

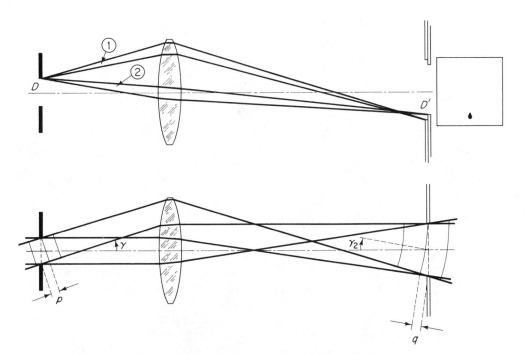

Figure 1.6-7 Different magnifications in coma (*top*) and deriving the sine condition for the elimination of coma (*bottom*).

$$\sin \gamma = \frac{\Delta p}{\Delta D}$$

At the image, the same rays subtend γ' and, neglecting the curvature of the wavefronts,

$$\sin \gamma' = \frac{\Delta q}{\Delta D'}$$

For the different images to coincide, Δp must equal Δq, and therefore, $\Delta D \sin \gamma$ must equal $\Delta D' \sin \gamma'$ or

$$\frac{\sin \gamma}{\sin \gamma'} = \text{constant} \qquad\qquad [1.6\text{-}19]$$

which is known as *Abbe's sine condition.*

Correction for coma is possible by bending, or by using a combination of lenses symmetric about a central stop. A system free of both spherical aberration and coma is called an *aplanat.*

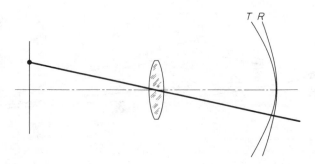

Figure 1.6-8   Tangential focal surface (T) and radial focal surface (R) as they occur in oblique astigmatism. In the example shown the astigmatism is *positive*.

Oblique Astigmatism

Oblique astigmatism, like coma, is another *off-axis* aberration.* Its main characteristic is that for off-axis points a focal *point* is drawn out into a complex, three-dimensional focal figure called a *conoid*. The conoid is limited by two *line* foci, oriented at right angles to each other. These line foci lie on (hypothetical) surfaces, one called the *tangential surface, T,* and the other the *radial surface, R* (Figure 1.6-8). *In* the axis, the two surfaces touch: For axial rays there is no astigmatism.

    The line focus closer to the lens is oriented *tangential* (tangent to a circle drawn around the optic axis) and the line focus farther away from the lens is oriented *radial*. Consequently, a point object, because of astigmatism, will be imaged as a tangential line at T and as a radial line at R (Figure 1.6-9a).

    If the object were a radial line, its image would be blurred at the tangential focus, T, and sharp at the radial focus, R (b). In contrast, if the object were a tangential line, its image would be sharp at T, and blurred at R (c). If the object were a combination of radial and tangential lines, such as a spoked wheel, the rim of the wheel (which is the sum of short tangential lines) will be in focus at T and the spokes (which are radial lines) at R (d).

    To understand why these lines are oriented tangential and radial, respectively, consider the following. Let a bundle of light originate at an axial point source. The light spreads out as a right-circular cone. If and when this cone falls off-axis on a lens surface, it forms an ellipse. The major axis of the ellipse is

---

*The term astigmatism comes from the Greek, $\acute{\alpha}$ = alpha privative, meaning "not," and $\sigma\tau\acute{\iota}\gamma\mu\alpha$ = mark or point, meaning that a point object is no longer imaged as a point. A system free of astigmatism is called an *anastigmat*.

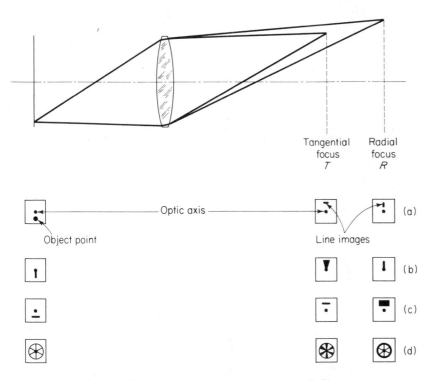

Figure 1.6-9 Astigmatic images of a sequence of objects.

always radial, the minor axis tangential. But rays that lie in the meridian of the major axis of the ellipse encounter part of a lens whose projected diameter, in this meridian, is *less*. Therefore, the power of the lens, in this meridian, is *higher* and the light will come to a focus *closer* to the lens. This light forms the tangential focus. Rays in the plane of the minor axis form the radial focus.

Axial Astigmatism (Cylinder Lenses). Cylinder lenses have properties that remind us of astigmatism. But *oblique* astigmatism is an off-axis aberration. Cylinder lenses have *axial astigmatism*. While oblique astigmatism is a nuisance, axial astigmatism is often induced on purpose; it is used in the correction of visual deficiencies and in anamorphic (wide-screen) motion picture cameras and projectors.

Consider the cylinder lens shown in Figure 1.6-10. Assume that two pieces of cardboard are held to either side of the lens, with the light grazing along their surfaces; on the cardboard we see a *ribbon* of light. In the orientation shown, the light continues straight through the lens with *no refraction,* because the thickness of the lens, in this *axis meridian,* is constant; the lens acts merely as a plane-parallel slab of glass.

But imagine that the cardboard is turned through 90°, as in Figure 1.6-11. The ribbon of light now lies in the other principal meridian, called *power meridian.*

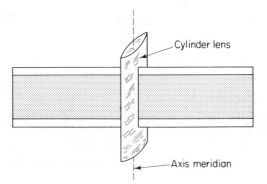

Figure 1.6-10 Biconvex cylinder. Ribbon of light in the axis meridian.

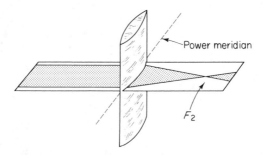

Figure 1.6-11 Cylinder with ribbon of light in the power meridian.

In this orientation the light *is* refracted, and comes to a focus, as with a conventional converging lens. A diverging, or *negative cylinder,* not shown, is comparable to a minus lens. Both types of cylinder lenses are *simple cylinders;* they have *no* power in their axis meridians.

In a compound cylinder lens or *spherocylinder* neither surface is plane: one surface is spherical, the other cylindric. Such a lens *has* power even in its axis meridian. To understand why, consider two separate lenses. The sphere alone, in the meridian shown in Figure 1.6-12, top left, projects a point object into a point image; the cylinder, with the ribbon of light in its *axis* meridian, *does not contribute.*

In the power meridian, however, *both lenses contribute;* the light comes to a focus *closer* to the lens (bottom). But note that the light *does not stop* at the foci; at the left-hand focus (lower diagram) it continues on and, by the time it has reached the right-hand focus, it has spread out into a *line,* oriented in direction of the power meridian. On the other hand (upper diagram), when the light is at the left-hand focus, it has *not yet* converged into a point; it is still a line, oriented in direction of the axis meridian. (Note that the first line focus is always parallel to the meridian of lesser power.) In reality, the two separate lenses shown in Figure 1.6-12 are fused into one lens, a *sphero-cylinder.*

The two lines that limit the conoid are a certain (axial) distance apart; this

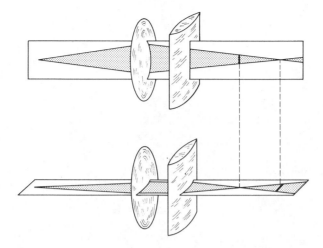

Figure 1.6-12   Spherocylinder (schematic), with ribbons of light in the axis meridian (*top*) and in the power meridian (*bottom*).

distance is called the *interval of Sturm*.* Within this interval there is one plane where the cross section of the light is circular; this is the *circle of least confusion.*

Example. *A spherocylinder 70 mm in diameter produces two line foci whose distances from the lens are 12.5 cm and 20 cm, respectively. Determine the position and size of the circle of least confusion.*

Solution. First change the two focal lengths into powers:

$$P = \frac{1}{0.125} = +8 \text{ diopters}$$

$$P' = \frac{1}{0.2} = +5 \text{ diopters}$$

The circle of least confusion is located at the "dioptric midpoint,"

$$\frac{8 + 5}{2} = +6.5 \text{ diopters}$$

Thus it is

---

*Charles-François Sturm (1803–1855). Born in Switzerland, Sturm became a French citizen, mathematician and physicist, and professor of analysis and mechanics at the École Polytechnique in Paris. He is known for the theorem that bears his name (referring to the number of roots of an algebraic equation) and for his contributions to projective geometry, propagation of sound in water, and the optics of cylinder lenses. Much of Sturm's scientific work is contained in lecture notes, published posthumously as two books, *Cours d'analyse de l'École polytechnique* (Paris, 1857–59) and *Cours de mécanique de l'École polytechnique* (Paris, 1861).

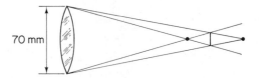

70 mm

Figure 1.6-13

$$\frac{1}{6.5} = 0.1538 \approx \boxed{15.4 \text{ cm}}$$

from the lens, always a little closer to the left-hand focus (Figure 1.6-13).

The diameter d of the circle of least confusion is found from similar triangles:

$$\frac{70}{200} = \frac{d}{200 - 154}$$

$$d = \boxed{16.1 \text{ mm}}$$

A *toric surface* has two radii of curvature (different in the two principal meridians), but in contrast to a simple cylinder, even the axis meridian has some power other than zero. Toric surfaces are used in ophthalmic lenses (which are menisci) whenever an additional cylinder component is needed. Usually, only one surface is toric, but *bitoric* lenses can be made also.

Curvature of Field

Curvature of field is a longitudinal defect; it causes the image of a two-dimensional *plane* object to become *curved*. As in oblique astigmatism, rays passing obliquely through a lens encounter a lens whose *projected* diameter is *less*. But a lesser diameter, all other parameters being equal, means higher power ("power error") and hence a shorter focal length. Therefore, the more oblique the rays, the more curved the image (Figure 1.6-14). Both curvature of field and oblique astigmatism vary directly with the tangent of the obliquity of the chief ray and inversely with the square of the focal length.

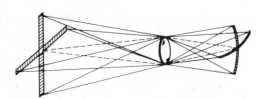

Figure 1.6-14   Curvature of field.

Distortion

Distortion is the transverse counterpart of curvature of field. Here, the size of the image, and therefore the magnification, vary as a function of distance from the optic axis. There are two types of distortion, *pincushion distortion,* where the corners of a square, for example, are drawn out like a pillow, and *barrel distortion,* where the corners have retracted (Figure 1.6-15). In pincushion distortion the transverse magnification increases with increasing obliquity of the rays; in barrel distortion it decreases.

Look at a sheet of graph paper from a distance of about 50 cm. Hold a high plus lens first fairly close to the paper and then farther away from it. You will have no difficulty distinguishing the two types of distortion.

Distortion is readily explained by ray tracing. The lens in Figure 1.6-16 forms an image of object 1–2 on the screen on the right. While point 1 is imaged (correctly) into 1', point 2, because of astigmatism and curvature of field, is imaged into 2', which lies to the left of the screen. But when the rays forming 2' have reached the screen, they have expanded into a blur circle with the chief ray at its center, C. If a circular aperture is placed to the right of the lens, the blur circle contracts and its center moves to D (solid line), which is farther away from the axis. This results in pincushion distortion.

On the other hand, if the stop is placed to the left of the lens, C moves to D', which is closer to the axis; this results in barrel distortion. To eliminate distortion, the correct position for a stop is *between* two lenses.

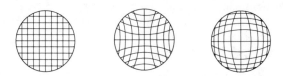

Figure 1.6-15 Undistorted image (*left*), pincushion distortion (*center*), and barrel distortion (*right*).

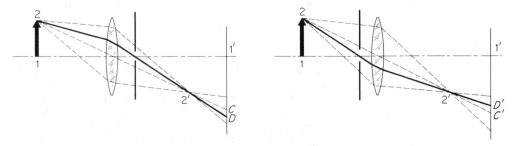

Figure 1.6-16 Rays causing pincushion distortion (*left*) and barrel distortion (*right*).

The two lenses must also meet the *Petzval condition**:

$$\frac{1}{n_1 f_1} + \frac{1}{n_2 f_2} = 0 \qquad [1.6\text{-}20]$$

If, in addition, the ratio of the slopes γ is constant,

$$\frac{\tan \gamma'}{\tan \gamma} = \text{constant} \qquad [1.6\text{-}21]$$

distortion will vanish as well. Systems built in this way are called *orthoscopic*.

> **Example.** *The Petzval condition can be met even if the two lenses have the same index of refraction. Consider a combination of two lenses that have the same power but opposite signs, and set them some distance d apart. Assume that $P_1 = +10 \ m^{-1}$, $P_2 = -10 \ m^{-1}$, $n = 1.55$, and $d = 31$ mm. This satisfies the Petzval condition,*
>
> $$\frac{+10}{1.55} + \frac{-10}{1.55} = 0$$
>
> *But is there any power left? (Work Problem 1.6-13 to find out.)*

Obviously, it would be better to make the plus lens of glass of higher index than the minus lens. But this is just the opposite of what is needed for correction of chromatic aberration. It is only by the use of newer types of high-index low-dispersion glass that the conditions for achromatism and flatness of field can be met at the same time.

Chromatic Aberration

Since the index of refraction of matter varies with wavelength, a single lens has different powers for different colors: *Blue* light comes to a focus *closer* to the lens than red light (Figure 1.6-17). The horizontal distance between the two images is called *longitudinal chromatic aberration*, or *"longitudinal color."*

---

*Josef Max Petzval (1807–1891), Hungarian mathematician and professor of mathematics at the University of Vienna. Petzval is known for his eloquent, colorful lectures; he wrote two volumes on the integration of linear differential equations and extended Gaussian optics beyond the paraxial approximation by including higher powers. For diversion he liked fencing and other forms of physical activity. Soon after Louis Jacques Maudé Daguerre (1787–1851) had announced, in 1839, his process of photography, Petzval set out to calculate, with the help of eight artillery men familiar with arithmetic and loaned to him by Archduke Ludwig of Austria, a portrait camera lens, of much larger aperture (to allow

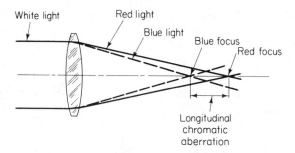

Figure 1.6-17 Longitudinal chromatic aberration.

In addition, the images produced by different colors are of different sizes; they have different transverse (lateral) magnifications. This is called *lateral chromatic aberration,* or *"lateral color."* An image formed in the "blue focus" is closer to the lens and smaller; its details are surrounded by a red halo. The "red focus" is farther away from the lens and larger; its details have a blue halo.

Correction for chromatic aberration.
There are several ways of correcting for chromatic aberration. The best known is to use two lenses in contact, one made of crown, the other of flint (Figure 1.6-18). The crown lens is given more plus power than necessary. Its dispersion is moderate. Flint, on the other hand, has high dispersion. The two dispersions are then made equal but of opposite sign so that they cancel. Therefore, because of the higher dispersion of flint, the flint component can have less minus power than the crown has plus power, the combination has excess plus power, and the result is a *positive achromat.**

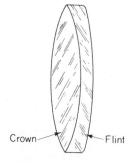

Figure 1.6-18 Achromatic doublet.

---

shorter exposures) and much better correction than any other lens known at that time (1840). Unfortunately, when in 1859 burglars broke into his country home on the Kahlenberg near Vienna looking for valuables, they destroyed part of the manuscript and only a short version remains: J. Petzval, *Bericht über die Ergebnisse einiger dioptrischer Untersuchungen* (Pest, 1843).

*This method of correcting for chromatic aberration goes back to 1729, when Chester Moor-Hall (1704–1771), British justice of the peace and amateur astronomer, designed the first achromatic contact doublet. Apparently, Moor-Hall had discovered that glass containing lead oxide, of the type used to make fine table glassware, had higher dispersion than window glass. To keep his discovery secret, he ordered one element for his doublet from a certain lens-maker in London, the other from another. As it happened, both men subcontracted the work to a third optician, who, on finding that both lenses were for the same customer and had one radius in common, placed them in contact and saw that the image was free of color. News of this discovery slowly spread to other opticians, among them John Dollond (1706–1761), whose son Peter (1739–1820) urged him to apply for a patent so that

Consider a combination of two lenses in contact, to be corrected for blue and red. We know that the approximate power of a combination is the sum of the powers of the elements:

$$\mathbf{P} = P_1 + P_2 \qquad [1.6\text{-}22]$$

We substitute for P the lens-makers formula, once for the crown element and once for the flint,

$$\mathbf{P} = \underbrace{(n-1)\left(\frac{1}{R_1} - \frac{1}{R_2}\right)}_{crown} + \underbrace{(n-1)\left(\frac{1}{R_1} - \frac{1}{R_2}\right)}_{flint}$$

For convenience we replace the terms containing the R's by A's, writing

$$\mathbf{P} = (n_1 - 1)A_1 + (n_2 - 1)A_2 \qquad [1.6\text{-}23]$$

In order to make the combination achromatic, its power at the two colors must be the same. Denoting blue by the subscript F and red by C, we obtain

$$(n_{1F} - 1)A_1 + (n_{2F} - 1)A_2 = (n_{1C} - 1)A_1 + (n_{2C} - 1)A_2$$

Multiplying out and canceling gives

$$\frac{A_1}{A_2} = -\frac{n_{2F} - n_{2C}}{n_{1F} - n_{1C}} \qquad [1.6\text{-}24]$$

We repeat the process for yellow (subscript D),

$$P_{1D} = (n_{1D} - 1)A_1 \quad \text{and} \quad P_{2D} = (n_{2D} - 1)A_2$$

Dividing one by the other gives

$$\frac{A_1}{A_2} = \frac{P_{1D}(n_{2D} - 1)}{P_{2D}(n_{1D} - 1)} \qquad [1.6\text{-}25]$$

Then setting Equation [1.6-25] equal to [1.6-24] and solving for P_{1D}/P_{2D} yields

$$\frac{P_{1D}}{P_{2D}} = -\frac{(n_{2F} - n_{2C})/(n_{2D} - 1)}{(n_{1F} - n_{1C})/(n_{1D} - 1)}$$

Both the numerator and the denominator in the right-hand term are the inverse of Abbe's number; therefore,

---

he could collect royalties. Naturally, the other London opticians objected and took the case to court, producing Moor-Hall as a witness. The court agreed that indeed Moor-Hall was the inventor, but in a much-quoted decision, the judge, Lord Camden, ruled in favor of Dollond, saying: "It is not the person who locked up his invention in his scritoire that ought to profit by a patent for such invention, but he who brought it forth for the benefit of the public."

$$\frac{P_1}{P_2} = -\frac{\nu_1}{\nu_2}$$

Note the minus sign. It means, since Abbe's number can only be positive, that if one of the lenses is positive, the other must be negative. Finally, we substitute the values of P_1 and P_2, respectively, from Equation [1.6-22] and obtain

$$P_1 = P\left(\frac{\nu_1}{\nu_1 - \nu_2}\right) \quad \text{and} \quad P_2 = -P\left(\frac{\nu_2}{\nu_1 - \nu_2}\right) \quad [1.6\text{-}26]$$

Example. *Design a crown-flint doublet of +10 cm focal length, achromatic for blue and red, using the refractive indices listed in Chapter 1.2, page 30.*

Solution. First, we determine Abbe's numbers. We find for crown

$$\nu_1 = \frac{n_D - 1}{n_F - n_C} = \frac{1.5230 - 1.0}{1.5293 - 1.5204} = 58.7640$$

and for flint

$$\nu_2 = \frac{1.7200 - 1.0}{1.7378 - 1.7130} = 29.0323$$

Inserting these figures in Equations [1.6-26] gives

$$P_1 = (+10)\left(\frac{58.7640}{58.7640 - 29.0323}\right) = +19.7648 \text{ m}^{-1}$$

and

$$P_2 = -(+10)\left(\frac{29.0323}{58.7640 - 29.0323}\right) = -9.7648 \text{ m}^{-1}$$

The combined power of the two lenses is $+10 \text{ m}^{-1}$, which serves as a check on our calculations so far.

Knowing the focal lengths required of the two lenses, we are ready to choose their radii. For reasons of economy, the converging lens is made symmetric, equiconvex. Also, the two lenses are to be in contact. Thus $R_1 = -R_2 = -R_3$. For the first lens, from the lens-makers formula,

$$P = (n_{\text{lens}} - 1)\left(\frac{1}{R_1} - \frac{1}{R_2}\right)$$

$$+19.7648 = (1.523 - 1.000)\left(\frac{2}{R_1}\right) = \frac{1.046}{R_1}$$

and thus

$$R_1 = 0.0529 = \boxed{+5.29 \text{ cm}}$$

The next two surfaces have radii of

$$R_2 = R_3 = \boxed{-5.29 \text{ cm}}$$

and for the last surface,

$$-9.7648 = (0.7200)\left(\frac{1}{-0.0529} - \frac{1}{R_4}\right)$$

$$R_4 = -\frac{0.7200}{3.8458} = -0.1872 = \boxed{-18.72 \text{ cm}}$$

Another method of making a system achromatic is to use two positive lenses, made of the same type of glass and separated by a distance equal to one-half the sum of their focal lengths. To see why this approach works, we start out from the equivalent power equation, Equation [1.4-9],

$$\mathbf{P} = P_1 + P_2 - P_1 P_2 d$$

Then, following Equation [1.6-23],

$$\mathbf{P} = (n - 1)A_1 + (n - 1)A_2 - (n - 1)A_1(n - 1)A_2 d$$

$$= (n - 1)(A_1 + A_2) - (n - 1)^2 A_1 A_2 d$$

For the combination to be achromatic, $\mathbf{P}$ must stay constant at different wavelengths; thus by differentiation,

$$\frac{d\mathbf{P}}{dn} = A_1 + A_2 - 2(n - 1)A_1 A_2 d = 0$$

We multiply by $(n - 1)$ and substitute for $(n - 1)A$ the corresponding P:

$$P_1 + P_2 = 2P_1 P_2 d$$

$$d = \frac{P_1 + P_2}{2P_1 P_2}$$

which is equal to

$$\boxed{d = \frac{1}{2}(f_1 + f_2)} \qquad [1.6\text{-}27]$$

Spaced doublets of this type are used as eyepieces (the Huygens and Ramsden eyepiece), discussed in Chapter 1.8.

A system corrected for chromatic aberration is called an *achromat*.

Such a system is corrected for two colors, and usually for spherical aberration too. *Apochromats* are corrected for three colors and *superachromats* for four.

Concluding remarks.
Are some aberrations more important than others? That depends on the application. For some purposes the highest resolution (no coma!) is required and distortion is not as critical; for others it may be just the opposite. There is no way to eliminate all aberrations at the same time; in fact, eliminating some of them will usually make others come up worse. All that we can do is *balance* them. A summary of the various aberrations is found in Table 1.6-1.

Nowadays, correcting for aberrations is nearly always done by computer, varying the power, shape, thickness, and separation of the different elements and choosing the right refractive indices and dispersive powers of the glasses. Still, despite all the publicity, the role of the computer is probably overemphasized. True, computers greatly help in any lens design. But it is the lens designer who develops the concept and who determines how much of each aberration he is willing to tolerate. The computer helps the designer make certain decisions and takes much of the tedium out of his work. Lens design as such still *remains an art*.

Table 1.6-1 SUMMARY OF ABERRATIONS

Aberration	*Character*	*Correction*
1. Spherical aberration	Monochromatic, on- and off-axis, image blur	Bending, high index, aspherics, gradient index, doublet
2. Coma	Monochromatic, off-axis only, blur	Bending, spaced doublet with central stop
3. Oblique astigmatism	Monochromatic, off-axis, blur	Spaced doublet with stop
4. Curvature of field	Monochromatic, off-axis	Spaced doublet
5. Distortion	Monochromatic, off-axis	Spaced doublet with stop
6. Chromatic aberration	Heterochromatic, on- and off-axis, blur	Contact doublet, spaced doublet

SUGGESTIONS FOR FURTHER READING

W. H. Price, "The Photographic Lens," *Scientific American* **235** (Aug. 1976), 72–83.

A. Nussbaum and R. A. Phillips, *Contemporary Optics for Scientists and Engineers* (Englewood Cliffs, NJ: Prentice-Hall, Inc., 1976).

R. Kingslake, *Lens Design Fundamentals* (New York: Academic Press, Inc., 1978).

PROBLEMS

1.6-1. If peripheral rays are traced through a +5.00-diopter lens of 39 mm diameter, it is found that these rays come to a focus 5 mm ahead of the paraxial rays. What is the diameter of the blur circle in the paraxial focus?

1.6-2. If a lens of 50 mm diameter and 25.4 cm focal length has a longitudinal spherical aberration of 4 mm, what is its transverse spherical aberration?

1.6-3. A thin meniscus of index 1.60 has radii $R_1 = +15$ cm and $R_2 = +30$ cm. Determine:
(a) The Coddington shape factor.
(b) The position factor for an object 1.2 m away.

1.6-4. A lens of index 1.66 forms of an object 40 cm away an image 8 cm away. The front surface of the lens has a radius of curvature of +120 mm. Find the shape and position factors.

1.6-5. A lens made of flint of index 1.72 has a focal length of +5 cm. For parallel incident light and for the lens to have minimum spherical aberration, determine the position and shape factors and the two radii of curvature necessary.

1.6-6. A tiny lens made of crown ($n = 1.523$) forms of an object 7 mm away an image 16 mm away. What is the best form to minimize spherical aberration?

1.6-7. If a target such as that shown in Figure 1.6-19 is imaged through a system not corrected for astigmatism, what changes will occur in the two principal foci?

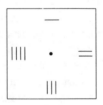

Figure 1.6-19

1.6-8. The letters ∃H⊥ serve as the object for a plus spherocylinder, axis vertical. Show what the image looks like in the two principal image planes.

1.6-9. Collimated light is focused by a spherocylinder into two line images 20 cm and 25 cm, respectively, away from the lens. If then a −1.00-diopter sphere is placed in contact with the spherocylinder, how long will the interval of Sturm be?

1.6-10. A large-diameter spherocylinder forms one line focus at a distance of 21.7 cm and another line focus at a distance of 23.8 cm. To what diameter must an iris diaphragm close to the lens be stopped down in order to produce a circle of least confusion 0.8 mm in diameter?

1.6-11. A spherocylinder forms with parallel light a circle of least confusion 6 mm in diameter on a screen located 30 cm from the lens. If a point source, placed 60 cm from the lens, gives a *line* image on the screen, what is the diameter of the lens?

1.6-12. A conventional thin lens forms of a pinhole 40 cm to the left of the lens a real image 20 cm to the right of the lens. On adding a cylinder to the lens, a horizontal line image is formed 10 cm to the right of the two lenses. What is the power and the orientation of the cylinder?

1.6-13. To meet the Petzval condition for the elimination of curvature of field, two lenses, both of +10.00 diopters power and $n = 1.55$, are placed 31 mm apart. What is the power of the combination?

1.6-14. If one lens of a combination of two is made of crown and has +17 cm focal length and if the combination is to be free from curvature of field, what should be the focal length of the other lens, provided that it is made of flint? Use the refractive indices for yellow, Table 1.2-2.

1.6-15. If a lens, made of ophthalmic crown ($n = 1.523$), has +8.50 diopters power, what is the power of a lens of the same shape but made of flint ($n = 1.720$)?

1.6-16. A certain lens has, for yellow light, an index of refraction of 1.6500 and a focal length of 62 cm. For red light, the focal length becomes 62.5 cm. What is the refractive index for that light?

1.6-17. Two lenses in contact, one made of crown, the other of flint, form a diverging (negative) achromatic doublet. Trace several rays representing two colors through the combination.

1.6-18. What are the powers of two thin lenses, one made of crown, the other of dense flint, that must be placed in contact to obtain a +6.00-diopter doublet achromatic for blue and red? Use the refractive indices from Table 1.2-2.

1.6-19. A +8.75-diopter crown lens is to be combined with a flint lens to make a contact doublet achromatic for blue and red. Using the data from Table 1.2-2, determine the power of the flint lens.

1.6-20. Design an achromatic doublet of 40 cm focal length, following the example in the text but using as the first lens a negative flint meniscus and as the second lens a convex-plane crown element, the plane surface being last.

1.6-21. An achromatic doublet consists of two positive lenses separated by 8 cm. The first lens has a focal length of 12 cm. What is the focal length:
(a) Of the second lens?
(b) Of the whole system?

1.6-22. Using two positive lenses, one of them of $f = 12$ cm, design an achromatic doublet of $f = 8$ cm. Find the focal length of the second lens and the distance between both.

1.7

Apertures and Stops

IN EVERY OPTICAL SYSTEM, the bundle of light passing through is limited in cross section, either by the finite size of the lenses or by additional diaphragms called *stops* (Figure 1.7-1). A lens that is too small will reduce the efficiency of others. A lens that is too large is wasteful. Therefore, the proper design of any system requires that the diameters of the various lenses are correctly chosen. The systematic study of such considerations is called the *theory of stops*.

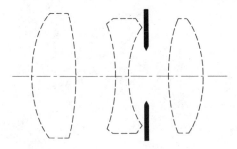

Figure 1.7-1   Aperture stop placed in three-element lens system.

Stops

Aperture stop.   An *aperture stop* is an opening, usually circular, in an otherwise opaque screen. Frequently, the aperture stop is simply the rim of a lens, or the edge of a mirror. In a camera the aperture stop is often an iris diaphragm (that can be varied in size). An aperture stop limits the cross section of the image-forming light. Making the aperture smaller increases the *depth of focus* and the depth of field (it also makes the image dimmer, without restricting the field of view).

The concept of depth of focus assumes that there exists a certain *blur circle* small enough that it does not affect the quality of the image. Generally, a blur circle 30 μm in diameter or $1/1000$ the focal length of the system, whatever is larger, is considered acceptable.

Consider Figure 1.7-2 and assume that the light comes from an axial point object and focuses precisely in the image plane. But if we can tolerate a given blur circle diameter, b, then for a given aperture diameter, D, we find that

$$\frac{D}{f + \Delta f} = \frac{b}{\Delta f}$$

where Δf is the depth of focus. Since $\Delta f \ll f$,

$$\frac{D}{f} = \frac{b}{\Delta f} \qquad\qquad [1.7\text{-}1]$$

The ratio D/f is called the *aperture ratio* of the system and its reciprocal the *f-stop number*, or "*speed*":

$$f\text{-stop} = \frac{\text{focal length}}{\text{diameter of aperture}} \qquad\qquad [1.7\text{-}2]$$

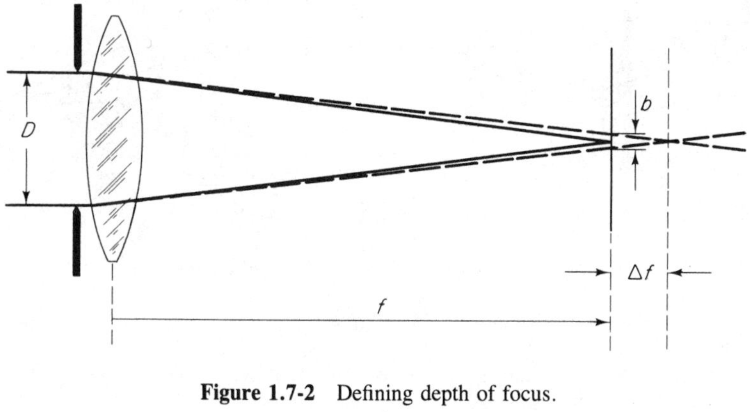

Figure 1.7-2   Defining depth of focus.

For example, if for a camera lens $f = 50$ mm and $D = 1.25$ cm, the f-stop number is 50 mm/12.5 mm $= 4$, which customarily is written $f/4$. (Camera lenses often have f-stop numbers such as $f/1.4, f/2, f/2.8, f/4, f/5.6, f/8$, and so on. They are chosen so that the next higher number requires twice the exposure time.) For objects nearby and image distances, i, appreciably longer than f, the *effective f-stop number* is i/D, rather than f/D.

Depending on the depth of focus, Δf or Δi, there is a conjugate distance, Δo, in the object space (not shown). That distance is called the *depth of field*. As before, a smaller aperture will increase the depth of field.

For the largest possible depth of field it would not be wise to focus at infinity because some depth of field, "beyond infinity," would be wasted. Instead, we focus at the *hyperfocal distance;* the depth of field then extends from one-half that distance out to infinity.

The term *angular aperture* refers to the (total) apex angle of the cone of light converging at the focus. For the same $f/4$ lens as before,

$$\sigma = 2 \tan^{-1} \left(\frac{6.25 \text{ mm}}{50 \text{ mm}} \right) = 14.3°$$

Field stop. A *field stop,* in contrast to an aperture stop, limits the angular field, or *field of view*. As the field stop is made smaller, the field of view becomes more narrow (as in Figure 1.7-3c) but the amount of light admitted remains the same. A field stop is often located in the image plane (as shown) but it may be located in the object plane (as in a slide projector where the opaque mount of the slide limits the field).

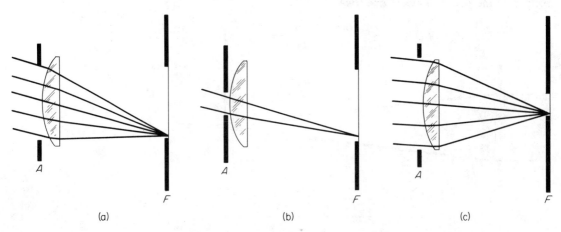

(a)   (b)   (c)

Figure 1.7-3 Aperture stop (A) and field stop (F). Making the aperture stop smaller, as in (b), limits the amount of light but does not restrict the field of view. Making the field stop smaller, as in (c), has the opposite effect.

Pupils

A stop is something tangible and real. A pupil is more conceptual. A pupil may be a real physical stop; or it may be a real or even a virtual image of a stop. Pupils are as important as lenses in the design of optical systems. A pupil is defined as the cross section of a bundle where light from all parts of the object passes through, in equal amounts, completely mixed, *with no preferential spatial separation*.

Examples

1. Assume that we have an extended object and that a lens forms an image of this object as shown in Figure 1.7-4. There is only one cross section common to all rays: *that cross section is the lens*. Of course, we could draw an infinite number of other cross sections through the beam, closer to the object or closer to the image. But these cross sections would all contain light that comes predominantly from certain parts of the object, more from the tip of the arrow or more from its foot, and thus they are not pupils.

2. If we place a stop between an (extended) object and a lens, that *stop will become the pupil* (Figure 1.7-5). Rays from all points in the object pass through the stop, completely mixed, with no spatial separation. If we partially occlude the stop with the point of a pencil and at the same time look at the image, the point can hardly be seen (except that the image becomes dimmer). In this case, the lens does *not* act as a pupil: rays from the tip of the arrow pass through the lower part of the lens, and rays from the foot pass through the upper part. If we hold the pencil close to the lens, the point *would* show as a shadow superimposed on the image.

3. So far, I have referred to *a* pupil. More specifically, the stop in Figure 1.7-5 is the *entrance pupil. Its image,* as formed by the lens, *is the exit pupil.* Note that there are two processes of image formation that go on side by side: (a) the lens transforms an object into its image, and (b) the lens transforms the entrance pupil into the exit pupil. Object and image are conjugate, and entrance pupil and exit pupil are conjugate also.

4. We could *move* the entrance pupil, that is, we could move the stop, back and forth along the axis. If we move the stop closer to the lens,

Figure 1.7-4   Lens acting as a pupil.

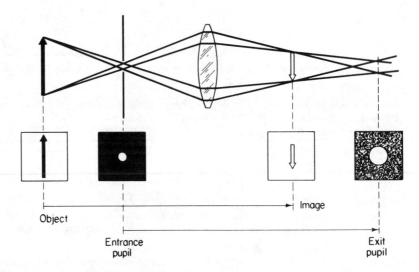

Figure 1.7-5   Entrance pupil and exit pupil. Long horizontal arrows show that there are two independent processes of image formation.

its image, the exit pupil, moves away. If the entrance pupil is moved into the left-hand focal plane, the exit pupil moves out to infinity on the right. The system is then called *telecentric on the image side* (Figure 1.7-6). Conversely, if the stop is moved into the right-hand focal plane, the system is *telecentric on the object side*. Such systems are useful in precision measurements, whenever the (transverse) size of an object is to be compared with a scale that cannot be brought into contact with the object.

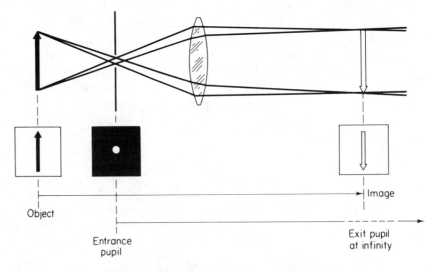

Figure 1.7-6   System telecentric on the image side.

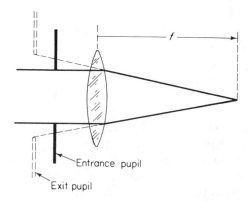

Figure 1.7-7   Entrance pupil located within first focal length of lens.

5. If the stop is moved even closer to the lens so that it comes to *within the focal length,* the entrance pupil and exit pupil lie on the same side (Figure 1.7-7). The exit pupil, like the image formed by a magnifying glass, has become a virtual, magnified image of the stop.

Example. *A stop 8 mm in diameter is placed halfway between an extended object and a large-diameter lens of 9 cm focal length. The lens projects an image of the object onto a screen 14 cm away. What is the diameter of the exit pupil?*

Solution. Refer to Figure 1.7-5. First, from the known focal length and the image distance, we find the object distance,

$$o = \frac{if}{f - i} = \frac{(14)(9)}{9 - 14} = -25.2 \text{ cm}$$

The stop is one-half that distance in front of the lens,

$$o' = \frac{-25.2 \text{ cm}}{2} = -12.6 \text{ cm}$$

Considering the diameters given, we have little doubt that the stop, rather than the lens, is the entrance pupil. Its image, the exit pupil, is formed at a distance

$$i' = \frac{o'f}{o' + f} = \frac{(-12.6)(9)}{(-12.6) + 9} = 31.5 \text{ cm}$$

to the right of the lens. The diameters of the two pupils, D_{EP} and D_{XP}, are proportional to their distances, o' and i'. Therefore, the diameter of the exit pupil is

$$D_{\text{XP}} = \frac{D_{\text{EP}}i'}{o'} = \frac{(0.8)(31.5)}{-12.6} = \boxed{2 \text{ cm}}$$

Pupils of combinations of lenses.

When a stop is placed ahead of a lens, that stop becomes the entrance pupil. But if another lens is placed ahead of the stop, the light does not reach the stop directly. Instead, the light "sees" the stop the same way we see newsprint through a magnifying glass. In other words, the light comes to an *image* of the stop, rather than to the stop itself, and *this image*, by definition, *is the entrance pupil.*

Therefore, the entrance pupil of a combination of lenses is the image of the stop formed by all lenses *preceding* the stop. Conversely, the exit pupil of a combination of lenses is the image of the stop formed by all lenses *following* the stop. Furthermore, the exit pupil may be located to the right of the entrance pupil or to the left of it, or it may coincide with it.

Example. *Two lenses, a +8.00-diopter lens and a minus lens of unknown power, are mounted coaxially and 8 cm apart. The system is afocal, that is, light entering the system parallel at one side emerges parallel at the other. If a stop 15 mm in diameter is placed halfway between the lenses:*
(a) Where is the entrance pupil?
(b) Where is the exit pupil?
(c) What are their diameters?

Solution. For the system to be afocal, the focal points of the two lenses must coincide (at F in Figure 1.7-8). Therefore, if the first lens has +8.00 diopters power, or $(1/8)(100) = 12.5$ cm focal length, and if the two lenses are 8 cm apart, the second lens must have $-(12.5 - 8) = -4.5$ cm focal length.

The *entrance pupil* is the image of the stop formed by the first lens. Considering that the stop lies to the right of the lens and the f_2 focal length consequently is on the left, we find the entrance pupil

$$i = \frac{of_2}{o + f_2} = \frac{(4)(-12.5)}{(4) + (-12.5)} = \boxed{+5.88 \text{ cm}}$$

to the right of the (first) lens.

The *exit pupil* is the image of the stop formed by the second lens. It is

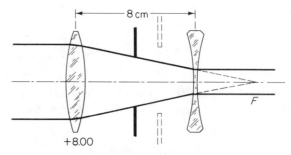

Figure 1.7-8

$$i' = \frac{(-4)(-4.5)}{(-4) + (-4.5)} = \boxed{-2.12 \text{ cm}}$$

to the left of the (second) lens. In other words, the two pupils coincide—which is typical of an afocal system.

However, the pupil *sizes* are different. The diameter of the entrance pupil is

$$D_{\text{EP}} = \frac{D_{\text{STOP}}i'}{o'} = \frac{(15)(58.8)}{40} = \boxed{22 \text{ mm}}$$

and the diameter of the exit pupil

$$D_{\text{XP}} = \frac{(15)(21.2)}{40} = \boxed{8 \text{ mm}}$$

Finally, we look at a combination of two lenses and *two stops*. First consider the *aperture stop*. If we use graphical ray tracing, drawing a parallel ray and a chief ray, we find that the image of the aperture stop, formed by the lens preceding the stop, is the *entrance pupil*, EP (Figure 1.7-9a). The image of the aperture stop, formed by the lens following the stop, is the *exit pupil*, XP (b).

Then consider the *field stop*. The image of the field stop, formed by the lens preceding the stop, is the *entrance window* or *entrance port*, EW (c). The image of the field stop, formed by the lens following the stop, is the *exit window* or *exit port*, XW (d).

Now we are ready to locate the *rays limiting the bundle*. There is a cone of light that comes from the center of the entrance pupil, touches the rim of the entrance window, and goes on to the center of the aperture stop. From there it proceeds to the rim of the field stop, the center of the exit pupil, and the rim of the exit window (full lines in Figure 1.7-9e). The other bundle, limited by dashed lines, begins at the rim of the entrance pupil, and goes on to the center of the entrance window. It then proceeds to the rim of the aperture stop, the center of the field stop, the rim of the exit pupil, and finally the center of the exit window. The two bundles are completely intertwined yet independent of each other. In particular, note that:

1. *Aperture stops relate to pupils,* both entrance pupil and exit pupil. *Field stops relate to windows,* both entrance window and exit window.

2. A *chief ray* (solid lines, Figure 1.7-9e) always enters a system through the center of the entrance pupil. And because entrance pupil and exit pupil are conjugate, the chief ray leaves the system through the center of the exit pupil.

3. At the exit pupil, the bundle has the *least diameter,* because an aperture stop is by necessity farther away from the last lens than a field stop. That has important consequences: In many optical instruments, such

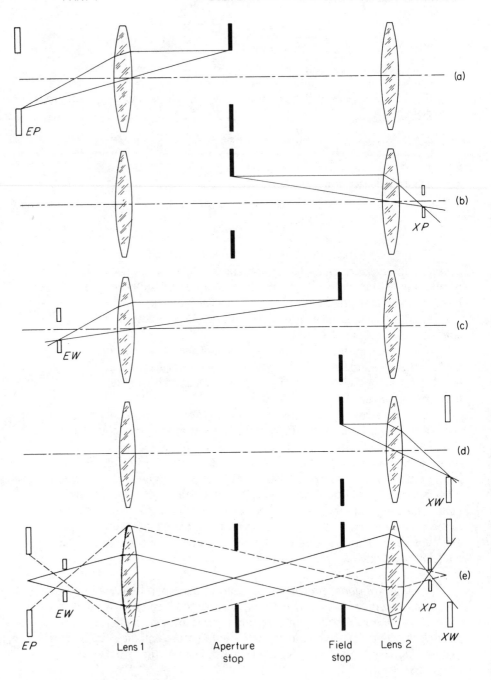

Figure 1.7-9 Entrance pupil, exit pupil, entrance window, and exit window of a system composed of two lenses and two stops.

as in a telescope or microscope, the entrance pupil is the objective lens (no separate stop is needed). The exit pupil is the image of this lens formed by the eyepiece. This image is where the density of the light is highest and where the pupil of the observer's eye should be placed to assure comfortable viewing and to avoid *vignetting*. (This aspect is discussed in numerical detail in the example on page 157).

    4. The light source should be large enough to fill the entrance pupil. A light source that does not fill the entrance pupil cannot fill the aperture stop either; and if it does not, the amount of light received in the image and the resolution of the image will be less than they could be. This is why it is desirable that the lenses, and the stops and other elements in an optical system, all have the right diameters.

PROBLEMS

1.7-1. An object is placed 30 cm in front of a lens of 50 mm diameter and 75 mm focal length. What is the *effective f*-stop number?

1.7-2. A camera lens of 50 mm focal length is set at $f/8$. If the camera is focused at an object 15 cm away, what will the effective f-stop number be?

1.7-3. Compare the depth of focus, as a percentage of focal length, of a 30-mm-focal-length wide-angle lens with the depth of focus of a 135-mm-focal-length telephoto lens, both set at $f/4$.

1.7-4. Continue with Problem 1.7-3 and determine the nearest conjugate object distance when focusing at infinity.

1.7-5. If a pinhole camera has the shape of a cube:
(a) How large a field of view will it cover?
(b) How large would be the field if the camera were built around a cube made out of glass of $n = 1.52$?

1.7-6. A photographic slide, taken on 35-mm film, has a usable area of 24×36 mm². What is the angular field, measured diagonally, that can be projected through a lens of 123 mm focal length?

1.7-7. A stop, placed a short distance to the rear of a converging lens, acts as the exit pupil. By ray tracing find the entrance pupil.

1.7-8. Two stops, of different diameters, are mounted a certain distance apart from each other. How must the stops be placed if for parallel light it is the smaller stop that limits the bundle while for divergent light it is the larger stop?

1.7-9. A thin lens of 50 mm focal length has a diameter of 4 cm. A stop 2 cm in diameter is placed 3 cm to the left of the lens, and an axial point object is located 20 cm to the left of the stop.
(a) Which of the two, the stop or the lens, limits the bundle?
(b) What is the effective f-stop number of the system;
(c) Where is the exit pupil located?

1.7-10. Two thin lenses, 5 cm in diameter each and of focal lengths $+10$ cm and $+6$ cm, respectively, are placed 4 cm apart. An aperture stop 2 cm in diameter is set halfway between the lenses. Find the diameters of the entrance pupil and the exit pupil.

1.8

Optical Systems

THERE ARE SO MANY OPTICAL SYSTEMS that it would be futile to try to list them. Instead, I present a few characteristic examples.

Camera Lenses

Camera lenses are a good subject for a study of evolution; they have gradually developed from very simple to highly complex systems, punctuated from time to time by fundamentally new inventions.

The earliest type of a camera, known to the Arabs 1000 years ago, was the pinhole camera or *camera obscura,* mentioned on page 16. In the Middle Ages, around 1568,* a *biconvex lens* was added to the camera obscura, allowing the image to be focused. In 1812 came the *meniscus,* its concave side facing the object. Such a lens has little curvature of field but because of spherical and chromatic aberration, its aperture can be no larger than about $f/11$.

Next came the *achromatic meniscus,* the so-called "landscape lens." But *two achromatic menisci,* their concave sides facing each other as in Figure 1.8-1, are much better. An example is the *Rapid Rectilinear,*

---

*According to J. M. Eder, *History of Photography* (New York: Columbia University Press, 1945).

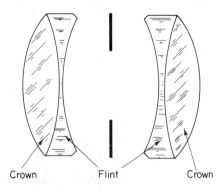

Figure 1.8-1 Orthoscopic doublet.

developed in 1866. It has very little coma, distortion, and lateral color but still has considerable spherical aberration, and either astigmatism or curvature of field. Its speed is about $f/8$. Several other symmetrical lenses followed, most notably the *Planar*, which has two additional plano-convex lenses, one on each side. The *Petzval portrait lens,* two separate positive achromats, has been mentioned earlier (page 120).

The design of camera lenses then took a dramatically different turn, due to the development of the *Taylor–Cooke triplet*.* This is the familiar three-lens system used here to introduce each major new topic.

The Taylor triplet consists of two positive lenses, in front and back, both made of crown, and a negative lens, in between, made of flint. The negative lens is the crucial element. Its distance from the first (positive) lens is relatively large. Thus the light, as it reaches the minus lens, has fairly high vergence. Hence, the power of the minus lens can also be kept high to correct for spherical and chromatic aberration, and to reduce coma, astigmatism, curvature of field, and distortion, and still leave sufficient power in the system. Because of its good correction, the Taylor triplet could be made with much higher speeds ($f/6.3$ and better) and a wider field than any other camera lens known at that time.

---

*Designed in 1893 by Harold Dennis Taylor (1862–1943), optical manager of the T. Cooke & Sons Optical Company, Bishophill, York, England. It seems that early camera lenses were often "corrected" for aberrations by placing stops to eliminate unwanted marginal rays (and making long exposure times a necessity). Taylor broke with tradition. In his invention disclosure he describes his "idea of eliminating the diaphragm corrections and throwing the whole burden of flattening the final image and correcting marginal astigmatism upon a negative lens." H. D. Taylor, *A Simplified Form and Improved Type of Photographic Lens,* Brit. Patent 22,607, 6th Oct. 1894, and *Lens,* U.S. Patent 540,122, May 28, 1895. In 1906, Taylor (who had started out to become an architect) published a book, *A System of Applied Optics,* where he set forth the principles of lens design he had used, an extension of Coddington's earlier work.

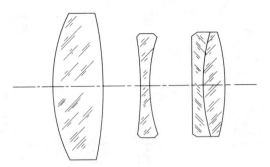

Figure 1.8-2   *Tessar* camera lens.

From this concept of the Taylor–Cooke triplet there derived numerous other camera lenses. The most successful of these is the *Zeiss Tessar*.* As shown in Figure 1.8-2, the rear element became an achromatic cemented doublet. The Tessar has excellent correction for spherical aberration, good achromatism, sufficiently high speed ($f/3.5$ and even $f/2.8$), very little astigmatism, and a field free of curvature and distortion out to about 60°. Today the Tessar and its derivatives are the most widely used high-quality camera lenses. Table 1.8-1 lists some typical parameters.

A *telephoto lens* has a focal length longer than the diagonal across the film negative. These lenses are usually of the Galilean type (see pages 152, 154). A *wide-angle lens* has a focal length shorter than the diagonal. A lens that covers a 180° field (a hemisphere) is called a *fisheye lens* or *sky lens*. An example is shown in Figure 1.8-3.

A *zoom lens* is a system whose focal length can be varied, without moving the image out of focus. A zoom lens, therefore, provides variable magnification. Many zoom systems are Taylor triplets, plus–minus–plus. In some types, the two plus lenses are mechanically linked and move as a unit while the minus lens remains in place. With the minus lens in midposition, the image has a certain size. Moving the two plus lenses forward makes the image smaller, moving them backward makes it larger. Such zoom lenses are called *"optically compensated."*

In other types, the lenses move in a more complex way, often nonlinear. The front lens in Figure 1.8-4, for example, is stationary. The

---

*Invented by Paul Rudolph (1858–1935), lens designer at the Carl Zeiss optical company in Jena, Germany. In 1895 Rudolph designed the Planar, in 1902 the Tessar. In his invention disclosures, *Sphärisch, chromatisch und astigmatisch korrigiertes Objektiv aus vier, durch die Blende in zwei Gruppen geteilte Linsen*, Dtsch. Reichspatent 142 294, 25 Apr. 1902, and *Photographic Objective*, U.S. Patent 721,240, Feb. 24, 1903, Rudolph describes how with a "comparatively small number of components" he developed a "spherically, chromatically, and astigmatically corrected objective" that is "fruitful in an extraordinary degree." For details on modern Tessars, see R. Kingslake, *Lens Design Fundamentals*, pp. 277–86 (New York: Academic Press, Inc., 1978).

Table 1.8-1 Tessar Camera Lens*

Axial thickness, d (mm) Refractive index, n	Radius of curvature (mm)	Air space (mm)
Lens 1 $d = 3.57$ $n = 1.6116$	$R_1 = +16.28$ $R_2 = -275.7$	1.89
Lens 2 $d = 0.81$ $n = 1.6053$	$R_3 = -34.57$ $R_4 = +15.82$	3.25
Lens 3 $d = 2.17$ $n = 1.5123$	$R_5 = \infty$ $R_6 = +19.20$	
Lens 4 $d = 3.96$ $n = 1.6116$	$R_7 = +19.20$ $R_8 = -24.00$	

*Focal length: 50.82 mm.

second lens and the third lens both move, the rates and directions of motion being governed mechanically by cams, or slots cut into a rotatable cylinder. Such zoom lenses are called *"mechanically compensated."*

Why does the upper example in Figure 1.8-4 give a wide-angle effect and the lower example a telephoto effect? Look at the angles subtended at the image by peripheral rays and consider the *Lagrange invariant, nhγ* (page 55). Since the product $h\gamma$ must remain constant, a *larger* angle (upper example) relates to a *smaller* image, as produced by a *wide-angle* lens, and a smaller angle (bottom example) to a larger image, as produced by a telephoto lens.

Camera lenses can be built to remarkable specifications. For example, cameras in high-flying aircraft can from a height of more than 12 km

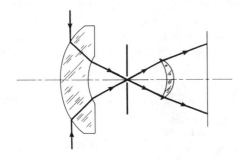

Figure 1.8-3 Sky Lens.

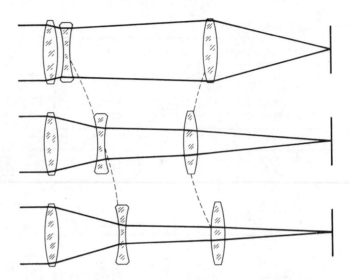

Figure 1.8-4   Mechanically compensated zoom system. (*Top*) Wide-angle and (*bottom*) telephoto effect.

(40,000 ft) resolve numbers as small as those on an automobile license plate.

The Eye

Image formation.   The human eye, like a camera, forms a real image on a light-sensitive surface, the *retina*. On entering the eye, the light first passes through a transparent layer, the *cornea* (see Figure 1.8-5). It then enters the *anterior chamber* which is filled with a liquid, the *aqueous humor*. The anterior chamber is bounded by an aperture stop, the *iris*. The image of the central opening, a virtual image because it is seen through the cornea, is called the *pupil* of the eye. The diameter of the pupil varies as a function of the amount (and wavelength) of the light received. Next comes the *crystalline lens,* made of an elastic, jellylike substance. The curvature of the two surfaces of this lens is controlled by the *ciliary muscle*. After refraction by the lens the light passes through the *vitreous humor* and then reaches the retina.

The optical properties of the eye can be represented either by the simpler *reduced eye* or by the more realistic *schematic eye*. The reduced eye (discussed on page 43) is assumed to have a single refractive surface, with a radius of curvature of $R = +5.7$ mm. The two focal lengths are $f_1 = -16.8$ mm and $f_2 = +22.5$ mm, and the power is $+60$ diopters.

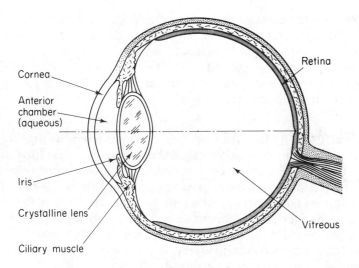

Figure 1.8-5   Cross section through the human eye.

Gullstrand's *schematic eye* is assumed to have three surfaces, the front surface of the cornea and the front and rear surfaces of the crystalline lens. Their dimensions are shown in Table 1.8-2. The total power is again 60 diopters. Of this, about 43 diopters is contributed by the cornea, the remainder by the lens.

The *crystalline lens* of the human eye is remarkable in that it is a *gradient-index lens,* the prototype of an entirely new class of optical elements that I will present in detail in Chapter 3.5. The refractive index of the crystalline lens is highest in the center (1.406), lowest at the equator (1.375), and only slightly higher (1.387 and 1.385) at the vertices, where it borders the aqueous and the vitreous, respectively.

Focusing on a nearby object is done by *accommodation*. Accommodation is accomplished by a change of power of the lens, resulting from a change of shape. The lens itself is nearly spherical. But it is suspended by

Table 1.8-2   SCHEMATIC EYE

Radii of curvature	
Cornea	+7.80 mm
Lens, anterior surface	+10.00 mm
posterior surface	−6.00 mm
Refractive indices	
Aqueous	1.336
Lens (cumulative)	1.413
Vitreous	1.336

ligaments and these ligaments are connected to the circumferential ciliary muscle. Thus the lens is stretched out into a flatter shape. During accommodation, the ciliary muscle contracts, relaxing the ligaments. This reduces the tension on the lens and allows it to become more spherical. The effect is most noticeable on the front surface; its radius of curvature changes from $+10.0$ mm to $+5.3$ mm.

Refractive anomalies. The two most common refractive errors of the eye are *myopia* (nearsightedness) and *hyperopia* (hypermetropia, farsightedness). In myopia, the image of a distant object is formed in front of the retina and thus is seen out of focus (Figure 1.8-6a). Only objects that are not farther away than the *far point* will focus on the retina (b). Correction for myopia requires a minus lens. Its second (left-hand) focal point should coincide with the far point (c). Then, parallel light incident on the eye becomes divergent, so that it appears to come from the far point and *will* focus on the retina (d).

In *hyperopia,* parallel light comes to a focus behind the eye (Figure 1.8-7a). The far point is to the right of the retina. It is defined as the point to which rays from an object must converge for a sharp image to be formed

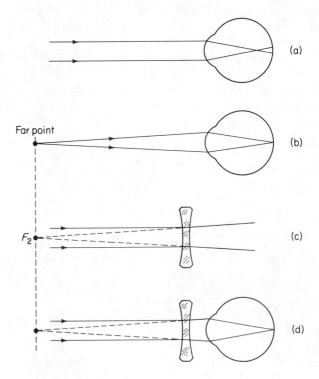

Figure 1.8-6 Myopia and its correction.

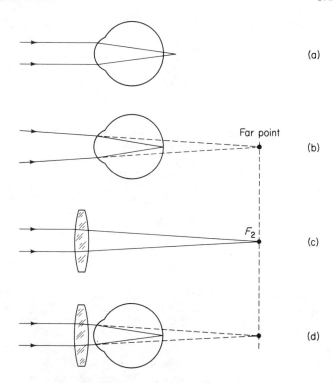

Figure 1.8-7 Hyperopia and its correction.

on the retina (b). For correction, again the second focal point of the correcting lens should coincide with the far point (c). Then rays entering the eye are already converging and will focus on the retina (d).

Example. *A person wants to look at the image of his or her own eyes, without accommodation, using a concave mirror of 60 cm radius of curvature. How far must the mirror be from the eye if the person has*
(a) Normal vision?
(b) 4.00 diopter myopia, without correction?
(c) 4.00 diopter hyperopia, also without correction?

Solution. First we determine the focal length of the mirror:

$$f = \frac{R}{-2} = \frac{-60}{-2} = +30 \text{ cm}$$

(a) For the light to come to a focus on the retina, it must enter the eye parallel, and hence leave the mirror parallel, and hence originate at the focal plane. Therefore, the distance between mirror and eye is

30 cm

(b) A 4.00-diopter myopia means that the light, when entering the eye, must have -4.00 m^{-1} vergence in order to focus on the retina. Such light comes from an object $1/-4.00$ m$^{-1} = -25$ cm in front of the eye. This object is the image formed by the mirror; its distance from the mirror is the image distance. The object distance (relative to the mirror) is the image distance *plus* 25 cm, $-o = i + 25$ cm, or $i = -o - 25$ cm. Then from the thin-lens equation,

$$o = \frac{if}{f - i} = \frac{(-o - 25)(30)}{(30) - (-o - 25)} = \frac{-30o - 750}{55 + o}$$

$$o^2 + 85o + 750 = 0$$

Since, if

$$ax^2 + bx + c = 0$$

$$x = \frac{-b \pm \sqrt{b^2 - 4ac}}{2a}$$

$$o = \frac{-85 \pm \sqrt{(85)^2 - 4(1)(750)}}{(2)(1)}$$

$$= \frac{-85 \pm \sqrt{4225}}{2} = \frac{-85 \pm 65}{2} = \boxed{-75 \text{ cm}}$$

which holds when using the $-$ sign in the numerator. The image formed of the eye by the mirror is *real* (why?). It appears at a distance

$$i = \frac{of}{o + f} = \frac{(-75)(+30)}{(-75) + (+30)} = -50 \text{ cm}$$

from the mirror or 25 cm from the eye, as it should.

   The alternative solution, when using the $+$ sign, is -10 cm, which means that if the mirror is held 10 cm from the eye, a *virtual* image is seen 15 cm behind the mirror.

(c) In 4.00-diopter hyperopia the far point is 25 cm behind the eye, thus that distance is *subtracted* from the image distance, $-o = i - 25$ cm, or $i = -o + 25$ cm. Then,

$$o = \frac{(-o + 25)(30)}{(30) - (-o + 25)} = \frac{-30o + 750}{5 + o}$$

$$o^2 + 35o - 750 = 0$$

$$o = \frac{-35 \pm \sqrt{(35)^2 - 4(1)(-750)}}{2} = \frac{-35 \pm 65}{2} = \boxed{-50 \text{ cm}}$$

In this case the alternative solution is $+15$ cm, that is, 15 cm behind the eye (inside the head), which is impossible.

Telescopes and Microscopes

Telescopes and microscopes have the same purpose: to make visible details than cannot be seen otherwise. Both telescopes and microscopes have two major components: an *objective lens* which forms an intermediate real image and an *eyepiece,* or ocular, which further magnifies this image.

Astronomical telescope. *Kepler's astronomical refractor**

contains two converging lenses or lens systems: an objective of long focal length and an eyepiece of short focal length. The lenses are spaced so that the second focus of the first lens coincides with the first focus of the second lens. The system is afocal: Parallel light entering the first lens emerges parallel from the second lens (Figure 1.8-8). In the normal eye, without accommodation, such light will come to a focus on the retina. Thus the observer will see a virtual image projected out into infinity.

But experience shows that often, when a person looks into an optical instrument, he or she involuntarily accommodates, a phenomenon called

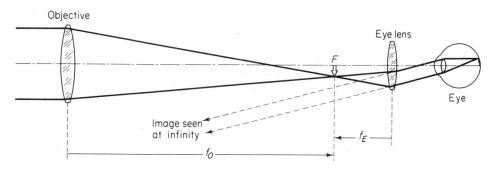

Figure 1.8-8   Astronomical telescope focused for infinity on both object and image sides.

---

*Johannes Kepler (1571–1630), German astronomer and physicist. Kepler's most important contributions were to astronomy. Analyzing Tycho Brahe's data he formulated his three laws of planetary motion; recognized the need for better optics in astronomy; investigated both theoretically and experimentally the formation of images by lenses, mirrors, and the eye; explained the rainbow; and was the first, using a pinhole camera, to observe sunspots. In one of his books, *Ad Vitellionem Paralipomena . . . Astronomiae Pars Optica,* 1604, he laid the groundwork for much of today's geometrical optics. In another, *Mathematici Dioptrice,* 1611, he described his telescope. Kepler believed in mysticism, cast horoscopes to finance his studies, once tried to find the sounds that belong to each planet producing the "music of the spheres."

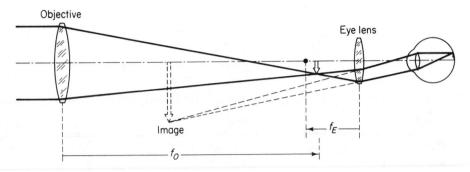

Figure 1.8-9　Astronomical telescope focused for object at infinity and *near point* viewing distance.

"instrument myopia."* The light entering the eye, then, is slightly divergent, as when reading. The intermediate image formed by the objective now lies *inside* the focal length of the eyepiece and the image is seen at the distance of most distinct vision, 25 cm in front of the eye (Figure 1.8-9).

Magnification of telescope.

Why do we see more stars with a larger telescope? Not because of higher *"magnification"*: fixed stars appear as points no matter how much they are magnified. Instead, a telescope has more *light-gathering power*.

Recall that *angular magnification* ("magnifying power") is the ratio of the angular size of the image to the angular size of the object:

$$M_\theta = \frac{\theta'}{\theta} \qquad [1.3\text{-}21]$$

From Figure 1.8-10, we find that $\theta = h/f_O$ and $\theta' = h/-f_E$. Substituting and canceling h gives

$$\boxed{M_\theta = -\frac{f_O}{f_E}} \qquad [1.8\text{-}1]$$

where the minus sign means that the image is inverted.

Angular magnification, from Equation [1.3-25], is also the inverse of transverse magnification, $M_\theta = 1/M_t$, and transverse magnification is the ratio of image size to object size, $M_t = h'/h$. Thus $M_\theta = h/h'$. This holds for *any* object distance. We could even make the object distance *zero;* that

---

*Instrument myopia can be avoided by first moving the eyepiece *out* of the telescope beyond the point of best focus, and then moving it in and *stopping* when the image is sharp. Viewing while accommodating is tiring; it also makes the lines of sight converge, which serves no useful purpose.

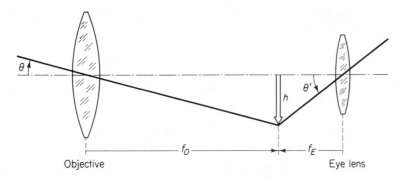

Figure 1.8-10 Deriving angular magnification of telescope.

is, we could make the objective lens the object and form an image of it by
the eyepiece. But since the objective is the entrance pupil and its image,
by definition, is the exit pupil, the angular magnification of a telescope is
also defined as

$$M_\theta = \frac{\text{diameter of entrance pupil}}{\text{diameter of exit pupil}} \qquad [1.8\text{-}2]$$

Based on this definition, there is a simple way for finding the angular
magnification of a telescope. Aim the telescope at the sky or some other diffuse
source. Hold a sheet of paper at some distance behind the eyepiece and move it
back and forth until the aperture of the objective is in focus. A pencil point held
close to the objective helps you do this. The image of the objective (lens) is the
exit pupil, or *Ramsden circle*. Divide one by the other, as in Equation [1.8-2]. This
gives the angular magnification of the telescope.

If a telescope has the specifications 6 × 30, this means that it has 6×
angular magnification and an objective 30 mm in diameter. The inverse ratio,
30/6, is the size of the exit pupil (in millimeters), and the square of that, 25, is
the relative light-gathering power (fully utilized only if the pupil of the eye is at
least of the same size).

When the eye accommodates, the angular magnification produced by
a telescope becomes a little higher, as with a magnifier. See the following
example.

Example. *A telescope may have a 40-cm-focal-length objective and a
2.5-cm-focal-length eyepiece.*
(a) Determine the angular magnification if the system is afocal.
(b) Determine the angular magnification if the object is 4.5 m away.
*(c) Determine the angular magnification if the object is 4.5 m away and the
eye accommodates.*
(d) Repeat (a) and (b), using the ratio of entrance pupil to exit pupil.

Solution. (a) With both object and image at infinity, the angular magnification is

$$M_\theta = \frac{f_O}{f_E} = \frac{+40}{-2.5} = \boxed{-16\times}$$

(b) If the telescope is focused at an object 4.5 m away, the intermediate image is formed farther away from the objective, at a distance of

$$i = \frac{of}{o+f} = \frac{(-4.5)(+0.4)}{-4.5+0.4} = +44 \text{ cm}$$

The angular magnification then is

$$M'_\theta = \frac{i_O}{f_E} = \frac{+44}{-2.5} = \boxed{-17.6\times}$$

(c) If the object is 4.5 m away and the eye accommodates, the image is seen at the near point, 25 cm from the eye, and the intermediate image, formed by the objective, is located

$$o = \frac{if_E}{f_E - i} = \frac{(-25)(+2.5)}{+2.5+25} = -2.27 \text{ cm}$$

in front of the eyepiece. The angular magnification then becomes still higher:

$$M''_\theta = \frac{i_O}{o_E} = \frac{+44}{-2.27} = \boxed{-19.4\times}$$

(d) For using the pupil ratio, Equation [1.8-2], first determine the distance between eyepiece and exit pupil. For both object and image at infinity, this distance is

$$i = \frac{-(40+2.5)(+2.5)}{-(40+2.5)+2.5} = +2.656 \text{ cm}$$

Then, if D is the diameter of the objective, the diameter, d, of the exit pupil is

$$d = \frac{(D)(+2.656)}{-(40+2.5)}$$

and the angular magnification

$$M_\theta = \frac{D}{d} = \boxed{-16\times}$$

the same as before. For the object at 4.5 m,

$$i' = \frac{-(44+2.5)(+2.5)}{-(44+2.5)+2.5} = +2.642 \text{ cm}$$

$$d' = \frac{(D)(+2.642)}{-(44+2.5)}$$

and

$$M'_\theta = \frac{D}{d'} = \boxed{-17.6\times}$$

again the same as before.

Eyepieces.   The field of view of a two-lens telescope is unnec-
essarily restricted because of *vignetting* (vin yet' ting). Vignetting is the
loss of light, especially in the corners of an image. As seen in Figure
1.8-11, top, some of the light (dashed ray) does not reach the eye at all.
Correction is possible by placing an additional lens, called a *field lens*, in,
or close to, the plane of the intermediate image, between objective and
eyepiece. This lens has no effect on magnification; it merely directs all rays
passing through into the *eye lens*. Furthermore, since the field lens is closer
to the entrance pupil, the field of view is larger than with the eye lens
alone.

Such considerations have led to the development of *eyepieces*. A
Huygens eyepiece consists of two plano-convex lenses, their convex sides
facing the object. The field lens has about twice the focal length of the eye
lens (for use in a microscope) or about $2\frac{1}{2}$ times the focal length (for use
in a telescope). The two lenses are separated by one-half the sum of their
focal lengths. A *Ramsden eyepiece* has two equal, or nearly equal, plano-
convex lenses, their convex sides facing each other. Their separation is
about two-thirds of their focal length. In contrast to the Huygens type, the
Ramsden eyepiece can have a reticle in front, to the left. A *Kellner
eyepiece* is essentially a Ramsden eyepiece with the eye lens made achromatic (Figure 1.8-12, left).

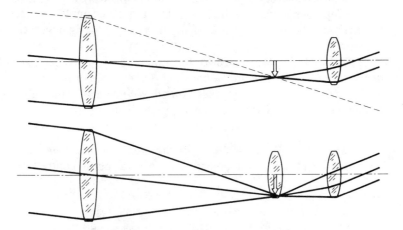

Figure 1.8-11   Astronomical telescope containing only two lenses (*top*) and with an
additional field lens (*bottom*).

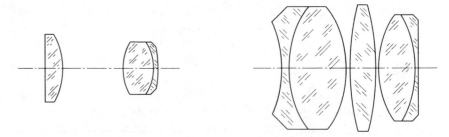

Figure 1.8-12   Kellner eyepiece (*left*) and Erfle eyepiece (*right*). To scale. (From *Optical Design* MIL-HDBK-141.)

An *orthoscopic eyepiece* is like a Ramsden eyepiece but has a triplet as the field lens; it has a wide field and good color correction. The *Erfle eyepiece* (Figure 1.8-12, right) has two or even three achromatic doublets; it gives a particularly wide field (up to 70°) and is often used in high-quality astronomical telescopes.* A *Barlow lens* is a negative lens placed just ahead of the focal plane of the objective, doubling and sometimes tripling the angular magnification of the telescope. A *Gauss eyepiece* contains a reticle, a beamsplitter, and a light bulb to illuminate the reticle (which helps in aiming at a target at night). In a *filar micrometer eyepiece,* a hair line can be moved across the field and the distance of motion read on a drum (which is useful for measuring the image size or the spacing of image details).

Terrestrial telescopes.   The drawback of the astronomical telescope is that the image is inverted. This is awkward for terrestrial observations. Correction is possible in three ways: by adding a *relay* or *erector lens* (which we will discuss in detail in the example on page 156), by inverting the image by use of a *prism,* or by designing a *Galilean telescope.*

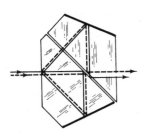

Figure 1.8-13 Pechan prism.

Prism telescopes, if used for binocular vision, are called *prism binoculars;* otherwise, they are *monoculars.* Older types of prism binoculars have two sets of Porro prisms. Newer models contain a Pechan, also called Malmros, Schmidt, Thompson, or Z prism (Figure 1.8-13).

Galileo's telescope.   In the *Galilean telescope,*† as in the astronomical telescope, the objective is positive, but in contrast to it, the

---

*H. Erfle, *Ocular,* U.S. Patent 1,478,704, Dec. 25, 1923.

†Named after Galileo Galilei (1564–1642), Italian scientist. While, as a medical student, attending mass in a cathedral in Pisa, Galileo is said to have watched a bronze

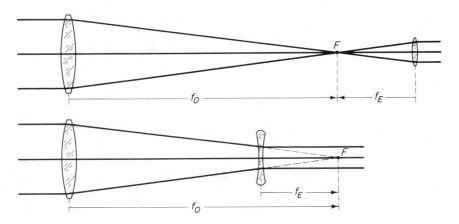

Figure 1.8-14 Focal points coincide in both the astronomical telescope (*top*) and Galileo's telescope (*bottom*).

eyepiece is negative. Still, the two focal points coincide. In Figure 1.8-14, top, F is both the second (right-hand) focal point of the first lens and the first (left-hand) focal point of the second lens. In Galileo's type (bottom), F is still the second focal point of the first lens and again the first (but now the *right*-hand) focal point of the (negative) second lens. Parallel light incident on the system is refracted toward F but before it reaches the focus, the light is intercepted by the negative lens and the emergent light is parallel again. The image, therefore, is seen at infinity. *With* accommodation, the image is seen at the distance of most distinct vision (not shown).

Telescopes of the Galilean type do not give much magnification (because the exit pupil lies to the left of the last lens and the eye cannot be brought there) but they are short, and hence are often used as opera glasses. Also, since there is no *real* intermediate image, there is no way to place a reticle.

---

chandelier swinging in the breeze, discovered that the frequency of oscillation, no matter what the amplitude, was constant. When in the early 1600s Dutch opticians had found that a combination of two lenses showed distant objects upright as well as magnified, Galileo quickly figured out the lenses necessary, stating in his *Il Saggiatore* (Roma: Appresso Giacomo Mascardi, MDCXXIII) that he had "discovered the same instrument, not by chance, but by the way of pure reasoning." In 1609 he demonstrated his "cannocchiale" in public, looking from the Campanile of San Marco in Venice at the Campanile San Giustinio in Padua (where he taught mathematics), 33 km away. The word *telescope* was coined later by a Greek poet and theologian, Ioannis Demisiani of Cephalonia, combining $\tau\tilde{\eta}\lambda\eta$ = far and $\sigma\kappa o\pi\grave{o}\varsigma$ = viewer. The second of more than 100 telescopes that Galileo made was built into a lead tube 70 cm long and had 3× angular magnification; the fifth, called *Discoverer*, had 30×. With it he resolved the Milky Way into a myriad of stars and discovered the moons of Jupiter and the phases of Venus—which proved its rotation around the sun.

Example 1. *Telephoto lenses for cameras are often built in the form of a Galilean telescope. Assume that the front lens has +50 mm focal length and the second lens −25 mm focal length. The distance between both is 30 mm. Determine:*
(a) The focal length.
(b) The actual physical length of the system.

Solution. (a) The term *focal length* refers, more precisely, to *equivalent focal length*. With the dimensions given, the equivalent focal length of the system, from Table 1.4-1, is

$$\mathbf{f} = \frac{f_1 f_2}{f_1 + f_2 - d} = \frac{(+50)(-25)}{+50 - 25 - 30} = \boxed{+250 \text{ mm}}$$

This focal length, by definition, is the distance from the second principal plane, $\mathbf{H}_2$, of the system to the film in the camera (Figure 1.8-15).

(b) In order to find the physical length of the system, we determine first the back vertex power (of the system), using Equation [1.4-14]. We convert the focal lengths given into powers, express the distance d between the lenses in meters, and set $n = 1$, because the medium between the lenses is air. Then

$$\mathbf{P}_2 = \frac{+20}{1 - (+20)(0.03)} + (-40) = 10 \text{ m}^{-1}$$

The reciprocal of $\mathbf{P}_2$ is the back vertex focal length, $\mathbf{a}_2$, and this, in our example, is the distance from the second lens to the plane of the film,

$$\mathbf{a}_2 = \frac{1}{10} = 0.1 \text{ m} = 100 \text{ mm}$$

The total length of the system, from the first lens to the film, therefore, is

$$30 + 100 = \boxed{130 \text{ mm}}$$

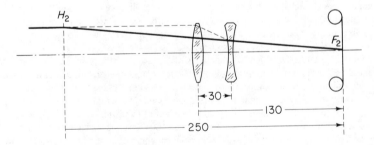

Figure 1.8-15 Example of telephoto lens. To scale.

considerably less than its focal length, 250 mm. This is very convenient when toting around such gear.

Example 2. *Design a telescopic sight, to be used on a hunting rifle.*

Solution. With no more information given than this, consider first the diameters of the various pupils and the magnification desired. The diameter of the pupil of the eye varies from 2 to 8 mm, depending on the ambient light. In daylight, the pupil is about 3 mm in size and not much is gained by making the exit pupil of the telescope larger, except that aligning the eye with the eyepiece becomes easier. Let us begin with an exit pupil 3.75 mm in diameter (Figure 1.8-16a).

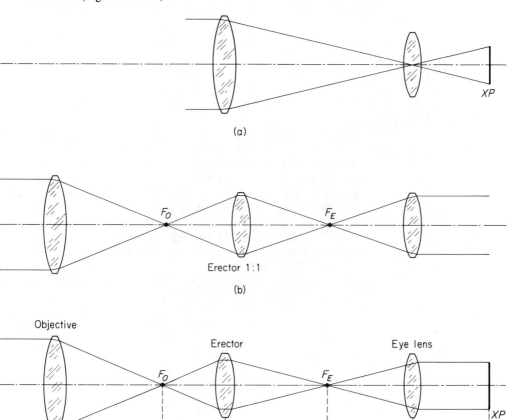

Figure 1.8-16 Successive steps in designing a telescope. For details see the text.

Since angular magnification is the ratio of the diameter of the objective to the diameter of the exit pupil, a higher magnification, for a given exit pupil, requires a larger-diameter objective. For daylight use, reasonable length, light weight, and moderate cost, we may settle for $M_\theta = 8\times$. The objective lens, then, should have a diameter of

$$D = (8)(3.75) = 30 \text{ mm}$$

Certainly, we must have an upright image. But for reasons explained earlier, an $8\times$ telescope cannot be built as a Galilean type. Instead, we use an astronomical telescope together with an *erector* (Figure 1.8-16b). We choose, somewhat arbitrarily, 12 cm as the focal length, f_O, of the objective. Since $M_\theta = f_O/f_E$, this requires an eyepiece of focal length

$$f_E = \frac{f_O}{M_\theta} = \frac{120 \text{ mm}}{8} = 15 \text{ mm}$$

With this focal length, the exit pupil is

$$i = \frac{of}{o + f} = \frac{-(120 + 15)(15)}{-120 - 15 + 15} = +16.9 \text{ mm}$$

to the right of the eyepiece. But this does not give enough *eye relief* (= distance from the last lens to the exit pupil which should coincide with the pupil of the eye). The pupil of the eye is about 3.6 mm behind the cornea and the lens is held in a mount; therefore, the actual clearance between the telescope and the eye might be less than 10 mm, not enough for comfort and to avoid the danger of recoil, and certainly not enough for people who wear glasses.

The solution lies in making the erector provide some magnification too. For example, if the eyepiece is given a more reasonable focal length, perhaps 30 mm, the erector can make up for the deficit and provide $2\times$. Assume that the erector has a focal length of 32 mm. Solving $i/o = -2$ for i and inserting this in $i = of/(o + f)$ gives

$$(-2)(o) = \frac{(o)(32)}{o + 32}$$

and therefore,

$$o = -48 \text{ mm}$$

and

$$i = 96 \text{ mm}$$

The design so far looks as shown in Figure 1.8-16c. The total length is now about 30 cm, acceptable for a riflescope.

The system still has too narrow a *field of view*. Draw a chief ray (*A* in Figure 1.8-17) through the center of the objective and just caught by the erector. Ray *B*, parallel to *A*, also goes through the erector; but *C*, also parallel to *A*, will not. In other words, the top half of the objective aperture

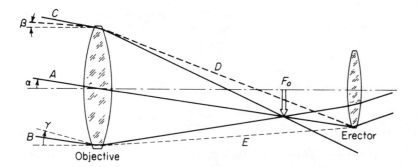

Figure 1.8-17 Various degrees of vignetting.

is not used; light entering there is lost, and vignetting is 50 percent. The angle at which this occurs is α, subtended by A and the axis.

At which angle does *all* the light go through, without vignetting? That happens when D falls within the erector ($\beta < \alpha$). [On the other hand, 100 percent vignetting occurs when E goes through like margins of both lenses ($\gamma > \alpha$).]

We could provide a larger field, and reduce vignetting, by making the erector larger in diameter. But a larger lens has more severe aberrations, and correcting them is expensive. Instead, we place a *field lens* at F_O. Its focal length is chosen so that the objective is imaged into the erector:

$$f_{\text{field lens}} = \frac{oi}{o - i} = \frac{(-120)(48)}{-120 - 48} = 34.3 \text{ mm}$$

This is an important step; it shows how by adding a simple lens (which does not even contribute to image formation) we get a wider field, less vignetting.

The diameter of the erector is determined by rays that come from the rim of the objective and pass through F_O (chief rays with respect to the field lens). In our example,

$$D_{\text{erector}} = (30)\left(\frac{48}{120}\right) = 12 \text{ mm}$$

The same result is obtained by making the erector 96/30 times larger than the exit pupil: (96/30)(3.75) = 12 mm. (All diameters refer to free apertures; the actual lenses may be 2 mm larger.) A 12-mm aperture would, *without* a field lens, give at 50 percent vignetting a (total) field of

$$2 \tan^{-1}\left(\frac{6}{120 + 48}\right) = 4.1°$$

But *with* a field lens at F_O (assuming its diameter is the same as that of the erector), the field of view is

$$2 \tan^{-1}\left(\frac{6}{120}\right) = 5.7°$$

There is another benefit besides the larger field. The lens placed at F_O limits the angular field for *all* rays, which means that up to this angle there is no vignetting: The boundary between no vignetting and complete vignetting is sharp.

Finally, we place another field lens at F_E (see page 151, design of eyepieces). This lens should have a focal length of

$$f = \frac{(-96)(30)}{-96 - 30} = 22.9 \text{ mm}$$

Its diameter, and the diameter of the eye lens, however, should be relatively large. The field of 5.7°, just mentioned, is the field of the telescope; when looking into the eyepiece, the field actually seen is $8 \times 5.7 = 45°$. This is a nice wide field, which should not be restricted by making the field lens (at F_E) and the eye lens too small.

We have now arrived at the *thin-lens solution*. Next we change to realistic, "thick" lenses, move the two field lenses slightly away from the foci (so that dust accumulating on them will not be seen), achromatize the system, trace a series of rays including skew rays through it, and correct for other aberrations. Finally, we place cross-hairs at either F_O or F_E. This completes our design.

Matrix example. *An astronomical telescope that has an objective of 80 cm focal length and gives 8× angular magnification is focused for infinity on both the object and image side. Determine the system matrix.*

Solution. First we determine the focal length of the eyepiece. Since

$$M_\theta = \mathsf{b} = 8\times = \frac{f_O}{f_E} = \frac{80 \text{ cm}}{f_E}$$

we find that

$$f_E = \frac{80}{8} = 10 \text{ cm}$$

The separation of the two lenses, then, is

$$d = f_O + f_E = 90 \text{ cm}$$

Next we determine the refractive powers,

$$P_O = \frac{1}{0.8} = +1.25 \text{ m}^{-1}$$

$$P_E = \frac{1}{0.1} = +10 \text{ m}^{-1}$$

Inserting these figures in the system matrix gives

$$S = R_2 T R_1 = \begin{bmatrix} 1 & 10 \\ 0 & 1 \end{bmatrix} \begin{bmatrix} 1 & 0 \\ -0.9 & 1 \end{bmatrix} \begin{bmatrix} 1 & 1.25 \\ 0 & 1 \end{bmatrix}$$

$$= \begin{bmatrix} 1 & 10 \\ 0 & 1 \end{bmatrix} \begin{bmatrix} 1 & 1.25 \\ -0.9 & -0.125 \end{bmatrix} = \begin{bmatrix} -8 & 0 \\ -0.9 & -0.125 \end{bmatrix}$$

For an afocal system the equivalent power a is zero, the angular magnification $b = -8\times$, and the distance between lenses $d = (-)0.9$ m.

Anamorphic Systems. Anamorphic systems, used in wide-screen motion picture cameras and projectors, provide different magnifications in the horizontal meridian than in the vertical. The anamorphic part is placed in front of the conventional camera lens, or projector lens (the "prime lens"); it is, in essence, a reversed Galilean telescope composed of cylinder lenses whose axes are vertical. Looking through a telescope in the wrong direction makes the image appear smaller and the field wider. But with an anamorphic system this effect is limited to the horizontal dimension; thus a wide horizon is *compressed* into the normal film format. In the vertical dimension the cylinders do not affect the focal length or the field of the prime lens. In the projector, the compressed horizon is expanded again to normal proportions using a similar attachment.

Microscopes.
The microscope is a good topic for learning about optics. For an introduction we need no more than geometric optics. But only later, after we have read about diffraction theory (Chapter 2.3) and optical transformations (4.3), will we fully understand how the microscope works and what its limitations are. For special types of microscopes we may also need polarization (3.4), interference (2.2), and perhaps physiological optics, photography, and photometry (3.6).

A microscope is an instrument for viewing objects that are very small (from the Greek $\mu\iota\kappa\rho\grave{o}\varsigma$ = small and $\sigma\kappa o\pi\epsilon\hat{\iota}\nu$ = to view). A *simple microscope* is merely a magnifying glass, of particularly short focal length. An example is van Leeuwenhoek's microscope* (Figure 1.8-18).

An eyepiece is added to change a simple microscope into a *compound microscope* (Figure 1.8-19). Thus we have *two* lens systems, an *objective* of short focal length, and the *eyepiece*. Like a telescope, the

---

*Antoni van Leeuwenhoek (1632–1723), Dutch clerk and biologist. In 401 publications, including 375 letters to the Royal Society of London and a book, *Arcana Naturae per Microscopum Detecta* (Secrets of Nature Discovered through the Microscope), van Leeuwenhoek described bacteria and yeast cells, the circulation of blood, red blood cells (which had been misinterpreted by Malpighi as fat cells), spermatozoa (though he fancied seeing tiny human heads inside), and the cross-striation of muscle fibers. He also explained that certain insects do not develop from mud, but from eggs laid there. See C. Dobell, *Antoni van Leeuwenhoek and His Little Animals* (New York: Russell & Russell, Publishers, 1958).

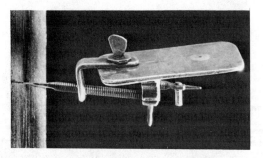

Figure 1.8-18   Van Leeuwenhoek's microscope (brass microscope, 125 ×). (Copyright Rijksmuseum voor de Geschiedenis der Natuurwetenschappen, Leiden, The Netherlands. Reproduced by permission.)

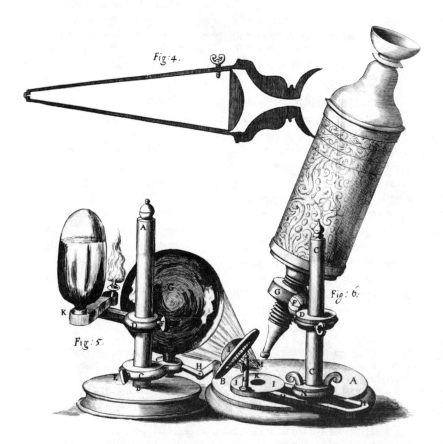

Figure 1.8-19   Robert Hooke's compound microscope (plate from Hooke's *Micrographica,* 1665; The Science Museum, London SW 7. Reproduced by permission). Note tiny objective lens, far left, in cross-sectional diagram on top.

objective forms a real, inverted, and magnified intermediate image. This image is further magnified by the eyepiece. Unlike a telescope, the right-hand focus of the objective and the left-hand focus of the eyepiece are a certain distance, T, apart. This distance, in essence Newton's image distance relative to the objective, is called the *optical tube length;* it is usually 160 mm.

Assume that the final image is seen without accommodation, as shown in Figure 1.8-20. The objective forms of object h an intermediate image h'. Since the shaded triangles are similar,

$$\frac{h}{h'} = \frac{f_O}{T}$$

But h'/h is by definition the transverse magnification, and therefore,

$$M_t = \frac{T}{f_O} \qquad\qquad [1.8\text{-}3]$$

The eyepiece acts as a magnifier and thus, from Equation [1.3-22], it provides an angular magnification of

$$M_\theta = \frac{25}{f_E} \qquad (f \text{ in cm}) \qquad [1.8\text{-}4]$$

The *total* magnification is the product of both:

$$\boxed{\mathbf{M} = M_t M_\theta = \frac{T}{f_O}\frac{25}{f_E}} \qquad [1.8\text{-}5]$$

where T, 25 cm, and f_E should be taken as negative.

The light follows a different path depending on whether the microscope is used for projection and microphotography or for viewing. Consider *projection* (Figure 1.8-21, left-hand side). With the specimen outside the focal length, the objective produces an intermediate real image. This image is located outside, in front of, the first focal plane of the eyepiece.

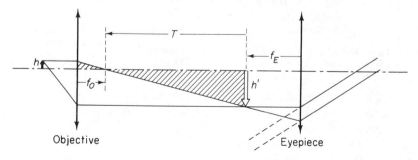

Figure 1.8-20 Deriving magnification of a compound microscope.

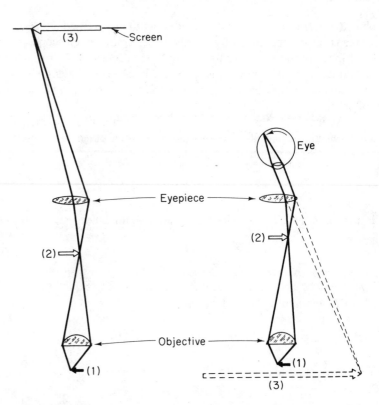

Figure 1.8-21   Ray diagram of a compound microscope used for projection (*left*) and for visual observation (*right*). Note positions of specimen (1), intermediate image (2), and final image (3).

The eyepiece relays this image to the screen or photographic film. When the microscope is used for *viewing,* the objective lens is brought closer to the object. This causes the intermediate image to move farther away, to a position inside the focal length of the eyepiece, which now acts as a magnifier (right). Since a magnifier does not invert an image, the final image, with respect to the original specimen, remains inverted.

Since a microscope is so similar to an astronomical telescope, what is the difference between the two? It is the *optical tube length,* the separation of the foci of the objective and the eyepiece. It explains why it is not possible to use the same instrument and focus simultaneously at close-by and faraway objects.

The *numerical aperture,* N.A., of a microscope depends on the refractive index of the medium between specimen and objective and on the half-angle of the cone of light entering the objective (Figure 1.8-22), N.A. $= n \sin \alpha$. The useful magnification of a microscope is about 1000 times the numerical aperture. Although it is possible to go beyond that

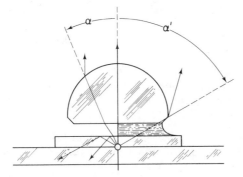

Figure 1.8-22 Half-angle of cone of light entering a conventional microscope objective (*left*) and an oil immersion objective (*right*).

limit by using a higher-power eyepiece or by projecting the image on a far screen, the result is a larger image but not the disclosure of more detail; any magnification beyond the useful magnification is an "empty magnification," as we will discuss further in Chapter 4.3 (page 426).

Example 1. *If the objective lens of a microscope bears the designation "16 mm" and the eyepiece "12.5×," by how much must the tube of the microscope be raised when changing from visual observation (without accommodation) to photography, assuming that the photographic film is 60 mm above the eyepiece?*

Solution. First we determine the focal length of the eyepiece,

$$f_E = \frac{25 \text{ cm}}{12.5} = 20 \text{ mm}$$

If the film, and therefore the image, is 60 mm away (from the eyepiece), the object distance (for the eyepiece) is

$$o_E = \frac{if}{f - i} = \frac{(60)(20)}{20 - 60} = -30 \text{ mm}$$

or **10** mm more than for visual observation (-20 mm).

Next, we use Gauss' thin-lens equation to find the two object distances for the objective assuming that, if not noted otherwise, $T = 160$ mm. For visual observation,

$$o = \frac{if}{f - i} = \frac{(16 + 160)(16)}{16 - (16 + 160)} = -17.6 \text{ mm}$$

while for photography,

$$o' = \frac{(16 + 160 - \mathbf{10})(16)}{16 - (16 + 160 - \mathbf{10})} = -17.7 \text{ mm}$$

Thus the tube must be raised

$$17.7 - 17.6 = \boxed{0.1 \text{ mm}}$$

Example 2. *Sometimes a microscope problem can be solved in some other way. Consider, for example, that the objective has a focal length of 4 mm, the eyepiece 30 mm, and that the distance between both is 184 mm. What is the total magnification?*

Solution. Knowing the focal lengths of two lenses and the distance, d, between them reminds us of the concept of *equivalent focal length,* **f.** Therefore,

$$\mathbf{f} = \frac{f_O f_E}{f_O + f_E - d} = \frac{(4)(30)}{4 + 30 - 184} = -0.8 \text{ mm}$$

Now that we have **f,** we consider the system a simple magnifier with an angular magnification of

$$M_\theta = \frac{25 \text{ cm}}{-0.8 \text{ mm}} = \boxed{-312.5\times}$$

The same result is obtained from

$$M_\theta = \frac{T}{f_O}\left(\frac{25}{f_E}\right) = \frac{-(184 - 4 - 30)}{4}\left(\frac{-250}{-30}\right) = \boxed{-312.5\times}$$

Reflecting Systems

Both telescopes and microscopes can be built in the form of *reflecting systems.* Conventional telescopes and microscopes are combinations of lenses; they are *dioptric systems* (from the Greek $\delta\iota\grave{\alpha}$ = through). In contrast, *catoptric systems* (from $\kappa\alpha\tau\grave{\alpha}$ = onto) are combinations of mirrors. *Catadioptric systems* are lens–mirror combinations.

Mirrors have no chromatic aberration. Also, mirrors let us extend the range of investigation into the UV, the IR, and even into the X-ray region, where no refractive materials are known.

Catoptric systems. Reflecting systems are widely used in astronomy. In principle, an astronomical reflector is similar to a refractor, the difference being that the objective lens is replaced by a mirror. In *Herschel's telescope,* not shown, the mirror is slightly tilted with respect to the axis and the eyepiece is placed alongside the tube of the instrument. In *Newton's telescope,* the light is reflected out of the tube by an additional mirror or prism (Figure 1.8-23).

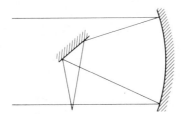

Figure 1.8-23 Newton's telescope.

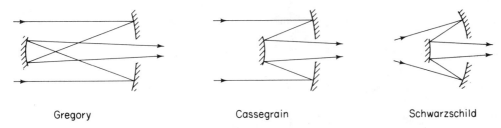

Gregory   Cassegrain   Schwarzschild

Figure 1.8-24 (*Left to right*) Gregory's telescope, Cassegrain's telescope, and Schwarzschild's microscope.

*Gregory's telescope** has two curved mirrors, a concave primary (mirror) with a central opening, and a concave secondary mounted in front of the primary (Figure 1.8-24, left). The distance between both mirrors is equal to, or slightly larger than, the sum of their focal lengths.

Cassegrain's telescope[†] is the equivalent of Galileo's type, with the secondary being convex (Figure 1.8-24, center). This mirror contributes significantly to the overall magnification, because its distance from the primary's focus is much less (about one-fourth) than that from the secondary to the image. Cassegrainian telescopes are widely used; they are much shorter than a Gregorian of the same focal length.

---

*James Gregory (1638–1675), Scottish mathematician and astronomer, suggested the system named after him in 1663, but opticians at that time could not grind the mirrors required.

[†]Guillaume Cassegrain, French. Little is known about Cassegrain's life; according to some, he was a professor of physics at the College of Chartres, according to others, a sculptor at the court of Louis XIV. In 1672, nine years after Gregory, Cassegrain combined a concave and a *convex* mirror, intercepting the rays before they came to a focus (an idea quickly dismissed by Newton as only a minor modification of Gregory's precedent). The real virtue of Cassegrain's design is that the spherical aberration introduced by one mirror is partially canceled by the other, a fact established by Ramsden a century later. Cassegrain's original system had a paraboloid primary and a hyperboloid secondary; since then the term "Cassegrainian" has broadened to mean any combination of a concave primary and a convex secondary. In the *Ritchey–Chrétien* type, for example, both mirrors are hyperboloid, completely eliminating spherical aberration and coma.

The *Schwarzschild type* (right), in principle similar to the Cassegrainian, is used mainly in microscopy.* Its focal length is easily derived from Gullstrand's equivalent power equation, Equation [1.4-7], $\mathbf{P} = P_1 + P_2 - P_1P_2d$. If we substitute $-2/R$ for the powers of the two mirrors and, because the light changes direction, change the sign in the R_2 term, then

$$\frac{1}{f} = \frac{-2}{R_1} + \frac{2}{R_2} + \frac{4d}{R_1R_2} \qquad [1.8\text{-}6]$$

where d is the distance between the mirrors.

Example 1. *Using matrix algebra, determine the angular magnification of a Cassegrainian telescope that has a primary of 40 cm radius of curvature and a secondary of 10 cm radius of curvature, the mirrors being separated by 15 cm.*

Solution. Converting these radii into powers gives

$$P_1 = \frac{-2}{R_1} = \frac{-2}{-0.4} = +5 \text{ m}^{-1}$$

$$P_2 = -\frac{-2}{-0.1} = -20 \text{ m}^{-1}$$

We then determine the system matrix:

$$\mathsf{S} = \begin{bmatrix} 1 & P_2 \\ 0 & 1 \end{bmatrix} \begin{bmatrix} 1 & 0 \\ -d & 1 \end{bmatrix} \begin{bmatrix} 1 & P_1 \\ 0 & 1 \end{bmatrix}$$

$$= \begin{bmatrix} 1 & -20 \\ 0 & 1 \end{bmatrix} \begin{bmatrix} 1 & 0 \\ -0.15 & 1 \end{bmatrix} \begin{bmatrix} 1 & 5 \\ 0 & 1 \end{bmatrix}$$

$$= \begin{bmatrix} 1 & -20 \\ 0 & 1 \end{bmatrix} \begin{bmatrix} 1 & 5 \\ -0.15 & 0.25 \end{bmatrix} = \begin{bmatrix} 4 & 0 \\ -0.15 & 0.25 \end{bmatrix}$$

The determinant is $(4 \times 0.25) - [0 \times (-0.15)] = 1$. From the Gaussian constants we find that the equivalent power, a, is zero (the system is afocal) and that the angular magnification is

$$\mathsf{b} = M_\theta = \boxed{4\times}$$

Example 2. *A microscope objective of the Schwarzschild type consists of a primary of 50 mm radius of curvature and a secondary of 10 mm radius of curvature. The distance between both is 40 mm. Where must the object be placed in order to produce an image 18 cm to the right of the primary?*

---

*D. S. Grey, "A New Series of Microscope Objectives: II. Preliminary Investigation of Catadioptric Schwarzschild Systems," *J. Opt. Soc. Am.* **39** (1949), 723–28.

Solution. While the light is reflected by the two mirrors, it is much easier to work such a problem if the system is *unfolded*, as shown in Figure 1.8-25.

This time we use the vergence method. We determine first the powers of the two mirrors:

$$P_1 = \frac{-2}{R_1} = \frac{-2}{-0.05} = +40 \text{ m}^{-1}$$

$$P_2 = \frac{-2}{+0.01} = -200 \text{ m}^{-1}$$

Since the image is 18 cm from the primary, it is $18 + 4 = 22$ cm from the secondary. Therefore, the exit vergence at the secondary must be

$$V_2' = \frac{1}{+0.22} = +4.545 \text{ m}^{-1}$$

The entrance vergence at the secondary, then, is

$$V_2 = V_2' - P_2 = +4.545 - (-200) = +204.545 \text{ m}^{-1}$$

Such wavefronts would come to a focus $1/204.545 = 4.888$ mm behind the secondary, or 44.888 mm behind the primary. The exit vergence at the primary, then, is

$$V_1' = \frac{1}{+0.044888} = +22.278 \text{ m}^{-1}$$

and the entrance vergence is

$$V_1 = 22.278 - (+40) = -17.722 \text{ m}^{-1}$$

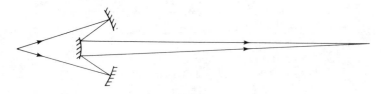

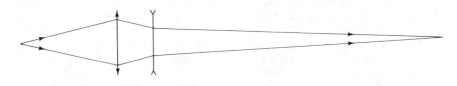

Figure 1.8-25 Schwarzschild-type microscope objective in its actual configuration (*top*) and unfolded (*bottom*).

which means that the object distance is

$$o = \frac{1}{-17.722} = -56.4 \text{ mm}$$

Therefore, the object must be placed

$$-56.4 + 40 = \boxed{-16.4 \text{ mm}}$$

in front of the secondary.

Catadioptric systems.   The two-mirror systems described so far have serious off-axis aberrations and their fields are small. A significant improvement is possible by adding a *corrector,* an odd-shaped glass plate, shown in Figure 1.8-26. *Kellner* had originally conceived this idea for use in searchlights (where light is to be projected well collimated, over long distances).* *Schmidt*[†] extended the idea to astronomical telescopes. His first reflector had a speed of $f/1.75$, unheard of at that time, and gave photographs of superb quality.

Schmidt's concept of a corrector plate revolutionized the design of large-aperture, wide-field reflectors. But correctors are difficult to design and produce. This difficulty was overcome by *Maksutov*[‡] who used *only spherical surfaces.* The two radii of curvature of the corrector, R_a and R_b

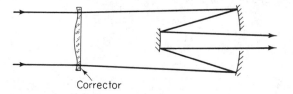

Corrector

Figure 1.8-26   Cassegrain–Schmidt reflector.

---

*G. A. H. Kellner, *Projecting Lamp,* U.S. Patent 969,785, Sept. 13, 1910.

[†]Bernhard Voldemar Schmidt (1879–1935), Estonian optician and instrument maker. A rather independent, self-taught, highly skilled craftsman, despite the loss of his right forearm, Schmidt set up shop in Mittweida near Jena, Germany, later joined the Bergedorf astronomical observatory near Hamburg, where in 1930 he designed and built the reflectors named after him.

[‡]D. D. Maksutov, "New Catadioptric Meniscus Systems," *J. Opt. Soc. Am.* **34** (1944), 270–84. In his original design, Maksutov used the following parameters (in millimeters): primary R_1 −823.2; meniscus diameter 100, thickness 10.0, radius R_a −152.8, R_b −158.6, $n = 1.5163$, $\nu = 64.1$; meniscus–mirror distance 539.1; f-stop number $f/4$.

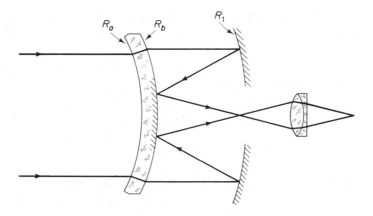

Figure 1.8-27   Cassegrain–Maksutov system showing meniscus corrector (*left*) and primary (*center right*). Doublet on the far right is a field lens.

in Figure 1.8-27, are used as variables to eliminate spherical aberration and coma. Furthermore, minimum longitudinal color results when

$$\frac{R_a - R_b}{t} = \frac{n^2 - 1}{n^2} \qquad [1.8\text{-}7]$$

where t is the thickness of the glass and n its index for yellow. Lateral color and curvature of field are compensated for by a field lens behind the primary. The secondary is either applied directly to the corrector, or a *Mangin mirror** is added. The Cassegrain–Maksutov type is short, sturdy, and lightweight; it has a large aperture and a wide field and is completely sealed, making it impervious to air turbulence. It is a very popular design.

The trend in the design of astronomical telescopes is toward *deformable* mirrors. Such mirrors consist of a thin, flexible face plate, often made of beryllium and supported on a rigid backplate by a series of thick piezoelectric disks. Telescopes of this kind include a sensor that monitors distortions of the incoming wavefronts (distorted because of air turbulence) and an elaborate real-time servo and feedback system to compensate for these distortions, eliminating the poor "seeing" (image quality) that so often afflicts astronomical observations. All of this goes by the name of *adaptive optics*.[†]

---

*A Mangin mirror is a thick negative meniscus, the rear surface made reflective. The negative lens introduces negative spherical aberration (overcorrection) which compensates for the positive aberration (undercorrection) of the mirror component. Compared with a parabola of the same focal length, a Mangin mirror has about half the coma.

[†]See R. H. Freeman and J. E. Pearson, "Deformable mirrors for all seasons and reasons," *Applied Optics* **21** (1982), 580–88.

SUGGESTIONS FOR FURTHER READING

A. Cox, *Photographic Optics,* 15th edition (New York: Focal Press, Inc., 1974).

R. Kingslake, "The Development of the Zoom Lens," *J. Soc. Motion Pict. Television Eng.* **69** (1960), 534–44.

S. Duke-Elder and D. Abrams, *Ophthalmic Optics and Refraction,* Vol. V of S. Duke-Elder, editor, *System of Ophthalmology* (St. Louis: The C. V. Mosby Company, 1970).

A. G. Ingalls, editor, *Amateur Telescope Making,* Vols. 1–3 (New York: Scientific American, Inc., 1955).

R. B. McLaughlin, *Special Methods in Light Microscopy* (London: Microscope Publications Ltd., 1977).

J. Maxwell, *Catadioptric Imaging Systems* (New York: American Elsevier Publishing Company, Inc., 1972).

D. F. Horne, *Optical Instruments and Their Applications* (Bristol, England: Adam Hilger Ltd., 1980).

PROBLEMS

1.8-1. Note the camera lens shown in Figure 1.8-28.

Figure 1.8-28

(a) From which basic type is it derived?
(b) How does it differ from a Tessar?
(c) What should be its focal length to serve as a "standard lens" (that is, neither telephoto nor wide-angle), assuming a film format of 24 mm × 36 mm?

1.8-2. A mechanically compensated zoom lens contains three elements which have focal lengths of +15 cm, −5 cm, and +13.5 cm,

respectively. In one zoom position, the distance between lenses 1 and 2 is 5 cm, and the distance between lenses 2 and 3 is 13.8 cm. In another position, the distance between lenses 1 and 2 is 7.5 cm, and the distance between lenses 2 and 3 is 6.65 cm. If the system is focused for infinity, by how much does the focal plane vary?

1.8-3. A target, in the form of a back-lighted cutout arrow pointing upward, is placed at some distance in front of the eye in order to view the various *Purkinje images* caused by reflection at the refracting surfaces of the eye. Show:
(a) The orientation and the relative sizes of the three images.
(b) The change that occurs in accommodation.

1.8-4. Assume that an *aphakic* eye (an eye from which the crystalline lens has been removed) has the radius of curvature of the cornea and the refractive index of a *schematic* eye and the length of a *reduced* eye. Find the location of the far point.

1.8-5. An astronomical telescope consists of two lenses separated by a distance of 1 m. If the angular magnification, when focusing at infinity, is 5×, what are the refractive powers of the two lenses?

1.8-6. If the objective of an astronomical telescope is 50 mm in diameter and has 40 cm focal length, and if the angular magnification is 8×, what is the diameter of the exit pupil and the power of the eyepiece.

1.8-7. The objective of a telescope has a diameter of 40 mm and a focal length of 32 cm. When focused at infinity, the exit pupil is 2.5 mm in diameter. What is:
(a) The angular magnification of the system?
(b) The focal length of the eyepiece?

1.8-8. A telescope has an objective of 50 cm focal length and an eyepiece of 5 cm focal length. With the telescope focused on a target 3 m away and used without accommodation, what is:
(a) The distance between the two lenses?
(b) The angular magnification?

1.8-9. What angular magnification is produced by a telescope that has an objective of 30 cm focal length and an eyepiece of 3.5 cm focal length and that is used for looking, *with* accommodation, at an object 13.12 m away?

1.8-10. A telescope has a +2.00-diopter objective and gives 10× angular magnification. In which direction and how much must a −5.00-diopter myope, not wearing corrective glasses, move the eyepiece in order to see the image in focus?

1.8-11. A telescope with an angular magnification of 10× is pointed at the sun and focused to give an image at infinity. The focal length of the objective is 60 cm. How far and in what direction must the eyepiece be moved to project an image of the sun on a screen 30 cm away from the eyepiece?

1.8-12. An astronomical telescope is focused for an object at infinity, without accommodation. If the telescope is used
(a) for taking photographs with a camera focused for infinity,
(b) for taking photographs using only the camera body, without the camera lens, and
(c) by an observer who does accommodate, how must the distance between the objective

and the eyepiece be changed?

1.8-13. A 2.5× Galilean telescope has a −25.00-diopter eye lens. When focused for infinity, what is the distance between the lenses?

1.8-14. A Galilean telescope, containing a +50-cm F.L. objective and a −8-cm F.L. eyepiece, is focused at an object 6.25 m away. What must be the distance between the two lenses if the image is to be at infinity? Solve by the vergence method.

1.8-15. A *low-vision aid* can be realized by combining a positive spectacle lens with a high-minus contact lens. For example, if +20.00-diopter spectacles are worn over −32.00-diopter contact lenses, what is:
(a) The angular magnification?
(b) The distance between the lenses?

1.8-16. A low-vision telescope consists of +20.00-diopter spectacles and −40.00-diopter contact lenses, separated by 25 mm. To read a book held 40 cm in front of the system, determine:
(a) The power of the *reading cap* (placed in front of the telescope, in contact with the spectacle lenses).
(b) Using matrix algebra, the angular magnification of the system.

1.8-17. Find the system matrix of a telescope that has an objective of 120 mm focal length and an eyepiece of 30 mm focal length. What is the angular magnification?

1.8-18. Refer to the telescope example discussed on page 158 and assume that the object is now 50 m away. Determine the system matrix and the angular magnification. Check by the determinant.

1.8-19. While a single frame of standard 35-mm movie film is 18 mm × 24 mm in size, a Cinemascope screen may be 5 m high and 20 m wide.
(a) By how much do the two formats differ?
(b) What kind of projector lenses are needed to expand the image and how should they be oriented?

1.8-20. Continue with Problem 1.8-19 and assume that a +15-cm-focal-length cylinder is mounted in front of the prime lens. What other lens is needed, and where should it be placed?

1.8-21. If a microscope has an objective of 20 mm focal length, an optical tube length of 16 cm, and an eyepiece marked 12.5×, what is the total magnification?

1.8-22. A microscope has an objective of 20 mm focal length and an eyepiece of 50 mm focal length, both components separated by a distance of 170 mm. Determine the equivalent focal length of the system and the magnification.

1.8-23. A microscope has three objectives, of focal lengths 16 mm, 4 mm, and 1.6 mm, an optical tube length of 16 cm, and two sets of eyepieces, marked 5× and 10×. What is:
(a) The highest magnification possible?
(b) The least magnification possible?

1.8-24. An oil-immersion microscope objective is marked 1.6 mm, 1.25 N.A. It is used, at 16 cm optical tube length, together with an eyepiece marked 25×. If the oil has an index of 1.515, is the microscope's total magnification within the useful limit?

1.8-25. A Gregorian telescope, focused for infinity, consists of a primary (mirror) of $R_1 = -25$ cm and a secondary of $R_2 = +10$ cm. Determine:
(a) The separation of the two mirrors.
(b) The system matrix.
(c) The angular magnification.
(Note that the light between mirrors changes direction; therefore, that distance must change signs too.)

1.8-26. A Cassegrainian telescope of $R_1 = -40$ cm, $R_2 = -13\frac{1}{3}$ cm, and $d = 15$ cm is focused at the moon. Where is the image located? Solve by the vergence method.

1.8-27. The Schwarzschild microscope problem presented in Example 2 on page 166 should now be solved by matrix algebra.

1.8-28. What should be the radius of curvature of the back surface of a Maksutov corrector if its front surface has a radius of −74 mm, its thickness is 5 mm, and the material spectacle crown ($n = 1.523$)?

Part **2**

Wave Optics

FROM OUR STUDY OF GEOMETRIC OPTICS we might conclude that the boundaries of a shadow are sharp and distinct. But close experimental observation shows that they are not. In fact, the boundaries of a shadow, seen with white light, are diffuse and, seen with monochromatic light, they consist of fringes. This can be explained by assuming that light is made up of *waves*.

Light waves behave much like other waves. They bend around obstacles, combine with like waves, and reinforce or weaken each other. They pass through apertures, slits, and thin films in characteristic ways that cannot be predicted from geometric optics. These properties, and others, are discussed in the following chapters, combined under the heading of *wave optics*.

2.1

Light as a Wave Phenomenon

WITHIN CERTAIN LIMITS, light can be considered a wave phenomenon. We begin with a study of wave motion. A motion that repeats itself in equal intervals of time is called a *periodic motion*. If the motion follows a sine function, or a cosine function, and the waveform therefore is either sinusoidal or cosinusoidal, it is called a *simple harmonic motion*.

Simple Harmonic Motion

A good way of describing simple harmonic motion is by considering it a projection of uniform circular motion. Think of a flywheel whose axis is horizontal, and which has a knob protruding from it near the rim. If light is incident on one side of the flywheel, normal to its axis of rotation, and a shadow is cast on a distant screen, the knob will be seen moving up and down.

Let us represent the flywheel by a *reference circle* (Figure 2.1-1, left). Point P_0 is the protruding knob. It travels in a circular path, counterclockwise, at a uniform angular velocity ω. At $t = 0$, the knob is at P_0. After a given interval of time, the knob has reached P_1. The projection of P_1 on the y axis gives point Q. Thus as P moves around the circle, Q will move up and down along the y axis.

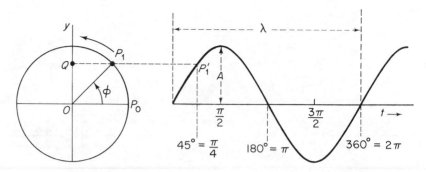

Figure 2.1-1   Reference circle (*left*) and simple harmonic motion (*right*).

It is customary to plot on the *x* axis the independent variable, *time*, and on the *y* axis the dependent variable, the *displacement*. Let point *P* begin its motion at the initial position P_0, where $t = 0$ and $y = 0$. When *P* has advanced through $\phi = 45°$, it has reached P_1. Since the total angle around a point is 360° or 2π radians, point *P* has moved through $2\pi(45°/360°) = \frac{1}{4}\pi$ rad.

After another equal interval of time, *P* has moved through a total of $\phi = 90°$, or $\frac{1}{2}\pi$ rad. The projection, *y*, has reached a maximum. After turning through $\phi = 180° = \pi$ rad, *P* is at a position where again $y = 0$. After $\phi = 270° = \frac{3}{2}\pi$ rad, *P* is at the lower maximum, and after $\phi = 360° = 2\pi$ rad, it is back at its initial position. Connecting all points gives the sinusoidal curve shown in Figure 2.1-1, right.

We call the radius of the reference circle **A**, for *amplitude*, and the angle ϕ the *phase*. Then

$$\sin \phi = \frac{y}{\mathbf{A}}$$

where *y* is the displacement of *Q* above *O*. Substituting the definition of angular velocity, $\omega = \phi/t$, solved for ϕ, gives

$$\boxed{y = \mathbf{A} \sin (\omega t)} \qquad \text{[2.1-1]}$$

which is the *equation of motion* of *Q* as a function of time.

Note that simple harmonic motion can be described in terms of a *rotating vector*, or *phasor*, *OP*. This is a very useful concept. Also note how the velocity of *Q* changes. At $y = 0$, the velocity, *u*, is highest. Toward the maximum value of the displacement, both $+y$ and $-y$, point *Q* slows down. At the maxima themselves, $u = 0$. Then *u* changes direction, increasing again toward maximum velocity at $y = 0$, and so on.

It is often convenient to use the cosine function of ωt, rather than the sine function. Also, point *P* may well begin its motion, not at P_0, but at a certain initial phase angle ϕ. Hence we may write, instead of Equation [2.1-1],

$$y = \mathbf{A} \sin (\omega t + \phi)$$

or   [2.1-2]

$$y = \mathbf{A} \cos (\omega t + \phi)$$

as the case may be.

Some characteristics of light have been introduced before (on page 14). For example,

---

Amplitude, $\mathbf{A}$, is the height of the wave above the axis of propagation.

Wavelength, λ, is the distance between consecutive equivalent points on the wave.

Frequency, ν, is the number of oscillations per second.

Period, T, is the time it takes a point on the wave to make one complete oscillation. Period and frequency, therefore, are reciprocal to one another; $T = 1/\nu$.

The *velocity* of propagation, v, of a wave is the product of wavelength and frequency;

$$\boxed{v = \lambda \nu}$$ [2.1-3]

Amplitude and power.   So far I have always referred to the *amplitude* of a wave. But in most practical applications the amount of light generated, propagated, or received is a matter of *energy*, or more precisely, it is a matter of energy per unit of time. Energy per time is *power*.* How are amplitude and energy related to each other?

To use a mechanical analog: Assume that some mass is oscillating according to Equation [2.1-1]:

$$y = \mathbf{A} \sin(\omega t)$$

Differentiating the displacement, y, with respect to time gives the velocity, u, of the oscillation:

---

*Often the counterpart to amplitude is called "intensity," but it should not be; "intensity" is a term that is much overworked. *Power* is the central theme. For a further discussion of energy, power, and "intensity," see Chapter 3.6.

$$u = \frac{dy}{dt} = \mathbf{A}\omega \cos(\omega t) \qquad [2.1\text{-}4]$$

A second differentiation (with respect to time) yields the acceleration, a:

$$a = \frac{du}{dt} = -\mathbf{A}\omega^2 \sin(\omega t) = -\omega^2 y \qquad [2.1\text{-}5]$$

From mechanics we recall that the kinetic energy, KE, of a mass m is

$$\text{KE} = \tfrac{1}{2} m\text{u}^2 \qquad [2.1\text{-}6]$$

Substituting Equation [2.1-4] into [2.1-6] gives

$$\text{KE} = \tfrac{1}{2} m\mathbf{A}^2 \omega^2 \cos^2(\omega t) \qquad [2.1\text{-}7]$$

At the points of reversal, the high point and the low point of the oscillation, the kinetic energy is zero. At the equilibrium position (where $y = 0$), the kinetic energy is at a maximum. In contrast, at the points of reversal the potential energy, PE, is at a maximum and at the equilibrium positions it is zero. At any given time,

$$\text{PE} = \tfrac{1}{2} may \qquad [2.1\text{-}8]$$

and thus, substituting Equation [2.1-5] and neglecting the minus sign,

$$\text{PE} = \tfrac{1}{2} m\omega^2 y^2 \qquad [2.1\text{-}9]$$

Inserting Equation [2.1-1] into [2.1-9] leads to

$$\text{PE} = \tfrac{1}{2} m\omega^2 \mathbf{A}^2 \sin^2(\omega t) \qquad [2.1\text{-}10]$$

Equations [2.1-7] and [2.1-10] together represent the *total* energy, Q:

$$Q = \text{KE} + \text{PE} = \tfrac{1}{2} m\mathbf{A}^2 \omega^2 \cos^2(\omega t) + \tfrac{1}{2} m\omega^2 \mathbf{A}^2 \sin^2(\omega t)$$

Since $\sin^2\alpha + \cos^2\alpha = 1$, this simplifies into

$$Q = \tfrac{1}{2} m\mathbf{A}^2 \omega^2 \qquad [2.1\text{-}11]$$

The *energy* ("intensity") *of the light, therefore, is proportional to the square of the amplitude:*

$$\boxed{Q \propto \mathbf{A}^2} \qquad [2.1\text{-}12]$$

Complex notation.   We have seen that simple harmonic motion can be represented by either a sine function or a cosine function:

$$y = \mathbf{A} \sin(\omega t + \phi)$$
$$y = \mathbf{A} \cos(\omega t + \phi)$$
$$[2.1\text{-}2]$$

Such functions can be combined into *complex numbers*. The concept of a complex number often instills unnecessary fears. In reality, it is quite simple. A complex number is a number that has both a real part and an imaginary part. The two parts can be drawn along orthogonal axes, a real axis, x, and an imaginary axis, y (Figure 2.1-2). Point P, which represents the "complex" number, then has the *rectangular* coordinates (x, y).

But **A** can also be represented by *polar* coordinates (A, ϕ):

$$\sin \phi = \frac{y}{A} \quad \text{and} \quad \cos \phi = \frac{x}{A} \qquad [2.1\text{-}13]$$

Furthermore, we see from Figure 2.1-2 that

$$\cos(90° - \phi) = \frac{y}{A} \qquad [2.1\text{-}14]$$

In short, the sine function and the cosine function (of a wave vector) are essentially the same, except that they *differ in phase by 90°*.

Assume now that we have two vectors, **a** and **b**, that differ by 90° (they are *orthogonal* to one another). In complex notation, we preface **b** by a symbol, i. This symbol means that **b**, as compared with **a**, has made a *counterclockwise rotation through 90°*. The resultant of **a** and i**b** is then

$$\mathbf{A} = \mathbf{a} + i\mathbf{b} = A \cos \phi + iA \sin \phi \qquad [2.1\text{-}15]$$

Whereas i stands for a (single) counterclockwise 90° rotation, we may repeat the process and consider two consecutive 90° rotations. The term $i(i\mathbf{b})$, or $i^2\mathbf{b}$, then means that **b** has turned through 180° and has become $-\mathbf{b}$:

$$i^2\mathbf{b} = -\mathbf{b}; \quad i^2 = -\frac{\mathbf{b}}{\mathbf{b}} = -1; \quad \text{thus} \quad i = \sqrt{-1} \quad [2.1\text{-}16]$$

We expand the sine and cosine terms in Equation [2.1-15]. This gives

$$\sin \phi = \phi - \frac{\phi^3}{3!} + \frac{\phi^5}{5!} \cdots \quad \text{and} \quad \cos \phi = 1 - \frac{\phi^2}{2!} + \frac{\phi^4}{4!} \cdots \quad [2.1\text{-}17]$$

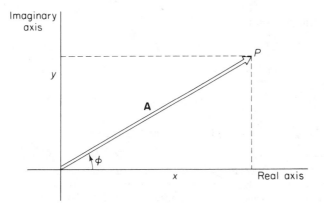

Figure 2.1-2 *Argand diagram* shows vector, **A,** extending from 0 to (complex) point P.

We add the two trigonometric terms in Equation [2.1-15], neglecting A for the time being:

$$\cos \phi + i \sin \phi = 1 + i\phi - \frac{\phi^2}{2!} - i\frac{\phi^3}{3!} + \frac{\phi^4}{4!} + \cdots \qquad [2.1\text{-}18]$$

Since from Equation [2.1-16] $-1 = i^2$, Equation [2.1-18] can be written as

$$\cos \phi + i \sin \phi = 1 + i\phi + \frac{(i\phi)^2}{2!} + \frac{(i\phi)^3}{3!} + \cdots + \frac{(i\phi)^N}{N!} + \cdots \qquad [2.1\text{-}19]$$

In short,

$$\cos \phi + i \sin \phi = e^{i\phi} \qquad [2.1\text{-}20]$$

which is *Euler's formula.* * Equation [2.1-15] then becomes

$$\mathbf{A} = Ae^{i\phi} \qquad [2.1\text{-}21]$$

where A is the absolute value of the (complex) *amplitude* $\mathbf{A}$ of the wave:

$$A = |\mathbf{a} + i\mathbf{b}| = \sqrt{a^2 + b^2} \qquad [2.1\text{-}22]$$

and ϕ is its *phase:*

$$\phi = \tan^{-1}\left(\frac{b}{a}\right) \qquad [2.1\text{-}23]$$

Term e is the base of natural logarithms, $2.718\,28 \ldots$.

Whenever we are concerned with the *energy* of light, we need the *square* of the amplitude. Using complex notation, this is easy to do. We merely multiply the complex number by its *complex conjugate*. The complex conjugate of $\mathbf{A}$, for example, is written $\mathbf{A}^*$. Conjugate terms, whenever they contain i, change signs; thus if

$$\mathbf{A} = \mathbf{a} + i\mathbf{b} \qquad [2.1\text{-}15a]$$

the complex conjugate is

$$\mathbf{A}^* = \mathbf{a} - i\mathbf{b} \qquad [2.1\text{-}24]$$

The complex conjugate of $\mathbf{A} = Ae^{i\phi}$, Equation [2.1-21], therefore, becomes

$$\mathbf{A}^* = Ae^{-i\phi} \qquad [2.1\text{-}25]$$

and the square, the product of $\mathbf{A}$ and $\mathbf{A}^*$, becomes

---

*Leonard Euler (1707–1783), Swiss mathematician. Euler worked on a wide range of topics, from astronomy to perturbation theory to harmonic wave motion, laying the groundwork for Thomas Young, who some 50 years later showed that light can be represented by waves. Euler's formula is most remarkable in that it connects geometry (trigonometric functions) with algebra (exponential functions).

$$\mathbf{AA}^* = Ae^{i\phi}Ae^{-i\phi} = A^2 \qquad [2.1\text{-}26]$$

which is (proportional to) the energy.

Finally, since the presence of i implies a 90° rotation, the $iA \sin \phi$ term in Equation [2.1-15] becomes redundant and we may write the *real part* in this equation,

$$y = A \cos(\omega t)$$

as

$$\boxed{y = Ae^{i\omega t}} \qquad [2.1\text{-}27]$$

This is a very useful expression (as we will see in the next several chapters).

Moving Waves

So far we have discussed the simple harmonic motion of a *single* point, and different ways of describing such motion. Now we proceed to a *series* of points, which together make up a wave, and see how this wave propagates through space.

There are two ways of looking at moving waves. For example, consider the waves on the surface of a lake at a given instant of time. If we take a snapshot of the waves, their motion is frozen at time $t = 0$. That means that we consider the shape of the wave *as a function of space,*

$$y = f(x) \qquad [2.1\text{-}28]$$

But we can also look at the wave as a *function of time*. For example, if we look through a narrow slot, as in Figure 2.1-3, we will see a given point on the water surface merely move up and down (the x position remaining constant). In contrast to Equation [2.1-28], we now have

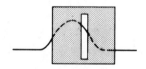

$$y = f(t) \qquad [2.1\text{-}29]$$

Figure 2.1-3 Moving wave as seen through narrow slot.

Now assume that the wave configuration moves forward, to the right. If v is the velocity of the wave, then after time t the wave has moved through distance $x = vt$. Therefore, in order to retain the equality in $y = f(x)$, vt must be *subtracted* from x_0:

$$y = f(x_0 - vt) \qquad [2.1\text{-}30]$$

If the wave were to move to the left, vt is *added* and

$$y = f(x_0 + vt) \qquad [2.1\text{-}31]$$

Since angular velocity $\omega = \phi/t$, we have for *one* oscillation,

$$\omega = \frac{2\pi}{T} \qquad\qquad [2.1\text{-}32]$$

where T is the period of the wave, the reciprocal of the frequency, $1/T = \nu$. Substituting this expression in Equation [2.1-32] gives

$$\omega = 2\pi\nu \qquad\qquad [2.1\text{-}33]$$

and substituting this in Equation [2.1-1],

$$y = \mathbf{A}\,\sin(\omega t)$$

leads to

$$y = \mathbf{A}\,\sin(2\pi\nu t) \qquad\qquad [2.1\text{-}34]$$

This describes the displacement in terms of *time*.

Then take the relationship

$$v = \lambda\nu = \frac{x}{t} \qquad\qquad [2.1\text{-}35]$$

Retain the two right-hand terms and solve for ν:

$$\nu = \frac{x}{\lambda t} \qquad\qquad [2.1\text{-}36]$$

Substitute Equation [2.1-36] in [2.1-34] and cancel t:

$$y = \mathbf{A}\,\sin\!\left(2\pi\frac{x}{\lambda}\right) \qquad\qquad [2.1\text{-}37]$$

This describes the displacement in terms of *space*.

We are now ready to represent the displacement in *space and time*, that is, we make the wave *move*. We combine Equations [2.1-30] and [2.1-37]. This gives

$$\boxed{\,y = \mathbf{A}\,\sin\!\left[\frac{2\pi}{\lambda}(x - vt)\right]\,} \qquad\qquad [2.1\text{-}38]$$

which is an important equation. It describes the displacement of any point on a (sinusoidal) wave in both space and time.

There are several other expressions, equivalent to Equation [2.1-38], that are useful, depending on which parameters of the wave are known. For example, multiplying out the term in square brackets in Equation [2.1-38], substituting in the last term $v = \lambda\nu$, and canceling λ gives

$$y = \mathbf{A}\,\sin\!\left[2\pi\!\left(\frac{x}{\lambda} - \nu t\right)\right] \qquad\qquad [2.1\text{-}39]$$

Eliminating λ by the same relationship solved for λ leads to

$$y = \mathbf{A}\,\sin\left[2\pi\nu\left(\frac{x}{v} - t\right)\right] \qquad\qquad [2.1\text{-}40]$$

and substituting $\nu = 1/T$ in Equation [2.1-39] results in

$$y = \mathbf{A}\,\sin\left[2\pi\left(\frac{x}{\lambda} - \frac{t}{T}\right)\right] \qquad\qquad [2.1\text{-}41]$$

Note that x/λ and t/T are dimensionless quantities, as they should be.

Example. *A particle is in simple harmonic motion with a period of 3 s and an amplitude of 6 cm. One-half second after the particle has passed through its equilibrium position, what is:*
(a) Its displacement?
(b) Its velocity?
(c) Its acceleration?

Solution. (a) Since from Equation [2.1-32] $\omega = 2\pi/T$, we can set Equation [2.1-1]:

$$y = \mathbf{A}\,\sin(\omega t) = \mathbf{A}\,\sin\left(\frac{2\pi}{T}t\right)$$

Then we replace 2π by 180° and insert the actual figures; hence the *displacement* is

$$y = (0.06)\sin\left[\frac{(2)(180°)}{3}(0.5)\right] = \boxed{0.052 \text{ m}}$$

(b) The *velocity* of the particle, from Equation [2.1-4], making the same substitution and neglecting the sign, is

$$u = \mathbf{A}\left(\frac{2\pi}{T}\right)\cos\left(\frac{2\pi}{T}t\right)$$

(Here the first 2π term must be expressed in radians while in the second, the cosine function, π should be converted into degrees.) Thus

$$u = 0.06\left(\frac{2\pi}{3}\right)\cos 60° \approx \boxed{0.063 \text{ m s}^{-1}}$$

(c) For the *acceleration* we find, from Equation [2.1-5],

$$a = \left(\frac{2\pi}{T}\right)^2 y = \left(\frac{2\pi}{3}\right)^2 (0.052) = \boxed{0.228 \text{ m s}^{-2}}$$

The General Wave Equation

Returning to Equation [2.1-38], how does the displacement, y, vary as a function of space, x, and time, t? This requires partial differentiation of y with respect to x and t:

$$\frac{\partial y}{\partial x} = \frac{2\pi}{\lambda} \mathbf{A} \cos\left[\frac{2\pi}{\lambda}(x - vt)\right] \qquad \text{[2.1-42]}$$

and

$$\frac{\partial y}{\partial t} = \frac{2\pi v}{\lambda} \mathbf{A} \cos\left[\frac{2\pi}{\lambda}(x - vt)\right] \qquad \text{[2.1-43]}$$

Combining both equations and eliminating equal factors gives

$$\frac{\partial y}{\partial x} = -\frac{1}{v}\frac{\partial y}{\partial t} \qquad \text{[2.1-44]}$$

which is the differential equation describing the wave. If the wave were to move to the left, in the $-x$ direction,

$$\frac{\partial y}{\partial x} = \frac{2\pi}{\lambda} \mathbf{A} \cos\left[\frac{2\pi}{\lambda}(x + vt)\right] \qquad \text{[2.1-45]}$$

$$\frac{\partial y}{\partial t} = \frac{2\pi v}{\lambda} \mathbf{A} \cos\left[\frac{2\pi}{\lambda}(x + vt)\right] \qquad \text{[2.1-46]}$$

and, therefore,

$$\frac{\partial y}{\partial x} = +\frac{1}{v}\frac{\partial y}{\partial t} \qquad \text{[2.1-47]}$$

The *second* derivative,

$$\frac{\partial^2 y}{\partial x^2} = \frac{1}{v^2}\frac{\partial^2 y}{\partial t^2} \qquad \text{[2.1-48]}$$

however, holds for any (sinusoidal) wave, independent of the direction of travel, either $-x$ or $+x$.

There are two ramifications that follow. First, I have referred to a displacement, y, which might give the impression that something material really *moves*. This, however, need not be so, especially not in the case of light (see Chapter 3.1). We replace y by the more general term ξ which stands for *any disturbance,* without prejudice as to its nature, and instead of using Equation [2.1-48] write

$$\boxed{\frac{\partial^2 \xi}{\partial t^2} = v^2\frac{\partial^2 \xi}{\partial x^2}} \qquad \text{[2.1-49]}$$

This is the *one-dimensional wave equation*. It connects variations in time and space to the velocity of propagation of the wave.

Second, we do not want to restrict ourselves to waves that propagate along the x-axis. But if we are to include waves propagating in *any* direction, we need to extend the right-hand term in Equation [2.1-49] to the y and z axes, and replace it by

$$\frac{\partial^2 \xi}{\partial x^2} + \frac{\partial^2 \xi}{\partial y^2} + \frac{\partial^2 \xi}{\partial z^2} \qquad [2.1\text{-}50]$$

Such an expression can be written, much shorter, using the *vector differential operator* ∇, read "del." The operator ∇ stands for

$$\nabla = \frac{\partial}{\partial x} + \frac{\partial}{\partial y} + \frac{\partial}{\partial z} \qquad [2.1\text{-}51]$$

If we multiply this *vector* operator ∇ by itself, we obtain a *scalar* (operator), ∇^2:

$$\nabla \cdot \nabla = \nabla^2 \qquad [2.1\text{-}52]$$

This is the *Laplacian operator;* it represents the sum of three second-order partial derivatives:

$$\nabla^2 = \frac{\partial^2}{\partial x^2} + \frac{\partial^2}{\partial y^2} + \frac{\partial^2}{\partial z^2} \qquad [2.1\text{-}53]$$

For a wave propagating in any direction in three-dimensional space, therefore, Equation [2.1-49] becomes

$$\boxed{\frac{\partial^2 \xi}{\partial t^2} = v^2 \nabla^2 \xi} \qquad [2.1\text{-}54]$$

which is the *general* or *three-dimensional scalar wave equation*.

At this point we will temporarily leave the theory of wave motion and defer discussion of what actually constitutes ξ to a later chapter (3.1), when we come to the electromagnetic nature of light. We turn now to some more tangible aspects of waves, interference (Chapters 2.2 and 2.5) and diffraction (2.3 and 2.4).

SUGGESTIONS FOR FURTHER READING

A. P. French, *Vibrations and Waves* (New York: W. W. Norton & Company, Inc., 1971).

J. R. Pierce, *Almost all about waves* (Cambridge, MA: The MIT Press, 1974).

I. G. Main, *Vibration and Waves in Physics* (New York: Cambridge University Press, 1978).

For an introduction to optics' close cousin, acoustics, see

J. Backus, *The Acoustical Foundations of Music*, 2nd edition (New York: W. W. Norton & Company, Inc., 1977).

PROBLEMS

2.1-1. If a young person can hear sound over a frequency range from 22 Hz to 15 kHz and if the velocity of sound is $v = 331$ m s^{-1}, what are the wavelengths corresponding to these limits?

2.1-2. If sound waves of 662 Hz frequency propagate at a velocity of 331 m s^{-1}, what is:
(a) Their wavelength?
(b) The phase angle difference (in radians) of two points on the wave 40 cm apart?

2.1-3. The *sum* (or difference) of two complex numbers is generally another complex number. As an example try $(5 + 4i) - (3 + 2i)$.

2.1-4. Determine the *product* of two complex numbers such as $(5 + 2i)(2 + 5i)$, remembering that $i \times i = -1$.

2.1-5. The displacement of a point P on a sinusoidal wave, at a certain time, is two-thirds of its amplitude, $y = \frac{2}{3}A$. If $\lambda = 10$ cm, how far has P advanced from $x = y = 0$?

2.1-6. What is the wavelength and the velocity of propagation of the wave that can be represented by $y = 4 \sin(10x - 20t)$?

2.1-7. If the amplitude of a certain wave is 2 cm, the period 0.02 s, and the velocity 40 cm s^{-1}, what is:
(a) The frequency?
(b) The wavelength?

(c) The equation describing the displacement as a function of space and time?

2.1-8. A given point is oscillating in simple harmonic motion with an amplitude of 5 cm and a period of 8 s. If the *initial phase angle* is $\epsilon = \pi/3$ rad, find the displacement y at time $t = 0$ and the displacement after a time interval of 2 s.

2.1-9. If surface waves on a lake can be described by $y = 0.4 \sin(2t)$, what is:
(a) The maximum displacement?
(b) The velocity and acceleration, at time $t = 0$, of a cork floating on the water?

2.1-10. A wave has a period of 0.2 s and an amplitude of 4 cm. What is the displacement and velocity of an element of the wave at the time the phase angle has increased from zero to 60°?

2.1-11. A point moves counterclockwise in a circle at a constant speed of 5 cm s^{-1}. If the period is 6 s and if at time $t = 0$ the point is 30° away from the $+x$ direction, what is:
(a) Its displacement, y?
(b) The displacement at $t = 4$ s?
(c) The equation of motion in terms of A, ω, and ϵ?

2.1-12. Continue with Problem 2.1-11 and determine the numerical value of the velocity, dy/dt, and of the acceleration, d^2y/dt^2, at time $t = 4$ s.

2.2

Interference

THE TERM INTERFERENCE REFERS to the phenomenon that waves, under certain conditions, intensify or weaken each other. Expressions such as "constructive" and "destructive" interference should not be used because they imply that sometimes, and somehow, there could be a "destruction" of light. Certainly this is not so: If less light reaches a given point, more light reaches some other point; interference merely causes a *redistribution* of the light.

The phenomenon of interference is inseparably tied to that of *diffraction*. In fact, diffraction is more inclusive; it contains interference and, in a sense, even refraction and reflection. It is only because diffraction is mathematically more complex that I treat interference and diffraction in separate chapters and discuss interference first.

Superposition of Waves

The prerequisite of all interference is the *superposition* of waves. Consider the following four cases.

1. Superposition of waves of equal phase and frequency. Assume that two sinusoidal waves of the same frequency are traveling side by side in the same medium. The waves are of *equal phase*,

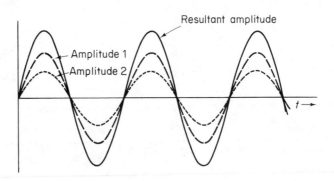

Figure 2.2-1 Superposition of waves of equal phase and frequency.

without any phase angle difference between them. We plot the amplitudes of the waves, as in Figure 2.2-1, and add them, point by point. The resultant amplitude then is the sum of the individual amplitudes,

$$\mathbf{A} = \mathbf{A}_1 + \mathbf{A}_2 + \mathbf{A}_3 + \cdots + \mathbf{A}_N \qquad [2.2\text{-}1]$$

and the resultant energy ("intensity"), from Equation [2.1-12], is proportional to the square of the sum of the amplitudes,

$$I \propto (\mathbf{A}_1 + \mathbf{A}_2 + \mathbf{A}_3 + \cdots + \mathbf{A}_N)^2 \qquad [2.2\text{-}2]$$

2. Superposition of waves of constant phase difference.

Now consider two waves that have the same frequency but that have a certain *constant phase angle difference* between them. The two waves have certain *different initial phase angles,* ϕ. The resultant wave can be found by several methods: by graphical construction, by trigonometry, by vector addition, and by the use of complex numbers.

(a) *Graphical construction.* As before, we draw the two wave amplitudes on top of each other and add corresponding points. This gives the resultant wave.

(b) *Trigonometric solution.* Again, the two waves have the same frequency (and therefore the same angular velocity ω) but different phase angles, ϕ_1 and ϕ_2:

$$y_1 = \mathbf{A}_1 \sin(\omega t + \phi_1)$$

and   [2.2-3]

$$y_2 = \mathbf{A}_2 \sin(\omega t + \phi_2)$$

Since

$$\sin(\alpha + \beta) = \sin \alpha \cos \beta + \cos \alpha \sin \beta \qquad [2.2\text{-}4]$$

Equations [2.2-3] can be written

$$y_1 = A_1(\sin \omega t \cos \phi_1 + \cos \omega t \sin \phi_1)$$

and   [2.2-5]

$$y_2 = A_2(\sin \omega t \cos \phi_2 + \cos \omega t \sin \phi_2)$$

The resultant displacement y is the algebraic sum of the two individual displacements:

$$y = y_1 + y_2$$

$$= A_1 \sin \omega t \cos \phi_1 + A_1 \cos \omega t \sin \phi_1 + A_2 \sin \omega t \cos \phi_2$$

$$+ A_2 \cos \omega t \sin \phi_2$$

$$= (A_1 \cos \phi_1 + A_2 \cos \phi_2)\sin \omega t$$

$$+ (A_1 \sin \phi_1 + A_2 \sin \phi_2)\cos \omega t \qquad [2.2\text{-}6]$$

The terms in parentheses are constant in time. Thus we can set

$$A_1 \cos \phi_1 + A_2 \cos \phi_2 = A \cos \epsilon \qquad [2.2\text{-}7]$$

and

$$A_1 \sin \phi_1 + A_2 \sin \phi_2 = A \sin \epsilon \qquad [2.2\text{-}8]$$

where **A** is the amplitude of the resultant wave and ϵ the new initial phase angle. In order to solve for **A** and ϵ, we square Equations [2.2-7] and [2.2-8] and add them. This gives

$$A^2 \cos^2 \epsilon + A^2 \sin^2 \epsilon = A_1^2(\cos^2 \phi_1 + \sin^2 \phi_1)$$

$$+ 2A_1 A_2(\cos \phi_1 \cos \phi_2 + \sin \phi_1 \sin \phi_2)$$

$$+ A_2^2(\cos^2 \phi_2 + \sin^2 \phi_2) \qquad [2.2\text{-}9]$$

Since

$$\sin^2 \alpha + \cos^2 \alpha = 1$$

and

$$\cos(\alpha - \beta) = \cos \alpha \cos \beta + \sin \alpha \sin \beta$$

this simplifies to

$$A^2 = A_1^2 + A_2^2 + 2A_1 A_2 \cos(\phi_1 - \phi_2) \qquad [2.2\text{-}10]$$

We can now substitute Equations [2.2-7] and [2.2-8] in [2.2-6]. This gives

$$y = A \cos \epsilon \sin \omega t + A \sin \epsilon \cos \omega t \qquad [2.2\text{-}11]$$

Angle ϵ is found by dividing Equation [2.2-8] by [2.2-7]:

$$\tan \epsilon = \frac{A_1 \sin \phi_1 + A_2 \sin \phi_2}{A_1 \cos \phi_1 + A_2 \cos \phi_2} \qquad [2.2\text{-}12]$$

Finally, using the same identity $\sin(\alpha + \beta) = \sin \alpha \cos \beta + \cos \alpha \sin \beta$ as before in Equation [2.2-4], we arrive at

$$y = \mathbf{A} \sin(\omega t + \epsilon) \qquad [2.2\text{-}13]$$

This agrees with our earlier first equation in [2.1-2] and shows that *the resultant of two sinusoidal waves is again a sinusoidal wave*, of the same frequency but with a new amplitude, $\mathbf{A}$, and a new phase angle, ϵ.

Example. *The two waves are represented by cosine functions, $y_1 = A \cos(\omega t + \phi_1)$ and $y_2 = A \cos(\omega t + \phi_2)$, respectively. Find the resultant wave and the magnitude of the new amplitude.*

Solution. Algebraic addition shows that the resultant displacement is

$$y = y_1 + y_2 = \mathbf{A} \cos(\omega t + \phi_1) + \mathbf{A} \cos(\omega t + \phi_2)$$

Then from the identity,

$$\cos \alpha + \cos \beta = 2 \cos \tfrac{1}{2}(\alpha + \beta)\cos \tfrac{1}{2}(\alpha - \beta)$$

we find that the resultant displacement

$$y = \mathbf{A}[2 \cos \tfrac{1}{2}(\omega t + \phi_1 + \omega t + \phi_2)\cos \tfrac{1}{2}(\omega t + \phi_1 - \omega t - \phi_2)]$$

$$= 2\mathbf{A}[\cos(\omega t + \tfrac{1}{2} \cos \phi_1 + \tfrac{1}{2} \cos \phi_2)\cos \tfrac{1}{2}(\phi_1 - \phi_2)] \qquad [2.2\text{-}14]$$

and the new amplitude

$$\mathbf{A} = 2\mathbf{A} \cos \tfrac{1}{2}(\phi_1 - \phi_2) \qquad [2.2\text{-}15]$$

(c) *Vector addition.* Three or more simple harmonic motions of the same frequency could be added in the same way, consecutively, each time giving an equation of the form of Equation [2.2-13]. *Vector addition*, though, is more convenient.

Assume that one of the waves follows the equation

$$y_1 = \mathbf{A}_1 \sin(\omega t) \qquad [2.2\text{-}16]$$

The phasor representing this wave at time t_0 is shown in Figure 2.2-2, left. As before, the phasor rotates counterclockwise and with angular velocity ω. The other wave has the same angular velocity but a different amplitude, $\mathbf{A}_2$, and lags behind the first wave by a (constant) phase difference δ (center). Thus

$$y_2 = \mathbf{A}_2 \sin(\omega t - \delta) \qquad [2.2\text{-}17]$$

The resultant of both phasors can be found by the parallelogram method of vector addition or, more easily, by a *vector triangle*. Note that both phasors, and the vector triangle formed by them, *rotate as a unit*.

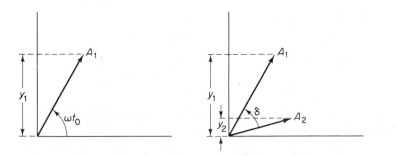
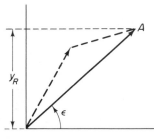

Figure 2.2-2 (*Left*) Phasor representing wave amplitude A_1; (*center*) two phasors representing waves of different amplitudes and with constant phase difference δ; (*right*) vector triangle showing resultant phasor.

If we then plot the displacements, y, versus time, t, for the two component waves and for the resultant, we arrive at Figure 2.2-3. As before, *the sum of two, or more, sinusoidal waves of the same frequency is a sinusoidal wave*.

(d) The advantage of using *complex numbers* is that the (vector) addition of real amplitudes can be written more easily in the form of an (algebraic) addition of complex amplitudes. For example, consider the *real* parts of two waves that follow the equations

$$\mathbf{A}_1 = A_1 e^{i(\omega t + \phi_1)}$$

and   [2.2-18]

$$\mathbf{A}_2 = A_2 e^{i(\omega t + \phi_2)}$$

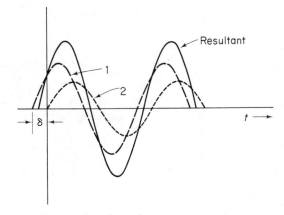

Figure 2.2-3 Two sine waves, 1 and 2, of constant phase difference, δ, and the *resultant*, also a sine wave.

Adding these two equations gives

$$\mathbf{A} = \mathbf{A}_1 + \mathbf{A}_2 = A_1 e^{i(\omega t + \phi_1)} + A_2 e^{i(\omega t + \phi_2)} \qquad [2.2\text{-}19]$$

We can now take out the common exponent $i\omega t$:

$$\mathbf{A} = e^{i\omega t}(A_1 e^{i\phi_1} + A_2 e^{i\phi_2}) \qquad [2.2\text{-}20]$$

The square of the resultant, $\mathbf{A}^2$, is found by multiplying the complex terms by their complex conjugates:

$$\mathbf{A}^2 = (A_1 e^{i\phi_1} + A_2 e^{i\phi_2})(A_1 e^{-i\phi_1} + A_2 e^{-i\phi_2})$$

$$= \mathbf{A}_1^2 + \mathbf{A}_2^2 + A_1 A_2 [e^{i(\phi_1 - \phi_2)} + e^{-i(\phi_1 - \phi_2)}] \qquad [2.2\text{-}21]$$

Then, from Euler's formula (page 180),

$$e^{i\phi} + e^{-i\phi} = \cos \phi + i \sin \phi + \cos \phi - i \sin \phi = 2 \cos \phi \qquad [2.2\text{-}22]$$

and therefore, Equation [2.2-21] becomes

$$\mathbf{A}^2 = \mathbf{A}_1^2 + \mathbf{A}_2^2 + 2\mathbf{A}_1\mathbf{A}_2 \cos(\phi_1 - \phi_2) \qquad [2.2\text{-}23]$$

the same as Equation [2.2-10].

Interference fringes.   Since the energy ("intensity") of a wave is proportional to the square of the amplitude, $I \propto \mathbf{A}^2$, and since $\phi_1 - \phi_2$ is the phase angle difference, δ, Equation [2.2-23] may be written

$$\mathbf{I} = I_1 + I_2 + 2\sqrt{I_1 I_2} \cos \delta \qquad [2.2\text{-}24]$$

Whenever the phase difference is zero, $\delta = 0$, we will have a *maximum* amount of light,

$$\mathbf{I}_{max} = I_1 + I_2 + 2\sqrt{I_1 I_2}$$

which, when $I_1 = I_2$, becomes

$$I_{max} = 4I_1 \qquad [2.2\text{-}25]$$

On the other hand, whenever the phase difference is 180°, $\delta = 180°$, $\cos 180° = -1$, and we will have a *minimum* amount of light:

$$I_{min} = I_1 + I_2 - 2\sqrt{I_1 I_2}$$

which, when $I_1 = I_2$, becomes

$$I_{min} = 0 \qquad [2.2\text{-}26]$$

At points that lie between the maxima and minima, we find, following Equation [2.2-24] and assuming that $I_1 = I_2$, that

$$I = I_1 + I_1 + 2I_1 \cos \delta$$
$$= 2I_1 (1 + \cos \delta)$$
$$= 4I_1 \cos^2 \tfrac{1}{2} \delta \qquad\qquad [2.2\text{-}27]$$

If the two contributions are not equal, there will be some excess light (from the more energetic contribution) and the minima will not be completely dark.

Consequently, the light distribution resulting from a superposition of waves will consist of alternately bright and dark bands called *interference fringes*. Such fringes can be observed visually, projected on a screen, or recorded photoelectrically.

Figure 2.2-4 shows the energy distribution across a few fringes projected on a screen. If only one beam of light illuminates the screen, the distribution is uniform throughout (lower horizontal line marked I_1). If we have two beams of light of equal energy, but lacking the property necessary to produce interference, the distribution is again uniform, but twice as much light will reach the screen (upper horizontal line marked $2I_1$). But if the two beams *are capable of producing interference,** they form alternate maxima and minima. If the initial contributions have equal amplitudes, and therefore equal energies, then the maxima, from Equation [2.2-25], contain *four times* the energy of the individual contribution; the minima, from Equation [2.2-26], contain no energy.

The result is the $\cos^2$ curve labeled $4I_1$. Note that the *integrated* values, that is, the areas under the curves $2I_1$ and $4I_1$, *are the same:* Energy can neither be created nor destroyed; interference merely causes it to be *redistributed*.

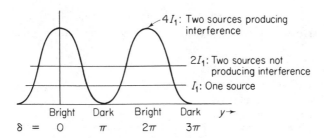

Figure 2.2-4 in figure:
4I_1: Two sources producing interference
2I_1: Two sources not producing interference
I_1: One source
Bright   Dark   Bright   Dark $y \rightarrow$
$\delta =$ 0 π 2π 3π

Figure 2.2-4 Energy distributions as produced by one source, two sources not causing interference, and two sources capable of causing interference.

---

*The property of light necessary to produce interference is called *coherence;* it will be discussed later in this chapter (beginning on page 206).

3. Superposition of waves of different frequencies.

So far we have assumed that the waves have the same, well-defined, frequency. But light is never truly "monochromatic"; even light that nominally, or predominantly, has one frequency, always contains a certain range of frequencies (*quasimonochromatic* light). When waves of different frequencies superimpose, the result is more complicated. This case and its ramifications are discussed in Chapter 4.1, under the heading *Fourier transform spectroscopy*.

4. Superposition of waves of random phase difference.

Finally, we consider waves that have random phase differences between them, or that have widely different frequencies. If such waves superimpose, they do not produce any discernible interference. The resultant energy is obtained by adding the individual energies,

$$\sum_{}^{N} \mathbf{A}^2 = \mathbf{A}_1^2 + \mathbf{A}_2^2 + \mathbf{A}_3^2 + \cdots + \mathbf{A}_N^2 \qquad [2.2\text{-}28]$$

rather than by adding the amplitudes and squaring their sum.

Young's Double-Slit Experiment

Let us now turn to a description of the most fundamental, classical experiment on the interference of light: Young's *double-slit experiment*. *
At first, Young had tried two pinholes, but quickly realized that the fringes were much brighter when two parallel slits were used (S_1 and S_2 in Figure 2.2-5).

In the center of the field, where the contributions from the two slits have traveled through equal distances and where the path difference is zero:

---

* Thomas Young (1773–1829), British physician and physicist. The son of a Quaker family, Young could read at age 2, at 6 began studying Latin, and at 13 had also mastered Greek, Hebrew, Italian, and French. At 19 he entered medical school, correctly explained the accommodation of the eye, and was elected Fellow of the Royal Society. In 1796, Young graduated from the University of Göttingen Medical School, opened a practice in London, and five years later became Professor of Natural Philosophy (physics) at the Royal Institution of London. That same year, 1801, he read before the Royal Society the first of several papers presenting the wave theory of light and the principle of interference, much to the opposition of Newton's followers. Young made noteworthy contributions also to acoustics, atmospheric refraction, elasticity, fluid dynamics, and color vision; later in his life, he helped decipher the Egyptian Rosetta Stone hieroglyphics. Th. Young, "On the Theory of Light and Colours," *Phil. Trans. Roy. Soc. London* **92** (1802), 12–48; and "Experiments and calculations relative to Physical Optics," *ibid.* **94** (1804), 1–16.

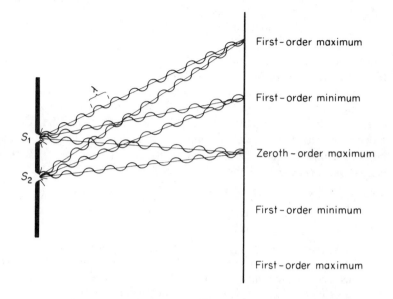

First - order maximum

First - order minimum

Zeroth - order maximum

First - order minimum

First - order maximum

Figure 2.2-5   Maxima and minima in Young's double-slit experiment. Vertical dimensions greatly exaggerated.

$$\Gamma = 0$$

we have the *zeroth-order maximum*. But maxima will also occur whenever the path difference is one wavelength, λ, or an integral multiple of a wavelength, $m\lambda$:

$$\Gamma = m\lambda \qquad\qquad [2.2\text{-}29]$$

The integer m is called the *order* of interference.

In order to calculate the positions of the maxima, we call d the distance between the centers of the two slits, θ the change in direction of the light, and Γ the path difference between the two contributions. Then from Figure 2.2-6,

$$\sin \theta = \frac{\Gamma}{d} \qquad\qquad [2.2\text{-}30]$$

Combining Equations [2.2-29] and [2.2-30] gives

$$\boxed{d \sin \theta = m\lambda} \qquad m = 0, 1, 2, \ldots \qquad [2.2\text{-}31]$$

which is *Young's double-slit equation for maxima*. Interference *minima* occur whenever one of the contributions has shifted in phase by $\frac{1}{2}\lambda$, that is, when

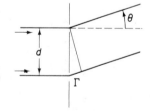

Figure 2.2-6 Deriving the double-slit equation.

$$\boxed{d \sin \theta = (m - \tfrac{1}{2})\lambda} \qquad m = 1, 2, 3, \ldots \qquad [2.2\text{-}32]$$

(Note that now $m = 1$ is the lowest possible order; in double-slit interference there is no zeroth-order minimum.)

In a typical double-slit experiment, the angles θ are small; thus their sines may be replaced by the angles themselves, measured in radians. Consequently, from Equation [2.2-31], the separation of the fringes is proportional to the wavelength and, for a given wavelength, inversely proportional to d. Since d and θ are easy to measure, Young's experiment can be used to determine the wavelength.

Example. *Light passes through two narrow slits of d = 0.8 mm. On a screen 1.6 m away the distance between the two second-order maxima is 5 mm. What is the wavelength of the light?*

Solution. For small angles, $\sin \theta$ can be set equal to y/x, where y is the distance of a given maximum, or minimum, from the optic axis and x is the distance from the slits to the screen. The double-slit equation $d \sin \theta = m\lambda$ can then be written $d(y/x) = m\lambda$ and, when solved for λ,

$$\lambda = \frac{dy}{mx} \qquad [2.2\text{-}33]$$

Since the distance given is that *between* the two maxima, rather than the distance of one maximum from the axis, we use one-half that distance, $y = 2.5$ mm, so that

$$\lambda = \frac{(0.8 \text{ mm})(2.5 \text{ mm})}{(2)(1600 \text{ mm})} = \boxed{625 \text{ nm}}$$

Variations on the theme of double-slit interference.

Ordinarily, the separation d of the two slits in Young's experiment is a few millimeters at most. If the two slits are placed farther apart, but still within the circumference of a converging lens, the lens makes the two contributions come together no matter what the separation of the slits. However, the wavefronts in the two contributions subtend a larger angle and, hence, the fringes become more closely spaced, requiring a magnifying glass to see them. Then, the lens near the double slits and the magnifier together form a telescope. Indeed, placing two (movable) slits in front of a telescope is a method well known in astronomy for determining the angular separation of binaries (double stars) or the diameter of fixed stars, dimensions too small to be resolved by direct (non-interferometric) observation.

Example. *Light comes from two stars, each generating its own fringe pattern. As before, y/x is the angular distance, as seen from the lens,*

*between adjacent maxima in either of the patterns. At one-half that angle,
the maxima of one pattern fall exactly between the maxima of the other,
obliterating both patterns. This occurs at an angle $\theta = \frac{1}{2}(y/x)$ or, from
Equation [2.2-33], at $\theta = \lambda/2d$. Thus the least separation d between the
slits at which the fringe pattern disappears determines the angular separation of the two sources.*

An extension of this principle is found in *Michelson's stellar interferometer.* Here the distance *d* is made still larger, exceeding the diameter
of the telescope objective. This is made possible by four mirrors, the outer
mirrors taking the place of the double slits (Figure 2.2-7).*

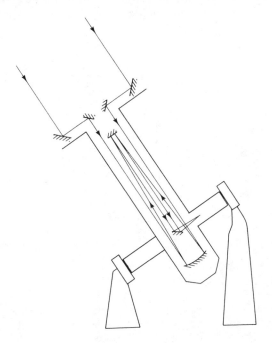

Figure 2.2-7   Michelson's stellar interferometer.

---

*On December 13, 1920, Michelson's coworker F. G. Pease, using this method,
made the first conclusive measurements on the star α Orionis (Betelgeuse). The interference
fringes disappeared (although fringes could still be seen in the images of other stars) when
the two outer mirrors were 307 cm apart. Using Rayleigh's criterion, which we will discuss
in the next chapter, and assuming a wavelength of 575 nm, the angular diameter of α Orionis
was found to be $\theta = 1.22(575 \times 10^{-9})/3.07 = 2.28 \times 10^{-7}$ rad $= (2.28 \times 10^{-7})$
$(57°/\text{rad})(3600 \text{ sec}/\text{deg}) = 0.047$ arc sec, corresponding to a linear diameter of 4.1×10^{8}
km, more than the orbit of the Earth $(3 \times 10^{8}$ km). A. A. Michelson and F. G. Pease,
"Measurement of the Diameter of α Orionis with the Interferometer," *Astrophys. J.* **53**
(1921), 249–59.

The Michelson Interferometer

General construction.
There is no interferometer that is as versatile as the *Michelson interferometer*.* It makes it possible to bring two optical planes into coincidence and to move them virtually through one another. In addition, Michelson's instrument can be used with an extended source; therefore, compared with a set of double slits, it gives much brighter fringes. The Michelson interferometer usually takes the form shown in Figure 2.2-8.

A bundle of light from an extended source—or from a point source and collimator—falls on a beamsplitter, A, which on its rear surface has a reflective coating. The beamsplitter divides the light into two bundles,

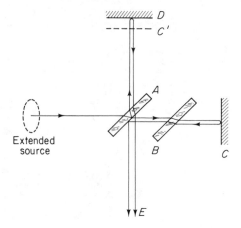

Figure 2.2-8   The Michelson interferometer.

---

* Albert Abraham Michelson (1852–1931). Born in Strelno, Prussia (now Strzelno, Poland), Michelson was two years old when his parents brought him to the United States. He graduated from, and taught at, the U.S. Naval Academy and later worked at the Case School of Applied Science in Cleveland, at Clark University in Worcester, Massachusetts, and at the University of Chicago. In 1907 he was awarded the Nobel prize in physics, the first American scientist to be so honored. Michelson is best known for his precise determination of the velocity of light, for inventing the interferometer that bears his name, for the Michelson–Morley aether drift experiment, and for establishing the length of the meter in terms of wavelength of light. He also made noteworthy contributions to astronomy, spectroscopy, and geophysics, was proficient in tennis and other sports, played the violin, and liked to paint landscapes. The interferometer was first described in A. A. Michelson, "Interference Phenomena in a new form of Refractometer," *Am. J. Sci.* (3) **23** (1882), 395–400, and *Phil. Mag.* (5) **13** (1882), 236–42. Also see: Dorothy Michelson Livingston, *The Master of Light: A Biography of Albert A. Michelson* (Chicago: University of Chicago Press, 1973).

one transmitted toward mirror C, the other reflected toward D. The two mirrors, C and D, return the light to A. There they recombine and proceed toward E, where interference is observed.

One of the mirrors is mounted so that it can be moved along the axis. But note that, if reflection at A occurs at the rear surface as shown, the light reflected at D will pass through A three times while the light reflected at C will pass through only once. For this reason, a compensating plate, B, of the same thickness and inclination as A, is inserted into the A–C path. If we look into the instrument from E, we see mirror D, and in addition we see a virtual image, C', of mirror C. Depending on the positions of the mirrors, image C' may be in front of, or behind, or exactly coincident with mirror D.

Practical considerations

1. If the two mirrors have the same axial distance from the rear face of A and if they are perpendicular to each other, then image C' is coincident with mirror D. At this coincidence position, the two paths are of equal length. Thus we expect the waves to reinforce each other and to form a maximum. But this is not so, because of a π *phase change* which *occurs on external* (air-to-glass) *reflection only*. No phase change occurs on internal (glass-to-air) reflection, and none occurs on transmission or refraction. (See page 305 for a discussion of phase changes on reflection.) Look again at Figure 2.2-8 and note that it is the light that comes from C and goes to E that is reflected, air-to-glass, at A, and therefore undergoes the π change.* This means that at the coincidence position there will be a *minimum: the center of the field will be dark*.

2. Now we *move* one of the mirrors. If the mirror is moved through a distance of a quarter of a wavelength, $d = \frac{1}{4}\lambda$, the path length (because the light passes through the distance twice) changes by $\frac{1}{2}\lambda$, the two bundles get out of phase by 180°, the phase change compensates, and we have a *maximum*. Moving the mirror by another $\frac{1}{4}$ wavelength gives another minimum, another $\frac{1}{4}\lambda$ another maximum, and so on. Thus

$$\boxed{2d = m\lambda} \qquad m = 0, 1, 2, \ldots \qquad [2.2\text{-}34]$$

which is the *Michelson interferometer equation*.

3. Next, assume that we look *obliquely* into the interferometer and that our line of sight makes an angle α with the axis. Ordinarily, the two

---

* Since the beamsplitter usually has a *metallic* coating, the situation is slightly more complicated. A phase change does occur, but in contrast to uncoated glass it is not of exactly π rad magnitude.

planes D and C' are a distance d apart and the two virtual images, I and I', are separated by $2d$. But for oblique incidence, as we see from Figure 2.2-9, the path difference between the two lines of sight becomes *less* and, instead of Equation [2.2-34], we have

$$2d \cos \alpha = m\lambda \qquad\qquad [2.2\text{-}35]$$

For a given mirror separation d, and a given order m and wavelength λ, angle α is constant. Therefore, the fringes are of rotational symmetry; they appear in the form of *circles,* concentric around the axis. They are *fringes of equal inclination.*

4. Now we move the mirrors farther *away from each other.* Consequently, as d becomes larger, a given ring—which has a certain order m—increases in size because the product $2d \cos \alpha$ must remain constant. The rings, therefore, expand and new rings appear in the center, one ring appearing each time one mirror is moved by one-half a wavelength (Figure 2.2-10). Also note that as d increases, the rings in the periphery disappear slower than new rings appear in the center; thus the field of view becomes more crowded with thinner rings. Conversely, as d is made smaller, the rings appear to contract and to disappear in the center.

5. Finally, if the mirrors are *tilted,* and are no longer precisely perpendicular to each other, the fringes will appear straight. Actually, they are sections of large hyperbolas. (Fringes that occur while C' and D

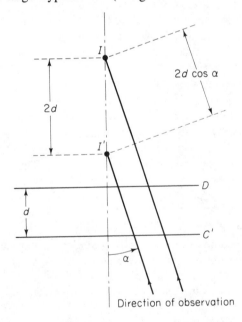

Figure 2.2-9 Looking off-axis into the Michelson interferometer.

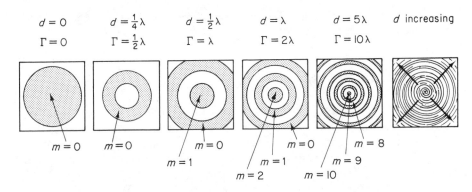

Figure 2.2-10 Appearance of fringes in the Michelson interferometer as the mirrors are moved away from each other. Arrows on the far right indicate motion of the fringes.

subtend a certain angle are called *fringes of equal thickness,* or *Fizeau fringes.*) Now, as one of the mirrors is moved, the fringes move across the field and may be counted as they pass a certain point. For each fringe that passes, the optical path length has changed by one wavelength, and one of the mirrors has moved through one-half wavelength. Therefore, if m fringes are counted as the mirror is moved through d, the wavelength of the light is $\lambda = 2d/m$. With care, $\frac{1}{100}$ of the width of a fringe can be measured.

Aligning the Michelson interferometer. Tape a black mark to a ground glass placed between light source and beamsplitter. Make the two paths equal to within 1 mm. Look into the interferometer from E. Three reflected images of the mark will be seen (Figure 2.2-11a). A slight turn of the mirror tilt controls will show which of these can be moved. Superimpose the movable image exactly on the right-hand image of the stationary pair of images, aligning the movable image first in *height* and then *sideways* (b). At this point, interference fringes should become visible (c).

Using the tilt controls, bring the center of the fringes into the center of the field of view (d). Move the other mirror along the axis so that the fringes move *inward* (e). Near the zeroth order, alignment becomes critical. Even placing your hand over and close to the paths—which will raise the temperature of the air and

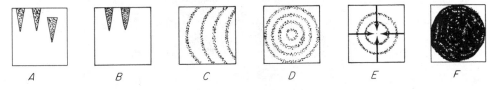

Figure 2.2-11 Aligning the Michelson interferometer.

lower its refractive index—will make the fringes wiggle and move in and out of the field. At the zeroth order, the center of the field will be dark (f).

Applications

The most important applications of interferometry are *precision measurements*. In principle almost any interferometer will do, but for particular purposes some types are better suited than others. We distinguish two groups, whether the instrument is used for *reflecting* or for *transparent* objects.

The first group applies to the *examination of surfaces,* to *metrology,* and to the *alignment* of optical and mechanical components. Here we use the Michelson interferometer, or one of its many derivatives. For example, if we want to know how flat a surface has been ground and polished, we use that surface in place of one of the mirrors. The other mirror serves as a reference. Any depression or elevation in the surface shows as characteristic distortions of the fringe pattern, as we see in Figure 2.2-12.

Figure 2.2-12   Interference effects caused by the weight of an insect resting on a water surface. [From Cagnet, Françon, Thrierr, *Atlas of Optical Phenomena.* p. 44. (Berlin: Springer-Verlag, 1962). Reproduced by permission.]

Example. *When examining the surface of a polished workpiece in thallium light (535 nm), some scratch marks are seen where the fringes are distorted by $\frac{4}{10}$ the distance between fringes (Figure 2.2-13). How deep are the scratches?*

Solution. As always in interferometry, consecutive fringes indicate a path difference of one wavelength, 1λ. But the light is reflected, and goes back and forth through its path; therefore, a depth, d, in the surface means one-half of a wavelength,

$$d = \tfrac{1}{2}\lambda$$

In our example, the distortion is $\frac{4}{10}$ of the fringe separation and hence the depth of the scratch is

$$d = (0.4)(\tfrac{1}{2})(535 \times 10^{-9}) = \boxed{0.1 \ \mu m}$$

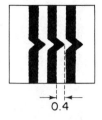

Figure 2.2-13

For transparent objects the *Mach–Zehnder interferometer* is preferred, mainly because the light passes through the sampling field only once and interpretation of the results is easier. The Mach–Zehnder interferometer, illustrated in Figure 2.2-14, contains two mirrors and two beamsplitters. The first beamsplitter divides the incoming light, the mirrors reflect the beams as shown, and the second beamsplitter brings the beams together again. The two paths may be widely separated, which permits testing large objects, as in a windtunnel.

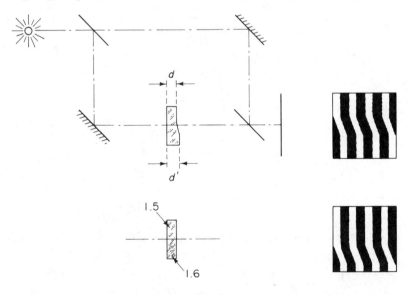

Figure 2.2-14 Fringe shifts in an interferogram (*right*) will result if the sample inserted in one arm of the interferometer has local variations of thickness (*top*) or variations of refractive index (*bottom*). Interferometer shown is Mach–Zehnder type.

Interferometers respond to differences in *optical path length*. From Equation [1.1-1], we recall that optical path length is the product of thickness and refractive index:

$$S = Ln \qquad [2.2\text{-}36]$$

The *path difference*, Γ, is simply the difference between two such path lengths:

$$\Gamma = S_2 - S_1 \qquad [2.2\text{-}37]$$

Also, since a wave of length λ can be represented by the full circumference, 2π rad, of the reference circle, path difference, Γ, and *phase difference*, δ, are connected by

$$\frac{\Gamma}{\lambda} = \frac{\delta}{2\pi} \qquad [2.2\text{-}38]$$

Note that any fringe shift seen in an interferometer may be due to either a change in thickness or a change in refractive index. In the upper example in Figure 2.2-14 we assume that the refractive index within the sample is constant but the thickness is not; thus from Equations [2.2-36] and [2.2-37], we have an optical path difference of

$$\Gamma = n(L_2 - L_1) \qquad [2.2\text{-}39]$$

In the lower example, the thickness is constant but the index is not; thus we have

$$\Gamma = L(n_2 - n_1) \qquad [2.2\text{-}40]$$

Since from Equation [2.2-29],

$$\Gamma = m\lambda$$

a given path difference will cause

$$m = \frac{1}{\lambda} n(L_2 - L_1) \qquad [2.2\text{-}41]$$

or

$$m = \frac{1}{\lambda} L(n_2 - n_1) \qquad [2.2\text{-}42]$$

additional wavelengths to be present within the thicker, or denser, medium and a fringe shift by m fringes will result. For media of irregularly changing refractive index we have

$$m = \frac{1}{\lambda} \int_A^B (n_2 - n_1) \, dL \qquad [2.2\text{-}43]$$

where the integral is taken from A to B, over the length of the path, L.

There are a few more types of interferometers worth noting for special applications. The *Twyman–Green interferometer** is derived from the Michelson type. It is particularly useful for the testing of optical components such as lenses, prisms, and mirrors. The element to be tested is placed in one of the arms of the interferometer as shown in Figure 2.2-15. The light must be well collimated. Then, if the lens under test, L, is optically perfect, the light will be returned collimated, with precisely plane wavefronts, and the fringe pattern will be uniform. If the lens is not perfect, any variations in optical thickness will show as distortions (of the fringes) that look like contour lines on a map.

At first sight, the Twyman–Green interferometer looks like the Michelson type. But there is an important difference: in the Twyman–Green instrument the light is collimated, whereas in the Michelson interferometer it is not. If there is a path difference, the Michelson interferometer would show concentric rings; these, as we have seen, are *fringes of equal inclination*. In the Twyman–Green instrument there is no variation in the angle of incidence and any fringes that are generated by the test object are *fringes of equal thickness*.

The *Kösters prism interferometer,*[†] likewise, is derived from the Michelson type, with the beamsplitter located inside a solid prism (Figure

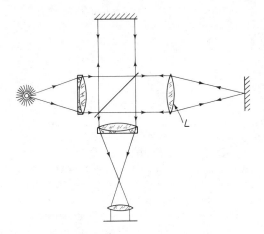

Figure 2.2-15 Twyman–Green interferometer set up for testing lens (L).

---

* Invented by Frank Twyman, manager, and Alfred Green, foreman, at Adam Hilger Ltd., astronomical and optical instrument makers in London, and first described in F. Twyman and A. Green, *Improvements in Finishing Prisms or Lenses or Combinations of the same and in Apparatus therefor*, Brit. Patent 103,832, Feb. 5, 1917.

† W. Kösters, *Interferenzdoppelprisma für Messzwecke*, Dtsch. Reichspatent 595 211 (1931).

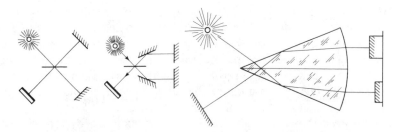

Figure 2.2-16 Folding the paths in a Michelson interferometer (*left*) leads to the Kösters prism interferometer (*right*).

2.2-16). This makes the system particularly compact and mechanically stable.

The *Jamin interferometer** is similar to the Mach–Zehnder type, except that the left-hand side beamsplitter and mirror and the right-hand side mirror and beamsplitter are combined into one thick plane-parallel plate each. This makes for mechanical stability and easier alignment. *Shearing interferometers* also use a beamsplitter to divide the light. The two beams are then recombined with a certain amount of lateral offset or *shear*. No perfect reference is needed since any given wavefront is compared with itself.

In a *triangular path*, or *circular, interferometer* the light is divided by a beamsplitter into two bundles that travel in opposite directions (Figure 2.2-17). They are recombined by the same beamsplitter.[†]

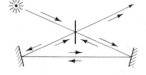

Figure   2.2-17 Triangular path interferometer.

Coherence

Coherence is sometimes defined as the condition necessary to produce interference, and interference is defined as an interaction of waves that are "coherent." Nothing much follows from such circular arguments. Moreover, coherence is occasionally thought to be either present or not present, as if light were *either* coherent *or* incoherent. In fact, no light is completely coherent, and none is completely incoherent.

Coherence is a *property* of light. Interference is the *process* of interaction. Coherence means that two or more waves in a radiation field are in a fixed and predictable phase relationship to each other.

For expediency, we distinguish two classes of coherence, *spatial coherence* and *temporal coherence*. Spatial coherence or, more precisely,

---

*J. Jamin, "Description d'un nouvel appareil de recherches, fondé sur les interferences," *Compt. rendu* **42** (1856), 482–85.

[†]G. Sagnac, "Sur les interférences de deux faisceaux superposés en seus inverses le long d'un circuit optique de grandes dimensions," *Compt. rendu* **150** (1910), 1302–5.

transverse spatial coherence refers to the phase relationship between waves traveling side by side, at a certain distance from one another. The farther apart the two waves, the less likely they will be in phase, and the less coherent the light will be. Temporal coherence, or *longitudinal* spatial coherence (often called monochromaticity), applies to waves traveling the same path. It refers to the constancy, and predictability, of phase as a function of time.

Spatial coherence.

Assume that the light is incident on a double slit as shown in Figure 2.2-18. On a distant screen (right) a system of interference fringes is seen. With highly (spatially) coherent light, the fringes are rather distinct; their *contrast* is high. As the two slits are moved farther apart, the fringes become more closely spaced and will lose contrast. We will see in detail shortly that the degree of contrast is a measure of the degree of spatial coherence.

Temporal coherence.

In principle, the arrangement in Figure 2.2-18 can be used to determine the degree of temporal coherence. We only need to *delay* one of the beams, that is, place a sheet of transparent material over one of the slits. If at a given thickness (of the sheet) no interference fringes are visible, the path difference introduced by the material exceeds the temporal coherence.

Temporal coherence requires discussion of the peculiar nature of light. Certain types of electromagnetic radiation, such as microwaves and radiowaves, and also sound waves and other mechanical waves, can be generated in the form of continuous waves of almost infinite length. But light waves come in *wavetrains*. These wavetrains are of finite length; each contains only a limited number of waves. The length of a wavetrain, Δs, is called the *coherence length*. It is the product of the number of waves, N, contained in the train and of their wavelength, λ:

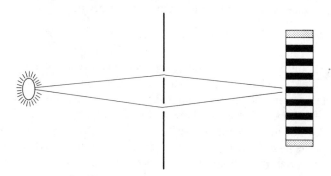

Figure 2.2-18   Young's double slits and spatial coherence.

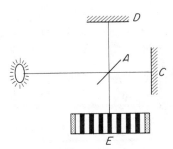

Figure 2.2-19   Michelson's interferometer and temporal coherence.

$$\Delta s = N\lambda \qquad\qquad [2.2\text{-}44]$$

Since velocity is defined as the distance traveled per unit of time, it takes a wavetrain of length Δs a certain length of time, Δt, to pass a given point:

$$\Delta t = \frac{\Delta s}{c} \qquad\qquad [2.2\text{-}45]$$

where c is the velocity of light. The length of time Δt is called the *coherence time*.

It is more convenient to measure the degree of temporal coherence using a Michelson interferometer. If the two paths in Figure 2.2-19, A–C–A and A–D–A, are equal in length, the fringes seen at E have maximum contrast, and hence maximum temporal coherence. If they are not of equal length, the contrast is less. Temporal coherence, then, is inversely proportional to the magnitude of the path difference and directly proportional to the length of the wavetrains.

Light from the green mercury line has a coherence length of about 11 mm, light from the orange krypton line about 80 cm, and light from a laser can have coherence lengths of hundreds of kilometers. The reason for this is a matter of generating such waves, rather than a matter of wavelength.

Partial coherence.   It is incorrect to think of light as *either* coherent *or* incoherent. Light can have different *degrees of coherence*, which introduces the concept of *partial coherence*. Assume that two wavetrains of light, each of finite length Δs, overlap to their full extent. Such complete overlap will result in distinct maxima and minima of the highest degree of contrast. But even if the wavetrains overlap *only in part*, as in Figure 2.2-20, interference is possible, although the degree of contrast of the fringes is less, depending on the degree of overlap. The question, then, is not how much the wavetrains must overlap to produce interference; the question is how much contrast we *need* to see any fringes.

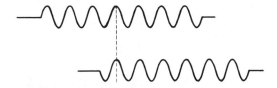

Figure 2.2-20   Partial overlap of two wavetrains.

The definition of contrast is a matter of comparison. The amount of power incident per unit area is called *areance* (see page 385). Contrast, γ, can be defined as the ratio of the difference between maximum areance, E_{max}, and minimum areance, E_{min}, to the sum of such areances:

$$\gamma \equiv \frac{E_{max} - E_{min}}{E_{max} + E_{min}} \qquad [2.2\text{-}46]$$

Let the maximum areance, E_{max}, assume any arbitrary value and let E_{min} become zero. The contrast then is $\gamma = 1$. On the other hand, if $E_{max} = E_{min}$, then $\gamma = 0$ and *no* fringes can be seen. The contrast may vary between 0 and 1 and assume any value in between. Generally, a contrast of 0.8 is considered high and fringes of contrast 0.2 are barely visible.

How does contrast relate to coherence? Assume that two points on a distant screen are illuminated by two bundles of light that produce equal areances E_0. Each of these bundles consists of two parts, *A* and *B*.* Parts *A* may be "completely coherent" and cause areances

$$E_A = \mathscr{C} E_0 \qquad [2.2\text{-}47]$$

Parts *B* may be "completely incoherent," both as to themselves and with respect to *A*, and cause areances

$$E_B = (1 - \mathscr{C})E_0 \qquad [2.2\text{-}48]$$

The quantity $\mathscr{C}$ is called the *degree of coherence*.

If interference results, it is because of parts *A*. These parts form fringes whose maxima, from Equation [2.2-25], have intensities, and cause areances, *four times* as high as the individual contributions. The maximum areance, thus, is $4\mathscr{C}E_0$ and the minimum areance is zero. On this pattern a uniform distribution is superimposed which, because it comes from two sources, has an areance *twice* that of Equation [2.2-48],

$$E_B = 2(1 - \mathscr{C})E_0$$

As a result, the areance in the maxima is

---

*Following P. H. van Cittert, "Degree of Coherence," *Physica* **24** (1958), 505–7.

$$E_{max} = 4\mathscr{C}E_0 + 2(1 - \mathscr{C})E_0 = 2(1 + \mathscr{C})E_0 \qquad [2.2\text{-}49]$$

and in the minima it is

$$E_{min} = 2(1 - \mathscr{C})E_0 \qquad [2.2\text{-}50]$$

If Equations [2.2-49] and [2.2-50] are substituted in [2.2-46],

$$\gamma = \frac{2(1 + \mathscr{C})E_0 - 2(1 - \mathscr{C})E_0}{2(1 + \mathscr{C})E_0 + 2(1 - \mathscr{C})E_0} = \frac{4\mathscr{C}E_0}{4E_0} = \mathscr{C} \qquad [2.2\text{-}51]$$

which shows that the *degree of contrast* of the fringes produced by interference of two waves *is equal to the degree of coherence* between these two waves.

The highest contrast, and therefore the highest degree of coherence, will result when the minimum areance in Equation [2.2-46] is zero. The contrast and the degree of coherence then become *unity*. Although conceivable in theory, this figure cannot be attained in practice because noise prevents the minima from having no light at all. *Complete* coherence is merely a theoretical limit.

But why can't the degree of coherence be *zero*; that is, why can't light be completely incoherent? Think of an object that consists of an array of luminous points (*pixels*) incoherent among themselves. The light is focused by a lens on a screen. One would expect that the light in the image would be completely incoherent too. But this is not so. The reason is the finite aperture of the lens. The lens forms, of each pixel, not a point image but a diffuse patch of light. The image, then, contains a multiplicity of such patches which *among* themselves are incoherent but, since each patch comes from one pixel, are coherent *within* each patch. Since the patches overlap, thus, there still exists a certain degree of coherence.

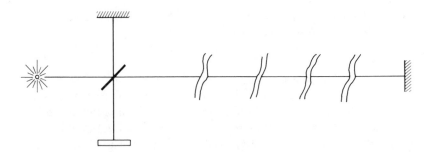

Figure 2.2-21   Unequal-path Michelson interferometer as used for determining the maximum path difference through a turbulent atmosphere at which fringes could still be seen. From R.B. Herrick and J.R. Meyer-Arendt, "Interferometry through the Turbulent Atmosphere at an Optical Path Difference of 354 m," *Applied Optics* **5** (1966), 981–83. Three years later, an even longer path difference was reached by V.V. Pokasov and S.S. Khmelevtsov, "Interferometry with a Path Difference of up to 500 m through the Turbulent Atmosphere" (in Russian), *Izvest. Vysh. Zav. Fizika (Tomsk U)* **8** (1969), 139–41.

The contrast becomes less also when the light passes through an *inhomogeneous medium*. The degree of contrast then is a function not only of the coherence length but is also proportional to the degree of homogeneity (Figure 2.2-21).

SUGGESTIONS FOR FURTHER READING

S. Tolansky, *An Introduction to Interferometry* (New York: John Wiley & Sons, Inc., 1973).

J. Dyson, *Interferometry as a Measuring Tool* (Brighton, England: The Machinery Publishing Co. Ltd., 1970).

M. J. Beran and G. B. Parrent, Jr., *Theory of Partial Coherence* (Englewood Cliffs, NJ: Prentice-Hall, Inc., 1964).

PROBLEMS

2.2-1. Two sinusoidal waves of equal amplitude are $\frac{1}{4}$ of a wavelength out of phase. What is the amplitude of the resultant wave?

2.2-2. Two harmonic waves, given by $y_1 = 7 \sin(\omega t + \pi/2)$ and $y_2 = 5 \sin(\omega t + \pi/3)$, superimpose on each other. Find the resultant amplitude.

2.2-3. By vector construction add two waves of equal wavelength that have an amplitude ratio of $5:3$ and a (constant) phase difference of $\frac{3}{4}\pi$ between them. What is:
(a) The resultant amplitude?
(b) The phase difference between the resultant and the first of the two contributing waves?

2.2-4. Add two sine waves that have wavelengths of a ratio $2:1$ and amplitudes of a ratio $2:3$. Assume that the waves start in phase.

2.2-5. Two waves, $y_1 = \sin x$ and $y_2 = \frac{1}{2} \sin 2x$, are added to one another. Find the equation of the resultant wave and draw plots of the three waves.

2.2-6. Two waves are given by $y_1 = 2 \sin x$ and $y_2 = \sin(x + \pi/3)$, respectively. Determine the amplitude, the new phase angle, and the equation of the resultant wave.

2.2-7. Monochromatic light passes through a double slit, producing interference. The distance between the slit centers is 1.2 mm and the distance between consecutive fringes on a screen 5 m away is 0.3 cm. What is:
(a) The wavelength?
(b) The color of the light?

2.2-8. Light passes through two narrow slits 0.8 mm apart. If on a screen 80 cm away the distance between the two second-order maxima is 2 mm, what is the wavelength of the light?

2.2-9. Light of 600 nm wavelength passes through a double slit and forms interference fringes on a screen 1.2 m away. If the slits are 0.2 mm apart, what is the distance between the zeroth and:
(a) A third-order maximum?
(b) A third-order minimum?

2.2-10. Two narrow slits 0.3 mm apart are illuminated by monochromatic light. On a screen 1.2 m away, the distance between the two fifth-order minima is found to be 22.78 mm. Determine the wavelength of the light.

2.2-11. If a double-slit experiment is performed *under water* ($n = \frac{4}{3}$), will the distance between fringes change? For example, how far from the axis will be the third-order maximum in Problem 2.2-9?

2.2-12. When one of the slits in Young's experiment is covered by a film of transparent material, the zeroth order is seen to shift by 2.2 fringes. If the refractive index of the material is 1.4 and the wavelength of the light 500 nm, how thick is the film?

2.2-13. A double slit is illuminated by light containing two wavelengths, 450 nm and 600 nm. What is the least order at which a maximum of one wavelength will fall exactly on a minimum of the other?

2.2-14. Two slits are illuminated by light that consists of two wavelengths. One wavelength is known to be 550 nm. On a screen, the fourth minimum of the 550 nm light coincides with the third maximum of the other light. What is the wavelength of the unknown light?

2.2-15. A series of double slits are cut into an opaque screen such that the separation, d, between slits is constant but the spacing between different double slits varies at random (Figure 2.2-22). Describe the resulting interference pattern.

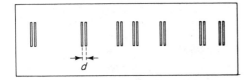

Figure 2.2-22

2.2-16. A satellite circling Earth is transmitting microwaves of 15 cm wavelength. When the satellite is above a ground station which has two antennas 100 m apart and located in the plane of the orbit, a signal is received that fluctuates with a period of $\frac{1}{10}$ s. If the satellite is known to be at an altitude of 400 km and if we neglect the curvature of Earth, what is the satellite's velocity?

2.2-17. Using red cadmium light, $\lambda = 643.8$ nm, Michelson in his original experiment could still see interference fringes after he had moved one of the mirrors 25 cm away from the coincidence position. How many fringes did he count?

2.2-18. How far must one of the mirrors of a Michelson interferometer be moved for 400 fringes of light of 500 nm wavelength to cross the center of the field of view?

2.2-19. As one of the mirrors in a Michelson interferometer is moved, 500 fringes are counted crossing the field of view. If the micrometer reads 5.070 mm at the beginning of the count and 5.245 mm at the end, what is the wavelength of the light?

2.2-20. A Michelson interferometer is used with blue light of 475 nm wavelength and set up for a path difference of 5μm. What is the angular radius of:
(a) The tenth-order ring?
(b) The twentieth-order ring?

2.2-21. If one arm of a Michelson interferometer contains a tube 2.5 cm long which is first evacuated and then slowly filled with air ($n = 1.0003$), how many fringes will cross the center? Assume light of 600 nm wavelength.

2.2-22. A Jamin interferometer has two test chambers placed side by side, each 30 cm long. One of these cells is left evacuated while the other is gradually filled with a certain gas. If, with light of 500 nm wavelength, 240 fringes are seen to cross the field, what is the refractive index of the gas?

2.2-23. The crystalline lens of the human eye has in its center a refractive index of 1.41 while near the periphery the index is 1.38. If a slice $\frac{1}{10}$ mm thick is cut out of the lens and examined in 600 nm light in a Mach–Zehnder interferometer, what is the distortion of the fringes?

2.2-24. A brine solution is placed in a cuvette 4 mm wide. When examined in light of 500 nm wavelength in an interferometer, the fringe pattern is distorted, from top to bottom, by 16 fringes. If the average refractive index of the solution is 1.462 and the variation is linear, what are the two refractive indices, at the top and bottom of the liquid?

2.2-25. The *Rayleigh interferometer* is derived from the double-slit design. The two bundles of light pass through two tubes that are placed side by side and are filled with different gases, or with the same gas at different pressures. If the tubes are 10 cm long, if one tube contains air at atmospheric pressure ($n = 1.0003$) and the other tube is evacuated and then air is slowly admitted to it, how many fringes will cross the field? Use $\lambda = 600$ nm.

2.2-26. Continue with Problem 2.2-25 and assume now that both tubes are filled with air and that the air in one tube is gradually replaced by a gas of unknown, but higher, refractive index. If, while this is done, 40 fringes cross the center of the field, what is the refractive index of the gas?

2.2-27. If light of $\lambda = 660$ nm has wavetrains 20λ long, what is its:
(a) Coherence length?
(b) Coherence time?

2.2-28. Determine the number of waves per wavetrain in light from:
(a) The green mercury line (546 nm).

(b) The orange krypton line (606 nm).
(c) A helium–neon laser (633 nm), assuming that its coherence length is 20 km?

2.2-29. If the contrast in an interference pattern is 50 percent, and if the maxima receive 15 units of light, how much do the minima receive?

2.2-30. Two wavetrains overlap to 29 percent of their length. If the maxima in the resulting interference pattern receive 20 units of light, how much do the minima receive?

2.3

Diffraction

WHEN LIGHT PASSES THROUGH A NARROW SLIT, it spreads out more than what could be accounted for by geometric construction. This is an example of *diffraction*. Diffraction can be defined as any departure from *the predictions of geometric optics*.

Introduction

Consider a point source of quasimonochromatic light and an obstacle placed halfway between the source and the screen. Following the rules of geometric optics, we expect to see a well-defined, distinct shadow (Figure 2.3-1, left). Careful examination, however, shows that the edges of the

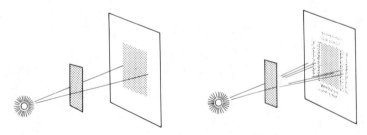

Figure 2.3-1 Shadow in geometric optics (*left*) and in wave optics (*right*).

shadow are not sharp. Inside the shadow the amount of light gradually decreases. Outside the shadow it increases, forming alternately brighter and darker fringes (right).*

We distinguish two major classes, *Fraunhofer diffraction* and *Fresnel diffraction*. Fraunhofer diffraction refers to parallel, collimated light ("far-field diffraction"). In Fresnel diffraction, the light need not be parallel ("near-field diffraction"). Fresnel diffraction is more general; it includes Fraunhofer diffraction as a special case. But Fraunhofer diffraction is so much easier to discuss that it is customarily presented first.

Fraunhofer Diffraction

Single slit.

We begin with *Fraunhofer diffraction*[†] on a single slit of width s (Figure 2.3-2). The light is collimated as the waves enter the

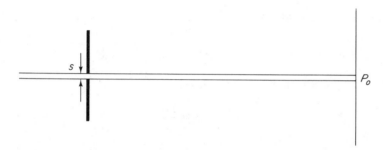

Figure 2.3-2   Central maximum in Fraunhofer diffraction.

---

*These fringes were first seen by Francesco Maria Grimaldi (1618–1663), Italian mathematician and teacher of rhetoric at the College of the Jesuit Order in Bologna. Grimaldi let sunlight enter a dark room through a hole in a curtain and found that the shadow cast by an object was surrounded by colored fringes. In his book *Physico-mathesis de Lumine, Coloribus et Iride* (*The Science of Physics of Light, Colors, and Vision*), published by a friend two years after his death, Grimaldi wrote: "Lumen propagatur seu diffunditur non solum directe, refracte ac reflexe, sed etiam quodam quarto modo diffracte" (Light propagates and diffuses not only directly or by refraction or reflection, but also, in a certain fourth manner, by diffraction).

[†]Joseph Fraunhofer (1787–1826), German. After working for a while as a lens grinder and apprentice optician, Fraunhofer became a partner in an optical company that made precision theodolites, professor at the University of Munich, and was knighted by King Maximilian of Bavaria. In his short life (he died of tuberculosis at age 39), Fraunhofer produced large-aperture telescope lenses, exceptionally well corrected for spherical and chromatic aberration, ruled precision gratings and discovered their use for spectroscopy, and found that the spectrum of the sun is crossed by dark lines since named *Fraunhofer lines*. His theory of diffraction appeared first in J. Fraunhofer, "Kurzer Bericht von den Resultaten neuerer Versuche über die Gesetze des Lichtes, und die Theorie derselben," *Ann. Physik* (3) **14** (1823), 337–78.

slit; the waves are plane and in phase (uniphase). On a distant screen, at point P_0, the waves are still in phase and form the *zeroth-order maximum.**

At certain angles, when point P moves off axis, there will be diffraction maxima and minima. Consider the *minima* first. Assume, as shown in Figure 2.3-3, that the light passing through the slit contains three rays, 1–2–3. After diffraction, the light subtends with the axis an angle θ. For a minimum to occur, this angle must be of a size such that rays 1 and 2 are $\lambda/2$ out of phase and, hence, cancel each other. Generally, though, *any* ray that passes through the *upper* half of the slit will cancel a corresponding ray, a distance $s/2$ away, that passes through the *lower* half; thus no light will reach point P.

The first minimum, therefore, will be where the path difference, Γ, between rays 1 and 2 is

$$\Gamma_{12} = \tfrac{1}{2}\lambda$$

which is the same as saying that the path difference between rays 1 and 3 is

$$\Gamma_{13} = \lambda \qquad\qquad [2.3\text{-}1]$$

From the construction it follows that

$$\sin\theta = \frac{\Gamma_{13}}{s} \qquad\qquad [2.3\text{-}2]$$

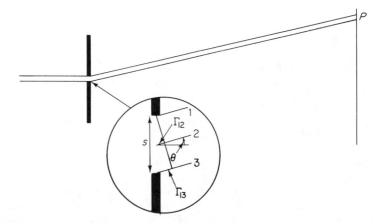

Figure 2.3-3 Fraunhofer diffraction, first minimum. Insert shows magnified view of path differences inside aperture.

---

*In Fraunhofer diffraction the distance from the source to the slit, and from the slit to the screen, is large, and the light parallel. This is easily accomplished by use of a laser.

and, combining Equations [2.3-1] and [2.3-2], we have

$$s \sin \theta = \lambda$$

But there will be other minima (whenever Γ_{13} is an integral multiple of λ); thus

$$\boxed{s \sin \theta = m\lambda} \qquad m = 1, 2, 3, \ldots \qquad [2.3\text{-}3]$$

which is the equation for *minima in Fraunhofer diffraction on a single slit*.

This equation shows, first, that if the slit is made narrower, the angle θ becomes larger and the light spreads out more. Second, as the wavelength increases, θ increases too: *red light is diffracted more than blue light,* the opposite of *re*fraction.

Vector construction.
Now we divide the width of the slit into several narrow strips of equal width Δs. Light passing through each of these strips has an amplitude $\mathbf{A}_i$, and each can be represented by a short phasor (Figure 2.3-4). Since the light is parallel and in phase, these phasors have the same length and direction.

At point P_0, the elemental waves arrive in phase. This is where the resultant amplitude, $\mathbf{A}$, has its *maximum*—the arithmetic sum of the phasors (Figure 2.3-5):

$$\mathbf{A} = \sum \mathbf{A}_i \qquad [2.3\text{-}4]$$

For other points P on the screen, waves passing through different Δs's will be out of phase. For example, at the first minimum there is a phase difference of 180° between the first and the central phasor. This means that if we have 12 phasors, any two adjacent phasors differ by

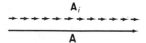

Figure 2.3-5 Vector representation of amplitudes in the zeroth-order maximum.

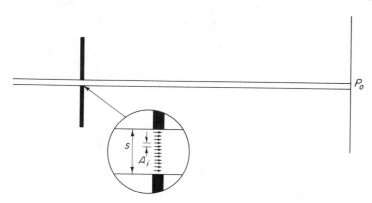

Figure 2.3-4 Fraunhofer diffraction by a single slit. Insert shows arrows inside aperture; these represent phasors.

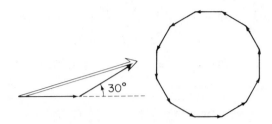

Figure 2.3-6 Vector addition of phasors, differing by 30° (*left*). Vibration curve contains 12 such phasors (*right*).

$360°/12 = 30°$; the *vibration curve* formed by these phasors is a (closed) circle and the resultant is *zero* (Figure 2.3-6).

Next assume that point P is located somewhere *between* the zeroth-order maximum and the first-order minimum, and that the phase difference between rays 1 and 3 is only 180°. The vibration curve will then be a *half-circle*. The length of the *arc* is always $\Sigma\, A_i$. But the resultant amplitude is the *chord*, the diameter of the half-circle,

$$A = 2\frac{\Sigma\, A_i}{\pi} \qquad\qquad [2.3\text{-}5]$$

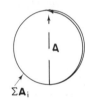

Figure 2.3-7 Vibration curve at first-order maximum.

Now consider that point P is *outside* the first-order minimum. The combined phasors then make more than one revolution. At the *first-order maximum*, for example, the vibration curve will have turned through (approximately) $1\frac{1}{2}$ revolutions (Figure 2.3-7). The circumference of the circle, therefore, is $\frac{2}{3}\,\Sigma\, A_i$ and

$$A \approx \frac{2}{3}\frac{\Sigma\, A_i}{\pi} \qquad\qquad [2.3\text{-}6]$$

At the next (second-order) minimum we again have $A = 0$, and at the next (second-order) maximum

$$A = \;\approx \frac{2}{5}\frac{\Sigma\, A_i}{\pi} \qquad\qquad [2.3\text{-}7]$$

and so on.

We call δ the phase angle difference between peripheral rays 1 and 3. But δ, as in Figure 2.3-8, is also the angle subtended by the extensions of the first and the last phasors. From the construction,

$$\sin \tfrac{1}{2}\delta = \frac{A/2}{R}$$

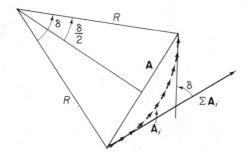

Figure 2.3-8 Vector construction used for finding intensity in single-slit diffraction.

or

$$\mathbf{A} = 2R \sin \tfrac{1}{2}\delta \qquad\qquad [2.3\text{-}8]$$

and furthermore,

$$\delta = \frac{\Sigma \, \mathbf{A}_i}{R} \qquad\qquad [2.3\text{-}9]$$

Solving Equation [2.3-9] for R and inserting it into [2.3-8] gives

$$\mathbf{A} = 2\frac{\Sigma \, \mathbf{A}_i}{\delta} \sin\frac{\delta}{2} = \Sigma \, \mathbf{A}_i \frac{\sin \delta/2}{\delta/2} \qquad\qquad [2.3\text{-}10]$$

Since, from Equation [2.1-12], the "intensity" is proportional to the square of the amplitude, the intensity at point P_0, at an angular distance θ from the axis, is

$$I_\theta = I_0 \left(\frac{\sin \delta/2}{\delta/2}\right)^2 \qquad\qquad [2.3\text{-}11]$$

where I_0 is the intensity in the zeroth-order maximum. This maximum is the central peak in Figure 2.3-9. At certain angular distances (from the axis), the amplitudes, and the intensities, fall to zero. This will occur whenever

$$\tfrac{1}{2}\delta = \pi,\ 2\pi,\ 3\pi,\ \ldots \qquad\qquad [2.3\text{-}12]$$

The minima, therefore, are equidistant, at least for small angles. The maxima, however, are not.

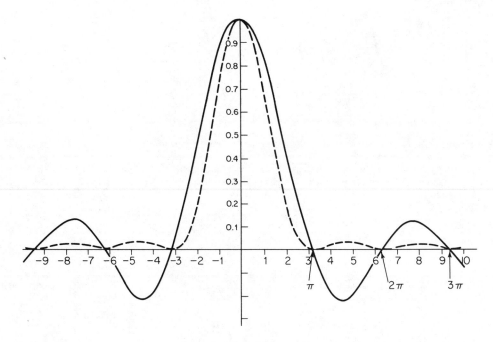

Figure 2.3-9 Amplitude distribution (solid line) and intensity distribution (dashed line) in Fraunhofer diffraction by a single slit.

Diffraction maxima. In contrast to double-slit interference, the *maxima* do not fall exactly halfway between the minima; they are slightly displaced toward the center. Their locations are found by differentiating Equation [2.3-11] with respect to δ and setting it equal to zero:

$$\frac{d}{d\delta}\left[I_0\left(\frac{\sin \delta/2}{\delta/2}\right)^2\right] = 2I_0\left(\frac{\sin \delta/2}{\delta/2}\right)\left[-\frac{\sin \delta/2}{(\delta/2)^2} + \frac{\cos \delta/2}{\delta/2}\right] = 0$$

[2.3-13]

Maxima, therefore, will occur whenever

$$\tan\left(\frac{\delta}{2}\right) = \frac{\delta}{2}$$

[2.3-14]

This equation is easiest to solve by graphical construction. Draw two plots, one of $y = \tan \delta/2$ and the other of $y = \delta/2$. The maxima are located where the two plots intersect. For the first four maxima we find that $y = 0$, 1.4303π, 2.4590π, and 3.4707π. The higher orders are even more nearly halfway between the minima.

The *intensities* in the maxima can be calculated to a good approximation by determining $(\sin^2 \delta/2)/(\delta/2)$ at the halfway positions, that is, where

$$\frac{\delta}{2} = \frac{3\pi}{2}, \frac{5\pi}{2}, \frac{7\pi}{2}, \cdots \qquad [2.3\text{-}15]$$

This gives for the first maximum $4/(9\pi^2)$ or approximately 4.5 percent, for the second maximum $4/(25\pi^2)$ or 1.6 percent, for the third maximum $4/(49\pi^2)$ or 0.8 percent, and so on, of the intensity in the zeroth-order maximum.

Circular aperture. Fraunhofer diffraction by a circular aperture is of considerable practical interest because most lenses and stops are round. The result is again a series of maxima and minima; but behind a circular aperture these maxima and minima take the form of concentric rings. The bright central maximum is known as *Airy's disk.* * The pattern looks similar to that formed behind a slit but the dimensions are different, as shown in Figure 2.3-10.

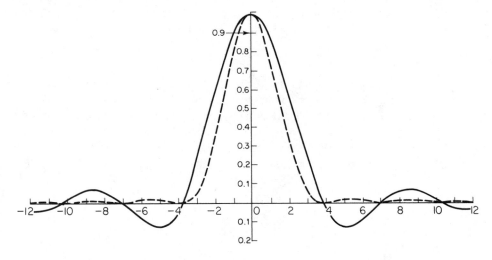

Figure 2.3-10 Amplitude and intensity distributions in the diffraction pattern behind a circular aperture. Note similar shape but different dimensions as those of Figure 2.3-9.

---

*Sir George Biddell Airy (1801–1892), British mathematician and Astronomer Royal. Airy is perhaps best known for the disk just mentioned. He described it in G. B. Airy, "On the Diffraction of an Object-glass with Circular Aperture," *Trans. Cambridge Phil. Soc.* **5** (1835), 283–91.

Table 2.3-1 POSITIONS OF MINIMA
IN FRAUNHOFER DIFFRACTION

Minimum	Single slit $m =$	Circular aperture $J =$
First-order	1	1.220
Second-order	2	2.233
Third-order	3	3.238
Fourth-order	4	4.241
Fifth-order	5	5.243

The mathematical analysis of diffraction behind a circular aperture is much more difficult than diffraction behind a slit. As before, the aperture is divided into a series of narrow strips of equal width. But since these strips are not of equal length, the amplitudes are unequal too. The resultant amplitude is found by integration.

The dimensions, as mentioned, are different. Whereas, for example, the positions of the *minima* behind a slit are given by the simple relationship $s \sin \theta = m\lambda$, we now have to replace m by another factor, J, that comes from *Bessel functions of the first kind*. Table 2.3-1 shows a comparison. For the first *maximum* (behind a circular aperture) set $J = 1.63$.

Rayleigh's criterion.
Diffraction limits the resolvance (resolving power) of an optical system. For example, consider two nearby stars. Only when the diffraction patterns of these two stars are separate will the stars appear separate. When the central maxima fuse, the two stars appear as one. When the central maximum of one star coincides with the first minimum of the other, resolution is marginal, a condition called *Rayleigh's criterion* (Figure 2.3-11).

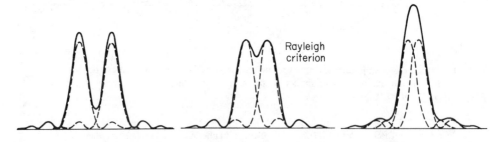

Figure 2.3-11 Rayleigh's criterion. Note the separation of the maxima in the left-hand plot and the close overlap on the right.

Therefore, from Table 2.3-1, the *minimum angle of resolution* provided by a lens of diameter D, at a wavelength λ, is

$$\theta_{min} \approx 1.22\frac{\lambda}{D} \qquad\qquad [2.3\text{-}16]$$

The lens, even if it could be fully corrected for all aberrations, is still *diffraction-limited*.

Rayleigh's criterion has been criticized for several reasons. Some people claim that it is overgenerous; under favorable conditions details can be resolved that are slightly smaller in size. Others believe in Rayleigh's criterion as if it were written in tablets of gold by the angels.* Rayleigh himself considered it to be no more than an approximation.

Example. *A diffraction-limited telescope with a 7.6-cm aperture is aimed at a target 12.5 km away. Assuming light of 500 nm wavelength and neglecting air turbulence, what size details can be resolved by the telescope?*

Solution. Using Rayleigh's criterion, we first determine the *angular resolution:*

$$\theta_{min} \approx 1.22\frac{\lambda}{D} = (1.22)\frac{500 \times 10^{-9}}{0.076} = 8 \times 10^{-6} \text{ rad}$$

At a distance of 12.5 km, this means a *linear* resolution of

$$d = (12.5 \times 10^3)(8 \times 10^{-6}) = \boxed{10 \text{ cm}}$$

Superresolution. Diffraction is the limit to all resolution. Or is it? Several approaches have been tried to overcome this limit. For example, we could superimpose *color* on a (black-and-white) object. Then the intensity of the light emitted from any one point in the object depends not only on the location but is also a function of wavelength; spatial information is represented, and superseded, by color information. The interesting result is that now images can be transmitted through fog, or other scattering media, at a resolution *higher* than otherwise possible.[†]

---

*As quoted from F. H. Perrin, "Methods of Appraising Photographic Systems," *J. Soc. Motion Pict. Telev. Eng.* **69** (1960), 152.

[†]A. I. Kartashev, "Optical Systems with Enhanced Resolving Power," *Optics and Spectroscopy* **9** (1960), 204–6.

Fresnel Diffraction

Fresnel integrals.   *Fresnel diffraction** is not restricted to parallel light. Consider the arrangement shown in Figure 2.3-12. A horizontal slit source, *S*, emits cylindrical wavefronts. Wavefront *W* has just reached the center of a slit, oriented parallel to the source and placed at distance *a* from the source and at distance *b* from a screen. As before, we divide the wavefront into a series of narrow strips, each strip of width *dW* and each parallel to the edges of the slit.

For reasons that will soon become apparent, it is practical to choose the width of these strips so that they correspond to path lengths that increase by one-half of a wavelength, $\frac{1}{2}\lambda$, from strip to strip. Thus, while the distance along the axis, from the center of the slit to P_0, is *b*, the distance from the next higher element is $b + \frac{1}{2}\lambda$, from the next higher, second element $b + 2(\frac{1}{2}\lambda)$, from the third element $b + 3(\frac{1}{2}\lambda)$, and so on. The wavefront elements defined this way are known as *Fresnel half-period*

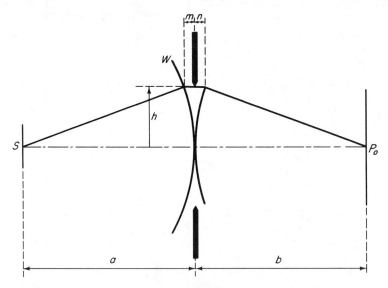

Figure 2.3-12   Fresnel diffraction by a slit. (Vertical dimensions exaggerated.)

---

*Augustin Jean Fresnel (1788–1827), French physicist. Fresnel studied mathematics, then civil engineering, at the French École des Ponts et Chaussées (School of Bridges and Roads); he went into optics later. In his dissertation, "Sur la Diffraction de la lumière, où l'on examine particulièrement le phénomène des franges colorées que présentent les ombres des corps éclairés par un point lumineux," *Ann. Chim. et Phys.* (2) **1** (1816), 239–81, Fresnel presented the first rigorous treatment of diffraction. He also worked on more mundane problems, invented flat, weight-saving lenses for use in lighthouses.

zones. These zones are discussed in more detail when we come to circular apertures and *zone plates* (page 236).

Consider first the disturbance, *dy*, caused by the element *dW* which is closest to the edge, acting on point P_0 on the screen. From Equation [2.1-34], this disturbance is

$$dy = A \sin(2\pi\nu t)\, dW \qquad\qquad [2.3\text{-}17]$$

Since $\nu = 1/T$, this is equal to

$$dy = A \sin\left(2\pi\frac{t}{T}\right) dW \qquad\qquad [2.3\text{-}18]$$

At another point *P* on the screen, some distance away from P_0, the disturbance is

$$dy = A \sin\left[2\pi\left(\frac{t}{T} - \frac{b}{\lambda}\right)\right] dW \qquad\qquad [2.3\text{-}19]$$

The disturbance due to some *other* element *dW*, farther away from the edge, is

$$dy = A \sin\left[2\pi\left(\frac{t}{T} - \frac{b}{\lambda}\right) + \delta\right] dW \qquad\qquad [2.3\text{-}20]$$

where δ is the phase difference between the light from these two elements.

The *total* disturbance at *P*, due to all elements *dW*, is found by integration:

$$y = A \int \sin\left[2\pi\left(\frac{t}{T} - \frac{b}{\lambda}\right) + \delta\right] dW \qquad\qquad [2.3\text{-}21]$$

If we use the identity $\sin(\alpha + \beta) = \sin\alpha\cos\beta + \cos\alpha\sin\beta$, as we did in Equation [2.2-4], we can write Equation [2.3-21] as

$$y = A \sin\left[2\pi\left(\frac{t}{T} - \frac{b}{\lambda}\right)\right] \int \cos\delta\, dW + A \cos\left[2\pi\left(\frac{t}{T} - \frac{b}{\lambda}\right)\right] \int \sin\delta\, dW$$

$$[2.3\text{-}22]$$

We then set

$$\mathbf{A}\cos\theta = A \int \cos\delta\, dW$$

and   [2.3-23]

$$\mathbf{A}\sin\theta = A \int \sin\delta\, dW$$

so that Equation [2.3-22] becomes

$$y = \mathbf{A} \sin\left[2\pi\left(\frac{t}{T} - \frac{b}{\lambda}\right) - \theta\right] \qquad [2.3\text{-}24]$$

This equation represents the disturbance caused by all contributions, at any point, P.

Before we proceed to a determination of the intensity, consider the *path difference*, Γ, and the *phase difference*, δ, between rays (from different wavefront elements). First, we find from Figure 2.3-13, and from the sagitta formula derived on page 63, that

$$2am = h^2 \qquad \text{and} \qquad 2bn = h^2 \qquad [2.3\text{-}25]$$

The path difference, then, is the sum of $m + n$;

$$\Gamma = \frac{h^2}{2a} + \frac{h^2}{2b} = h^2\left(\frac{a + b}{2ab}\right) \qquad [2.3\text{-}26]$$

Path difference and phase difference are connected as

$$\frac{\Gamma}{\lambda} = \frac{\delta}{2\pi} \qquad [2.2\text{-}38]$$

Solving for δ and substituting Equation [2.3-26] gives

$$\delta = 2\pi\frac{\Gamma}{\lambda} = \pi h^2\left(\frac{a + b}{ab\lambda}\right) \qquad [2.3\text{-}27]$$

In order to become independent of the specific values of a, b, and λ (which may vary from experiment to experiment), we introduce a variable, v. This variable is part of the vibration curve, representing the amplitude vector (phasor) of an element of light passing through one of the half-period zones. Numerically, v is defined as

$$\tfrac{1}{2}\pi v^2 = \delta \qquad [2.3\text{-}28]$$

so that by substitution in Equation [2.3-27],

$$\pi h^2\left(\frac{a + b}{ab\lambda}\right) = \tfrac{1}{2}\pi v^2 \qquad [2.3\text{-}29]$$

or

$$v = h\sqrt{\frac{2(a + b)}{ab\lambda}} \qquad [2.3\text{-}30]$$

The direction of v is given by the angle, δ, it subtends with the $+x$ axis (Figure 2.3-13). But v is only a *short length* on what may be a long and complicated curve and, therefore,

$$\sin\delta = \frac{dy}{dv} \qquad \text{and} \qquad \cos\delta = \frac{dx}{dv} \qquad [2.3\text{-}31]$$

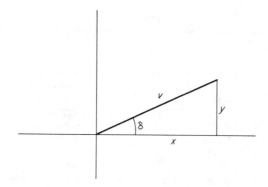

Figure 2.3-13   Part of a vibration curve.

Solving for dx and dy, respectively, and substituting Equation [2.3-28] gives

$$dx = \cos \delta \, dv = \cos \tfrac{1}{2} \pi v^2 \, dv$$

and   [2.3-32]

$$dy = \sin \delta \, dv = \sin \tfrac{1}{2} \pi v^2 \, dv$$

Now, in order to obtain the *intensity*, we square, and add, Equations [2.3-23]. This gives

$$I \propto \mathbf{A}^2 = \left[\int \cos \delta \, dW \right]^2 + \left[\int \sin \delta \, dW \right]^2 \qquad [2.3\text{-}33]$$

We can then substitute Equation [2.3-27] in [2.3-33]:

$$I = \left[\int \cos\left(\pi h^2 \frac{a+b}{ab\lambda} \right) dW \right]^2 + \left[\int \sin\left(\pi h^2 \frac{a+b}{ab\lambda} \right) dW \right]^2 \qquad [2.3\text{-}34]$$

and, using Equation [2.3-29], take the $ab\lambda$ term out in front of the integral signs. Equation [2.3-34] then reduces to

$$I = \tfrac{1}{2}\left(\frac{ab\lambda}{a+b} \right)\left[\left(\int \cos \tfrac{1}{2} \pi v^2 \, dv \right)^2 + \left(\int \sin \tfrac{1}{2} \pi v^2 \, dv \right)^2 \right] \qquad [2.3\text{-}35]$$

The two integrals that occur in this equation are known as *Fresnel's integrals*:

$$\boxed{\begin{aligned} x &= \int \cos \tfrac{1}{2} \pi v^2 \, dv \\[2mm] y &= \int \sin \tfrac{1}{2} \pi v^2 \, dv \end{aligned}} \qquad [2.3\text{-}36]$$

Table 2.3-2 FRESNEL INTEGRALS

v	x	y	v	x	y
0.00	0.0000	0.0000	4.50	0.5261	0.4342
0.10	0.1000	0.0005	4.60	0.5673	0.5162
0.20	0.1999	0.0042	4.70	0.4914	0.5672
0.30	0.2994	0.0141	4.80	0.4338	0.4968
0.40	0.3975	0.0334	4.90	0.5002	0.4350
0.50	0.4923	0.0647	5.00	0.5637	0.4992
0.60	0.5811	0.1105	5.05	0.5450	0.5442
0.70	0.6597	0.1721	5.10	0.4998	0.5624
0.80	0.7230	0.2493	5.15	0.4553	0.5427
0.90	0.7648	0.3398	5.20	0.4389	0.4969
1.00	0.7799	0.4383	5.25	0.4610	0.4536
1.10	0.7638	0.5365	5.30	0.5078	0.4405
1.20	0.7154	0.6234	5.35	0.5490	0.4662
1.30	0.6386	0.6863	5.40	0.5573	0.5140
1.40	0.5431	0.7135	5.45	0.5269	0.5519
1.50	0.4453	0.6975	5.50	0.4784	0.5537
1.60	0.3655	0.6389	5.55	0.4456	0.5181
1.70	0.3238	0.5492	5.60	0.4517	0.4700
1.80	0.3336	0.4508	5.65	0.4926	0.4441
1.90	0.3944	0.3734	5.70	0.5385	0.4595
2.00	0.4882	0.3434	5.75	0.5551	0.5049
2.10	0.5815	0.3743	5.80	0.5298	0.5461
2.20	0.6363	0.4557	5.85	0.4819	0.5513
2.30	0.6266	0.5531	5.90	0.4486	0.5163
2.40	0.5550	0.6197	5.95	0.4566	0.4688
2.50	0.4574	0.6192	6.00	0.4995	0.4470
2.60	0.3890	0.5500	6.05	0.5424	0.4689
2.70	0.3925	0.4529	6.10	0.5495	0.5165
2.80	0.4675	0.3915	6.15	0.5146	0.5496
2.90	0.5624	0.4101	6.20	0.4676	0.5398
3.00	0.6058	0.4963	6.25	0.4493	0.4954
3.10	0.5616	0.5818	6.30	0.4760	0.4555
3.20	0.4664	0.5933	6.35	0.5240	0.4560
3.30	0.4058	0.5192	6.40	0.5496	0.4965
3.40	0.4385	0.4296	6.45	0.5292	0.5398
3.50	0.5326	0.4152	6.50	0.4816	0.5454
3.60	0.5880	0.4923	6.55	0.4520	0.5078
3.70	0.5420	0.5750	6.60	0.4690	0.4631
3.80	0.4481	0.5656	6.65	0.5161	0.4549
3.90	0.4223	0.4752	6.70	0.5467	0.4915
4.00	0.4984	0.4204	6.75	0.5302	0.5362
4.10	0.5738	0.4758	6.80	0.4831	0.5436
4.20	0.5418	0.5633	6.85	0.4539	0.5060
4.30	0.4494	0.5540	6.90	0.4732	0.4624
4.40	0.4383	0.4622	6.95	0.5207	0.4591

If these integrals are solved for certain values, and between certain limits, of v, they yield the resultant of the disturbances produced by corresponding elements of the wavefront. In short, we can predict the amount of light reaching the screen.

Cornu spiral.   If Fresnel's integrals are solved between a lower limit of $v_1 = 0$ and an upper limit of $v_2 = \infty$,

$$x = \int_0^\infty \cos \tfrac{1}{2}\pi v^2 \, dv$$

[2.3-37]

$$y = \int_0^\infty \sin \tfrac{1}{2}\pi v^2 \, dv$$

then x and y assume the values listed in Table 2.3-2. If we now plot x versus y, we obtain a curve known as *Cornu's spiral** (Figure 2.3-14). This spiral provides us with a most elegant means for the graphical solution of diffraction problems.

The magnitude of v is the length measured *along the spiral.* Each short fraction of v, dv, represents the amplitude of the light in one wavefront element, dW. In Figure 2.3-13 we saw that the direction of dv is given by the tangent of the angle δ it makes with the $+x$ axis,

$$\tan \delta = \frac{dy}{dx}$$

[2.3-38]

But this angle, $\delta = \tfrac{1}{2}\pi v^2$, Equation [2.3-28], is proportional to the *square* of v. Therefore, as we move away from the origin and v *increases,* the *steepness of the curve increases even faster*—which accounts for the spiral shape.

At the origin, where $v = 0$, the curve is parallel to the x axis. At $v = 1$, it is parallel to the y axis. At $v = \sqrt{2}$, and then at $v = 2$, the curve is again parallel to the x axis, and at $v = 3$, not shown, again parallel to the y axis.

Of particular interest are the two endpoints, the *eyes* of the spiral. At these points $v \to \pm\infty$. Using the identity

---

*Marie Alfred Cornu (1841–1902), French physicist, professor of experimental physics at the École Polytechnique in Paris. While working earlier at the École des Mines, Cornu became interested in optics, read Felix Billet's book *Traité d'optique,* repeated all the experiments described therein. He determined the speed of light by Fizeau's method and made precise measurements of the wavelengths of certain lines in the spectrum of hydrogen. He is probably best known for the spiral he described in A. Cornu, "Méthode nouvelle pour la discussion des problèmes de diffraction dans le cas d'une onde cylindrique," *J. de Physique* **3** (1874), 5–15, 44–52.

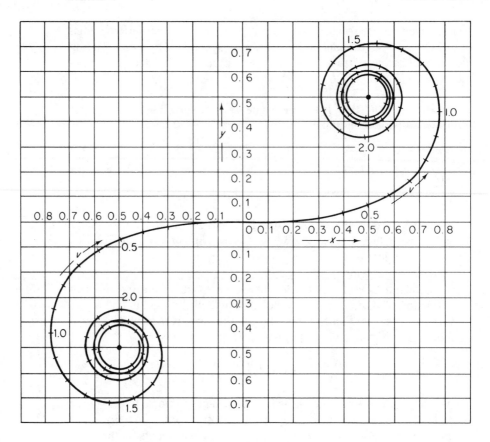

Figure 2.3-14 Cornu spiral.

$$\int_0^\infty \sin ax^2 \, dx = \int_0^\infty \cos ax^2 \, dx = \frac{1}{2}\sqrt{\frac{\frac{1}{2}\pi}{a}} \qquad [2.3\text{-}39]$$

we substitute $\frac{1}{2}\pi$ for a, and v for x, and obtain

$$\int_0^\infty \sin \tfrac{1}{2}\pi v^2 \, dv = \int_0^\infty \cos \tfrac{1}{2}\pi v^2 \, dv = \frac{1}{2}\sqrt{\frac{\frac{1}{2}\pi}{\frac{1}{2}\pi}} = \frac{1}{2} \qquad [2.3\text{-}40]$$

which means that the upper eye of the spiral has the coordinates

$$x = y = (+0.5, \, +0.5) \qquad\qquad [2.3\text{-}41]$$

and the lower eye,

$$x = y = (-0.5, \, -0.5) \qquad\qquad [2.3\text{-}42]$$

As with any vibration curve, the amplitude due to a wavefront element, dW, corresponds to an element of length, dv, *on* the curve. But the total amplitude corresponds to the shortest distance *between* points; it corresponds to the *chord*. The square of this chord represents the intensity.

Note that with Cornu's construction the numerical value of the intensity of an unobstructed wave is *two*, rather than unity. Since both eyes have the same coordinates, except for the signs, the distance from the origin to either eye is

$$A = \sqrt{0.5^2 + 0.5^2} = \frac{1}{\sqrt{2}}$$

The distance between both eyes, therefore, is twice that and the intensity is the square of it:

$$I_0 = \left(\frac{2}{\sqrt{2}}\right)^2 = 2 \qquad\qquad [2.3\text{-}43]$$

which is why the factor $\frac{1}{2}$ appears in Equation [2.3-35].

Now we apply Cornu's spiral to *diffraction at a single slit*. Consider first the central point P_0. This point lies on a straight line extending from the source through the center of the slit. The slit has a width s. This is twice the height, h, that we used earlier and, therefore, from Equation [2.3-30],

$$v_0 = (\tfrac{1}{2}s)\sqrt{\frac{2(a + b)}{ab\lambda}} \qquad\qquad [2.3\text{-}44]$$

where $-v$ and $+v$, respectively, are the limits of the Fresnel integrals.

Now imagine that the slit is gradually opened. Since s is proportional to v, the two points $-v$ and $+v$ move farther into the spirals. In Figure 2.3-15 we assume that the two points are initially 5 units apart (line 1). But at a certain width (of the slit), the distance between the two points (the chord) becomes line 2, that is, it reaches a *maximum*. This occurs when the slit is opened so that the two central half-period zones contribute. As the slit is opened farther, the length v (measured along the spiral) increases but the chord becomes shorter, until it reaches a *minimum* (not shown). In other words, as an aperture is gradually opened, the center of the fringe system alternates between maxima and minima.

Next consider the amount of light reaching a point P, some distance q down from P_0. Remember that in deriving the Fresnel integrals the construction in Figure 2.3-16, top, was used to show that Γ is the path difference introduced by taking the indirect path $c + d$ instead of the direct path $a + b$. This path difference, from Equation [2.3-26], therefore, is

$$\Gamma = (c + d) - (a + b) = h^2\left(\frac{a + b}{2ab}\right) \qquad\qquad [2.3\text{-}45]$$

But if point P is off center, as shown in Figure 2.3-16, bottom, then an additional term, k, must be added:

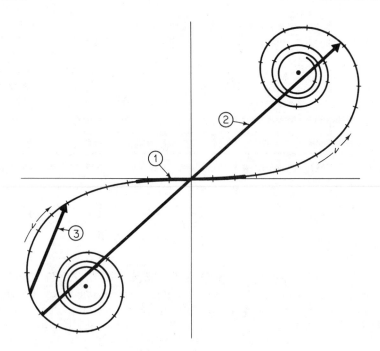

Figure 2.3-15 Cornu's spiral and its application to diffraction at a single slit.

$$k = q\left(\frac{a}{a+b}\right)$$ [2.3-46]

In contrast to Equation [2.3-44], the limits of the Fresnel integrals are now

$$v_1 = \left(\frac{s}{2} + k\right)\sqrt{\frac{2(a+b)}{ab\lambda}}$$

and [2.3-47]

$$v_2 = \left(-\frac{s}{2} + k\right)\sqrt{\frac{2(a+b)}{ab\lambda}}$$

The midpoint between v_1 and v_2 is the average of both:

$$\bar{v} = \frac{v_1 + v_2}{2}$$ [2.3-48]

We now solve Equation [2.3-46] for q, set $k = h$, and substitute for h Equation [2.3-30]. This gives us the distance q of the point on the screen from P_0:

$$q = \left(\frac{a+b}{a}\right)\bar{v}\sqrt{\frac{ab\lambda}{2(a+b)}}$$ [2.3-49]

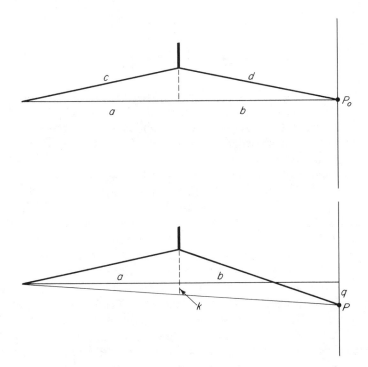

Figure 2.3-16 Geometry of different ray paths reaching a screen.

We now find v_0. This gives us the length *on* the spiral, $v = 2v_0$. Then read the corresponding length of the chord off the curve with a ruler, using either the x axis or the y axis as a scale. The square of this length is the intensity.

Example. *A single slit, 0.25 mm wide, is placed halfway between a source and a screen which are 4 m apart. Assuming a wavelength of 500 nm, determine the amplitude and the intensity at the center point, P_0, on the screen.*

Solution. From Equation [2.3-30] and setting $h = 0.25$ mm and $a = b = 2$ m, we find the length segment Δv on the spiral to be

$$v = h\sqrt{\frac{2(a + b)}{ab\lambda}} = 0.25 \times 10^{-3}\sqrt{\frac{2(2 + 2)}{(2)(2)(500 \times 10^{-9})}} = 0.5$$

If the center part of the spiral can be considered linear, then both the curve and the chord coincide, and extend from -0.25 to $+0.25$; therefore,

$$\mathbf{A} \approx v \approx \boxed{0.5}$$

and

$$I \propto \mathbf{A}^2 \approx \boxed{0.25}$$

We now turn to Fresnel diffraction by a *half-plane* or *knife edge*. Again we consider first point P_0, located at the geometric limit of the shadow (Figure 2.3-17). As before, the amplitude at P_0 is proportional to the length of the chord which extends from the lower eye of the spiral to the origin, as shown.

As point P moves *into* the shadow, the chord on the spiral retracts (Figure 2.3-18). The lower end of the chord remains anchored to the lower eye, but the upper end retraces the path of the spiral downward, becoming gradually shorter.

Conversely, as P moves *outside* the shadow, the upper end of the chord moves into the upper branch of the spiral (Figure 2.3-19), while its lower end remains anchored to the lower eye. But the *length of the chord undergoes periodic variations,* passing through a series of maxima and minima, rather than changing monotonically as before.

At certain points the amplitude is even higher than the amplitude with no obstacle in the path. In short: *Inside* the shadow the intensity drops off gradually, and *outside* the shadow there are fringes (Figure 2.3-20). At the edge of the shadow, only one-half of the wave contributes; thus the amplitude has fallen to one-half, and the intensity to one-fourth, of the unobstructed wave.

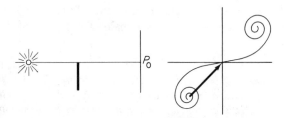

Figure 2.3-17   Fresnel diffraction at a knife edge (knife edge is heavy line in left-hand diagram); Cornu spiral (*right*) shows amplitude of light at point P_0.

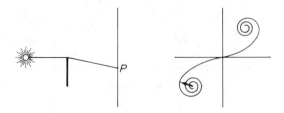

Figure 2.3-18   Amplitude vector *inside* shadow.

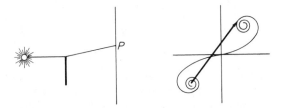

Figure 2.3-19   Amplitude vector *outside* shadow.

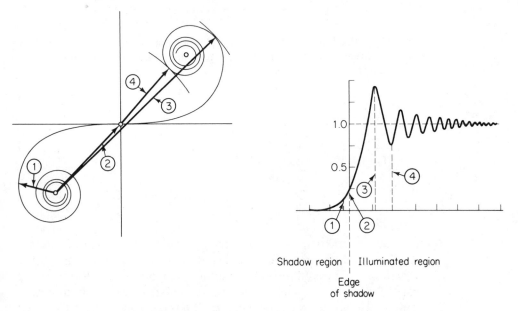

Shadow region ¦ Illuminated region

Edge
of shadow

Figure 2.3-20   Fresnel diffraction at a knife edge. Amplitudes in Cornu spiral (*left*) and intensity distribution on screen (*right*). Note corresponding numbers on left and right.

Example. *Whereas Cornu's spiral illustrates the principle very well, higher accuracy is obtained when the tabulated values of Fresnel's integrals are used directly. For example, calculate the relative intensity at a point:*
(a) Where $v = -1.00$, which is inside *the shadow.*
(b) Where $v = +1.00$, which is outside *the shadow.*

Solution. From Table 2.3-2,

$$v = 1.00, \quad x = 0.7799, \quad y = 0.4383$$

(a) *Inside* the shadow, the top of the phasor lies in the same (lower left-hand) quadrant as the lower eye of the spiral to which it is anchored. Thus the phasor extends from the coordinates $(-0.5; -0.5)$ to $(-0.7799; -0.4383)$ and the absolute values of the x and y components, respectively, must be *subtracted*:

$$\Delta x = 0.5 - 0.7799 = |0.2799|$$

and

$$\Delta y = 0.5 - 0.4383 = 0.0617$$

We need not solve for the actual length of the phasor (which has to be squared anyway to obtain the intensity). However, we recall that Cornu's construction gives a value of two for the unobstructed wave, and hence the relative intensity at the specified point is

$$I = \tfrac{1}{2}(0.2799^2 + 0.0617^2)I_0 = \boxed{0.041I_0}$$

(b) *Outside* the shadow, the phasor extends from the lower eye to the first (upper right-hand) quadrant; thus the x and y components must be *added*:

$$\Delta x = 0.5 + 0.7799 = 1.2799$$

and

$$\Delta y = 0.5 + 0.4383 = 0.9383$$

The relative intensity then is

$$I = \tfrac{1}{2}(1.2799^2 + 0.9383^2)I_0 = \boxed{1.26I_0}$$

Zone plate.   We now come to Fresnel diffraction by a *circular aperture*. For simplicity, assume that the light comes from infinity. The wavefronts that pass through the aperture in Figure 2.3-21, therefore, are *plane*. We divide the aperture into half-period zones, which from zone to zone are $\tfrac{1}{2}\lambda$ farther away from point P_0. If b is the distance from the center of the aperture to P_0, then $b + \tfrac{1}{2}\lambda$ corresponds to a (first) ring around the center of the aperture, $b + 2(\tfrac{1}{2}\lambda)$ to a second, wider ring, $b + 3(\tfrac{1}{2}\lambda)$ to the third ring, and $b + m(\tfrac{1}{2}\lambda)$ to the mth ring. These distances refer to *boundaries* between zones, rather than to the zones as such.

From Pythagoras' theorem we find that the radius of the first boundary is

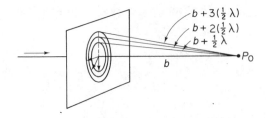

Figure 2.3-21   Fresnel's half-period zones on a plane wavefront.

$$R_1 = \sqrt{(b + \tfrac{1}{2}\lambda)^2 - b^2} \qquad [2.3\text{-}50]$$

and the radius of the mth boundary

$$R_m = \sqrt{(b + \tfrac{1}{2}m\lambda)^2 - b^2} \qquad [2.3\text{-}51]$$

Furthermore, the *area* of the first *zone* is

$$A_1 = \pi R_1^2 = \pi[(b + \tfrac{1}{2}\lambda)^2 - b^2] = \pi b\lambda + \tfrac{1}{4}\pi\lambda^2 \qquad [2.3\text{-}52]$$

or, since λ is small compared with b,

$$A_1 \approx \pi b\lambda \qquad [2.3\text{-}53]$$

The area of the second zone is

$$A_2 = \pi[(b + \lambda)^2 - b^2] - \pi b\lambda \approx 2\pi b\lambda - \pi b\lambda = \pi b\lambda \qquad [2.3\text{-}54]$$

and so on. The individual zones all have approximately the same area. Therefore, the contributions from these zones are very nearly equal; but since the zones differ by one-half of a wavelength, the resultant at P_0 is zero. However, any odd number of zones will give a bright center (while any even number will give a dark center).

If we now block out all odd-numbered zones, then only zones 2, 4, 6, . . . will contribute to the light reaching P_0, and P_0 will be bright; it will become the *focus* of a *zone plate** (Fig. 2.3-22).

In order to derive the focal length of a zone plate, we square Equations [2.3-51] and [2.3-50] and divide one by the other:

Figure 2.3-22   Center portion of a zone plate.

---

*A zone plate, or *zone lens* because of its focusing properties, is the only image-forming element for which there is no prototype in nature. The first to make a zone plate, in 1871, was Lord Rayleigh. Four years later, J. L. Soret described its properties, "Ueber die durch Kreisgitter erzeugten Diffractionsphänomene," *Ann. Physik* (6) **6** (1875), 99–113. Today, zone plates are more than a curiosity. They are used for image formation in regions of the spectrum (X rays, microwaves) where conventional lenses do not exist.

$$\frac{R_m^2}{R_1^2} = \frac{(b + \frac{1}{2}m\lambda)^2 - b^2}{(b + \frac{1}{2}\lambda)^2 - b^2}$$

Canceling b and λ gives

$$R_m^2 + \tfrac{1}{2}R_m^2\lambda = m(R_1^2 b + \tfrac{1}{2}R_1^2\lambda)$$

and, since $\lambda \ll R$,

$$R_m = R_1\sqrt{m} \qquad\qquad [2.3\text{-}55]$$

which means that the boundaries between zones have radii that are proportional to the square root of the natural numbers.

If we again square Equation [2.3-51] and rearrange the terms, we obtain

$$b^2 = (b + \tfrac{1}{2}m\lambda)^2 - R_m^2 = b^2 + bm\lambda + (\tfrac{1}{2}m\lambda)^2 - R_m^2 \quad [2.3\text{-}56]$$

and, since $\lambda \ll b$,

$$bm\lambda \approx R_m^2$$

The *focal length,* therefore, is

$$f = b \approx \frac{R_m^2}{m\lambda} \qquad\qquad [2.3\text{-}57]$$

Note the wavelength λ in this equation. It shows that a zone plate has severe chromatic aberration. (A conventional lens has chromatic aberration too, because its refractive index is a function of wavelength. This variation, though, is much smaller because n varies very slowly.)

A zone plate acts both as a positive and a negative lens; its focal length is actually $\pm f$. This is because light diffracted at the boundaries is directed not only *toward* the axis but also *away* from it. Also unlike an ordinary lens, a zone plate has several foci. This is because of the higher orders that occur in addition to the first order. The most intense, first-order focus is farthest away. The higher-order foci are located at distances $\frac{1}{3}, \frac{1}{5}, \frac{1}{7}, \ldots$ of that of the first order: If P were placed at a distance of $f/2$ from the zone plate, *two* successive boundaries would contribute, and cancel. But at a distance of $f/3$, the contributions would come from *three* boundaries: two cancel and the third causes the focus.

If a zone plate has completely opaque and completely clear zones, it is called a *Fresnel zone plate.* * If the zone plate has a sinusoidal transmissivity distribution, it is a *Gabor zone plate*—which, oddly enough, has no higher-order foci.

---

*A *Fresnel lens* is completely different from a *Fresnel zone plate*. A Fresnel lens contains a series of concentric grooves of prismatic profile that focus light, just as an ordinary lens does. Fresnel lenses are based on geometric optics, rather than on diffraction; they are used in lighthouses, automobile headlamps, and projectors, and have ophthalmic applications.

Gauss' thin-lens equation $1/o + 1/f = 1/i$, applies as well to a zone plate. Rayleigh's criterion, $\theta_{min} \approx 1.22\lambda/D$, also applies. If we replace θ by d/f and f by Equation [2.3-57], then

$$d_{min} \approx 1.22\frac{R_m^2}{mD}$$

But D is twice the radius, and thus

$$d_{min} \approx 1.22\frac{R_m}{2m} \qquad [2.3\text{-}58]$$

Interestingly, λ has dropped out: The wavelength, although it causes much chromatic aberration, has no effect on the resolution. Finally, we see from Equation [2.3-58] that to make a better zone plate (which will resolve smaller details), it (a) must be as small as possible and (b) must have as many zones as possible, two requirements that call for high-resolution photographic material on which to reproduce such zone plates.

Concluding phenomena.   There are a few more diffraction
phenomena worth mentioning.

If instead of an (open) aperture we have a (solid) circular obstacle, Fresnel's theory and experimental observation show that in the center of the shadow there is a bright point of light, called *Poisson's spot** (Figure 2.3-23). The reason is that in diffraction by a circular disk or circular aperture, there are only two rays (from diametrically opposite points) diffracted to any given off-axis point. But there are an infinite number of rays (from the whole circumference) that converge on the axis.

If the disk is illuminated by two point sources, two Poisson spots appear. These, in a way, are images of the two point sources. If there is an array of many point sources, there are as many Poisson spots. And if the many point sources are replaced by a complex object, such as a photographic slide illuminated from behind, a real image of the slide results, having used as the "objective lens" nothing but a solid metal sphere.

---

*Named after Siméon Denis Poisson (1781–1840), French physicist. As a member of the committee which was to judge Fresnel's dissertation, Poisson concluded that, if Fresnel were right, a central bright spot would appear in the shadow of a circular obstacle, a result which he offered as a *reductio ad absurdum,* that is, evidence that Fresnel's theory is absurd. He did not know that the bright spot in question had been discovered by Maraldi and Deslisle some 100 years earlier. After the objection raised by Poisson, Arago immediately tried the experiment using a disk 2 mm in diameter and rediscovered Maraldi's spot. Poisson's name became immortal, either despite his objection, or because of it, and the phenomenon is since called *Poisson's spot.*

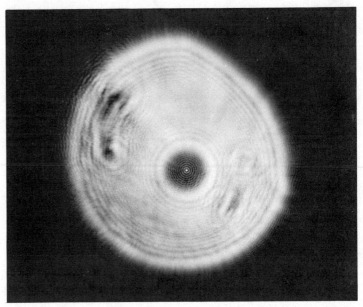

Figure 2.3-23 Poisson's spot.

Two diffracting objects are said to be complementary to each other if details in one object are opaque while the same details in the other object are transparent, and vice versa. The (Fraunhofer) diffraction patterns generated by two such objects, except for the zeroth order, are the same, an effect known as *Babinet's principle*.

Neither Huygens' principle nor Fresnel's theory considers the fact that the secondary Huygens waves cannot have equal amplitudes in all directions. If they did, they would move forward as well as backward. This dilemma was solved by Kirchhoff, who went beyond Huygens and Fresnel, introducing the *obliquity factor*.*

The obliquity factor makes the amplitude vary as a function of the angle θ by which the light departs from the forward direction. Using the identity $\cos^2 \alpha = \frac{1}{2}(1 + \cos 2\alpha)$, the amplitude then takes the form

$$\mathbf{A} = \mathbf{A}_0 \cos^2(\tfrac{1}{2}\theta) = \tfrac{1}{2}\mathbf{A}_0(1 + \cos\theta) \qquad [2.3\text{-}59]$$

In the forward direction, where $\theta = 0$, the obliquity factor reaches its highest value, unity. At $\theta = 90°$, the amplitude falls off to $\frac{1}{2}$, and the intensity to $\frac{1}{4}$. In the backward direction, $180°$, the obliquity factor and the

---

*For Kirchhoff footnote, see page 472. The fundamental paper on Kirchhoff's diffraction theory is G. Kirchhoff, "Zur Theorie der Lichtstrahlen," *Ann. Physik* (Neue Folge) **18** (1883), 663–95.

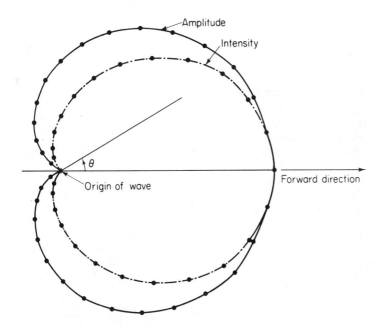

Figure 2.3-24   Obliquity factor in Huygens' construction.

amplitude and intensity all become zero. If the amplitude and intensity are plotted in three-dimensional space, a figure results that has the shape of a cardioid (Figure 2.3-24).

SUGGESTIONS FOR FURTHER READING

R. B. Hoover and F. S. Harris, Jr., "Die Beugungserscheinungen: a Tribute to F. M. Schwerd's Monumental Work on Fraunhofer Diffraction," *Applied Optics* **8** (1969), 2161–64.

B. Rossi, *Optics,* pp. 175–88 (Reading, MA: Addison-Wesley Publishing Company, Inc., 1957).

A. Sommerfeld, *Optics, Lectures on Theoretical Physics,* Vol. IV (New York: Academic Press, Inc., 1954).

J. M. Stone, *Radiation and Optics,* pp. 107–209 (New York: McGraw-Hill Book Company, 1963).

J. B. Keller, "Geometrical Theory of Diffraction," *J. Opt. Soc. Am.* **52** (1962), 116–30.

D. K. Lynch, *Atmospheric Phenomena.* Readings from *Scientific American* (San Francisco: W. H. Freeman and Company Publishers, 1980).

R. Greenler, *Rainbows, Halos, and Glories* (New York: Cambridge University Press, 1980).

PROBLEMS

2.3-1. A single slit, illuminated by red light of 600 nm wavelength, gives first-order Fraunhofer diffraction minima that subtend angles of 4° with the optic axis. How wide is the slit?

2.3-2. A slit 0.168 mm wide is illuminated by monochromatic light. If on a screen 2 m away from the slit, the two second-order minima are 3 cm apart, what is the wavelength of the light?

2.3-3. If the amplitude in the zeroth-order maximum in Fraunhofer diffraction at a single slit is 6 arbitrary units, what is the amplitude:
(a) In the first minimum?
(b) In the first, second, and fifth maxima?

2.3-4. Consider the first-order and second-order minima in Fraunhofer diffraction at a single slit. Find the phase angle difference between the *central* and *peripheral* (marginal) rays that pass through the slit and show what the vibration curve looks like.

2.3-5. Whenever the sky is slightly overcast, you may see the sun, or the moon, surrounded by a ring which is red inside and blue outside (Figure 2.3-25, left). This is called a *halo*.

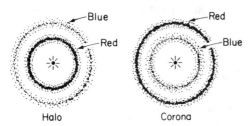

Figure 2.3-25

(a) How does it come about?
(b) The opposite sequence (right) is called a *corona;* here blue is inside and red outside. Why is this so?

2.3-6. When you are driving at night, the headlight of a motorcycle following you, seen through the fogged rear window of your car, appears to be surrounded by colored rings. Which color is inside, which outside?

2.3-7. If white light is passing through a small aperture that has the shape of an equilateral triangle, a regular pentagon, or a regular hexagon, respectively, how many light *fans* will occur because of diffraction?

2.3-8. In Figure 2.3-26 we have a variety of small apertures of different shapes to be inserted into a beam of light. For each aperture, show what the diffraction pattern will look like. Assume white light, except for the second example, which should also be drawn for monochromatic light.

Figure 2.3-26

2.3-9. Lycopodium seeds, which have an average diameter of 30 μm, are dusted on a glass plate. If parallel light of 589 nm wavelength is passed through the plate, what is the angular radius of the first diffraction maximum?

2.3-10. If, due to turbidity within the eye, a light source 3 m away appears surrounded by a red ring 50 cm in diameter, how large are the structural details in the eye that cause the ring? Assume that $\lambda = 630$ nm.

2.3-11. If the pupil of the eye is 3 mm in diameter, and if $\lambda = 525$ nm, what is the eye's angular resolution, in arc sec?

2.3-12. What is the limit of resolution (in arc sec) of a telescope that has an objective 40 mm in diameter, assuming that it is used with light of 550 nm wavelength?

2.3-13. (a) What is the angular distance that can be resolved by a lens that is 25 mm in diameter and has 130.4 cm focal length?
(b) If the lens is focused for infinity, what is the corresponding distance in the image plane? Assume that $\lambda = 550$ nm.

2.3-14. If the headlights of an approaching automobile are 1.22 m apart, what is the maximum distance at which the eye could resolve them? Assume pupils 4 mm in diameter and light of 500 nm wavelength.

2.3-15. A single slit is placed 2 m from a point source and a screen 4 m from the slit. If light of 750 nm wavelength, passing through the slit, is to have no more than $\lambda/10$ path difference between central and peripheral (marginal) rays, how wide can the slit be?

2.3-16. If a knife edge is 2 m away from a (slit)

source on one side and a screen on the other and if $\lambda = 500$ nm, how far from the geometrical limit of the shadow will a point be located for which $v = 1.00$ on the Cornu spiral?

2.3-17. From the table of Fresnel integrals, determine the relative intensity at a point:
(a) $v = -0.7$ *inside* the shadow.
(b) $v = +1.4$ *outside* the shadow behind a knife edge.

2.3-18. A wire, 0.3 mm in diameter, is placed 50 cm away from a source and an equal distance in front of a screen. If $\lambda = 500$ nm, determine the value of v on Cornu's spiral and the relative intensity at the edge of the geometrical shadow.

2.3-19. The eighth boundary of a zone plate has a diameter of 6 mm.
(a) What is its principal focal length for light of 500 nm wavelength?
(b) What are the next two focal lengths?

2.3-20. If a zone plate of the same focal length as in Problem 2.3-19 has 50 zones (100 bound-

aries), what is the least distance that it can resolve?

2.3-21. (a) What should be the diameter of a zone plate that has 55 zones and can resolve details 10 μm in size?
(b) What wavelength is necessary?

2.3-22. Determine the longitudinal chromatic aberration for two wavelengths, 450 nm and 600 nm, of the zone lens described in Problem 2.3-21.

2.3-23. In the lecture hall we show *Poisson's spot* on a screen 2.3 m away from an obstacle 2 mm in diameter. Why can't we see Poisson's spot during a solar eclipse? (Angular size of moon $\approx \frac{1}{2}$ degree.)

2.3-24. A wavefront, which has an amplitude of $\mathbf{A} = 1$ arbitrary unit, is advancing due east. Considering the *obliquity factor,* what is the ratio of the amplitudes, and of the intensities, of secondary Huygens' wavelets traveling west, northwest, north, northeast, and east?

2.4

Diffraction Gratings

A DIFFRACTION GRATING IS A DIRECT EXTENSION of Young's double slit and as such is based on both diffraction and interference. Gratings, and their applications, have developed very much into a technology all their own.

The Grating Equation

In its most elementary form, a diffraction grating consists of multiple parallel, equidistant slits in an opaque screen. As in double-slit interference (see Figure 2.2-6), d is the distance between the centers of any two adjacent slits, θ the angle through which the light is diffracted, m the order, and λ the wavelength. Then

$$\boxed{d \sin \theta = m\lambda} \qquad m = 0, 1, 2, \ldots \qquad \text{[2.4-1]}$$

the same equation derived for double-slit maxima.

If the light is incident at an angle α, then

$$d(\sin \theta - \sin \alpha) = m\lambda \qquad \text{[2.4-2]}$$

which is the (more complete) *grating equation*. (The minus sign is used to conform to our sign convention, which, as mentioned in Chapter 1.3, refers to angles measured clockwise, beginning at the axis, as positive, and to angles measured counterclockwise as negative.)

The principal use of diffraction gratings is in spectroscopy. A *grating spectrograph,* as we see from Figure 2.4-1, is similar to a prism spectrograph. The light passes first through a combination of a slit and a collimating lens. It then reaches the grating, set with its rulings parallel to the slit. Another lens, L_2, focuses the light on the screen. Light of different wavelengths, as a consequence of Equation [2.4-1], is diffracted at different angles, θ, and for *each* order, m, is drawn out into a spectrum, the blues nearer to the axis and the reds farther away from it. The zeroth order retains the composite color of the source and therefore is easy to identify.

The number of orders possible is limited by the separation of the slits, d, because θ cannot exceed 90°. A coarse grating, with only a few slits per unit width, will give many orders. A finely ruled grating, with many slits per unit width, may give only one or two orders, but their spectra are spread out much wider than those formed by the coarse grating.

Example. *A grating has 8000 slits ruled across a width of 4 cm. What is the wavelength, and the color, of the light whose two fifth-order maxima subtend an angle of 90°?*

Solution. Solving Equation [2.4-1] for λ and noting that 90° is the angle subtended by *two* maxima, one on each side, and thus $\theta = 45°$, we find that

$$\lambda = d\left(\frac{\sin \theta}{m}\right) = \frac{0.04}{8000}\left(\frac{\sin 45°}{5}\right) = \boxed{707 \text{ nm}}$$

Since 630 nm is considered the peak wavelength of red, 707 nm corresponds

to a color of $\boxed{\text{dark red.}}$

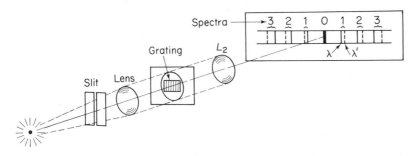

Figure 2.4-1 Schematic diagram of grating spectrograph. Light source is assumed to emit two wavelengths, $\lambda < \lambda'$.

Grating theory (cont'd).   So far we have considered only the separation, d, of the slits. But the slits also have a *finite width*, s (Figure 2.4-2).

This significantly changes the intensity distribution behind the grating. Consider the light that passes through a *single slit*. The path difference between peripheral rays, touching the edges of the slit, from Equation [2.3-2], is

$$\Gamma = s \sin \theta$$

Using Equation [2.2-38], this is equivalent to a phase difference

$$\delta = \frac{2\pi}{\lambda}\Gamma = \frac{2\pi}{\lambda} s \sin \theta \qquad [2.4\text{-}3]$$

The intensity distribution behind a single slit, from Equation [2.3-11], is

$$I_\theta = I_0 \left(\frac{\sin \delta/2}{\delta/2} \right)^2 \qquad [2.4\text{-}4]$$

We now substitute Equation [2.4-3] in [2.4-4],

$$I_\theta = I_0 \frac{\sin^2 \left(\frac{\pi}{\lambda} s \sin \theta \right)}{\left(\frac{\pi}{\lambda} s \sin \theta \right)^2} \qquad [2.4\text{-}5]$$

which is the *diffraction contribution*. It is due to the *finite width* of the slits.

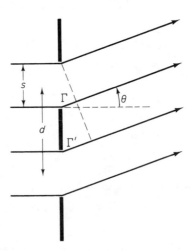

Figure 2.4-2   Diagram of diffraction grating showing slit width, s, and slit separation, d.

Now we consider *several, N, consecutive slits*. The path difference between rays that pass through (equivalent points in) adjacent slits, from Equation [2.2-30], is

$$\Gamma' = d \sin \theta$$

and the phase difference

$$\phi = \frac{2\pi}{\lambda} d \sin \theta \qquad \text{[2.4-6]}$$

The intensity distribution behind N slits could be found by integration; however, adding the complex amplitudes is more convenient. The amplitude of the light passing through each slit, from Equation [2.1-21], is

$$\mathbf{A} = A e^{i\phi}$$

But since, from slit to slit, there is a constant phase difference δ', consecutive amplitudes vary as

$$\mathbf{A}_1 = A e^{i\delta'}, \quad \mathbf{A}_2 = A e^{i2\delta'}, \quad \mathbf{A}_3 = A e^{i3\delta'}, \quad \dots \qquad \text{[2.4-7]}$$

The sum of these amplitudes is that of a *geometric series*. In general,

$$a + ar + ar^2 + ar^3 + \cdots + ar^{N-1} = a \frac{1 - r^N}{1 - r} \qquad \text{[2.4-8]}$$

and therefore,

$$A e^{i\delta'} = A[1 + e^{i\delta'} + e^{i2\delta'} + \cdots + e^{i(N-1)\delta'}] = A \frac{1 - e^{iN\delta'}}{1 - e^{i\delta'}} \qquad \text{[2.4-9]}$$

To find the intensity, we multiply the amplitude with its complex conjugate. This gives

$$\mathbf{A}^2 = A^2 \left[\frac{(1 - e^{iN\delta'})(1 - e^{-iN\delta'})}{(1 - e^{i\delta'})(1 - e^{-i\delta'})} \right] \qquad \text{[2.4-10]}$$

But

$$(1 - e^{iN\delta'})(1 - e^{-iN\delta'}) = 1 - \cos N\delta' \qquad \text{[2.4-11]}$$

and, likewise,

$$(1 - e^{i\delta'})(1 - e^{-i\delta'}) = 1 - \cos \delta' \qquad \text{[2.4-11a]}$$

Using the trigonometric relationship

$$1 - \cos \alpha = 2 \sin^2 (\tfrac{1}{2}\alpha) \qquad \text{[2.4-12]}$$

we obtain

$$\mathbf{A}^2 = A^2 \frac{\sin^2 (N\frac{1}{2}\delta')}{\sin^2 (\frac{1}{2}\delta')} \qquad [2.4\text{-}13]$$

and, substituting Equation [2.4-6] in [2.4-13], we have

$$I_\theta = I_0 \frac{\sin^2 \left(N\frac{\pi}{\lambda} d \sin \theta \right)}{\sin^2 \left(\frac{\pi}{\lambda} d \sin \theta \right)} \qquad [2.4\text{-}14]$$

which is the *interference contribution*. It is due to the *multiplicity* of slits.

The total, actual pattern behind the grating is found by multiplying Equations [2.4-5] and [2.4-14]:

$$I_\theta = I_0 \frac{\sin^2 \left(\frac{\pi}{\lambda} s \sin \theta \right) \sin^2 \left(N\frac{\pi}{\lambda} d \sin \theta \right)}{\left(\frac{\pi}{\lambda} s \sin \theta \right)^2 \sin^2 \left(\frac{\pi}{\lambda} d \sin \theta \right)} \qquad [2.4\text{-}15]$$

For simplicity, call D the pertinent term $(\pi s \sin \theta)/\lambda$ in the diffraction contribution and I the term $(\pi d \sin \theta)/\lambda$ in the interference contribution. The total pattern produced by the grating is then

$$I_\theta = I_0 \frac{\sin^2 D}{D^2} \frac{\sin^2 NI}{\sin^2 I} \qquad [2.4\text{-}16]$$

which means that one contribution is being *modulated* by the other contribution.

The numerator, $\sin^2 NI$, in the interference contribution becomes zero whenever

$$NI = k\pi \qquad k = 1, 2, 3, \ldots \qquad [2.4\text{-}17]$$

The denominator, $\sin^2 I$, similarly, becomes zero whenever

$$I = 0, \pi, 2\pi, \ldots \qquad [2.4\text{-}18]$$

Since the ratio 0/0 is indeterminate, Equation [2.4-17] is the condition for minima for all values of k *except* where $k = 0, N, 2N, 3N, \ldots, mN$. Substituting $k = mN$ in Equation [2.4-17], and replacing I by $(\pi d \sin \theta)/\lambda$, gives

$$N\frac{\pi d \sin \theta}{\lambda} = mN\pi \qquad [2.4\text{-}19]$$

This reduces to

$$d \sin \theta = m\lambda \qquad [2.4\text{-}1]$$

which again is the grating equation (for normal incidence). It refers to the *principal maxima*. Between any two adjacent (principal) maxima, there will be $N - 1$ loci of zero intensity (see Figure 2.4-3). Between these minima there are *secondary maxima*, but their intensities are much less than those in the principal maxima.

Finally, note the $\sin^2 N$ term in Equation [2.4-16]. It shows that the intensity in the principal maxima is proportional to the *square* of the number of slits. If there are *more* slits, the principal maxima become higher and more narrow and the secondary maxima between them are suppressed; these orders are "missing."

> **Example.** *A grating has slits that are 0.15 mm wide and separated, center to center, by 0.6 mm. Which of the higher-order maxima are missing?*
>
> **Solution.** We divide the grating equation (which relates to interference *maxima*),

$$d \sin \theta = m\lambda$$

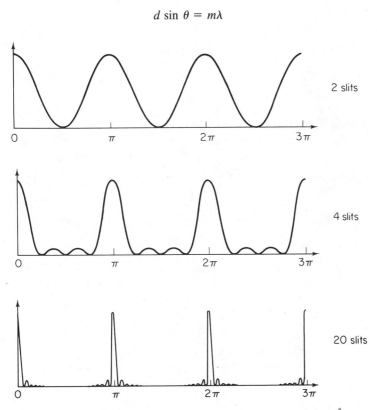

Figure 2.4-3   Intensity distribution behind a grating with 2 slits (*top*), 4 slits (*center*), and 20 slits (*bottom*).

by the single-slit equation for *minima*,

$$s \sin \theta = m'\lambda$$

This gives

$$\frac{d}{s} = \frac{m}{m'} = \frac{0.6 \text{ mm}}{0.15 \text{ mm}} = 4$$

Setting m' successively equal to 1, 2, 3, . . . , the missing orders are

$$m = (m')(4) = \boxed{4, 8, 12, \ldots}$$

Resolvance of a grating.

The resolvance ("resolving power") of a grating, like that of a prism, is a measure of its capacity to resolve two closely spaced spectrum lines. We take the elementary grating equation, $d \sin \theta = m\lambda$, and differentiate θ with respect to λ. Since $\Delta\lambda \ll \lambda$, this gives

$$\frac{d\theta}{d\lambda} = \frac{m}{d \cos \theta} \qquad\qquad [2.4\text{-}20]$$

The first term, $d\theta/d\lambda$, is the *angular dispersion* of the grating; it increases as the order, m, increases.

Now consider light that contains two wavelengths, λ and $(\lambda + \Delta\lambda)$. In order to resolve these two wavelengths, in accordance with Rayleigh's criterion, the line of wavelength $(\lambda + \Delta\lambda)$ should be no closer to the other line than the minimum adjoining the line of wavelength λ. We recall from Equation [2.2-29] that a line of order m relates to a path difference of

$$\Gamma = m\lambda$$

In the line of order $(m + 1)$ the path difference has increased by λ,

$$\Gamma' = (m + 1)\lambda \qquad\qquad [2.4\text{-}21]$$

But these two lines (which are *maxima*) are separated by $(N - 1)$ *minima*. So, in the $(N - 1)$th minimum, the path difference $\Gamma = m\lambda$ has increased to

$$\Gamma'' = m\lambda + \frac{\lambda}{N} \qquad\qquad [2.4\text{-}22]$$

The least wavelength difference, $\Delta\lambda$, that can be resolved is

$$m(\lambda + \Delta\lambda) = m\lambda + \frac{\lambda}{N} \qquad\qquad [2.4\text{-}23]$$

or

$$\boxed{\frac{\lambda}{\Delta\lambda} = Nm}$$ [2.4-24]

which defines the *resolvance* of the grating. As before, λ is the wavelength of the light, N the number of lines in the grating, and m the order in which the grating is used. The width of the grating is immaterial.

The resolvance, $\lambda/\Delta\lambda$, is proportional to N and m. To obtain higher resolution, we may either use a grating with more lines or go to a higher order. But note from Equation [2.4-24] that the resolvance (unlike the angular dispersion) is independent of the separation of the slits. In practice, grating lines are ruled very close to each other to keep the aperture size of the system within reasonable limits.

Example. *The sodium D doublet has wavelengths of 589 nm and 589.6 nm.*
(a) If only a grating with 400 rulings is available, what is the lowest order possible in which the D lines are resolved?
(b) How wide does the grating have to be?

Solution. (a) From Equation [2.4-24],

$$m = \frac{\lambda}{\Delta\lambda\,N} = \frac{589 \times 10^{-9}}{(0.6 \times 10^{-9})(4 \times 10^{2})} = 2.45$$

But orders must be *whole numbers* and therefore the lowest possible order in which the D lines are resolved is

$$m = \boxed{3}$$

(b) This is $\boxed{\text{immaterial.}}$

Types of gratings. We distinguish *transmission gratings* (which let the light pass through, as in Figures 2.4-1 and 2.4-2) from *reflection gratings* (which reflect it; Figures 2.4-4 and 2.4-5). Either type can be made with more than 1000 lines per millimeter. In contrast, *Ronchi grids** have only a few black, absorbing lines per millimeter. (Such grids

---

*Named after Vasco Ronchi (1897–), Italian physicist, long-time director of the Istituto Nazionale di Ottica in Florence and founder and president of the Fondazione "Giorgio Ronchi." After graduating from the University of Pisa, Ronchi devoted his life to optics, wrote 20 books and more than 700 papers on various topics, most notably optical testing and interferometry, the history of science, and the evolution of scientific thought. He received many honors.

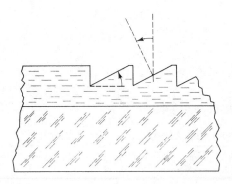

Figure 2.4-4 Profile of a blazed reflection grating. Angle shown is called *blaze angle*.

are used to obtain moiré patterns, to be discussed soon, and are based on geometric optics; the wavelength plays no role.)

A grating with completely opaque bars and clear intervals is called an *amplitude grating*. If the bars do not block the light but merely retard its phase, we have a *phase grating*. If the light distribution across the bars and intervals is sinusoidal, rather than "square-wave," all the light is diffracted into the two first-order spectra.

Most often, gratings are used that have many lines and observations made in a spectrum of low order. But high resolution can be achieved also by using few lines and high orders. This is possible, for example, by placing increasingly thicker glass plates in front of successive slits, retarding the light, from slit to slit, by many wavelengths. This is called an *echelon grating*.

Actually, there are no "slits" or "lines" in a grating. Modern gratings are *blazed*. First a plate of low-expansion glass is coated with a suitable metal such as aluminum or gold. Then the coating is ruled with a diamond tool, moved across in repetitive, parallel, equidistant strokes and with just enough pressure that the metal flows aside. After the diamond has passed, the metal hardens into the desired profile (Figure 2.4-4).

Gratings can be made with high precision, in sizes up to 50×75 cm^2, using interferometrically and electronically controlled ruling engines.* From original *masters,* plastic *replicas* can be produced. But even the best gratings sometimes have periodic errors that give rise to spurious lines in the spectrum, called *ghosts*.

Diffraction gratings can be mounted, like prisms, between a collimating lens and a focusing lens. If the grating is of the reflecting type,

*See G. R. Harrison, "The Diffraction Grating—An Opinionated Appraisal," *Applied Optics* **12** (1973), 2039–49. For building an intriguingly simple grating, see T. B. Greenslade, Jr., "Wire Diffraction Gratings," *Am. J. Physics* **41** (1973), 730–31.

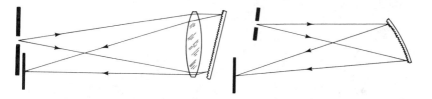

Figure 2.4-5   (*From left to right*) Plane reflection grating, Littrow mount; concave grating, no lens; plane grating, Fastie–Ebert mount. Note that right-hand system is folded: exit beam is in line with entrance beam, system is used as a monochromator.

one lens is sufficient for both functions. The result is the *Littrow mount* (Figure 2.4-5, left). No lens is required with a *concave reflection grating* (center). But curved gratings are difficult to rule. So it is preferable, after all, to use a plane grating, in conjunction with either two concave mirrors (*Czerny–Turner mount,* not shown) or with only one mirror (*Fastie–Ebert mount,* Figure 2.4-5, right).

Three-Dimensional Gratings

The gratings discussed so far are one-dimensional. A two-dimensional or *cross-grating* results if two gratings are laid on top of each other, their rulings oriented at right angles. A zone plate could also be considered a two-dimensional grating, of rotational symmetry (and space-varying frequency). I will refer again to two-dimensional gratings in Chapter 4.3. For now, we proceed to *three-dimensional gratings.*

X-ray diffraction.

For a while after Röntgen's discovery of X rays it was not known whether X rays were particles (like cathode rays) or waves of exceedingly short length. It occurred to Max von Laue that a crystal may act on X rays in a way similar to a grating on light. Indeed, when a collimated, and preferably monochromatic, beam of X rays is incident on a crystal, as shown in Figure 2.4-6, it generates a pattern of maxima symmetrical about the axis, a *Laue diagram.**

A crystal is a three-dimensional lattice of atoms, or groups of atoms, that is built out of repeating fundamental units of structure, called *unit cells.* X-ray diffraction on such units may be visualized as reflections

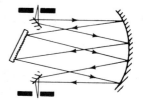

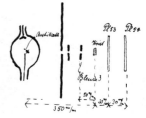

Figure 2.4-6   Reproduction from Friedrich, Knipping, and von Laue's notebook, dated 4 May 1912, showing X-ray tube (*left*), three collimating pinholes, crystal, and photographic plate (in two different positions). W. Friedrich, P. Knipping, and M. von Laue, *Sitzungsber. Bayer. Akad. Wiss., Math.-phys. Klasse* **1912,** 303, and "Interferenzerscheinungen bei Röntgenstrahlen," *Ann. Physik* (4) **41** (1913), 971–88. (Photo Deutsches Museum München. Reproduced by permission.)

*Max Theodor Felix von Laue (1879–1960), German physicist and professor at the University of Munich, later professor of theoretical physics and director of the Kaiser Wilhelm Institute, the present Max Planck Institute, in Berlin. Together with Friedrich and Knipping, von Laue tried crystals of zinc sulfide and copper sulfate. Both gave fine X-ray diffraction patterns, which established the wave nature of X rays. In 1914, von Laue received the Nobel prize.

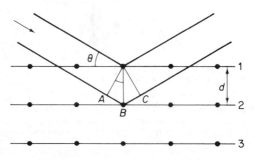

Figure 2.4-7   Deriving Bragg's law.

(Figure 2.4-7). Let a beam of monochromatic X rays incident on a crystal make an angle θ with the surface.* Some of the incident rays are reflected at the top layer, 1, but other rays penetrate to the deeper layers, 2, 3, and so on. The planes have an interplanar spacing d. If the wave reflected at 1 is reinforced by the wave reflected at 2, the path difference Γ between the two rays must be $\Gamma = AB + BC = \lambda$ or it must be a multiple of λ, $\Gamma = m\lambda$.

Since

$$\sin \theta = \frac{AB}{d}$$

$$AB = BC = d \sin \theta$$

Therefore, in order to produce a maximum, the path difference,

$$\Gamma = 2d \sin \theta \qquad [2.4\text{-}25]$$

must be equal to an integral multiple of a wavelength,

$$\boxed{2d \sin \theta = m\lambda} \qquad [2.4\text{-}26]$$

which is *Bragg's law*.[†] The same relationship holds if the X rays are transmitted through the crystal, rather than reflected by it.

So far we have considered lattice planes *parallel* to the surface. But there are other planes, in other orientations. Three out of many others possible are shown in Figure 2.4-8. These planes not only differ in slope

---

*Note that angles in X-ray crystallography are measured from the surface and not from the surface *normal* as elsewhere in optics.

[†]Named after the younger Bragg, son of Sir William (Henry) Bragg (1862–1942), British crystallographer and physicist. Sir (William) Lawrence Bragg (1890–1971), was a crystallographer, professor of physics at the University of Manchester, successor of Rutherford as head of the Cavendish Laboratory, and like his father before him, director of the Royal Institution. Both father and son shared the Nobel prize for physics in 1915. W. L. Bragg, "The Structure of Some Crystals as Indicated by their Diffraction," *Proc. Roy. Soc. London* **A89** (1914), 248–77.

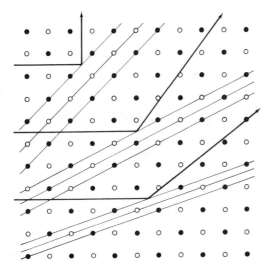

Figure 2.4-8   Three sets of lattice planes in a crystal and their reflections.

but also in the number of scatterers they contain per unit area; that is, they differ in population density. Different densities, in turn, cause maxima of different sizes ("intensities") in the Laue diagram.

Moiré Patterns

When two window screens are held close together, intersection of their lines produces another repetitive sequence of lines called a *moiré pattern.**

Assume that the spacing of the lines in the two original grids is d and that one of the grids has been turned through an angle θ with respect to the other grid. From the construction in Figure 2.4-9 and assuming that θ is small, it follows that

$$\theta \approx \frac{d}{m} \quad \text{or} \quad m \approx \frac{d}{\theta} \qquad [2.4\text{-}27]$$

Thus the more *parallel* the grids, the wider the spacing m of the moiré lines.

---

*The term "moiré" comes from the French; it refers to the wavy finish of textiles made from the wool of angora goats, in English it is "mohair." The moiré effect in optics was first noted by Lord Rayleigh, "On the Manufacture and Theory of Diffraction-gratings," *Phil. Mag.* (4) **47** (1874), 81–93. See also Y. Nishijima and G. Oster, "Moiré Patterns: Their Application to Refractive Index and Refractive Index Gradient Measurements," *J. Opt. Soc. Am.* **54** (1964), 1–5; and G. Oster and Y. Nishijima, "Moiré Patterns," *Scientific American* **208** (May 1963), 54–63.

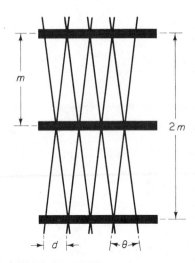

Figure 2.4-9   Moiré pattern formed by two grids of spacing d, subtending an angle θ. Distance m is spacing of moiré lines.

From this derivation we recognize one of the applications of moiré techniques: the *resolution* of lines that could not be resolved otherwise. Perhaps even more interesting is *moiré topography*. Its purpose is to map the surface of an irregular object, that is, to generate *contour lines,* similar to those on a topographic map. A practical arrangement is shown in Figure 2.4-10, the result in Figure 2.4-11.

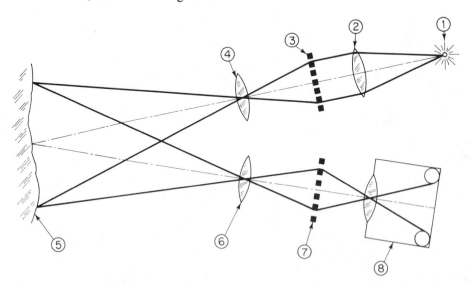

Figure 2.4-10   Moiré topography, projection method. 1, Light source; 2, condenser lens; 3, first grid; 4, projection lens; 5, object; 6, relay lens; 7, second grid; 8, camera.

Figure 2.4-11 Contour lines as produced by moiré topography. [From H. Takasaki, "Moiré Topography," *Applied Optics* **12** (1973), 845–50. Reproduced by permission.]

SUGGESTIONS FOR FURTHER READING

S. P. Davis, *Diffraction Grating Spectrographs* (New York: Holt, Rinehart and Winston, Inc., 1970).

M. C. Hutley, *Diffraction Gratings* (New York: Academic Press, Inc., 1982).

V. Ronchi, "Forty Years of History of a Grating Interferometer," *Applied Optics* **3** (1964), 437–51.

P. S. Theocaris, *Moiré Fringes in Strain Analysis* (Oxford: Pergamon Press Ltd., 1969).

G. Oster, "Optical Art," *Applied Optics* **4** (1965), 1359–69.

PROBLEMS

2.4-1. A grating, used in the second order, diffracts light of 632.8 nm wavelength through an angle of 30°. How many lines per millimeter does the grating have?

2.4-2. Green light of 520 nm wavelength is diffracted by a grating ruled with 2000 lines per centimeter. What is the angular distance of:
(a) The third order?
(b) The tenth order?

2.4-3. Collimated light containing the wavelengths 600 nm and 610 nm is diffracted by a plane grating ruled with 60 lines to the millimeter. If a lens of 2 m focal length is used to focus the light on a screen, what is the linear distance between these two lines:
(a) In the first order?
(b) In the second order?

2.4-4. Following up on our sign convention, show in the form of three ray diagrams how light is diffracted through an angle θ if the light is incident on a grating at an angle α that is:
(a) Zero.
(b) Positive.
(c) Negative.

2.4-5. A spectrum of white light is obtained with a grating ruled with 2500 lines/cm. What is the angular separation between the green (520 nm) and red (630 nm) in the second order?

2.4-6. The spectrum of mercury contains, among others, a blue line of 435.8 nm and a green line of 546.1 nm. If these two lines are to have an angular separation of at least 7°, and if the grating available has 500 lines/mm, in which order must the grating be used?

2.4-7. Red light of 630 nm wavelength, diffracted by a grating in a given order, overlaps blue light of 475 nm wavelength, diffracted by the same grating in the next higher order. What are the two orders?

2.4-8. Two slits 0.05 mm wide are moved apart until the third maximum produced by interference from the two slits on a screen 2 m away coincides with the first minimum in the diffraction pattern produced by either one of the slits. These lines of coincidence are 1.8 cm on each side from the zeroth-order maximum. What is the wavelength of the light and what is the separation of the slits?

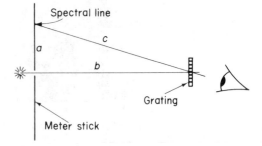

Figure 2.4-12

2.4-9. A grating is mounted at distance b from a meter stick on which a certain spectrum line is observed at distance a from the center (Figure 2.4-12). Derive a general equation for the wavelength of the light.

2.4-10. If you compare photographs of a certain line spectrum, one taken with a prism spectrograph and the other with a grating spectrograph, how can you tell which is which?

2.4-11. What is the least separation between wavelengths that can be resolved near 640 nm in the second order, using a diffraction grating that is 5 cm wide and ruled with 32 lines per millimeter?

2.4-12. How many lines must a grating have, used in the second order and near 550 nm, to resolve two lines 0.1 Å apart?

2.4-13. X rays of wavelength $\lambda = 1.3$ Å are diffracted in a Bragg spectrograph at an angle, in the first order, of 22°. What is the interplanar spacing of the crystal?

2.4-14. Find the angles at which 1.44-Å X rays glance off the polished face of a crystal of rock salt ($d = 2.8$ Å), forming the first- and second-order maxima.

2.4-15. Two identical grids are laid on top of each other. What angle must one grid subtend with the other so that moiré lines result that have twice the spacing in the original grids?

2.4-16. Continuing with Problem 2.4-15, what angle will give 13 times the spacing?

2.5

Thin Films

IN CONTRAST TO DOUBLE-SLIT and Michelson interference where two beams interact, interference in thin films is a matter of *multiple-beam interference*. Half the light is reflected by one side of the film, the other half by the other side. Thin films are a good example but only a special case. Interference in thick films, and in "air films" between surfaces, is more general.

Plane-Parallel Plates

Consider first a plane-parallel plate such as that shown in Figure 2.5-1 and assume that the light is incident at A. Part of the light is reflected toward B and another part is refracted into the plate toward C. This second part is reflected at C and emerges at D, continuing parallel to the first part.

At normal incidence, the path difference Γ between rays 1 and 2 is twice the optical thickness of the plate,

$$\Gamma = 2nd \qquad\qquad [2.5\text{-}1]$$

At oblique incidence,

$$\Gamma = n(AC + CD) - AB = \frac{2nd}{\cos \beta} - AB$$

We replace AB by $2AE \sin \alpha$ and substitute Snell's law, $\sin \alpha = n \sin \beta$, and $AE = d \tan \beta$. This gives

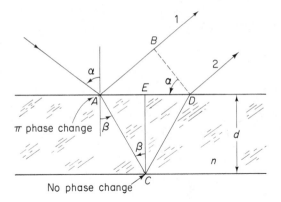

Figure 2.5-1 Notation used for interference in plane-parallel plate.

$$\Gamma = \frac{2nd}{\cos \beta} - 2d \tan \beta \, n \sin \beta$$

$$= 2nd \left(\frac{1}{\cos \beta} - \tan \beta \sin \beta \right)$$

$$= 2nd \left(\frac{1 - \sin^2 \beta}{\cos \beta} \right)$$

where n is the refractive index of the medium *between* the surfaces. Since for air $n \approx 1$, and since

$$\sin^2 \beta + \cos^2 \beta = 1$$

the path difference between rays 1 and 2 is

$$\Gamma = 2nd \cos \beta \qquad\qquad\qquad [2.5\text{-}2]$$

(which is similar to Bragg's law except that the angles θ and α are complementary).

But the light need not be *reflected*, as shown; it may be *transmitted*. Furthermore, the interval between the two surfaces may be an actual plate or film, or it may be a gap between plates. We have four possibilities, as illustrated in Figure 2.5-2. Note that in the two examples on the left, the refractive index *between* the surfaces is *higher* than outside; in the two examples on the right, it is *lower*. This determines whether or not there is a *phase change*.

We have seen earlier (on page 199), and will discuss in more detail later (on page 305), that on external (air-to-glass) reflection there is a π phase change. On internal (glass-to-air) reflection, and in refraction, there is none. In Figure 2.5-2a there is no π phase change involving any of the rays. Therefore, if the path difference between the two rays is an integral number of wavelengths, then from Equation [2.5-2],

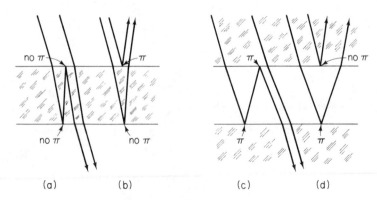

Figure 2.5-2   Interference in a plane-parallel plate (*left*) and in the air gap between two plates (*right*); π indicates phase change.

$$2nd \cos \beta = m\lambda \qquad m = 1, 2, 3, \ldots \qquad [2.5\text{-}3]$$

which is the condition for *maxima*.

But a maximum will result also in (c) because two π phase changes are equivalent to no phase change. (Again I use $m = 1$ as the lowest order because $m = 0$ would imply that either n or d is zero or that $\beta = 90°$, none of which is possible.)

Minima in transmission, as in (a) and (c), will result whenever

$$2nd \cos \beta = (m - \tfrac{1}{2})\lambda \qquad m = 1, 2, 3, \ldots \qquad [2.5\text{-}4]$$

With *reflected* light, there is generally *one* π phase shift. Then the two equations change places: Equation [2.5-3] describes the minima and Equation [2.5-4] the maxima.

The Fabry–Perot Interferometer

The most important application of multiple-beam interference between plane-parallel surfaces is that of the *Fabry–Perot interferometer*.*

---

*Marie Paul Auguste Charles Fabry (1867–1945), French physicist. While at the University of Marseilles, Fabry taught medical students, later became director of the French Institut d'Optique and professor of physics at the École Polytechnique (Sorbonne) in Paris, and is remembered as a brilliant lecturer of sometimes caustic wit. Besides his work on electricity, spectroscopy, atmospheric ozone, and astrophysics, Fabry today is best known for the instrument he developed together with Jean Baptiste Gaspard Gustave Alfred Perot (1863–1925), which he referred to as "this interferometer whose name I have the honor to bear myself." Ch. Fabry and A. Perot, "Théorie et applications d'une nouvelle méthode de spectroscopie interférentielle," *Ann. Chim. et Phys.* (7) **16** (1899), 115–44.

Its fringes can be made exceedingly narrow ("sharp"), much more so than in a Michelson interferometer. The resolvance, therefore, is very high, which makes the Fabry–Perot instrument a very powerful *spectrometer*. Also, as we will see later, reflection between plane-parallel plates is a precondition of *laser oscillation*.

As before, the light is reflected between two plane, parallel, highly reflective surfaces, separated by a distance d (Figure 2.5-3). This distance is generally of the order of several millimeters, up to a few centimeters at most; but when the system is used as a laser cavity, d may be considerably longer.

The medium between the two surfaces is usually air. Thus π phase changes occur on both of these (air-to-glass) surfaces and Equation [2.5-3] becomes

$$2d \cos \alpha = m\lambda \qquad\qquad [2.5\text{-}5]$$

which holds for *maxima*. Lens L may be a separate lens, as shown, or it may simply be the eye of the observer, in which case the retina serves as the screen. As in Michelson's interferometer, the maxima and minima, for given values of α and m, appear in the form of rings, concentric around the axis of the lens. The maximum nearest the center has the highest value of m. Maxima farther away have lower m's. If d is large, the rings are close together and a telescope may be needed to see them.

In the earlier Fabry–Perot interferometers, the distance between the plates could be varied, much as in the Michelson instrument. But keeping the plates exactly parallel during movement is very difficult. Today these systems have become obsolete and have been replaced by plates kept at *fixed distances* from each other. They are called *etalons*.

Variable spacing, though, is still used in the *scanning Fabry–Perot interferometer*. Here the plates are mounted on a spacer made out of barium titanate or some other piezoelectric material, driven by an alternating voltage of about 1 kV. Such spectrometers are generally used in the "central-spot scanning" mode.

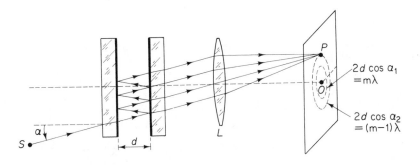

Figure 2.5-3   Fabry–Perot interferometer. S is part of an extended light source.

With an etalon we can determine precisely the relative wavelengths of different spectrum lines. Of course, if the system is mounted as shown in Figure 2.5-3, the rings formed by the different wavelengths are concentric and difficult to distinguish. But if a dispersing prism is placed between the etalon and lens L, the different ring systems are displayed side by side, separated according to wavelength. If, in addition, a slit is mounted in front of the source, with the slit oriented parallel to the apex of the prism, only narrow rectangular sections out of the various ring systems show, rather than the full rings; corresponding fringes can then be easily compared, using a single photograph.

Resolvance of the Fabry–Perot instrument. As before, we call $\Delta\lambda$ the difference between two wavelengths and $\Delta\alpha$ the angular separation between maxima formed by these wavelengths. The problem then becomes one of expressing $\Delta\lambda$ in terms of $\Delta\alpha$. To do so, we solve Equation [2.5-5],

$$2d \cos \alpha = m\lambda$$

for λ and differentiate it with respect to α:

$$\frac{d\lambda}{d\alpha} = -\frac{2d}{m} \sin \alpha \qquad [2.5\text{-}6]$$

For paraxial rays, and for small changes in wavelength and angle,

$$\frac{d\lambda}{d\alpha} = \frac{\Delta\lambda}{\Delta\alpha}$$

so that

$$\Delta\lambda = -\frac{2d}{m} \alpha \, \Delta\alpha \qquad [2.5\text{-}7]$$

(The minus sign is of no consequence because it is immaterial whether $\Delta\lambda$ is measured upward or downward from the mean wavelength.)

For paraxial rays α is small, $\cos \alpha \approx 1$, and Equation [2.5-5] reduces to

$$2d = m\lambda \qquad [2.5\text{-}8]$$

Now, if for a given ring

$$2d \cos \alpha = m\lambda$$

then for the next larger ring,

$$2d \cos \alpha' = (m' - 1)\lambda$$

Substituting for $\cos \alpha$ the approximation $\cos \alpha \approx 1 - \alpha^2/2$ yields

$$2d(1 - \alpha^2/2) = (m' - 1)\lambda \qquad [2.5\text{-}9]$$

Subtracting Equation [2.5-8] from [2.5-9] gives

$$\alpha^2 d = \lambda$$

and dividing this by Equation [2.5-7] yields

$$\frac{\lambda}{\Delta\lambda} = \frac{m}{2}\frac{\alpha}{\Delta\alpha} \qquad [2.5\text{-}10]$$

which relates chromatic resolvance to *angular* resolvance.

Next we consider the *phase difference* between adjacent rays. We recall from Equation [2.5-2] that the *path* difference between rays, traversing free space, is

$$\Gamma = 2d\cos\alpha$$

Then the *phase* difference between these rays, from Equation [2.2-38], is

$$\delta = 2\pi\frac{\Gamma}{\lambda} = \frac{2\pi}{\lambda}2d\cos\alpha = \frac{4\pi d}{\lambda}\cos\alpha \qquad [2.5\text{-}11]$$

Differentiation with respect to α gives

$$\frac{d\delta}{d\alpha} = -\frac{4\pi d}{\lambda}\sin\alpha$$

and, again assuming small angles,

$$\Delta\delta = -\frac{4\pi d}{\lambda}\alpha\,\Delta\alpha \qquad [2.5\text{-}12]$$

To determine how much light is reflected and transmitted at the two surfaces we use complex notation. We call A_0 the amplitude of the light incident on the first surface. A certain fraction of this light is reflected, $A_0\rho$, and another fraction, $A_0\tau$, is transmitted. The factors ρ and τ are called the *amplitude reflection coefficient* and the *amplitude transmission coefficient*. If there is no loss, the sum of the reflected and transmitted energies (that is, the squared values of the amplitudes) must equal the incident energy,

$$(A_0\rho)^2 + (A_0\tau)^2 = A_0^2$$

and thus

$$\rho^2 + \tau^2 = 1 \qquad [2.5\text{-}13]$$

Again, at the second surface part of the light is reflected (with an amplitude $A_0\rho^2$) and part is transmitted (with an amplitude $A_0\rho\tau$). The next ray is transmitted with an amplitude $A_0\rho^2\tau^2$, the next one after that with $A_0\rho^4\tau^2$, and so on (Figure 2.5-4). In general,

$$A_N = A_0\rho^{2(N-1)}\tau^2 \qquad [2.5\text{-}14]$$

In complex notation, the incident amplitude

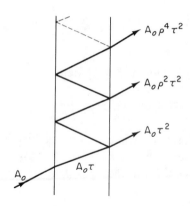

Figure 2.5-4   Amplitudes of successive rays passing through Fabry–Perot etalon.

$$\mathbf{A_0} \quad \text{becomes} \quad A_0 e^{i\omega t}$$

and

$$\mathbf{A_1} \rightarrow A_0 \tau^2 e^{i\omega t},$$

$$\mathbf{A_2} \rightarrow A_0 \rho^2 \tau^2 e^{i(\omega t - \delta)}, \; \ldots$$

$$\mathbf{A_N} \rightarrow A_0 \rho^{2(N-1)} \tau^2 e^{i[\omega t - (N-1)\delta]} \qquad [2.5\text{-}15]$$

If the number of terms approaches infinity, the series converges, and the transmitted amplitude reduces to

$$\mathbf{A_T} = A_0 e^{i\omega t} \left(\frac{\tau^2}{1 - \rho^2 e^{-i\delta}} \right) \qquad [2.5\text{-}16]$$

Multiplying this equation by its complex conjugate yields the (transmitted) *energy*,

$$I_T = I_0 \frac{\tau^2}{1 - 2\rho^2 \cos \delta + \rho^4} \qquad [2.5\text{-}17]$$

Using the identity $\cos \delta = 1 - 2 \sin^2 (\delta/2)$, Equation [2.5-17] can be transformed into

$$I_T = \frac{I_0}{1 + [4r \sin^2 (\delta/2)]/(1 - r)^2} \qquad [2.5\text{-}18]$$

Now we plot the angular position of the maxima (Figure 2.5-5). The minimum angular separation, $\Delta\alpha$, necessary to see *two* maxima must be such that the curves cross at the half-intensity point, $I_T = \frac{1}{2}I_0$. If they do, the denominator in Equation [2.5-18] must be 2.0 for each. This condition is satisfied when the second term in the denominator is unity,

$$\frac{4r \sin^2(\delta/2)}{(1 - r)^2} = 1$$

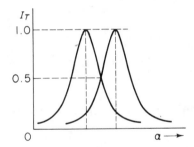

Figure 2.5-5 Plot of intensity versus angular position of two fringes just resolved according to Rayleigh's criterion.

or

$$\sin^2\left(\frac{\delta}{2}\right) = \frac{(1-r)^2}{4r} \qquad [2.5\text{-}19]$$

For the fringes to be narrow, $\delta/2$ must assume values of either near zero or near a multiple of π; any change of $\delta/2$ from these values is small. Then the sine may be set equal to the angle. Furthermore, if $\Delta\delta$ is the phase angle difference between adjacent maxima, Equation [2.5-19] becomes

$$\sin\tfrac{1}{2}\left(\frac{\Delta\delta}{2}\right) \approx \frac{\Delta\delta}{4} = \frac{1-r}{2\sqrt{r}} \qquad [2.5\text{-}20]$$

Solving Equation [2.5-12] for $\Delta\delta/4$ and combining it with Equation [2.5-20] gives

$$-\left(\frac{\pi d}{\lambda}\right)\alpha\,\Delta\alpha = \frac{1-r}{2\sqrt{r}} \qquad [2.5\text{-}21]$$

and solving Equation [2.5-7] for $\alpha\,\Delta\alpha$, inserting it into Equation [2.5-21], and eliminating $2d$, we have

$$\boxed{\frac{\lambda}{\Delta\lambda} = \frac{m\pi\sqrt{r}}{1-r}} \qquad [2.5\text{-}22]$$

The quantity $\lambda/\Delta\lambda$, as before, is the *resolvance* (resolving power) of the etalon. The resolvance depends on the wavelength, λ, on the order, m (and indirectly, therefore, on d), and on the reflectance, r, of the two mirrors. If the reflectance is high (close to unity), the resolvance is high. For adjacent fringes, that is, if $m = 1$, the right-hand term in Equation [2.5-22] is called the *finesse* of the interferometer; it may reach values between 30 and 1000.

As the two wavelengths, λ_1 and λ_2, become more and more different, the rings formed by them separate. Of course, rings formed by any wavelength will satisfy Equation [2.5-5]. But since we have two wavelengths, λ_1 and λ_2 (where $\lambda_2 > \lambda_1$), light of λ_2 forms rings *smaller* than those formed by λ_1, and, at some wavelength difference, an mth-order λ_2 ring coincides with an $(m + 1)$th-order λ_1 ring,

$$m\lambda_2 = (m + 1)\lambda_1$$

The wavelength difference at which this occurs is

$$\delta\lambda = \frac{\lambda_1}{m} \qquad\qquad [2.5\text{-}23]$$

or, substituting Equation [2.5-5] solved for m and assuming normal incidence, we obtain

$$\delta\lambda = \frac{\lambda^2}{2d} \qquad\qquad [2.5\text{-}24]$$

The interval $\delta\lambda$ is called the *free spectral range;* it is the change in wavelength necessary to shift the fringe system by one fringe.

Example. *Determine the least separation of two spectrum lines, near 500 nm, that can be resolved by an etalon that is 10 mm long (deep) and has a reflectance of 90 percent.*

Solution. For small angles, $\alpha \approx 0$ and $\cos \alpha \approx 1$. Then from Equation [2.5-5] we find that

$$m = \frac{2d}{\lambda} = \frac{(2)(0.01)}{5 \times 10^{-7}} = 40,000$$

Inserting this in Equation [2.5-22],

$$\frac{\lambda}{\Delta\lambda} = \frac{m\pi\sqrt{r}}{1 - r}$$

shows that

$$\Delta\lambda = \frac{\lambda(1 - r)}{m\pi\sqrt{r}}$$

$$= \frac{(500 \times 10^{-9})(1 - 0.9)}{(40,000)(\pi)\sqrt{0.9}} = \boxed{0.0042 \text{ Å}}$$

Hence, two lines only 0.0042 Å apart can be resolved—a much better result than from any other type of interferometer/spectrometer.

Clearly, if the reflectivities are still higher, the maxima are even narrower, and the resolvance even higher. At lower reflectivities the opposite is true, as shown in Figure 2.5-6.

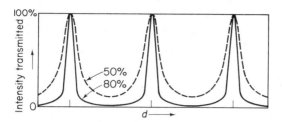

Figure 2.5-6 Plot of transmission versus plate separation in a Fabry–Perot interferometer. Percentages refer to reflectance of surfaces.

Newton's Rings

Newton's rings* result when a convex surface of long radius of curvature is placed in contact with a plane surface. As illustrated in Figure 2.5-7, R is the radius of curvature of the surface, r the radius of a given Newton ring, and d the width of the air gap at this ring. From the vertex depth formula, Equation [1.3-28],

$$d \approx \frac{r^2}{2R} \qquad\qquad [2.5\text{-}25]$$

If we insert this expression in the equation for minima in reflected light, Equation [2.5-3], we find that for dark Newton rings, at about normal incidence,

$$r^2 \approx m\lambda R \qquad\qquad [2.5\text{-}26]$$

Hence, if λ is known, the radius of curvature of the surface can be determined by measuring the diameter of one ring.

Example. *Assume that Newton's rings are formed between a plano-convex lens of 50 cm radius of curvature and a plano-concave lens, placed*

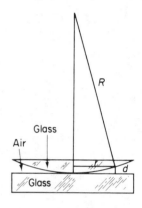

Figure 2.5-7 Notation as used with Newton's rings.

---

*Sir Isaac Newton (1642–1727), British scientist, Lucasian professor of mathematics at Cambridge University, and long-time president of the Royal Society. After graduating from Trinity College, the bubonic plague then raging forced Newton into seclusion for two years, during which time he invented calculus, discovered the composition of white light, formulated his three laws of motion, and conceived the idea of universal gravitation. In 1704 he published his *Opticks: or, a Treatise of the Reflexions, Refractions, Inflexions and Colours of Light*. A Latin version came out two years later. The English edition begins with these words: "My Design in this Book is not to explain the Properties of Light by Hypotheses, but to propose and prove them by Reason and Experiments: In order to which, I shall premise the following Definitions and Axioms." In 1705, Newton was knighted by Queen Anne. Good source material is found in R. S. Westfall, *Never at Rest: A Biography of Sir Isaac Newton* (New York: Cambridge University Press, 1981).

Figure 2.5-8

as shown in Figure 2.5-8. If the twentieth dark ring, seen in reflected light of 550 nm wavelength, has a diameter of 12 mm, what is the radius of curvature of the concave surface?

Solution. First, we *assume* that the lower surface is *plane*. Then from Equation [2.5-26] the radius of curvature (of a different lens on top) would have to be

$$R = \frac{r^2}{m\lambda} = \frac{(6 \times 10^{-3})^2}{(20)(550 \times 10^{-9})} = 3.273 \text{ m}$$

At the twentieth ring, the air gap between the actual lens and a plane surface, from Equation [2.5-25], is

$$d = \frac{r^2}{2R} = \frac{(6 \times 10^{-3})^2}{(2)(0.5)} = 0.036 \text{ mm}$$

But the air gap *needed* to produce the Newton rings, from the same equation, is

$$d = \frac{(6 \times 10^{-3})^2}{(2)(3.273)} = 0.0055 \text{ mm}$$

Thus, at the twentieth ring the air gap must be *less,* and the lower surface be *raised,* by

$$0.036 - 0.0055 \approx 0.03 \text{ mm}$$

Finally, we again use Equation [2.5-25] and find the radius needed:

$$R = \frac{r^2}{2d} = \frac{(6 \times 10^{-3})^2}{(2)(0.03 \times 10^{-3})} = \boxed{60 \text{ cm}}$$

An interesting variation of Newton's rings was discovered by Thomas Young. The support plate is made out of glass of a refractive index *higher* than that of the lens, and part of the air gap is filled with a liquid of *intermediate* index. Now that both boundaries are of the rare-to-dense type, π phase changes occur at both and the bright and dark fringes change places, but only within the area of the liquid (Figure 2.5-9).

In general, we distinguish two groups of fringes, *localized* and *non-localized*. Newton's rings are a good example of localized fringes; they appear at the surfaces that cause them. Fabry–Perot rings are localized fringes also; they are formed at infinity and therefore require a focusing lens to project them onto a screen. Localized fringes are produced by either extended or point sources.

Nonlocalized fringes are not restricted to a given surface. A good example are the fringes formed behind a double slit; they fill all (three-dimensional) space between the double slit and the screen. Nonlocalized fringes are produced by point, or line, sources.

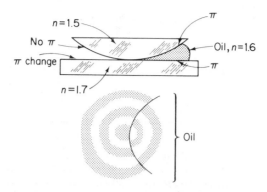

Figure 2.5-9 Newton's rings with part of gap filled with oil.

Fringes in thin films, seen in white light, often appear in brilliant colors. Soap bubbles and oil slicks on wet pavement are familiar examples. Some of the iridescent colors of insect wings are due to interference in thin films.

Interference Filters

Color filters are based either on absorption or on interference. Both types are characterized by three parameters: the wavelength of maximum transmission or *peak wavelength*, λ_{max}, in nm; the efficiency or *peak transmittance*, T_{max}, in units of relative power transmitted; and the spectral purity or *halfwidth*, HW, that is, the width of the *passband* at the level of one-half the peak transmittance, in nm (Figure 2.5-10).

Transmission-type *interference filters** consist of a spacer layer, for example a thin transparent layer of cryolite, Na_3AlF_6, sandwiched between two reflective coatings. (In older bandpass filters these coatings were metallic; in modern filters they are *dielectric,* that is, nonmetallic; they do not conduct electricity. Dielectric coatings are generally more efficient.) The coatings and the spacer are vacuum-deposited, in the proper sequence, on a plate of glass, and then another plate is cemented on top for protection (Figure 2.5-11). The separation between the two coatings, usually one-half of a wavelength or a multiple thereof, determines the wavelength. Interference filters can be made for any wavelength desired and to specifications much more rigid than those met by other filters. Table 2.5-1 shows a comparison.

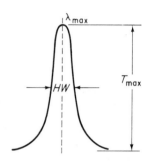

Figure 2.5-10 Characteristics of a color filter; peak wavelength, λ_{max}; peak transmittance, T_{max}; and halfwidth, HW.

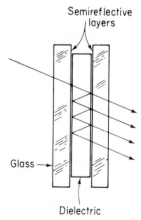

Figure 2.5-11 Interference filter.

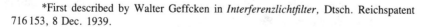

*First described by Walter Geffcken in *Interferenzlichtfilter,* Dtsch. Reichspatent 716153, 8 Dec. 1939.

Table 2.5-1 TYPICAL VALUES FOR TWO TYPES OF COLOR
FILTERS

	Absorption filter	Interference filter
T_{max}	30%	90%
HW	50 nm	1 Å
Cost	$10	$250

Example. *(a) Find the minimum thickness required of a layer of cryolite
(n = 1.35) in an interference filter designed to isolate light of 594 nm
wavelength. (b) How will the peak transmittance change if the filter is tilted
by 10°?*

Solution. (a) From the condition for maxima in transmitted light, Equation
[2.5-3], we find that

$$d = \frac{m\lambda}{2n \cos \beta} = \frac{(1)(594)}{(2)(1.35)(\cos 0)} = \boxed{220 \text{ nm}}$$

(b) In principle, the same equation applies. However, since Equation
[2.5-3] contains β (the angle of refraction), rather than α (the angle of
incidence, which is the same as the angle of tilt), we have to calculate β first,
using Snell's law, $\sin \alpha / \sin \beta = n_2/n_1$. Solving for $\sin \beta$, setting $n_1 = 1$,
and squaring both sides gives

$$\sin^2 \beta = \frac{\sin^2 \alpha}{n^2} \qquad [2.5\text{-}27]$$

Then, since

$$\sin^2 \beta + \cos^2 \beta = 1$$

or

$$\cos \beta = \sqrt{1 - \sin^2 \beta}$$

substituting Equation [2.5-27] yields

$$\cos \beta = \sqrt{1 - \frac{\sin^2 \alpha}{n^2}}$$

We then substitute this term for $\cos \beta$ in Equation [2.5-3]. Since the initial
transmission, at normal incidence, is at $\lambda_0 = 2nd$, the new peak wavelength, after tilting the filter, is

$$\lambda' = \lambda_0 \sqrt{1 - \frac{\sin^2 \alpha}{n^2}} \qquad [2.5\text{-}28]$$

(Interestingly, a higher index makes the wavelength change less.) Inserting
the actual figures gives

$$\lambda' = 594\sqrt{1 - \frac{\sin^2 10°}{1.35^2}} = 589 \text{ nm}$$

so that

$$\Delta\lambda = \boxed{5 \text{ nm}}$$

Note that the transmittance of the filter, when tilted, always changes toward the *shorter wavelengths*.

Interference filters can also be made for reflection. Such filters are called *dichroic mirrors;* they reflect certain wavelengths and transmit others. Heat reflectors or *hot mirrors* reflect infrared (which on absorption converts into heat) and transmit the visible; *cold mirrors,* used in movie projectors and Klieg lights, transmit infrared and reflect the visible. *Wedge-type interference filters* vary in spacing across the filter. This causes a whole spectrum to be transmitted, from blue at one end of the filter to red at the other. Such filters, combined with a slit for wavelength selection, are a "poor man's monochromator." *Rejection* or *minus filters* eliminate given wavelengths from the spectrum; they are used, for example, in laser safety goggles. Multilayer filters are discussed later in conjunction with antireflection coatings.

Antireflection Coatings

Part of the light passing through a boundary is lost due to reflection. Although at a single surface the loss is small, for multilens systems it may be considerable. Generally, the *reflectivity, r,* at a boundary between two media of refractive indices n_1 and n_2, and at normal incidence, is

$$r = \left(\frac{n_2 - n_1}{n_2 + n_1}\right)^2 \qquad\qquad [2.5\text{-}29]$$

a relationship that we will discuss in more detail in the next chapter (page 303). For example, at a boundary between air and glass ($n = 1.5$) the reflectivity is $r = [(1.5 - 1)/(1.5 + 1)]^2 = 0.04$, which means that 4 percent of the light is reflected and lost.

There are two ways of reducing, or even eliminating, such losses:

1. Reflection will not occur if $n_1 = n_2$. But then no refraction will occur either. Also, reflection will not occur if the transition between the two refractive indices is gradual rather than abrupt. This can be accomplished, to some extent, by etching.

2. Another way of eliminating reflection is by interference, applying a suitable coating to the surface.*

Assume that light is incident on a glass surface coated with a thin film of transparent material (Figure 2.5-12, top). Generally, some of the light is reflected on both sides of the film. These reflections can be made to cancel if the two reflected waves are out of phase by 180°, and if they have the same amplitude, $A_1 = A_2$.

(a) The amplitude condition, $A_1 = A_2$, is met if

$$\left(\frac{n_1 - n_0}{n_1 + n_0}\right)^2 = \left(\frac{n_g - n_1}{n_g + n_1}\right)^2 \qquad\qquad [2.5\text{-}30]$$

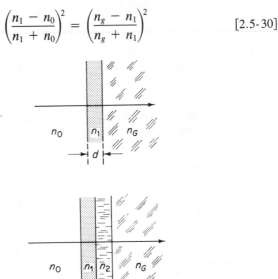

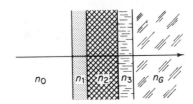

Figure 2.5-12 Antireflection coatings comprising one layer (*top*), two layers (*center*), and three layers (*bottom*).

---

*This discovery was made by Alexander Smakula (1900–), German-born physicist, at that time head of the research department of the Carl Zeiss Optical Company in Jena. Smakula later came to the United States to work at Ft. Belvoir, then at MIT. Fa. Carl Zeiss, *Verfahren zur Erhöhung der Lichtdurchlässigkeit optischer Teile durch Erniedrigung des Brechungsexponenten an den Grenzflächen dieser optischen Teile*, Dtsch. Reichspatent 685 767, 1 Nov. 1935.

where n_0 is the outside refractive index, n_1 the index of the coating, and n_g the index of the glass. For $n_0 = 1$,

$$\frac{n_1^2 - 2n_1 + 1}{n_1^2 + 2n_1 + 1} = \frac{n_g^2 - 2n_1 n_g + n_1^2}{n_g^2 + 2n_1 n_g + n_1^2}$$

which reduces to

$$4n_1^3 n_g + 4n_1 n_g = 4n_1^3 + 4n_1 n_g^2$$

Dividing both sides by $4n_1$ and rearranging gives

$$n_1^2 - n_1^2 n_g + n_g^2 - n_g = 0$$

from which

$$\boxed{n_{coating} \approx \sqrt{n_{glass}}} \qquad [2.5\text{-}31]$$

This means that the *refractive index* of the coating should be approximately equal to the square root of the index of the glass.

(b) The phase condition is met if, for normal incidence and following Equation [2.5-4],

$$2n_1 d = (m - \tfrac{1}{2})\lambda \qquad [2.5\text{-}32]$$

Thus, for $m = 1$,

$$\boxed{n_{coating} d = \tfrac{1}{4}\lambda} \qquad [2.5\text{-}33]$$

which means that the *optical thickness* of the coating should be equal to one-quarter of a wavelength. A coating of such index and thickness does not cause any light to be reflected. Then, if the coating is transparent and no light is lost to absorption or scattering, it follows, from the law of conservation of energy, that all the light must go through. Of course, the coating material not only must be transparent but also should be insoluble, scratch resistant, and in general impervious to damage. The best material known is magnesium fluoride, MgF_2; its refractive index is 1.38.

Multilayer antireflection coatings.

A single-layer coating is particularly effective only for one wavelength. The wavelength chosen is usually near the center of the visible spectrum, causing part of the blue and red to be reflected, which makes the coating on a camera lens appear purple. A much wider coverage across the spectrum is possible with multiple coatings, called *multilayers*. In a *two-layer antireflection (AR) coating*, for example, each layer is made $\tfrac{1}{4}\lambda$ thick. This is called a *quarter-quarter coating* (Figure 2.5-12, center). No reflection will occur if

$$\frac{n_1^2 n_3}{n_2^2} = n_0 \qquad [2.5\text{-}34]$$

Example. *If the first layer of a two-layer AR coating is magnesium fluoride and the substrate is ophthalmic crown (n = 1.523), what index should the second layer have?*

Solution. From Equation [2.5-34],

$$n_2 = \sqrt{\frac{n_1^2 n_3}{n_0}} = \sqrt{\frac{(1.38)^2(1.523)}{1.00}} = \boxed{1.70}$$

In many *three-layer AR coatings* the center, or "absentee," layer is made $\frac{1}{2}\lambda$ thick, the other two layers $\frac{1}{4}\lambda$ each. This is called a *quarter-half-quarter coating* (Figure 2.5-12, bottom). Such coatings are widely used; they are effective over most of the visible spectrum. The first, outside layer is often a $\lambda/4$ coating of magnesium fluoride, the next is $\lambda/2$ zirconium dioxide (ZrO_2, index 2.10), and the layer next to the substrate is $\lambda/4$ cerium fluoride (CeF_3, index 1.63) or aluminum oxide (Al_2O_3, index 1.76). Modern interference filters as well as antireflection coatings have up to 100 layers of alternating high- and low-index materials.

Antireflection coatings are of historic interest. During World War II the problem arose of finding an AR coating for German submarines as a countermeasure against enemy radar. Reflection can be eliminated if the ratio of the dielectric constants of air and of the coating applied to the conning tower equals the ratio of their magnetic permeabilities. If, in addition, the AR layer is absorptive for microwaves of the right wavelength, the submarine becomes undetectable by radar. (Of course, it is easy to change the wavelength of the radar but difficult to apply a new coating each time the wavelength is changed. So, although theoretically sound, in practice this method did not have much success.)

SUGGESTIONS FOR FURTHER READING

O. S. Heavens, *Thin Film Physics* (New York: Barnes & Noble, 1970).

Z. Knittl, *Optics of Thin Films, an Optical Multilayer Theory* (New York: John Wiley & Sons, Inc., 1976).

H. A. Macleod, *Thin-Film Optical Filters* (New York: American Elsevier Publishing Company, Inc., 1969).

P. Baumeister and G. Pincus, "Optical Interference Coatings," *Scientific American* **223** (Dec. 1970), 58–75.

PROBLEMS

2.5-1. A soap bubble, seen in white light, shows a particularly strong reflection of first-order red (630 nm). If the light is incident normally and the soapy water has a refractive index of 1.38, how thick is the wall of the bubble?

2.5-2. Light of wavelength 651.2 nm is reflected at right angles from a thin film of index 1.48. What is the least thickness of the film that will appear:
(a) Black?
(b) Bright?

2.5-3. An oil film (index = 1.47, thickness 0.12 μm) rests on a pool of water. If light strikes the film at an angle of 60°, what is the wavelength reflected in the first order?

2.5-4. A soap bubble ($n = 1.38$) has a wall 0.354 μm thick. If seen in reflected white light at normal incidence, what color is it?

2.5-5. Two rectangular plates of glass, each 12 cm long, are held in contact along one end while being separated by a hair located three-fourths of the way to the other end. When seen in light of 546 nm wavelength, 22 fringes are counted per centimeter. How thick is the hair?

2.5-6. Two circular disks of plane glass are laid on top of each other. At one point on their circumference the disks are separated by 0.0027 mm.
(a) How many dark fringes will be seen in reflected blue light (450 nm)?
(b) What shape will the fringes have?

2.5-7. Two glass plates are in contact on one side and two wavelengths apart on the other. Looking down at the plates, what do you see:
(a) In monochromatic light?
(b) In white light?

2.5-8. A wedge-shaped space between plane plates is filled with water ($n = \frac{4}{3}$) in such a way that a few bubbles of air are trapped between the plates. If 18 fringes are counted within a given distance inside an air bubble, how many fringes, within the same distance, are seen in the water?

2.5-9. What plate separation is needed in a Fabry–Perot interferometer in order to resolve two spectrum lines 0.05 Å apart if the average wavelength is 633 nm and the reflectance 80 percent?

2.5-10. (a) What is the order of the center fringe, seen with an etalon 2.5 cm deep and using the green 546-nm mercury line?
(b) If $r = 0.9$, what is the resolvance?

2.5-11. If in transmitted light a Fabry–Perot interferometer produces very narrow (bright) maxima, what do these maxima look like in *reflected* light?

2.5-12. The two surfaces in a scanning Fabry–Perot interferometer are kept 10 mm apart by a barium titanate spacer which, when a voltage of 380 V is applied, changes in length by one-half of 546 nm.

(a) When 1.14 kV is applied, and assuming that the response of the spacer is linear, how many fringes pass through the center of the field?
(b) If $r = 0.8$, how close a doublet (near 546 nm) can be resolved?

2.5-13. If two waves of equal amplitude and frequency are reflected between parallel surfaces and, hence, travel in opposite directions, they form a *standing-wave* pattern. The standing wave illustrated in Figure 2.5-13 is a *time exposure*. Show what the wave looks like:

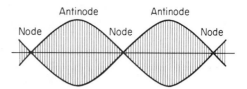

Figure 2.5-13

(a) In a *snapshot* photograph, at an instant of maximum displacement.
(b) $\frac{1}{4}$, $\frac{1}{2}$, and 1 period later.

2.5-14. Standing sound waves (sound has a velocity of 331 m s^{-1}) are set up in a cavity 20 cm × 30 cm × 50 cm in size. If only *nodes* can be reflected at the walls of the cavity, what are the lowest frequencies possible in each of the three dimensions?

2.5-15. In an experiment on Newton's rings, the diameter of the tenth dark ring formed by yellow sodium light (589 nm) and seen in reflection is 3 mm. What is the radius of curvature of the lens surface?

2.5-16. Newton's rings are observed in reflected light of 534 nm wavelength. If the twentieth bright ring has a diameter of 1 cm, what is the radius of curvature of the lens surface?

2.5-17. In a given Newton ring arrangement the diameter of the twentieth ring is 4 mm. What is the diameter of the thirtieth ring?

2.5-18. The diameter of a certain Newton ring is 10 mm. When an unknown liquid is poured into the gap between lens and support, the diameter of this ring shrinks to 8.45 mm. Calculate the liquid's index.

2.5-19. A single-layer interference filter, containing cryolite ($n = 1.35$) as the spacer material, has its peak transmittance at 540 nm. If the thickness of the cryolite changes by 1 percent, by how much does the peak wavelength change?

2.5-20. Plot the transmittance as a function of wavelength of:
(a) A *hot mirror* of the dichroic type.
(b) A *cold mirror* of the same type.

2.5-21. A certain interference filter *transmits* the green 546-nm mercury line only. Seen in *reflected* white light, however, the filter looks metallic (silvery).
(a) Why doesn't it look red (complementary to green)?
(b) Plot the transmittance and the reflectance of the filter as a function of wavelength.

2.5-22. If a narrow-band interference filter containing cryolite ($n = 1.35$) is precisely "tuned" to transmit one line of the sodium D doublet ($\lambda \approx 589$ nm), by how much must the

filter be tilted to transmit the other line, known to be 6 Å apart?

2.5-23. What percentage of light is reflected at normal incidence on a surface between air and glass ($n = \frac{5}{3}$)?

2.5-24. The antireflection coating on a lens is made of magnesium fluoride ($n = 1.38$). How thick a coating is needed to produce minimum reflection at 552 nm?

2.5-25. A camera lens consists of four uncoated, air-spaced elements, all of $n = 1.58$. How much light is lost due to surface reflections?

2.5-26. What percentage of the incident light is lost:
(a) On one side of an uncoated glass lens of $n = 1.9044$?
(b) After a coating of optimum refractive index has been applied?

Part *3*

Properties of Light and Matter

VIRTUALLY ALL INTERFERENCE can be explained by the wave nature of light. It is immaterial whether the light waves are longitudinal or transverse. Early in the history of physics, sound waves were found to be longitudinal. Perhaps because seeing and hearing are both sensory functions, light waves were thought to be longitudinal too. But polarization-of-light experiments showed that they are not; they are, in fact, transverse.

Light is made up of continually varying electric and magnetic fields. This concept of fields helps us to understand the refraction of light as it passes through boundaries and through nonhomogeneous media, and to formulate its velocity, scattering, state of polarization, and other *properties of light and matter*.

3.1

Light as an Electromagnetic Phenomenon

LIGHT WAVES ARE OFTEN COMPARED to other, mechanical waves. Certainly, all types of waves have some parameters in common. But light is fundamentally different. For instance, light requires no medium for propagation. This is, of course, in contrast to sound, and for a long time posed much difficulty when trying to explain the nature of light. The problem was solved by Clerk Maxwell, who showed that light is not a mechanical but an electromagnetic phenomenon, a result that can be expressed in the form of four fundamental equations known as *Maxwell's equations*.

In this chapter I will present the mathematical formalism of *Maxwell's equations* as the basis of the electromagnetic nature of light and *Fresnel's equations* as the basis of reflectivity. Readers who want to go on directly to the more tangible aspects of the velocity, scattering, and polarization of light may do so without much loss of continuity.

Maxwell's Equations

Light, and other electromagnetic waves, can be described by two vectors, the amplitude of the *electric field strength,* **E,** and the amplitude of the *magnetic field strength,* **H.** Both vectors are functions of space and time; they oscillate at right angles to each other and to the direction of propagation (Figure 3.1-1). They cannot be separated. A rigorous description of these *vector fields* is provided by Clerk Maxwell's equations.*

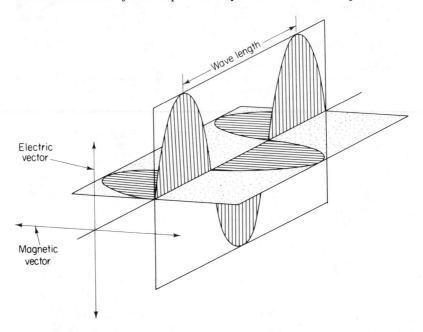

Figure 3.1-1   Electromagnetic wave represented by electric and magnetic field vectors.

*James Clerk Maxwell (1831–1879), Scottish mathematician and physicist. At age 15, Clerk Maxwell (his last name!) presented a paper before the Royal Society, at 16 became interested in optics when he had the chance of visiting Nicol, who gave him a pair of polarizing prisms. Using these, Clerk Maxwell went to work on photoelasticity, made major contributions also to refraction in nonhomogeneous media, color vision, gas dynamics, heat transfer, the theory of servomechanisms (governors), and correctly explained the rings of Saturn. In 1864 he read a paper before the Royal Society, published a year later: J. Clerk Maxwell, "A Dynamical Theory of the Electromagnetic Field," *Phil. Trans. Roy. Soc. London* **155** (1865), 459–512. Near the end of part I, on page 466, Clerk Maxwell says that "light itself . . . is an electromagnetic disturbance in the form of waves propagated through the electromagnetic field according to electromagnetic laws." In 1887, Heinrich Rudolf Hertz (1857–1894), German physicist, found the waves predicted by Clerk Maxwell 22 years earlier. In 1901, Guglielmo Marconi (1874–1937), Italian scientist and businessman, transmitted them across the Atlantic Ocean.

Clerk Maxwell, and others, tried to "explain" the concept of electromagnetic fields in terms of mechanical models. Gradually, all these attempts have been abandoned; today we realize that these fields cannot be reduced to anything simpler.

Maxwell's first equation.   Maxwell's first equation can be derived from *Coulomb's law,*

$$\mathbf{F} = \frac{1}{4\pi\varepsilon_0} \frac{q_1 q_2}{R^2} \hat{\mathbf{R}}$$ [3.1-1]

where vector $\mathbf{F}$ is the force, in newtons, N, between two point charges, q_1 and q_2, both given in coulombs, C; R is the distance, in meters; and $\hat{\mathbf{R}}$ is a *unit vector* indicating the direction from one charge to the other. The constant ε_0 is the *electric permittivity of free space*. The currently accepted best value of ε_0 in the Système International des Unités, the *International System of Units*, SI, is $8.854\,187\,82 \times 10^{-12}$, or approximately 8.85×10^{-12}, $C^2 N^{-1} m^{-2}$.

For an accumulation of many charges, Coulomb's law becomes

$$\mathbf{F} = \frac{1}{4\pi\varepsilon_0} \sum_i \frac{q_1 q_2}{R_1^2} \hat{\mathbf{R}}$$ [3.1-2]

and for an accumulation of charges on an irregular surface,

$$\mathbf{F} = \frac{1}{4\pi\varepsilon_0} \int \frac{q_0 dq}{R^2} \hat{\mathbf{R}} = \frac{q_0}{4\pi\varepsilon_0} \int \frac{dq}{R^2} \hat{\mathbf{R}}$$ [3.1-3]

where q_0 is the test charge and dq the fraction of charge on a surface element.

The *electric field strength*, **E,** also a vector, is the force per unit charge,

$$\mathbf{E} = \frac{\mathbf{F}}{q}$$ [3.1-4]

It has the units newtons/coulomb, $N\,C^{-1}$, or volts/meter, $V\,m^{-1}$. If we substitute for $\mathbf{F}$ Equation [3.1-3], then

$$\mathbf{E} = \frac{1}{4\pi\varepsilon_0} \int \frac{dq}{R^2} \hat{\mathbf{R}}$$ [3.1-5]

from which we see that a (stationary) *electric charge produces a* (steady) *electric field*.

The electric field strength may be visualized as lines originating at a certain charge and terminating at a charge of the opposite sign or at infinity. The more lines that intersect a given surface element, dA, the greater the field strength; hence, the magnitude of **E** is proportional to the *density* of these lines. The total number of lines intersecting the area is called the (electric) *flux*, ϕ; thus, if dA is oriented normal to the lines of flux, the fraction $d\phi$ passing through dA is

$$d\phi = E \, dA$$

But dA need not be normal to $d\phi$. For example, if, as in Figure 3.1-2, θ is the angle subtended by the surface normal $d\mathbf{S}$ and $\hat{\mathbf{R}}$ (or $\mathbf{E}$), then

$$d\phi = E \, dA \cos \theta$$

or, in vector notation,

$$d\phi = \mathbf{E} \cdot d\mathbf{S} \qquad\qquad [3.1\text{-}6]$$

where $\mathbf{E}$ and $d\mathbf{S}$ are multiplied in the form of a *vector dot product* (the result of which is a *scalar*).

But an actual surface is composed of many surface elements. The *total* flux passing through the surface is found by integration:

$$\phi = \frac{1}{4\pi\varepsilon_0} \int_S \frac{q}{R^2} \hat{\mathbf{R}} \cdot d\mathbf{S} \qquad\qquad [3.1\text{-}7]$$

If the surface is a *closed* surface, it is useful to state this explicitly by writing the integral sign in the form of $\oint$, as in Equation [3.1-9].

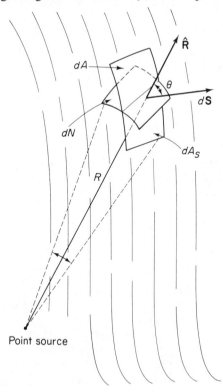

Figure 3.1-2 Element of actual surface, dA, and its projection, dN, as seen from point source.

At this point I introduce the term *solid angle*. The solid angle, ω, at a point P is subtended by an area A on the surface of a sphere of radius R and with its center at P. The unit of solid angle is the *steradian*, sr. One steradian, 1 sr, is defined as the solid (three-dimensional) angle subtended at the center of a sphere of 1 m radius by an area on its surface 1 m^2 in size, without regard to the *shape* of the area. But since *sr* is a *dimensionless* quantity, the distance R must be *squared*, $\omega = A/R^2$. Furthermore, since the surface of a sphere has an area $A = 4\pi R^2$, the *total* solid angle about a point is

$$\omega = \frac{4\pi R^2}{R^2} = 4\pi \qquad \text{sr} \qquad\qquad [3.1\text{-}8]$$

or

$$\oint_S \frac{1}{R^2} \hat{\mathbf{R}} \cdot d\mathbf{S} = 4\pi \qquad\qquad [3.1\text{-}9]$$

Substituting Equation [3.1-9] in [3.1-7] and canceling 4π and R shows that the total flux through the surface is

$$\phi = \frac{q}{\varepsilon_0} \qquad\qquad [3.1\text{-}10]$$

and combining Equation [3.1-10] with [3.1-6], integrated to sum over all contributions, that

$$\oint_S \mathbf{E} \cdot d\mathbf{S} = \frac{q}{\varepsilon_0} \qquad\qquad [3.1\text{-}11]$$

This is the *integral form of Gauss' law of electrostatics*. It says that the total flux of electric field strength through a closed surface is proportional to the charge enclosed by that surface.

If there are many charges and if these charges are distributed throughout a volume V, the charge distribution is better represented by the *volume charge density*, ρ, which has the unit coulombs/meter^3, C m^{-3}. Equation [3.1-11] now becomes

$$\oint_S \mathbf{E} \cdot d\mathbf{S} = \frac{1}{\varepsilon_0} \int_V \rho \, dV \qquad\qquad [3.1\text{-}12]$$

where the left-hand integral is taken over the closed surface, S, and the right-hand integral is taken over the volume, V.*

---

*Although I have written single integrals in Equation [3.1-12], and elsewhere, we should realize that with a surface integral the integration must be carried out twice, once for each dimension. With a volume integral, the integration must be carried out three times. Thus $\int_S dS$ is only a short-hand notation for the double integral $\iint_S dx \, dy$, and $\int_V dV$ stands for the triple integral $\iiint_V dx \, dy \, dz$.

The flux out of a volume element, a *voxel*, ΔV can be visualized as a *divergence* of flux lines. If the voxel has the dimensions Δx, Δy, Δz, the flux passing through the left-hand face, in direction of $-x$, is

$$d\phi_{-x} = E_x \, \Delta y \, \Delta z$$

and the flux passing through the right-hand face, in the direction of $+x$,

$$d\phi_{+x} = E_{x+\Delta x} \, \Delta y \, \Delta z$$

The difference between the two fluxes can be written in the form of a partial derivative,

$$d\phi_x = \left(\frac{\partial E_x}{\partial x}\right)\Delta x$$

The same argument applies to the other faces; thus the total flux ϕ out of the voxel is the sum of three partial derivatives:

$$\phi = \frac{\partial E_x}{\partial x} + \frac{\partial E_y}{\partial y} + \frac{\partial E_z}{\partial z} \qquad [3.1\text{-}13]$$

In order to be independent of the coordinate system, we use unit vectors, $\hat{\mathbf{i}}$, $\hat{\mathbf{j}}$, $\hat{\mathbf{k}}$. Equation [3.1-13] then becomes

$$\phi = \frac{\partial \mathbf{E}}{\partial x}\cdot\hat{\mathbf{i}} + \frac{\partial \mathbf{E}}{\partial y}\cdot\hat{\mathbf{j}} + \frac{\partial \mathbf{E}}{\partial z}\cdot\hat{\mathbf{k}} \qquad [3.1\text{-}14]$$

As on page 185, we use the vector operator del, $\boldsymbol{\nabla}$. But since we are considering the divergence (rather than a gradient), we add to it the dot symbol. This gives us a new operator, $\boldsymbol{\nabla}\cdot$, the *divergence operator,* and simplifies Equation [3.1-14] to

$$\phi = \boldsymbol{\nabla}\cdot\mathbf{E} \qquad [3.1\text{-}15]$$

The product, as before, is a scalar; it is the *divergence of* **E.** Substituting Equation [3.1-10] in this equation yields

$$\boldsymbol{\nabla}\cdot\mathbf{E} = \frac{q}{\varepsilon_0} \qquad [3.1\text{-}16]$$

which is the *differential form of Gauss' law*.

So far we have discussed charges in free space. There are some differences in matter. An external electric field causes (positive) nuclei and (negative) electrons to become displaced with respect to each other. This phenomenon is called *electric polarization,* **P,** or, more precisely, *polarization of electric charges in dielectric matter*. The magnitude of this polarization can be described by the *electric displacement,* **D.**

Generally, the displacement **D** and the field strength **E** are proportional to each other:

$$\mathbf{D} = \varepsilon\mathbf{E}$$ [3.1-17]

where ε is now the *electric permittivity of matter,* rather than the permittivity of free space, ε_0. The ratio $\varepsilon/\varepsilon_0$ is called the *relative permittivity,* or *dielectric constant, K.*

Replacing ε_0 by ε in Equation [3.1-16] and substituting Equation [3.1-17] gives

$$\boxed{\nabla\cdot\mathbf{D} = \rho}$$ [3.1-18]

which is *Maxwell's first equation in differential form.* It is essentially the same as the differential form of Gauss' law, and, in a way, equivalent to Coulomb's law. If the charge density in the region is zero,

$$\nabla\cdot\mathbf{E} = 0$$ [3.1-19]

We now take Gauss' law,

$$\oint_S \mathbf{E}\cdot d\mathbf{S} = \frac{1}{\varepsilon}\int_V \rho\,dV$$ [3.1-12]

replace $\varepsilon\mathbf{E}$ by **D,** and ρ by Equation [3.1-18], integrated over the volume V:

$$\oint_S \mathbf{D}\cdot d\mathbf{S} = \int_V \nabla\cdot\mathbf{D}\,dV$$ [3.1-20]

This is an important relationship known as *Gauss' divergence theorem.* It relates the surface integral of a vector function to the volume integral of the divergence of this same vector function. The surface integral is taken over the closed surface, S, bounding the volume, V. If we then integrate both sides of Equation [3.1-18] over V and substitute into it Equation [3.1-20], we obtain

$$\boxed{\oint_S \mathbf{D}\cdot d\mathbf{S} = \int_V \rho\,dV}$$ [3.1-21]

which is *Maxwell's first equation in integral form.* It shows that the flux of the displacement **D** through a closed surface is equal to the (free) charge enclosed by that surface. In free space where $\rho = 0$, again

$$\oint_S \mathbf{E}\cdot d\mathbf{S} = 0$$ [3.1-22]

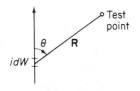

Figure 3.1-3 Measuring the magnetic field produced by a steady current.

Maxwell's second equation. Coulomb's law, which forms the basis of Maxwell's first equation, describes the force between charges *at rest*. Now we turn to charges *in motion*. Moving charges cause a *magnetic induction field*, **B**. Ordinarily, these charges are simply a current, i, in an element of wire, dW. As before, R is the distance from dW to the test point (Figure 3.1-3).

Individual contributions, dB, to the induction field cannot be observed because a moving charge by necessity requires at least part of a circuit. But it is helpful to describe such contributions as

$$dB = \mu_0 \frac{1}{4\pi} \frac{1}{R^2} i \, dW \sin \theta$$

or, in vector notation, as

$$d\mathbf{B} = \mu_0 \frac{1}{4\pi} \frac{1}{R^2} i \, d\mathbf{W} \times \hat{\mathbf{R}} \qquad [3.1\text{-}23]$$

The total induction, **B,** in a realistic circuit, is then found by integration,

$$\mathbf{B} = \mu_0 \frac{1}{4\pi} i \int \frac{1}{R^2} d\mathbf{W} \times \hat{\mathbf{R}} \qquad [3.1\text{-}24]$$

which is *Biot–Savart's law*. (The formal solution of this integral is given in Chapter 3.6, page 386).

The unit of magnetic induction, **B,** is the *tesla,* T, the same as webers/meter^2, Wb m^{-2}, or volt-sec/meter^2, V s m^{-2}. The constant μ_0 is called the *magnetic permeability of free space*. Its magnitude is $4\pi \times 10^{-7}$ Wb A^{-1} m^{-1}, by definition.

It is often preferable to replace the current i by the current *density j*, the current per unit area, $j = i/A$. Then,

$$\mathbf{B} = \mu_0 \frac{1}{4\pi} \int_V \frac{1}{R^2} \mathbf{j} \times \hat{\mathbf{R}} \, dV \qquad [3.1\text{-}25]$$

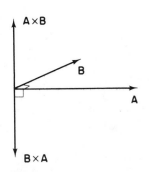

Figure 3.1-4 The vector cross product. The direction of **A × B** is found from the right-hand rule: Rotate **A** toward **B** with the fingers of the right hand; the thumb will point along **A × B**. The cross product **B × A** points in the opposite direction. **A × B** = **−B × A.**

The direction of **B,** in contrast to **D** and **E,** is given as a *vector cross product*. The product $d\mathbf{W} \times \mathbf{R}$, therefore, is again a vector, pointing in a direction *normal* to the plane defined by **W** and **R** (Figure 3.1-4).

Consider again the divergence of magnetic flux lines out of a voxel. When we apply the divergence operator to the **B** field, then

$$\nabla \cdot \mathbf{B} = \mu_0 \frac{1}{4\pi} \int_V \nabla \cdot \left(\frac{1}{R^2} \mathbf{j} \times \hat{\mathbf{R}} \right) dV \qquad [3.1\text{-}26]$$

The current density **j** is assumed to be constant; thus $\nabla \cdot \mathbf{j} = 0$. All that is left is the term $\mathbf{j} \cdot \mathbf{\nabla} \times (\hat{\mathbf{R}}/R)$, which becomes zero, so that

$$\boxed{\mathbf{\nabla} \cdot \mathbf{B} = 0}$$ [3.1-27]

which is *Maxwell's second equation in differential form.* One consequence of this equation is that magnetic flux lines form closed loops, in contrast to electric flux lines. Although predicted, magnetic monopoles (isolated magnetic "charges") have not yet been found.

In free space, the magnetic induction, **B**, and the *magnetic field strength,* **H**, are related as $\mathbf{B} = \mu_0\mathbf{H}$, with **H** given in units of A m^{-1}. In matter,

$$\mathbf{B} = \mu\mathbf{H}$$ [3.1-28]

where μ is now the *magnetic permeability of matter.* The ratio of both, μ/μ_0, is the *relative permeability.*

As before, we integrate the magnetic induction over a closed surface, S, enclosing a volume, V, of arbitrary size and shape,

$$\boxed{\oint_S \mathbf{B} \cdot d\mathbf{S} = 0}$$ [3.1-29]

which is *Maxwell's second equation in integral form.* In a way, this is the magnetic equivalent of Gauss' law,

$$\oint_S \mathbf{D} \cdot d\mathbf{S} = q$$ [3.1-11a]

Again, the integral form and the differential form of Maxwell's second equation may be transformed into each other using Gauss' divergence theorem,

$$\oint_S \mathbf{B} \cdot d\mathbf{S} = \int_V \mathbf{\nabla} \cdot \mathbf{B} \, dV$$ [3.1-30]

This is the magnetic equivalent of Equation [3.1-20].

Maxwell's third equation.   While the first Maxwell equation deals with a *steady* electric field, and the second equation with a steady magnetic field, we now come to *fields that vary as a function of time.* Assume that a bar magnet is thrust through a loop of wire (see footnote, page 354). This will cause an *electromotive force,* EMF, to occur in the wire. The magnitude of this force is proportional to the time rate of change of the magnetic flux, ϕ,

$$\text{EMF} = -\frac{d\phi}{dt} \qquad\qquad [3.1\text{-}31]$$

which is *Faraday's law of induced electricity.*

The *elementary linear generator,* shown in Figure 3.1-5, consists of two parallel wires that are a distance L apart. They form a track along which a metal rod can be moved. If the rod is moved through distance dx, at a velocity v, it cuts through a magnetic field, **B,** oriented at right angles to the plane defined by the track.

But B is the magnetic flux *density,* $B = \phi/A$. Then, since $A = L\, dx$, Faraday's law becomes

$$-\text{EMF} = \frac{d\phi}{dt} = \frac{B\, dA}{dt} = \frac{BL\, dx}{dt}$$

and, since $dx/dt = v$,

$$-\text{EMF} = BLv \qquad\qquad [3.1\text{-}32]$$

Consider now the *work* done on a charge, moving it from a point A, along a fraction of path dW, to a point B,

$$w = \int_A^B \mathbf{E} \cdot d\mathbf{W} \qquad\qquad [3.1\text{-}33]$$

which is a *line integral.* If the work is done instead along a *closed* path, for example if it is done around an elemental surface area dA, then the EMF is the *circulation* of **E,** along dW,

$$\text{EMF} = \oint_W \mathbf{E} \cdot d\mathbf{W} \qquad\qquad [3.1\text{-}34]$$

A real surface, of course, contains a great many surface elements, all lying side by side and each surrounded by a circulating current. Since all of the circulations are in the same sense, the circulations *between* elements

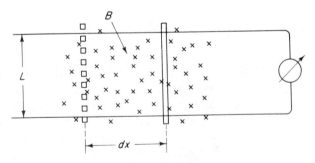

Figure 3.1-5 Elementary generator.

cancel, and all that is left is the circulation around the periphery, the closed path bounding the surface. We now wish to relate, in general terms, the line integral of a vector function along a closed path, $\oint_W \mathbf{E} \cdot d\mathbf{W}$, the right-hand part of Equation [3.1-34], to the surface integral of the *normal* component, the vector $\nabla \times \mathbf{E}$, of the enclosed surface. This is possible by the use of *Stokes' theorem*,

$$\oint_W \mathbf{E} \cdot d\mathbf{W} = \int_S (\nabla \times \mathbf{E}) \cdot d\mathbf{S} \qquad [3.1\text{-}35]$$

which means that the total work required to move $\mathbf{E}$ around the closed path (left-hand integral) is equal to the sum of the individual $(\nabla \times \mathbf{E}) \cdot d\mathbf{S}$ terms (right-hand integral). The surface need not have any particular shape; it need only be bounded by a closed path.

Now consider the magnetic flux, ϕ, of the induction $\mathbf{B}$ through an element $d\mathbf{S}$,

$$\phi = \int_S \mathbf{B} \cdot d\mathbf{S} \qquad [3.1\text{-}36]$$

By substituting Equation [3.1-36] in [3.1-31], we obtain

$$\text{EMF} = -\frac{d\phi}{dt} = -\frac{d}{dt} \left[\int_S \mathbf{B} \cdot d\mathbf{S} \right]$$

and hence

$$\text{EMF} = -\int_S \frac{d\mathbf{B}}{dt} \cdot d\mathbf{S} \qquad [3.1\text{-}37]$$

Combining then Equations [3.1-34] and [3.1-37] yields

$$\boxed{\oint_W \mathbf{E} \cdot d\mathbf{W} = -\int_S \frac{\partial \mathbf{B}}{\partial t} \cdot d\mathbf{S}} \qquad [3.1\text{-}38]$$

which is *Maxwell's third equation in integral form.*

Next we substitute Stokes' theorem in this equation. Since S represents the same surface, the integrands of the two surface integrals must be equal, and

$$\boxed{\nabla \times \mathbf{E} = -\frac{\partial \mathbf{B}}{\partial t}} \qquad [3.1\text{-}39]$$

which is *Maxwell's third equation in differential form.*

Both forms of Maxwell's third equation tell us that whenever a magnetic field changes with time, it generates an electric field.

Maxwell's fourth equation.   Maxwell's fourth equation is the final link. It can be considered the reverse of Faraday's law. According to Faraday's law, a changing magnetic field generates an electric field; and so does a changing electric field generate a magnetic field.

We derive the integral form of Maxwell's fourth equation from Biot–Savart's law, extending it to fluctuating currents. Follow a circular path L of radius R, concentric and located in a plane normal to a long straight wire that carries a current i. Then evaluate the line integral $\oint \mathbf{B} \cdot d\mathbf{L}$. The vector field $\mathbf{B}$ is everywhere parallel (tangent) to L. If we start at a given point on the path and return to the same point, we find that

$$\oint_L \mathbf{B} \cdot d\mathbf{L} = \mu_0 \frac{1}{2\pi R} i \oint dL \qquad [3.1\text{-}40]$$

But since path L is a closed loop, $L = 2\pi R$ and

$$\oint_L \mathbf{B} \cdot d\mathbf{L} = \int_0^{2\pi} \mu_0 \frac{1}{2\pi} i \frac{1}{R} R \, d\theta = \mu_0 \frac{1}{2\pi} i \frac{1}{R} 2\pi R$$

so that

$$\oint_L \mathbf{B} \cdot d\mathbf{L} = \mu_0 i \qquad [3.1\text{-}41]$$

which is *Ampère's circuital law*.

Current is the flow of charge per unit time, $i = dq/dt$, and current density, j, as we have seen, is $j = i/A$. *Surface* charge density, σ, furthermore, is $\sigma = q/A$. Therefore,

$$j = \frac{i}{A} = \frac{dq/dt}{A} = \frac{d\sigma}{dt} \qquad [3.1\text{-}42]$$

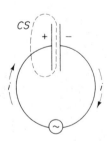

Figure 3.1-6   Deriving concept of displacement current.

Now comes the essential point. While Ampère's law is adequate for steady currents, such currents are not the only source of a magnetic field. This is seen from the fact that when a capacitor is being charged, a magnetic field is set up inside the capacitor, between the two plates.

Consider first a capacitor without dielectric. As charge flows in, it accumulates on one of the plates until the capacitor is fully charged to the voltage of the source. Assume that the charged plate is surrounded by a closed surface, *CS*, as shown in Figure 3.1-6. Then, while the capacitor is being charged, current i goes in but apparently none is coming out. For reasons of continuity, Clerk Maxwell thought that as much *displacement current* should exist between the plates, and go out across the closed surface, as real current is going in.

Displacement current, in short, exists whenever the electric field changes, but it ceases to exist as soon as $\mathbf{E}$ becomes constant. The displacement current *density*, $\mathbf{J}$, likewise, varies with $\mathbf{E}$ as a function of time, $\mathbf{J} = \epsilon_0\,\partial\mathbf{E}/\partial t$, or, since in free space $\epsilon_0\mathbf{E} = \mathbf{D}$,

$$\mathbf{J} = \frac{\partial\mathbf{D}}{\partial t} = \epsilon_0\frac{\partial\mathbf{E}}{\partial t} \qquad [3.1\text{-}43]$$

If the two plates are separated by a dielectric, then, as before, the electric field between the plates will cause the dielectric to become (electrically) polarized. This polarization, to some extent, is like a current, although the polarization charges never break away from the dielectric. Adding the polarization contribution $\partial\mathbf{P}/\partial t$, the *total* displacement current density is

$$\frac{\partial\mathbf{D}}{\partial t} = \epsilon\frac{\partial\mathbf{E}}{\partial t} + \frac{\partial\mathbf{P}}{\partial t} \qquad [3.1\text{-}44]$$

We return to Ampère's circuital law and replace $\mathbf{B}$ by $\mu_0\mathbf{H}$:

$$\oint_L \mathbf{H}\cdot d\mathbf{L} = i \qquad [3.1\text{-}45]$$

But from $j = i/A$ we see that current is also the sum of the current densities across all surface elements,

$$i = \int_S \mathbf{J}\cdot d\mathbf{S} \qquad [3.1\text{-}46]$$

Combining these last two equations, eliminating i, and replacing the time-varying part of $\mathbf{J}$ by Equation [3.1-43], we obtain

$$\oint_L \mathbf{H}\cdot d\mathbf{L} = \int_S \mathbf{J}\cdot d\mathbf{S} + \int_S \frac{\partial\mathbf{D}}{\partial t}\cdot d\mathbf{S} \qquad [3.1\text{-}47]$$

which is *Maxwell's fourth equation in integral form*. For steady currents, the last term drops out.

The left-hand side of this equation may be transformed into a surface integral using Stokes' theorem,

$$\oint_L \mathbf{H}\cdot d\mathbf{L} = \int_S (\mathbf{\nabla}\times\mathbf{H})\cdot d\mathbf{S} \qquad [3.1\text{-}48]$$

Since all integrals in Equations [3.1-47] and [3.1-48] refer to the same surface, the integrands are equal and

$$\mathbf{\nabla}\times\mathbf{H} = \mathbf{J} + \frac{\partial\mathbf{D}}{\partial t} \qquad [3.1\text{-}49]$$

which is *Maxwell's fourth equation in differential form*.

Consequences of Maxwell's equations. The terms $\partial \mathbf{D}/\partial t$ in both Equations [3.1-47] and [3.1-49] are one of Clerk Maxwell's major contributions to electromagnetic theory. These terms show that the variation in time of the electric field generates (part of) the magnetic field, just as Maxwell's third equations show that a time-varying magnetic field gives rise to an electric field. Moreover, from the vector cross products in Equations [3.1-39] and [3.1-49] we see that these two fields must be normal to each other and, as I will show shortly, normal to the direction of propagation, in free space and in isotropic, homogeneous media. And most significantly, the electric field arises directly from the (time-varying) magnetic field, and the magnetic field directly from the (time-varying) electric field. No loop of wire, or any other medium, is necessary: *Electromagnetic waves propagate of their own through space.*

In addition, there is another sequel to Maxwell's equations. It connects the last two Maxwell equations to the general wave equation, derived in Chapter 2.1, and to the velocity of light, discussed further in the next chapter. We begin with the differential form of Maxwell's third equation,

$$\nabla \times \mathbf{E} = -\frac{\partial \mathbf{B}}{\partial t}$$

and substitute $\mathbf{B} = \mu_0 \mathbf{H}$:

$$\nabla \times \mathbf{E} = -\mu_0 \frac{\partial \mathbf{H}}{\partial t} \qquad [3.1\text{-}50]$$

We then take the vector cross product (the curl) of this equation,

$$\nabla \times (\nabla \times \mathbf{E}) = -\mu_0 \frac{\partial}{\partial t}(\nabla \times \mathbf{H}) \qquad [3.1\text{-}51]$$

and substitute for $\nabla \times \mathbf{H}$ the differential form of Maxwell's fourth equation,

$$\nabla \times \mathbf{H} = \mathbf{J} + \frac{\partial \mathbf{D}}{\partial t}$$

Setting $\mathbf{D} = \varepsilon_0 \mathbf{E}$, we obtain

$$\nabla \times (\nabla \times \mathbf{E}) = -\mu_0 \frac{\partial}{\partial t}\left(\mathbf{J} + \varepsilon_0 \frac{\partial \mathbf{E}}{\partial t}\right) \qquad [3.1\text{-}52]$$

In this way, $\mathbf{H}$ has been eliminated and $\mathbf{E}$ remains. The curl of a curl of a vector function is

$$\nabla \times (\nabla \times \mathbf{E}) = \nabla(\nabla \cdot \mathbf{E}) - (\nabla \cdot \nabla)\mathbf{E} = \nabla\nabla \cdot \mathbf{E} - \nabla^2 \mathbf{E} \qquad [3.1\text{-}53]$$

Then, from Maxwell's first equation for a charge-free region, $\nabla \cdot \mathbf{E} = 0$, and from the general form of Ohm's law, $\mathbf{J} = \sigma \mathbf{E}$,

$$\mathbf{\nabla} \times (\mathbf{\nabla} \times \mathbf{E}) = -\mu_0 \frac{\partial}{\partial t}\left(\sigma \mathbf{E} + \varepsilon_0 \frac{\partial \mathbf{E}}{\partial t}\right) = \mathbf{\nabla}(0) - \nabla^2 \mathbf{E} \qquad [3.1\text{-}54]$$

and

$$\nabla^2 \mathbf{E} = \mu_0 \sigma \frac{\partial \mathbf{E}}{\partial t} + \varepsilon_0 \mu_0 \frac{\partial^2 \mathbf{E}}{\partial t^2} \qquad [3.1\text{-}55]$$

In free space, the middle term drops out:

$$\nabla^2 \mathbf{E} = \varepsilon_0 \mu_0 \frac{\partial^2 \mathbf{E}}{\partial t^2} \qquad [3.1\text{-}56]$$

Also, light propagating along the $+x$ axis must be independent of y and z, and thus $\nabla^2 \mathbf{E}$ can be replaced:

$$\frac{\partial^2 \mathbf{E}}{\partial x^2} = \varepsilon_0 \mu_0 \frac{\partial^2 \mathbf{E}}{\partial t^2} \qquad [3.1\text{-}57]$$

Now we set $\mathbf{E} = \xi$ and divide Equation [3.1-57] by the one-dimensional wave equation, Equation [2.1-49]:

$$\frac{\partial^2 \xi}{\partial t^2} = v^2 \frac{\partial^2 \xi}{\partial x^2}$$

In free space this yields

$$\boxed{c = \frac{1}{\sqrt{\epsilon_0 \mu_0}}} \qquad [3.1\text{-}58]$$

an important equation that comes up again in the next chapter.

Energy transfer.   One of the interesting aspects of electromagnetic fields is that they carry *energy*. Consider again the two field vectors, $\mathbf{E}$ and $\mathbf{H}$. First we determine the divergence of the cross product $\mathbf{E} \times \mathbf{H}$. From the identity

$$\mathbf{\nabla} \cdot (\mathbf{A} \times \mathbf{B}) = \mathbf{B} \cdot (\mathbf{\nabla} \times \mathbf{A}) - \mathbf{A} \cdot (\mathbf{\nabla} \times \mathbf{B}) \qquad [3.1\text{-}59]$$

it follows that

$$\mathbf{\nabla} \cdot (\mathbf{E} \times \mathbf{H}) = -\mathbf{E} \cdot (\mathbf{\nabla} \times \mathbf{H}) + \mathbf{H} \cdot (\mathbf{\nabla} \times \mathbf{E}) \qquad [3.1\text{-}60]$$

We substitute for the cross products on the right-hand side Maxwell's fourth and third equations, respectively. This gives

$$\mathbf{\nabla} \cdot (\mathbf{E} \times \mathbf{H}) = -\left(\mathbf{E} \cdot \frac{\partial \mathbf{D}}{\partial t} + \mathbf{H} \cdot \frac{\partial \mathbf{B}}{\partial t}\right) \qquad [3.1\text{-}61]$$

In free space, or in a medium that is linear and nondispersive (i.e., whose ε and μ are independent of the field strengths) and setting $\mathbf{D} = \epsilon_0 \mathbf{E}$ and $\mathbf{B} = \mu_0 \mathbf{H}$, the time derivatives on the right-hand side can be written

$$\mathbf{E} \cdot \frac{\partial \mathbf{D}}{\partial t} = \mathbf{E} \cdot \frac{\partial}{\partial t} \varepsilon_0 \mathbf{E} = \frac{1}{2} \varepsilon_0 \frac{\partial}{\partial t} E^2 \qquad [3.1\text{-}62]$$

and

$$\mathbf{H} \cdot \frac{\partial \mathbf{B}}{\partial t} = \mathbf{H} \cdot \frac{\partial}{\partial t} \mu_0 \mathbf{H} = \frac{1}{2} \mu_0 \frac{\partial}{\partial t} H^2 \qquad [3.1\text{-}63]$$

so that

$$\nabla \cdot (\mathbf{E} \times \mathbf{H}) = -\frac{\partial}{\partial t} \frac{1}{2} (\varepsilon_0 E^2 + \mu_0 H^2) \qquad [3.1\text{-}64]$$

where E and H are the instantaneous values of the fields. Integrating over the volume V bounded by the surface S yields

$$\oint_S (\mathbf{E} \times \mathbf{H}) \cdot d\mathbf{S} = -\frac{\partial}{\partial t} \int_V \frac{1}{2} (\varepsilon_0 E + \mu_0 H) \, dV \qquad [3.1\text{-}65]$$

The integrand on the right-hand side refers to the time rate or *flow of electromagnetic energy*, called the *Poynting vector*, **S***:

$$\boxed{\mathbf{S} = \mathbf{E} \times \mathbf{H}} \qquad [3.1\text{-}66]$$

It has the unit watts per square meter, W m^{-2}.

From the vector cross product we conclude that **S**, **E**, and **H** form a mutually orthogonal triad;[†] thus, since electromagnetic waves are transverse waves, **S** points in the direction of propagation of the wave (Figure 3.1-7).

Radiation pressure. Electromagnetic waves also carry *momentum*, which shows as *radiation pressure*, p. If the wave is incident on an absorbing (black) surface, the rate of change of momentum is equal to the force on the surface, p = F/A. Since *power*, ϕ, is a measure of how fast work can be done, $\phi = w/t$, and since work is force × distance, $w = Fs$, and velocity $v = s/t$, we find that $\phi = Fv$. Solving for F and substituting in the definition of pressure gives

---

*Named after John Henry Poynting (1852–1914), British physicist, professor of physics at the University of Birmingham.

[†]Both **E** and **H** each contribute one-half the total energy. However, the **E** field is considerably more *effective* than **H**; thus **E** is often referred to as the *optical field* or *light vector*. Most effects caused by light, from the photoelectric effect to the dissociation of silver halides in photographic film to the response of the retina, are due to the **E** field; the only exception is the *pressure* exerted by electromagnetic radiation.

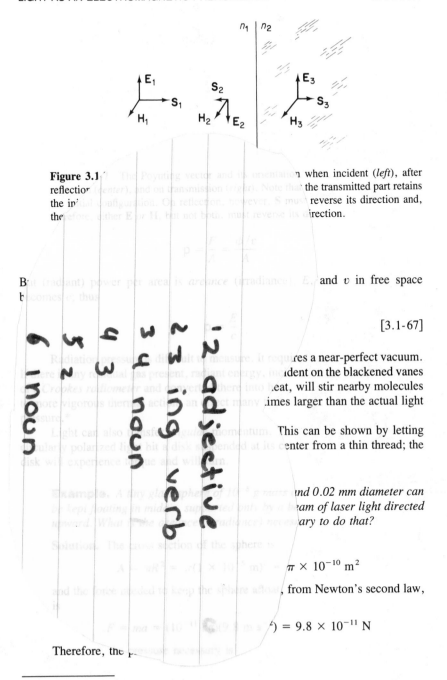

Figure 3.1 The Poynting vector and its orientation when incident (*left*), after reflection (*center*), and on transmission (*right*). Note that the transmitted part retains the initial configuration. On reflection, however, S must reverse its direction and, therefore, either E or H, but not both, must reverse its direction.

But (radiant) power per area is *irradiance* (irradiance) E, and v in free space becomes c; thus

$$[3.1-67]$$

Radiation pressure is difficult to measure. It requires a near-perfect vacuum. If there is any residual gas present, radiant energy, incident on the blackened vanes of a *Crookes radiometer* and converted there into heat, will stir nearby molecules into more vigorous thermal action than an effect many times larger than the actual light pressure.*

Light can also transfer *angular* momentum. This can be shown by letting circularly polarized light hit a disk suspended at its center from a thin thread; the disk will experience a torque and will turn.

Example. A tiny glass sphere of 10^{-8} grams and 0.02 mm diameter can be kept floating in midair, supported only by a beam of laser light directed upward. What is the necessary irradiance necessary to do that?

Solution. The cross section of the sphere is

$$A = \pi R^2 = \pi(1 \times 10^{-5} \text{ m})^2 = \pi \times 10^{-10} \text{ m}^2$$

and the force needed to keep the sphere afloat, from Newton's second law, is

$$F = ma = (10^{-11} \text{ kg})(9.8 \text{ m s}^{-2}) = 9.8 \times 10^{-11} \text{ N}$$

Therefore, the pressure necessary is

*Radiation pressure was measured first by Edward Leamington Nichols and Gordon Ferrie Hull in the United States and by Pyotr Nikolaievich Lebedev in the USSR, about 30 years after such effect had been predicted by Clerk Maxwell.

$$p = \frac{F}{A} = \frac{9.8 \times 10^{-11}\text{N}}{\pi \times 10^{-10} \text{ m}^2} \approx 0.3 \text{ pascal, Pa (N m}^{-2})$$

and the areance, from Equation [3.1-67], is

$$E = pc = 0.3 \times 3 \times 10^8 \approx \boxed{10^8 \text{ W m}^{-2}}$$

This is only five orders of magnitude higher than the areance provided by the sun, which, outside the atmosphere, amounts to 1.4 kW m^{-2}.

Fresnel's Equations

Boundary conditions.
We now turn to (linearly polarized) light incident on a *boundary*. The electric field **E** may oscillate in one of two directions: either *parallel* to the plane of incidence (as in Figure 3.1-8), this is called *p* polarization (for parallel) and the field given the subscript ∥; or the vector may oscillate *normal* to the plane of incidence (not shown in Figure 3.1-8), this is called *s* polarization (for the German "senkrecht" meaning perpendicular) and the field given the subscript ⊥.

As before, the angle of incidence is α and the angle of refraction β. We draw a closed rectangular path, *A–B–C–D*, that includes the boundary, and make the paths *AD* and *BC* vanishingly short (Figure 3.1-9). We then take the line integrals of the electric field around the closed rectangular

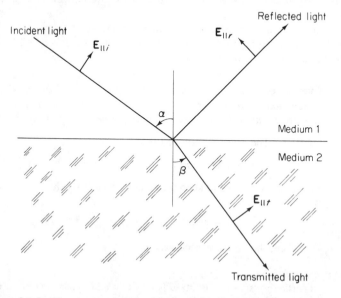

Figure 3.1-8 Notation used in conjunction with Fresnel's equations. Plane of incidence is the plane of the paper.

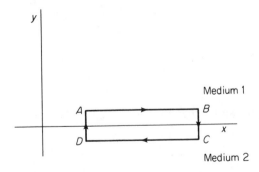

Figure 3.1-9 Closed path across boundary.

path and consider only the x component of the field, that is, the component that is *parallel* to the boundary,

$$\int_A^B \mathbf{E} \cdot d\mathbf{L} + \int_C^D \mathbf{E} \cdot d\mathbf{L} = \int_A^B E_{1x} \, dL - \int_C^D E_{2x} \, dL \qquad [3.1\text{-}68]$$

Since there are no free charges, we can set Equation [3.1-68] equal to zero so that

$$(E_{1x} - E_{2x})L = 0$$

and

$$E_{1x} = E_{2x} \qquad\qquad [3.1\text{-}69]$$

which means that the parallel components of the electric field are *equal and continuous* across the boundary. By a similar argument we find that the parallel components of the magnetic field are equal and continuous also:

$$H_{1x} = H_{2x} \qquad\qquad [3.1\text{-}70]$$

Now we consider the components of the two fields that are *normal* to the boundary (the y components). We construct a pillbox containing the boundary (Figure 3.1-10). Then, from Maxwell's first equation,

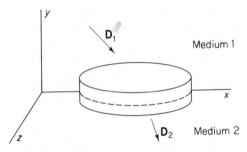

Figure 3.1-10 Pillbox straddling boundary.

$$\oint_S \mathbf{D} \cdot d\mathbf{S} = \int_V \rho \, dV$$

$$\oint_S \mathbf{D} \cdot d\mathbf{S} = \int_{S_1} \mathbf{D}_1 \cdot d\mathbf{S} + \int_{S_2} \mathbf{D}_2 \cdot d\mathbf{S} = -\int_{S_1} D_{1y} \, dS + \int_{S_2} D_{2y} \, dS$$

$$[3.1\text{-}71]$$

As long as there are no free charges present, this reduces to

$$(-D_{1y} + D_{2y})A = 0$$

and therefore

$$D_{1y} = D_{2y} \qquad\qquad [3.1\text{-}72]$$

which shows that the *normal* components of the electric displacement are equal and continuous across the boundary. The same applies to the normal components of the magnetic induction, **B**, using Maxwell's second equation,

$$\oint_S \mathbf{B} \cdot d\mathbf{S} = 0$$

and hence,

$$B_{1y} = B_{2y} \qquad\qquad [3.1\text{-}73]$$

It is only the *normal* components of **D**, or **B**, that are equal on both sides of the boundary. The total magnitudes of **D**, or **B**, need not be equal, nor do they have to point in the same direction; in fact, these fields may well be *reflected* or *refracted*, and may change direction.

Fresnel's equations.
In conjunction with the Fabry–Perot interferometer (page 265), I referred to the coefficients of reflection and transmission. Actually, there are four coefficients, two each for reflection and for transmission, depending on whether the **E** field oscillates parallel to the plane of incidence or normal to it. Thus we have two amplitude reflection coefficients, $\rho_\|$ and $\rho_\perp$, and two amplitude transmission coefficients, $\tau_\|$ and $\tau_\perp$. The squares of these coefficients are the *reflectivities*, $r_\|$ and $r_\perp$, and the transmissivities, $t_\|$ and $t_\perp$, respectively.*

Assume that there is no loss of energy. Then the flow of energy, the Poynting vector, of the incident light must equal the sum of the Poynting vectors of the reflected and the transmitted light:

$$\mathbf{S}_i = \mathbf{S}_r + \mathbf{S}_t \qquad\qquad [3.1\text{-}74]$$

---

*Terms ending in *-ion*, such as *reflection*, describe a process. Words ending in *-ivity*, such as *reflectivity*, refer to the general property of a material. Terms ending in *-ance*, such as *reflectance*, refer to the properties of a given object.

where the subscripts refer to the notation in Figure 3.1-8. Vectors $\mathbf{S}_i$ and $\mathbf{S}_r$ are in medium 1 (index n_1) and vector $\mathbf{S}_t$ is in medium 2 (n_2); thus, since energy is proportional to the square of the amplitude,

$$n_1 \mathbf{E}_i^2 = n_1 \mathbf{E}_r^2 + n_2 \mathbf{E}_t^2$$

or

$$n_1(\mathbf{E}_i^2 - \mathbf{E}_r^2) = n_2 \mathbf{E}_t^2 \qquad [3.1\text{-}75]$$

If A is the area of the boundary intercepted by the bundle of light, then the cross sections, C, of the individual bundles are

$$C_i = A \cos \alpha, \qquad C_r = A \cos \alpha, \qquad C_t = A \cos \beta$$

The ratio of the cross section of the transmitted light to that of the incident light, for example, is

$$\frac{C_t}{C_i} = \frac{\cos \beta}{\cos \alpha} \qquad [3.1\text{-}76]$$

so that Equation [3.1-75] becomes

$$n_1(\mathbf{E}_i^2 - \mathbf{E}_r^2)\cos \alpha = n_2 \mathbf{E}_t^2 \cos \beta \qquad [3.1\text{-}77]$$

Now consider separately the $\parallel$ and $\perp$ components. First we replace the term in parentheses on the left-hand side of Equation [3.1-75] by $(\mathbf{E}_i + \mathbf{E}_r)(\mathbf{E}_i - \mathbf{E}_r)$. Then we recall from Equation [3.1-69] that the $\parallel$ components are equal and continuous across the boundary. Hence, we can divide Equation [3.1-75] by

$$\mathbf{E}_{\parallel i} - \mathbf{E}_{\parallel r} = \mathbf{E}_{\parallel t} \qquad [3.1\text{-}78]$$

This gives

$$n_1(\mathbf{E}_{\parallel i} + \mathbf{E}_{\parallel r})\cos \alpha = n_2 \mathbf{E}_{\parallel t} \cos \beta \qquad [3.1\text{-}79]$$

We then eliminate $\mathbf{E}_{\parallel t}$, and use Snell's law to eliminate n_2/n_1:

$$\left(\frac{\mathbf{E}_r}{\mathbf{E}_i}\right)_{\parallel} = \frac{n_2 \cos \alpha - n_1 \cos \beta}{n_2 \cos \alpha + n_1 \cos \beta} = \frac{\sin \alpha \cos \alpha - \sin \beta \cos \beta}{\sin \alpha \cos \alpha + \sin \beta \cos \beta}$$

Thus

$$\boxed{\left(\frac{\mathbf{E}_r}{\mathbf{E}_i}\right)_{\parallel} = \frac{\tan(\alpha - \beta)}{\tan(\alpha + \beta)} = \rho_{\parallel}} \qquad [3.1\text{-}80]$$

which is *Fresnel's first equation*. It describes the amplitude reflection coefficient for $\parallel$ light, $\rho_{\parallel}$.

If instead we eliminate $E_{\parallel r}$ in Equation [3.1-79], we obtain

$$\left(\frac{E_t}{E_i}\right)_{\parallel} = \frac{2n_1 \cos \alpha}{n_1 \cos \beta + n_2 \cos \alpha} = \frac{2 \cos \alpha \sin \beta}{\sin \alpha \cos \alpha + \sin \beta \cos \beta}$$

or

$$\boxed{\left(\frac{E_t}{E_i}\right)_{\parallel} = \frac{2 \cos \alpha \sin \beta}{\sin(\alpha + \beta)\cos(\alpha - \beta)} = \tau_{\parallel}} \qquad [3.1\text{-}81]$$

which is *Fresnel's second equation*. It describes the transmission coefficient for $\parallel$ light, $\tau_{\parallel}$.

If Equation [3.1-77] is divided by

$$E_{\perp i} - E_{\perp r} = E_{\perp t} \qquad [3.1\text{-}82]$$

we obtain

$$n_1(E_{\perp i} - E_{\perp r})\cos \alpha = n_2 E_{\perp t} \cos \beta \qquad [3.1\text{-}83]$$

Eliminating $E_{\perp t}$ yields

$$\left(\frac{E_r}{E_i}\right)_{\perp} = \frac{n_1 \cos \alpha - n_2 \cos \beta}{n_1 \cos \alpha + n_2 \cos \beta} = -\frac{\sin \alpha \cos \beta - \sin \beta \cos \alpha}{\sin \alpha \cos \beta - \sin \beta \cos \alpha}$$

or

$$\boxed{\left(\frac{E_r}{E_i}\right)_{\perp} = -\frac{\sin(\alpha - \beta)}{\sin(\alpha + \beta)} = \rho_{\perp}} \qquad [3.1\text{-}84]$$

This is *Fresnel's third equation*. It describes $\rho_{\perp}$.

If we eliminate $E_{\perp r}$, we get

$$\left(\frac{E_t}{E_i}\right)_{\perp} = \frac{2n_1 \cos \alpha}{n_1 \cos \alpha + n_2 \cos \beta} = \frac{2 \cos \alpha \sin \beta}{\cos \alpha \sin \beta + \sin \alpha \cos \beta}$$

and

$$\boxed{\left(\frac{E_t}{E_i}\right)_{\perp} = \frac{2 \cos \alpha \sin \beta}{\sin(\alpha + \beta)} = \tau_{\perp}} \qquad [3.1\text{-}85]$$

which is *Fresnel's fourth equation*. It describes $\tau_{\perp}$.

The first two Fresnel equations, therefore, refer to light whose E vector oscillates parallel to the plane of incidence. In the third and fourth equations, E is normal to it. The first and third Fresnel equations apply to the reflection of light, the second and fourth to transmission.

Reflectivity.   We now apply Fresnel's equations to several practical problems. Let us consider first the *reflectivity at normal incidence*. At normal incidence, there is no difference between ‖ and ⊥ light. But we cannot simply set $\alpha = 0$, because substituting this in Fresnel's first or third equation would give an indeterminate result. Instead, we combine Equations [3.1-80] and [3.1-84],

$$\left(\frac{\mathbf{E}_r}{\mathbf{E}_i}\right)_{\|} = \frac{\tan(\alpha - \beta)}{\tan(\alpha + \beta)} = -\left(\frac{\mathbf{E}_r}{\mathbf{E}_i}\right)_{\perp} = \frac{\sin(\alpha - \beta)}{\sin(\alpha + \beta)}$$

$$= \frac{\sin\alpha\cos\beta - \cos\alpha\sin\beta}{\sin\alpha\cos\beta + \cos\alpha\sin\beta} \qquad [3.1\text{-}86]$$

Then, dividing both numerator and denominator by $\sin\beta$ and, applying Snell's law, replacing $\sin\alpha/\sin\beta$ by n_2/n_1, we obtain

$$\frac{\mathbf{E}_r}{\mathbf{E}_i} = \frac{n_2\cos\beta - n_1\cos\alpha}{n_2\cos\beta + n_1\cos\alpha} \qquad [3.1\text{-}87]$$

At normal incidence, $\alpha = \beta = 0$, and therefore the cosine terms drop out. Furthermore, since $\mathbf{E}_r/\mathbf{E}_i$ squared is the reflectivity, r,

$$\boxed{r = \left(\frac{n_2 - n_1}{n_2 + n_1}\right)^2} \qquad [3.1\text{-}88]$$

an equation that we have seen before in conjunction with antireflection coatings (page 273).

Now we turn to reflection at oblique incidence. How does the *reflectivity change as a function of the angle of incidence?* We distinguish between light incident from the side of the rarer medium (external reflection) and light incident from the side of the denser medium (internal reflection). Of course, we also distinguish between ‖ and ⊥ light.

Let us discuss *external reflection* first. From Equation [3.1-88] we note that for normal incidence and for a boundary between air and glass ($n = 1.5$) the reflectivity is $[(1.5 - 1.0)/(1.5 + 1.0)]^2 = 4$ percent. If we square Fresnel's third equation, we see that the reflectivity of ⊥ light varies as

$$r = \frac{\sin^2(\alpha - \beta)}{\sin^2(\alpha + \beta)} \qquad [3.1\text{-}89]$$

This means that, as α increases, the reflectivity increases, beginning at 4 percent (for the example given) until at grazing incidence it reaches 100 percent. This is shown in Figure 3.1-11, left-hand plot, upper curve.

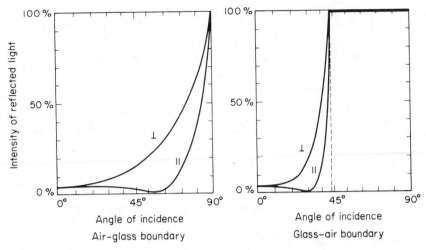

Figure 3.1-11   Reflectivity as a function of the angle of incidence.

The reflectivity of the $\parallel$ light is found from Fresnel's first equation. As the incident light reaches an angle such that $\alpha + \beta = 90°$, the denominator in Equation [3.1-80] becomes infinite and

$$r_\parallel = \frac{\tan^2(\alpha - \beta)}{\tan^2 90°} = 0 \qquad [3.1\text{-}90]$$

This means that at that angle no $\parallel$ light is being reflected.

Continuing with our example where $n = 1.5$ and replacing $\sin \beta$ in Snell's law by $\sin(90° - \alpha) = \cos \alpha$,

$$\frac{\sin \alpha}{\sin \beta} = \frac{\sin \alpha}{\cos \alpha} = \tan \alpha = \frac{n_2}{n_1}$$

$$\alpha = \tan^{-1}\left(\frac{n_2}{n_1}\right) = \tan^{-1}\left(\frac{1.5}{1.0}\right) = 56.3° \qquad [3.1\text{-}91]$$

At this angle, then, all of the reflected light is in the $\perp$ orientation, and none is in $\parallel$, as shown by the dip to zero in the left-hand plot, lower curve. The angle at which this occurs is known as *Brewster's angle* (which we will discuss in more detail in Chapter 3.4).

In *internal reflection* the $\perp$ reflectivity again increases monotonically. However, it reaches 100 percent sooner, at the minimum angle of total internal reflection. For glass of $n = 1.5$, we have

$$\alpha' = \sin^{-1}\left(\frac{n_2}{n_1}\right) = \sin^{-1}\left(\frac{1.0}{1.5}\right) = 41.8° \qquad [3.1\text{-}92]$$

The ∥ light reaches the Brewster angle sooner too:

$$\alpha'' = \tan^{-1}\left(\frac{n_2}{n_1}\right) = \tan^{-1}\left(\frac{1.0}{1.5}\right) = 33.7° \qquad [3.1\text{-}93]$$

At normal incidence the reflectivity is the same for both external and internal reflection. At increasingly oblique incidence, 100 percent reflection is reached sooner for internal reflection, later for external reflection. And finally, Brewster's angle affects only the ∥ component, both in external and internal reflection (see again Figure 3.1-11).

Phase change. Consider first ⊥ light. The third Fresnel equation has a minus sign: the ratio $\mathbf{E}_r/\mathbf{E}_i$ is negative. This means that for external reflection where $n_1 < n_2$ and $\alpha > \beta$, the direction of the electric vector of the reflected light is opposite that of the incident light (see also Figure 3.1-7). In other words, in external reflection, no matter what the angle of incidence, there is a phase change of 180° or π radians (Figure 3.1-12). In internal reflection there is none.

For ∥ light, the situation is a little more complex. As I said before, at normal incidence there is no difference between ∥ and ⊥ light. Thus we might be tempted to apply either Fresnel's first or third equation, wondering why the third equation has a minus sign and the first equation does not. The third equation deals with electric vectors that are *normal* to the plane of incidence, but with ∥ light (Fresnel's first equation) **E** lies *in* the plane of incidence. If the angle of incidence is small, both $\mathbf{E}_\parallel$ vectors point in *opposite* directions, toward the *y* axis, which by definition means a π phase change. Then imagine the rays being folded apart, increasing to 180° the angle between the incident and the reflected light. This makes the electric vectors point in the *same* direction, and would mean no phase change. Clearly, this cannot be. If we stipulate that at normal incidence the **E** fields are opposite, and a π phase change occurs, Fresnel's first equation must have a sign opposite to that of his third equation.

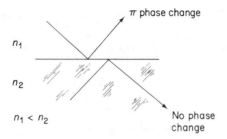

Figure 3.1-12   Phase change (for ⊥ light) occurs only at a rare-to-dense boundary.

Metallic reflection. The reflectivity of most metals is high and depends less on the angle of incidence than on the wavelength. So metals make good reflectors, and it is this property that makes metal reflection of interest in optics.

Refraction by a dielectric such as glass can be described by one figure, the refractive index n, as a function of wavelength. But since metals contain free electrons, their electric conductivity as well as their absorptivity are high, and another parameter is needed, called the *extinction coefficient*, k. These two terms, n and k, together describe the *complex refractive index*, $\mathbf{n}$:

$$\mathbf{n} = n - ik \qquad [3.1\text{-}94]$$

In most transparent materials the value for k is small, almost zero, but in highly absorbing materials such as metals it reaches values of around 3. Note that the complex notation applies merely to the refractive index; the angles α and β are real, as they are in conventional, nonmetallic refraction.

The refraction of light into a metal is very difficult to measure directly because of this high absorptivity. The refractive index of metals can only be inferred indirectly, by determining the reflectivity. At normal incidence, and in contrast to Equation [3.1-88], the reflectivity takes the form

$$r = \frac{(n-1)^2 + k^2}{(n+1)^2 + k^2} \qquad [3.1\text{-}95]$$

In practice, linearly polarized light is reflected at oblique incidence from the polished surface of a block of metal. The $\parallel$ and $\perp$ components of the light then have a certain phase difference between them, causing elliptical polarization. By measuring the ellipticity and the azimuth of the ellipse, the complex refractive index can be determined, a method known as *ellipsometry*.

SUGGESTIONS FOR FURTHER READING

D. J. Griffiths, *Introduction to Electrodynamics* (Englewood Cliffs, NJ: Prentice-Hall, Inc., 1981).

J. R. Reitz, F. J. Milford, and R. W. Christy, *Foundations of Electromagnetic Theory,* 3rd edition (Reading, MA: Addison-Wesley Publishing Company, Inc., 1979).

R. K. Wangsness, *Electromagnetic Fields* (New York: John Wiley & Sons, Inc., 1979).

J. B. Marion and M. A. Heald, *Classical Electromagnetic Radiation,* 2nd edition (New York: Academic Press, Inc., 1980).

R. M. A. Azzam and N. M. Bashara, *Ellipsometry and Polarized Light* (New York: North-Holland Publishing Company, 1977).

PROBLEMS

3.1-1. (a) What is the force between two small spheres separated by 2 cm, one sphere being charged to +8 nC, the other to −4 nC? (b) What is the force if the spheres are first brought into contact and then are separated and moved 2 cm apart?

3.1-2. An electric charge of −4 arbitrary units is located 20 cm to the left of a charge of +2 units (Figure 3.1-13). At a point P the electric field strength is found to be zero. How far is P from the +2 charge?

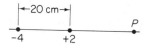

Figure 3.1-13

3.1-3. Two point charges, of 4.2 μC and 6.4 μC, act on each other in free space with a force of 6 N. By how much must one of the charges be moved if the two charges are immersed in a liquid with a relative permittivity of $K=2$ and the force between the charges is to remain constant?

3.1-4. Find the divergence of an electric field that has the x component x^2yz, the y component $-2y^3z$, and the z component x^3y^2z, at a point $(1, 2, -3)$.

3.1-5. A long straight wire carries a steady current of 6 A. Using Biot–Savart's law, determine:

(a) the magnetic induction, and
(b) the magnetic field strength,
assuming that both are caused by the current within 1 cm length of wire, at a point 20 cm from the wire.

3.1-6. An aircraft with a wing span of 20 m travels horizontally at a velocity of 900 km/h. If the vertical component of the Earth's magnetic induction is 50 μT, what is the potential difference generated between the wing tips?

3.1-7. Express the electric field strength, **E**, in terms of charge q, surface area A, and electric permittivity ε_0 and, using the definitions of current and current density, derive Equation [3.1-43].

3.1-8. From Maxwell's third equation we know that a time-varying magnetic field induces an electric field, and from Maxwell's fourth equation that a time-varying electric field induces a magnetic field. Why then are electromagnetic waves often represented such that the **E** vector and the **H** vector both go through zero at the same time, instead of being offset by 90°?

3.1-9. Sunlight, which on reaching the surface of the Earth has an areance of 1.35 kW m^{-2}, falls on a mirror 1 m^2 in size, exerting a certain radiation pressure. How large (or small) a mass, in the gravitational field of the Earth, would exert the same pressure?

3.1-10. A point light source is placed:
(a) In the focus of a collimating lens.
(b) In the focus of a paraboloid mirror.
(c) In one focus of a closed ellipsoid cavity, concentrating the light into the other focus.
Does the system, because of radiation pressure, experience any recoil?

3.1-11. Linearly polarized light is incident at 30° on a boundary between glass ($n = 1.5$) and air. Determine the ratio of the amplitudes, and energies, of the reflected light to those of the incident light, assuming that the **E** fields are oriented normal to the plane of incidence.

3.1-12. Light linearly polarized in the p orientation is internally reflected at an angle of 35° at the boundary between heavy flint ($n = 1.72$) and air ($n = 1.00$). Find the amplitude reflection coefficient and indicate whether or not there will be a phase change.

3.1-13. Calculate the Brewster angle for a glass–oil boundary (glass, $n = 1.52$; oil, $n = 1.81$) for:
(a) External reflection.
(b) Internal reflection.

3.1-14. What is the least angle at which a π phase change occurs in linearly polarized light, the **E** vector oscillating parallel to the plane of incidence and the light internally reflected at a boundary between crown ($n = 1.52$) and carbon disulfide ($n = 1.63$)?

3.1-15. Light falls at normal incidence on a boundary between water ($n = \frac{4}{3}$) and glass. What is the reflectivity if the glass is:
(a) Crown, $n = 1.52$?
(b) Flint, $n = 1.72$?

3.1-16. When light strikes a water surface ($n = \frac{4}{3}$) covered with a thick film of oil ($n = 1.65$), how much light is lost due to surface reflections?

3.1-17. A *cuvette*, made of glass of $n = 1.6$, is filled with a liquid of $n = 1.4$.
(a) What percentage of light, passing through the cuvette, is lost due to reflection on each of the four boundaries?
(b) What percentage finally emerges?

3.1-18. If two mirrors of 99 percent reflectance each are facing each other, and if a beam of light is reflected 50 times back and forth between the mirrors (= 100 reflections), what percentage of the light is lost?

3.1-19. (a) If the refractive index n_2 in Figure 3.1-14 is higher than both n_1 and n_3, or, in shorthand notation, if $1 < 2 > 3$, and if the light is incident from the left, at which of the two boundaries, 1–2 and 2–3, will a 180° phase change occur on reflection?

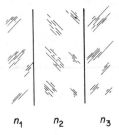

n_1 n_2 n_3

Figure 3.1-14

(b) At which boundaries will a phase change occur if $1 < 2 < 3$?
(c) At which boundaries will a phase change occur if $1 > 2 < 3$?
(d) At which boundaries will a phase change occur if $1 > 2 > 3$?

3.1-20. Prepare a table listing (A) external reflection and (B) internal reflection, each with respect to *p*- and *s*-oriented linearly polarized light, and indicate whether or not there is a π phase change for angles:
(a) Less than Brewster's angle.
(b) Larger than Brewster's angle.

3.2

Velocity
of Light

LIGHT TRAVELS SO FAST that there is nothing in our daily experience
to suggest that the speed of light is not infinite. Damianus, the son of
Heliodor of Larissa, thought that the propagation of light is instantaneous
because "at the same moment as the sun breaks through a cloud the light
reaches us." Naturally, we cannot see the breaking through of the sun
before the light reaches us; so, nothing follows from this argument. Kepler, in agreement with Damianus, stipulated that "light has no mass or
weight."

Galileo was one of the first to suggest that it may take light a finite
time to travel from one point to another. In his book *Discorsi,* he discussed
his theory on the velocity of light, using two fictitious characters, Sagredo
and a somewhat foolish man, Simplicio. Here is part of what they say:

Simplicio: Everyday experience shows that the propagation of light is instantaneous; for when we see a piece of artillery fired, at a great distance, the flash
reaches our eyes without lapse of time; but the sound reaches the ear only after a
noticeable interval.
Sagredo: Well, Simplicio, the only thing I am able to infer from this is that sound,
in reaching our ear, travels more slowly than light; it does not inform me, whether
light travels instantaneous or whether, although extremely rapid, it still takes time.

It is interesting how, over several centuries, the science of measuring
the velocity of light has progressed. The most significant steps along this
way are the subject of this chapter.

Direct Methods

All direct methods of determining the velocity of light are essentially *time-of-flight* measurements. The first of these was *Roemer's method.** Roemer recorded the times when the innermost moon of Jupiter, Io, reemerged from the shadow of Jupiter—which would occur (if Earth were standing still) every 42.5 h. But Earth revolves around the sun and in so doing assumes positions 1–2–3–4 in Figure 3.2-1.

Roemer noticed that when Earth was close to Jupiter (position 2), Io reappeared 10 min early, and when Earth was opposite (position 4) it reappeared 10 min late, compared with the times predicted on the basis of yearly averages. He concluded that the light took twice that time to traverse the diameter of Earth's orbit (at that time believed to be 2.57×10^8 km), and from that he calculated the velocity of light as $214\,000$ km s^{-1}.

Fizeau's method.   The first terrestrial determination of the speed of light was made by Fizeau.[†] The essential part of Fizeau's experiment, the forerunner of many others, is a shutter that chops the light into pulses which then travel, and return, along a given path.

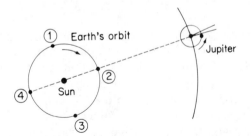

Figure 3.2-1   Roemer's method of measuring the speed of light.

---

*Ole Christensen Roemer (1644–1710), Danish astronomer. An assistant to the French astronomer Picard, Roemer measured the velocity of sound; later, at the Paris observatory, he made his famous determination of the speed of light, announcing the result at a meeting of the French Academy of Sciences in 1676. Five years later he was called back to Denmark by King Christian V to serve as Astronomer Royal, professor of astronomy, director of the observatory, and later as mayor of Copenhagen. C. Roemer, "Demonstration touchant le mouvement de la Lumiere trouvé," *Memoires de l'Acad. Roy. Sci. Paris* **10** (1676), 575–77.

[†]Armand Hippolyte Louis Fizeau (1819–1896), French physicist. The son of a physician, Fizeau started out studying medicine, became interested in optics and, being independently wealthy, could devote all of his time to research. He made noteworthy contributions to photography, used interferometry to measure the expansion coefficient of crystals, determined the velocity of light both in air and in running water. In his velocity-of-light experiment, Fizeau used a path between two Paris suburbs, Montmartre and Sures-

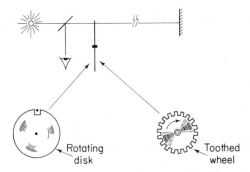

Figure 3.2-2 Fizeau's toothed-wheel experiment. Left-hand rotating disk is for illustration only; actual experiment has toothed wheel at bottom right.

Imagine a rotating disk with one notch cut into it (Figure 3.2-2, lower left). When the disk is standing still or rotating slowly, the observer, looking by way of a beamsplitter toward the distant mirror, sees the light source. But when the disk is made to turn faster and faster, the solid part of the disk cuts off the pulse on its return until the source disappears from sight. In the actual experiment, a toothed wheel takes the place of the disk (lower right).

Other experimenters, most notably Anderson and later Bergstrand, have improved on Fizeau's design, replacing the toothed wheel with an electronically driven Kerr cell. The same process is used extensively today for *electronic distance measurements, EDM.**

Foucault's method.   Wheatstone was the first to suggest that the toothed wheel be replaced by a rotating mirror, an idea that was then taken up by Foucault.[†] If mirror R in Figure 3.2-3 is standing still or turning slowly, the light is reflected to beamsplitter B following the path

nes, 8633 m long, and a notched wheel with 720 teeth, turning at 12.68 rev s^{-1}, found $c = 315\,300$ km s^{-1}, with estimated error limits of ± 500 km s^{-1}. H. Fizeau, "Sur une expérience relative à la vitesse de propagation de la lumière," *Compt. rendu* **29** (1849), 90–92.

  *W. C. Anderson, "Final Measurements of the Velocity of Light," *J. Opt. Soc. Am.* **31** (1941), 187–97. E. Bergstrand, "Velocity of Light and Measurement of Distances by High-Frequency Light Signalling," *Nature, London* **163** (1949), 338.

  [†]Jean Bernard Léon Foucault (1819–1868), French physicist. While a student in medical school, Foucault served as a lab assistant in an histology course and through work on a carbon arc became interested in optics. He built aspheric telescope mirrors and devised the knife-edge method to test them, invented the gyroscope and the pendulum named after him, determined the velocity of light by the rotating-mirror method, and discovered that light travels faster in air than in gas or water. L. Foucault, "Détermination expérimentale de la vitesse de la lumière; description des appareils," *Compt. rendu* **55** (1862), 792–96.

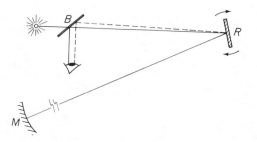

Figure 3.2-3   Foucault's rotating-mirror experiment.

of the solid line. But if the mirror is spinning, the light, reflected by the distant mirror, M, reaches R after it has turned a little; the light then follows the dashed line, the observer sees the image of the source slightly displaced, and this displacement is easy to measure. The mirror in Foucault's 1862 experiment made 830 turns per second. Then from the distance from B to R (1 m), the distance from R to M (with multiple reflections amounting to 20 m), from the displacement observed (0.7 mm), and from the formula

$$c \approx \frac{4\pi(830)(1)(20)}{0.7 \times 10^{-3}} \qquad [3.2\text{-}1]$$

Foucault found $c = 298\,000$ km s^{-1}, also with an estimated error of ± 500 km s^{-1}, increasingly close to today's value.

Michelson's method.   Precise measurements of the velocity of light were made by Michelson with an apparatus similar to Foucault's. The rotating mirror, driven by a jet of air, had eight (later 12 and then 16) reflecting surfaces (facets). Although Figure 3.2-4 is somewhat simplified, it illustrates the essential points. Light from source S falls on facet 1 of mirror M and is reflected toward assembly D. From there the light is returned to M and reflected into telescope T. As M rotates increasingly faster, light returning from D will no longer find facet 3 exactly at 45° and thus will not enter the telescope. However, if M rotates very fast so that facet 2 is brought into the position formerly occupied by 3, the light will be received through T. Thus, from the angular velocity of M and from

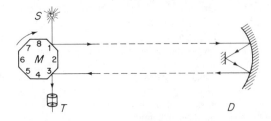

Figure 3.2-4   Michelson's method of measuring the speed of light.

the distance M to D the velocity of light can be calculated. After more than 1700 measurements, Michelson found that $c = 299\,796 \pm 4$ km s^{-1}.*

Indirect Methods

Stellar aberration.   Some 50 years after Roemer, Bradley[†] used an entirely different approach. Think of raindrops falling on the side windows of a car. When the car is standing still, the drops run straight down. But when the car is moving, the drops run down obliquely. Likewise, if we look at a star overhead (Bradley used Gamma Draconis, at the zenith), we could aim exactly at the star *if* Earth were standing still. But Earth moves in its orbit at a velocity **v**. Hence, the telescope must be tilted slightly, through an angle α, in order to let light that enters the objective pass through the center of the eyepiece. Half a year later, Earth is at the opposite side of its orbit, the direction of **v** is reversed, and the apparent position of the star has changed by 2α (Figure 3.2-5).

Bradley measured α, the *aberration constant,* and found 20.25 arc sec, very close to today's value (20.47). Since $\tan \alpha = \mathbf{v}/c$ and $\mathbf{v} \approx 2.98 \times 10^4 \mathrm{m\,s}^{-1}$,

$$c = \frac{\mathbf{v}}{\tan \alpha} = \frac{2.98 \times 10^4}{0.000\,098\,2} \approx 303\,000 \text{ km s}^{-1} \qquad [3.2\text{-}2]$$

with error limits of $\pm$ 5000 km s^{-1}.

---

*These experiments were performed in 1924 through 1926, with the rotating mirror located at the Mt. Wilson observatory and D on San Antonio Peak, 35 385.53 m away. The distance had been determined by the U.S. Coast and Geodetic Survey to an accuracy of 1 part in 6 800 000 (5.2 mm in 35.4 km). A. A. Michelson, "Measurement of the Velocity of Light between Mount Wilson and Mount San Antonio," *Astrophys. J.* **65** (1927), 1–22. Later, Michelson attempted to measure the velocity of light directly in an evacuated pipe 1.6 km long. That rotating mirror had 32 facets; the total path, with multiple reflections within the pipe, was 16 km. Michelson died in 1931 before the project was completed. Measurements taken after his death gave $c = 299\,774 \pm 2$ km s^{-1}. In 1933 an earthquake destroyed the apparatus.

[†]James Bradley (1693–1762), British astronomer, professor of astronomy at Oxford. Introduced to astronomy by an amateur astronomer uncle, Bradley served for a while as a vicar, later went into astronomy and became a most meticulous observer, devoting much of his time looking for stellar parallax, the apparent displacement of stars resulting from changes in the point of observation as Earth moves through its orbit. Although it was much later that the distances of stars were actually measured by parallax, Bradley together with his coworker Samuel Molyneaux, found a way to determine the speed of light, almost as a by-product. J. Bradley, "Account of a new discovered Motion of the Fixd Stars," *Phil. Trans. Roy. Soc. London* **35** (1728), #406, 637–60. See also A. B. Stewart, "The Discovery of Stellar Aberration," *Scientific American* **210** (Mar. 1964), 100–108.

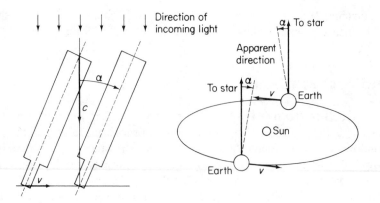

Figure 3.2-5   Aberration of starlight.

Ratio of electric constants.   A different approach is based on electromagnetic theory. By using Maxwell's third and fourth equations the velocity of light (page 295) is found to be

$$c = \frac{1}{\sqrt{\varepsilon_0 \mu_0}}$$ [3.2-3]

This equation connects the velocity of light to two electric constants, the electric permittivity, ε_0, and the magnetic permeability, μ_0, which are both of free space. Inserting $\varepsilon_0 = 8.85 \times 10^{-12} C^2 N^{-1} m^{-2}$ and $\mu_0 = 4\pi \times 10^{-7} Wb A^{-1} m^{-1}$ gives $c = 2.999 \times 10^8 m s^{-1}$, very close to the most recent figure. In matter, ε_0 and μ_0 become ε and μ, respectively. For most optical materials, μ is so nearly equal to unity that for all practical purposes the index of refraction,

$$n = \frac{c}{v} = \sqrt{\varepsilon}$$ [3.2-4]

Wavelength and frequency.   The velocity of propagation of a wave, from Equation [1.1-7], is equal to the product of wavelength and frequency, $v = \lambda \nu$. This relationship has led to the most accurate speed-of-light determination yet. The wavelength and frequency used were those of a highly stabilized helium–neon laser.* The wavelength, nominally 3.39 μm, was compared to the krypton-86 length standard, using a Fabry–Perot interferometer. The result was $\lambda = 3.392\,231\,376$ μm, with an error of less than 1 part in 10^{11}. The frequency, nominally 88 THz

*K. M. Evenson et al., "Speed of Light from Direct Frequency and Wavelength Measurements of the Methane-Stabilized Laser," *Phys. Rev. Lett.* **29** (1972), 1346–49.

(terahertz, 10^{12} Hz), was measured using a chain of phase-locked lasers and klystrons, comparing it to the cesium atomic-clock frequency standard. This gave $\nu = 88.376\,181\,627$ THz, with an error of 6 parts in 10^{10}, equivalent to 1 part in 300 million. From these measurements, the currently best estimate of the velocity of light is $3.392\,231\,376 \times 10^{-6}$ m $\times 88.376\,181\,627 \times 10^{12}$ Hz or

$$c = 299\,792\,456.2 \pm 1.1 \text{ m s}^{-1} \qquad [3.2\text{-}5]$$

The velocity of light is independent of the frequency or color and is the same for X rays, visible light, and radio waves, that is, the same throughout the whole electromagnetic spectrum: *The velocity of light is a natural constant.*

This fact may have far-reaching consequences, mainly because the special theory of relativity (Chapter 6.1) *requires* the speed of light to be constant. The velocity of light would then become a *defined* quantity rather than a measured one. The value likely to be adopted is

$$\boxed{c = 299\,792\,458 \text{ m s}^{-1}} \qquad [3.2\text{-}6]$$

If this proposal is adopted by the International Bureau of Weights and Measures, the meter would be defined as "the distance traveled in a time interval of $1/299\,792\,458$ of a second by plane electromagnetic waves in a vacuum." Any refinement in the accuracy of measurement would then alter, not the velocity of light, but the length of the meter.

The definition of the meter is another good example of evolution in optics. In 1789 the prototype meter was established as $1/40\,000\,000$ of the length of Earth's meridian passing through Paris, and set equal to the distance between two lines engraved on a platinum–iridium bar kept at Sèvres near Paris. But in 1827 Babinet proposed to use instead the wavelength of light as a reference. First, the red line of cadmium was selected for this purpose and then, because of its high monochromaticity, the orange line of krypton-86. And now it seems that the meter will no longer be a defined but rather a *derived* quantity, tied to the unit of time, the second, by the velocity of electromagnetic radiation.

Velocity of Light in Moving Matter

The question whether light is "dragged" along in moving matter, likewise, is of interest in the theory of relativity. We will pursue this question on page 537.

Phase Velocity and Group Velocity

There are some aspects of the velocity of light that I could have considered earlier in conjunction with interference. Assume that two waves are superimposed, both having the same amplitude but slightly different frequencies (Figure 3.2-6, top).

In free space, these two waves move along at the same velocity, called *phase velocity*. In matter, because of dispersion, different frequencies move along at different velocities. The wave of lower frequency moves at a velocity v and the wave of higher frequency at $v - dv$. The resultant of the two waves is another wave (dashed envelope, bottom), which has a frequency of its own, the *beat frequency*, and a velocity called *group velocity*.

To find the group velocity, assume that at time $t = 0$ two wave crests coincide, forming a maximum. After a time interval dt, both waves have advanced (to the right), the lower-frequency wave through distance $s = v\,dt$, and the higher-frequency wave through $s - ds = (v - dv)dt$. Since the two wave velocities (phase velocities) are different, the wave crests that form the next successive maximum are different too: one is the next adjacent crest. Thus the maximum has advanced, not by $v\,dt$, but by $v\,dt - \lambda$. Therefore, the group velocity is

$$\mathsf{v} = \frac{v\,dt - \lambda}{dt}$$

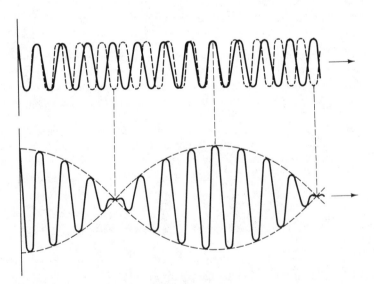

Figure 3.2-6   Two sinusoidal waves of different wavelengths (*top*) and their resultant (dashed lines, *bottom*)

and, because $dt = d\lambda/dv$,

$$\mathbf{v} = \frac{v\,dt}{dt} - \frac{\lambda}{dt} = v - \frac{\lambda\,dv}{d\lambda} \qquad [3.2\text{-}7]$$

The essential point is that in free space phase velocity and group velocity are the same. In matter they are different. And whereas the velocities of the individual component waves (in matter) are not much different, the group velocity is much less. Experimentally, we always measure group velocity; it is the wave *group* that carries the energy and, therefore, the group velocity is often referred to as *signal* velocity.

SUGGESTIONS FOR FURTHER READING

L. Essen and K. D. Froome, *The Velocity of Light* (New York: Academic Press, Inc., 1969).

E. Bergstrand, "Determination of the Velocity of Light," in *Fundamentals of Optics,* Vol. **24** of S. Flügge, editor, *Encyclopedia of Physics,* pp. 1–43 (Berlin: Springer-Verlag, 1956).

B. Jaffe, *Michelson and the Speed of Light* (Garden City, NY: Doubleday & Company, Inc., 1960).

PROBLEMS

3.2-1. Following Roemer's idea of determining the speed of light, plot a curve showing the time differences at which the eclipses of Jupiter's moon Io were seen, on the y axis, versus time, on the x axis. On the curve indicate four equally spaced positions of Earth in its orbit.

3.2-2. When Roemer's method was repeated recently, the light was found to traverse Earth's orbit in 16 min 38 s. If the radius of the orbit is assumed to be 1.49×10^{11}m, what result does this yield?

3.2-3. Refer to the data that Fizeau used in his original toothed-wheel experiment and determine the general equation.

3.2-4. A toothed wheel has 360 teeth. If the light is returned at a mirror 620 m away, what is the least speed of rotation that will cause the reflected light to be blocked out?

3.2-5. How fast did Michelson's octagon turn to make the return light *reappear* for the first time? Use the data referred to on page 313.

3.2-6. Assume that the light sent out by an electronic distance meter is modulated at a frequency of 25 MHz. If a phase difference of $0.3°$ is detected between the emitted waves and the waves returned by a distant retroreflector, to what precision (uncertainty of distance) does this correspond?

3.2-7. An oscillator that generates electromagnetic waves of 0.26 GHz is set up in front of a plane metal reflector. If standing waves, formed between source and reflector, have nodes every 57.5 cm, what is the velocity of the initial (moving) waves?

3.2-8. Since the magnetic permeability of free space is *defined* as $4\pi \times 10^{-7}$Wb A^{-1} m^{-1}, use the velocity of light, Equation [3.2-6], and determine the electric permittivity.

3.3

Light Scattering

"IT IS NOW, I BELIEVE, GENERALLY ADMITTED that the light which we receive from the clear sky is due in one way or another to small suspended particles which divert the light from its regular course." With these words John William Strutt, later Lord Rayleigh, began his first paper on *light scattering*. Light scattering is a phenomenon that occurs widely in nature and is of great utility. It accounts for the blue of the sky, as well as the white of the clouds, and is used as an analytical tool in chemistry and biology. We distinguish two major classes of light scattering, *Rayleigh scattering* and *Mie scattering*.

Rayleigh Scattering

Phenomenology.
When a beam of white light is passed through a gas freed of dust and other contaminants, observation in a dark room shows a faint hazy blue light coming sideways out of the beam. This is an example of *light scattering*. One of the early investigators of light scattering, Tyndall, let the light pass through a mixture of butyl nitrite and hydrochloric acid vapors.* Tyndall thought that the effect, since named

---

*John Tyndall (1820–1893), Irish physicist. Working as a surveyor, railroad engineer, and science teacher, Tyndall obtained a Ph.D. in mathematics, studied chemistry

after him, was due to the presence of small particles. He considered pure gases "optically empty."

At first Rayleigh agreed. Later he recognized that the molecules (of the air), rather than particulate matter, account for the blue of the sky. Since then, Rayleigh scattering has become synonymous with *molecular scattering*. The upper limit of the size of the particles that cause such scattering is about *one-tenth the wavelength of light.* *

Scattering as dipole radiation.
Light scattering is a process in which light interacts with matter. In the course of this process, energy is removed ("absorbed") from a beam of light and reemitted without appreciable change in wavelength. For simplicity we assume that the light interacts with only a single molecule.[†]

A molecule, in the absence of ionization, contains an equal number of positive and negative charges. Under the influence of an external electric field, the positive charges are displaced in one direction and the negative charges are displaced in the opposite direction, forming an *electric dipole* (Figure 3.3-1).

It is convenient to describe the properties of a dipole in terms of its *dipole moment, p*:

$$p = qd \qquad [3.3\text{-}1]$$

Figure 3.3-1 Electric dipole consists of two charges of opposite sign, separated by distance *d*.

under Bunsen, met Faraday, became professor of physics at the Royal Institution, and, when Faraday retired, succeeded him as superintendent. Tyndall wrote some 30 books, is noted for his work on light scattering, designed foghorns and both emission and absorption spectrophotometers, discovered the effect of penicillin on bacteria, and studied the mechanics of glacier motion, the latter no doubt because he was an accomplished mountain climber who, in 1861, made the first ascent of the Weisshorn, near Zermatt, in the Swiss Alps. J. Tyndall, "On the Blue Colour of the Sky, the Polarization of Skylight, and on the Polarization of Light by Cloudy matter generally," *Phil. Mag.* (4) **37** (1869), 384–94.

*John William Strutt, third Baron Rayleigh (1842–1919), British physicist, professor and later chancellor of Cambridge University, and successor of Clerk Maxwell as director of the Cavendish Laboratory. His first papers on light scattering were J. W. Strutt, "On the Light from the Sky, its Polarization and Colour," *Phil. Mag.* (4) **41** (1871), 107–120, 274–79; and "On the Scattering of Light by small Particles," *ibid.* (4) **41** (1871), 447–54. Later, in Lord Rayleigh, "On the Transmission of Light through an Atmosphere containing Small Particles in Suspension, and on the Origin of the Blue of the Sky," *Phil Mag.* (5) **47** (1899), 375–84, he corrected saying that "even in the absence of foreign particles we should still have a blue sky." On his estate at Terling, Lord Rayleigh had several darkrooms, one of them painted black with a mixture of soot and beer. It was here that he discovered argon, which brought him the 1904 Nobel prize in physics. In 1899, Rayleigh was admiring the sight of Mt. Everest from the terrace of a hotel in Darjeeling, 160 km away. At this distance, the outline of the mountain could barely be seen in the haze. From the degree of visibility and the refractive index of air, Rayleigh calculated the number of molecules per milliliter of air, and found 3×10^{19}, a figure close to today's value.

†In reality, light scattering is usually a case of *multiple scattering*, successive interactions with two or more molecules. The fundamental process, though, is the same.

where q is the charge and d the distance between the charges. With an external electric field $\mathbf{E}$, the induced dipole moment is

$$\mathbf{p} = \alpha\mathbf{E} \qquad [3.3\text{-}2]$$

where α is called the *polarizability* of the molecule. If the field varies sinusoidally as a function of time, the dipole moment will vary as well:

$$\mathbf{p}_t = \mathbf{p}_0 \sin(\omega t) \qquad [3.3\text{-}3]$$

The external field, in short, will make the dipole *oscillate,* in synchrony with the impinging field.

Since the positive charges in a molecule are relatively heavy (nuclei), they will, under the influence of an electromagnetic field, move very little. The negative charges (the electrons) will move much farther. Thus the induced oscillations will mainly affect the electrons. But electrons that undergo oscillatory accelerations are themselves radiators of electromagnetic waves, just as a dipole antenna is a radiator of radio waves. It is this *reradiation* that is fundamental to the process of light scattering.

1. Now assume that the incident electric field is oscillating up and down, along the y axis, as shown in Figure 3.3-2. The dipole consequently forms, and oscillates, also along the y axis. To an observer at S these oscillations appear in the form of *light*. If the observer changes position, moving around the dipole in an arc with the *dipole axis* as the pivot, he or she will see scattered light *from any direction,* as long as he or she remains in a plane *at right angles* to the axis of the dipole.

2. Next consider the plane *containing* the axis of the dipole (the paper plane). As before, the light is scattered in directions forward and backward. However, since electromagnetic waves are transverse, rather than longitudinal, an antenna does not radiate in the direction of its own length. Hence, an observer looking at the dipole from directly above or from directly below will see *no light*.

We call θ the angle subtended by the forward direction and the direction of the scattered light. In the forward direction where $\theta = 0$, and in the backward direction where $\theta = 180°$, the amplitude of the scattered light will reach its highest value. In directions straight up or down where $\theta = 90°$, the amplitude will drop to zero. The *amplitude* of the scattered

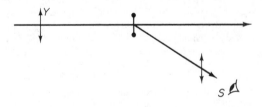

Figure 3.3-2 Linearly polarized light incident from the left, forming dipole. Observer at S.

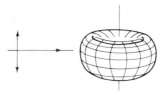

Figure 3.3-3 Three-dimensional plot of scattered light. Linearly polarized light incident from the left.

light varies as a function of the cosine of θ and the *energy*, following $Q \propto A$, varies as $\cos^2 \theta$. A three-dimensional plot of the scattered light looks like a doughnut (Figure 3.3-3).

3. Finally, consider that *unpolarized light* is incident on a volume of scatterers. Unpolarized light is the sum of two components polarized in orthogonal directions, oscillating at right angles to each other. The total energy of the scattered light is, likewise, the sum of two contributions; it follows a $1 + \cos^2 \theta$ distribution. This distribution, therefore, is (a) symmetric, with equal energies scattered forward and backward, and (b) the forward and back scatter components contain twice as much energy as scatter in the $\theta = 90°$ directions.

Rayleigh scattering as a function of wavelength.

Now let a bundle of light pass through a volume of scatterers of thickness Δx. We call I_0 the intensity (energy) of the light incident on the volume and I' the intensity passing through. A certain fraction of the light, i, is scattered out of the path and lost:

$$I_0 - I' = i \qquad\qquad [3.3\text{-}4]$$

The magnitude of this loss is proportional to I_0, to the thickness Δx, and to a constant of proportionality called *turbidity*, τ:

$$I' - I_0 = -i = I_0 \tau \, \Delta x \qquad\qquad [3.3\text{-}5]$$

Turbidity, in other words, is the *fractional loss* of light, the ratio i/I_0, per unit thickness; it is a term comparable to *absorptivity* (see Chapter 5.3) although I hasten to add that in pure scattering there is no absorption, or irretrievable loss of light: all the light removed from the incident beam is reemitted again almost instantaneously in the form of scattered light.

The scattering medium may be a gas that contains N molecules per unit volume. If the distances between the molecules are large relative to their size, these molecules will all scatter independently and the light scattered out of the beam, at $\theta = 90°$, will be *incoherent*. Therefore, the energy scattered by N molecules is N times the energy scattered by a single molecule. In the forward direction, at $\theta = 0$, however, the phases of the light are the same and, provided that the separation of the molecules is less than the coherence length of the light, the scattering is *coherent*. This means that, instead of adding the energies, we now add the amplitudes and square their sum.

The *total* turbidity, that is, the total amount of light scattered forward or backward, is found by integrating over the surface of a hemisphere of radius R:

$$\tau = 2\pi \int_0^\pi \frac{i}{I_0} R^2 \sin\theta \, d\theta = \frac{8\pi}{3} \frac{i}{I_0} R^2 \qquad [3.3\text{-}6]$$

The amplitude of the induced oscillation (of a dipole) increases as the driving frequency (of the incident light) approaches the natural frequency of oscillation of the molecule. Now, the natural frequency of the electric charges in a typical molecule is comparable to the frequency of ultraviolet radiation. Hence, more light is scattered at shorter wavelengths.

Rayleigh has shown that the amplitude of the scattered light is inversely proportional to the *square* of the wavelength. Furthermore, if we assume that the dipole is small compared with the wavelength of the light, then from Equation [3.3-6] and including the $(1 + \cos^2\theta)$ term, it follows that the intensity, i, of the scattered light is

$$i = I_0 \frac{8\pi^4 N\alpha^2}{\lambda^4 R^2} (1 + \cos^2\theta) \qquad [3.3\text{-}7]$$

where I_0 is again the intensity of the incident light, N the number of scatterers, α the polarizability, λ the wavelength, R the distance from the scatterers to the point of observation, and θ the angle subtended by the scattered light and the forward direction.

The essential point is that the intensity of the scattered light is *inversely proportional to the fourth power of the wavelength:*

$$\boxed{i \propto \frac{1}{\lambda^4}} \qquad [3.3\text{-}8]$$

Light of shorter wavelength is scattered much more than light of longer wavelength (Figure 3.3-4). Indeed, blue is *scattered out* of the light we receive from the sun and therefore the sun appears yellow. At sunset, when the light traverses an even longer path through the atmosphere, all but red is scattered out and the sun appears red. Conversely, light that has undergone multiple scattering (on the molecules of the air) is *scattered in;* that is, it is deflected *into* the line of sight of an observer who looks at the sky in any direction but at the sun. The sky, therefore, appears blue. As seen from high mountains, where there is not much air, or from the moon, where there is none, the sky is black.

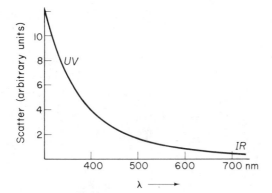

Figure 3.3-4 Rayleigh scattering: plot of intensity as a function of wavelength.

Example. *At a wavelength of 546 nm the scattering intensity of pure carbon tetrachloride is found to be 5.9 arbitrary units. What is the intensity at 436 nm?*

Solution. From Equation [3.3-8] we determine that

$$\frac{i_2}{i_1} = \left(\frac{\lambda_1}{\lambda_2}\right)^4 = \left(\frac{546}{436}\right)^4 = \frac{i_2}{5.9}$$

and thus

$$i_2 = \left(\frac{546}{436}\right)^4 (5.9) = \boxed{14.51}$$

Applications.

Light scattering can be used to determine particle weight of almost any magnitude, from the molecular weight of simple gases to the molecular weights of colloids, high polymers, and proteins and other substances of biochemical interest. Conversely, if the composition is known, the number of molecules per unit volume can be found.

Observations are usually made of the intensity of the scattered light as a function of the angle θ of scattering relative to the forward direction. The quantity actually measured is the *Rayleigh ratio, R_θ,* a parameter that is defined as

$$R_\theta = \frac{i_\theta}{I_0} R^2 \qquad\qquad [3.3\text{-}9]$$

and that, as I have indicated by the subscripts, is a function of θ. But, from Equation [3.3-5], $i/I_0 \propto \tau$, and hence R_θ is also proportional to τ_θ.

Most instruments for measuring light scattering follow the outline shown in Figure 3.3-5. Light comes from a source (left) and is collimated by a lens, L. After passing through an optional color or polarizing filter,

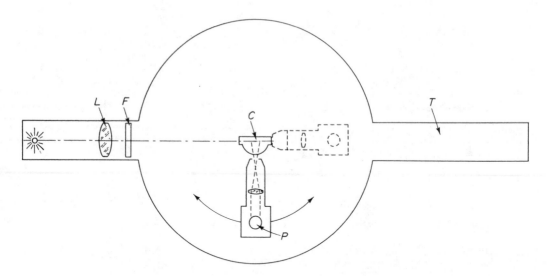

Figure 3.3-5 Schematic diagram of light-scattering apparatus.

F, the light is incident on a cell, C, containing the scattering medium. Preferably the cell can be quickly exchanged by an opal-glass reference standard, to permit calibration. After passing through the cell, the primary light is absorbed in a light trap, T.

Scattered light from the cell or from the reference standard goes through another slit-lens assembly and perhaps another polarizing filter to a photomultiplier tube, P. The phototube is mounted so that it can be swung around in an arc with the cell as the pivot. Its output signal is plotted versus the angle of rotation, the dashed outline of P in Figure 3.3-5 indicating the forward position, $\theta = 0$.

Mie Scattering

Rayleigh scattering versus Mie scattering. When the particles are larger than about one-tenth of a wavelength, the light scattered from one point on the particle may well be out of phase with light scattered from another point. The two contributions then interfere and the scattered-light distribution is no longer symmetrical as in the Rayleigh case. With increasing particle size, in fact, the scattered light becomes more concentrated in the forward direction (Figure 3.3-6). The difference between forward and back scatter is called *dissymmetry*, which is often defined as the ratio $i_{45°}/i_{135°}$. Dissymmetry can be used to determine the average size of particles, provided that their shape is known. For larger particles, however, the angular dependence of the scattering becomes quite

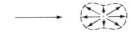

Small particles: Rayleigh scattering

Large particles: Mie scattering

Figure 3.3-6   Angular distribution in Rayleigh and Mie scattering.

complicated, showing a number of maxima and minima, and cannot be adequately described by a single parameter such as the dissymmetry.

Mie's theory* takes into account the size of the particles but also their refractive index, and the refractive index of the surrounding medium, as well as the shape, dielectric constant, and absorptivity of the particles. It is based on a formal solution of Clerk Maxwell's equations, leading to a series, in theory an infinite number, of partial waves. The amplitudes of these waves decrease rapidly as the particles become smaller. In Rayleigh scattering, we need to consider only the first few terms, and may neglect all others. Mie scattering, by contrast, requires many more terms; it is more general and includes Rayleigh scattering as a special case.

Mie scattering parameters.
Assume that light of cross section A is incident on a volume V of thickness Δx, $V = A\,\Delta x$. If there are N particles per unit volume, the actual volume contains $NA\,\Delta x$ particles. If r is the radius of a particle, each particle has a cross section πr^2 and together they have a cross section

$$\Sigma = NA\,\Delta x\,\pi r^2 \qquad\qquad [3.3\text{-}10]$$

provided that no one particle lies in the shadow of another particle; otherwise, the cross section is less. In most cases, the distances between particles are much larger than their radii, so the actual cross section is about the same as the maximum possible.

But then it turns out that the cross section of the light affected by a particle is not necessarily equal to the cross section of the particle. The reason is the refractive index difference between particles and the medium between them. The ratio of these two cross sections is called the *extinction factor, K*. This factor is generally larger than unity, which means that the

---

* Gustav Mie (1868–1957), German physicist, professor of physics at the University of Freiburg. G. Mie, "Beiträge zur Optik trüber Medien, speziell kolloidaler Metallösungen," *Ann. Physik* (4) **25** (1908), 377–445.

extinction by a particle is *higher* than what could be expected from its physical size. The ratio i/I_0 in Mie scattering, hence, is

$$\frac{-i}{I_0} = \frac{NA \, \Delta x \, K \pi r^2}{A} \qquad [3.3\text{-}11]$$

Combining this equation with our earlier Equation [3.3-5], canceling A and Δx, and replacing τ by the *Mie extinction coefficient*, μ, gives

$$\boxed{\mu = NK\pi r^2} \qquad [3.3\text{-}12]$$

which shows that μ is equal to the product of the number of particles times $K\pi r^2$, a term called the *extinction cross section*. If, as in most cases, the particles are of continuous size distribution, with radii ranging from r_1 to r_2, then

$$\mu = \pi \int_{r_1}^{r_2} N(r)K(r, n)r^2 \, dr \qquad [3.3\text{-}13]$$

The essential point is that there is *no λ^{-4} relationship*. Indeed, with increasing particle size the exponent 4 in the denominator in Equations [3.3-7] and [3.3-8] becomes less and gradually approaches zero. For sufficiently large particles there is no wavelength dependency at all. This is one of the characteristics that distinguish Mie scattering from Rayleigh scattering; it is, of course, the reason why most clouds are white. These clouds, and fog and mist and aerosol sprays, are composed of droplets at least 10 μm in diameter, 200 times the $\lambda/10$ limit referred to before.

The pertinent size of the particles is sometimes described by their ratio of radius to wavelength, r/λ. Instead of r/λ, however, more often the *size parameter, α,* is used, a term defined as the ratio of particle circumference to wavelength, $2\pi r/\lambda$. Particles that have the same α have similar scattering properties. If we plot the extinction factor, K, versus α, we obtain a curve that has the shape of a *damped oscillation* (Figure 3.3-7). With increasing particle size the value of K rises from 0 to nearly 4 and asymptotically approaches 2 for very large particles.

Under certain conditions, Mie scattering leads to results quite opposite to those that occur in Rayleigh scattering. In fact, some time during September 1950 over large parts of North America and Europe the sun was seen to be *blue,* an effect caused by oil droplets, remarkably uniform in size, that had been carried by winds in the upper atmosphere from forest fires burning in Alberta.*

> **Example.** *Sometimes, when the sun sets, a phenomenon is seen known as the* green flash. *How does it come about?*

---

*R. Wilson, "The Blue Sun of 1950 September," *Monthly Notices Roy. Astron. Soc.* **111** (1951), 478–89.

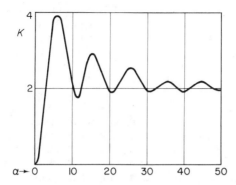

Figure 3.3-7 Mie scattering: plot of extinction factor, K, versus size parameter, α, for water droplets of $n = 1.33$.

Answer. This effect is due to a combination of dispersion and scattering. First, light from the setting sun, because of refraction in the stratified atmosphere, reaches the observer on Earth in a curved path.* Red is refracted least, and, after the sun has set, blue and green persist. Much of the blue is lost to Rayleigh scattering and green remains, the *green flash*.

SUGGESTIONS FOR FURTHER READING

H. C. van de Hulst, *Light Scattering by Small Particles* (New York: John Wiley & Sons, Inc., 1957).

M. Kerker, *The Scattering of Light and Other Electromagnetic Radiation* (New York: Academic Press, Inc., 1969).

H. C. van de Hulst, *Multiple Light Scattering, Tables, Formulas, and Applications* (New York: Academic Press, Inc., 1980).

E. J. McCartney, *Optics of the Atmosphere, Scattering by Molecules and Particles* (New York: Wiley-Interscience, 1976).

R. A. R. Tricker, *Introduction to Meteorological Optics* (New York: American Elsevier Publishing Company, Inc., 1970).

PROBLEMS

3.3-1. Determine the electric field strength, **E**, at a (horizontal) distance R from the center of a dipole of length d, similar to that shown in Figure 3.3-1. Recall that $\mathbf{E} = \mathbf{F}/q$ and that **F** is given by Equation [3.1-1].

3.3-2. Continue with Problem 3.3-1 and now assume that the charge separation, d, is small compared to R so that $d^2/4$ can be dropped.

3.3-3. Unpolarized light is incident on a medium causing molecular scattering. If the tur-

---

* See Chapter 3.5 for details on refraction in stratified, *gradient-index* media.

bidity at right angles to the incident light is 5 arbitrary units, what is the turbidity at 45°?

3.3-4. Plot the angular distribution of Rayleigh scattering for incident light that is:
(a) Linearly polarized *in* the plane of the drawing.
(b) Linearly polarized *normal* to the plane of the drawing.
(c) Unpolarized.

3.3-5. How much more Rayleigh scatter is produced by the 587-nm krypton line than by the 760-nm line?

3.3-6. If light of 635 nm wavelength causes a certain amount of Rayleigh scatter, light of what wavelength will give 10 times as much scatter?

3.3-7. If Rayleigh scattering causes 1.85 percent of 1.06 μm radiation to be lost when passing through 10 km of atmosphere, what per-

centage is lost of radiation of 694.3 nm?

3.3-8. Light of 760 nm and 810 nm wavelength is passed through a turbid medium. If 20 percent of the shorter wavelength is lost due to Rayleigh scattering, what percentage is lost of the longer wavelength?

3.3-9. A certain amount of light of 503.0 nm wavelength is scattered out of a beam passing through a volume of air at atmospheric pressure. What percentage of that amount is scattered at 632.8 nm wavelength at one-half the pressure?

3.3-10. Light scattering is one of several methods that can be used for building a *smoke detector* for the home. If you were given a small light source of low power consumption, such as a light-emitting diode, and a photocell as the receiver, how would you arrange these components?

3.4

Polarization
of Light

Types of Polarized Light

1. Most light as it occurs in nature is *unpolarized*. It is a mixture of light polarized in different ways and to different degrees. These differences result because such light is emitted by atoms, or groups of atoms, most of them oscillating independently. While the individual wavetrains (each with a coherence time usually no longer than about 10^{-8} s) oscillate in some constant mode and direction, the modes and directions of oscillation vary from train to train; thus the result is a sequence of oscillations all oriented *at random*.

2. If the electric field oscillates in a certain, constant orientation, the light is said to be *linearly polarized* (Figure 3.4-1). The term "plane polarized" is deprecated, for reasons that will become apparent shortly.

Figure 3.4-1 Linearly polarized light. Projection of wave on a plane intercepting the axis of propagation gives a *line*, hence the term *linear polarization*.

Figure 3.4-2 Partially linearly polarized light seen head-on.

3. If linearly polarized light contains an additional component of natural (unpolarized) light, it is called *partially* (linearly) *polarized* (Figure 3.4-2). Such light could come from a perfectly polarizing filter with some holes in it or from a low-quality filter.

4. In *circular polarization,* the electric vector no longer oscillates in a plane, as in linear polarization. Instead, the vector is constant in magnitude and proceeds in the form of a helix, *around* rather than *through,* the

Figure 3.4-3   Right-circularly polarized light.

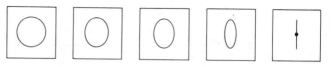

Figure 3.4-4   Circular, elliptical, and linear polarization.

axis of propagation (Figure 3.4-3). Within one wavelength, the vector completes one revolution. If the tip of the vector, when seen looking toward the light source, rotates clockwise, the light is said to be right-circularly polarized; if counterclockwise, left.

5. *Elliptical polarization* stands between circular and linear polarization. It is the most general type of polarization; linear and circular are the two extremes of elliptical polarization (Figure 3.4-4). The tip of the vector now proceeds in the form of a *flattened* helix; the vector rotates and, at the same time, it changes in magnitude.

The term "plane polarization," as opposed to "linear," is based on a three-dimensional convention. Circularly polarized light need then be called "helically polarized." To avoid awkward terms such as "circular helical" and "elliptical helical," I prefer a two-dimensional convention with the terms *linear, elliptical,* and *circular.*

6. *Optical rotation* (optical activity) is entirely different. It is a property of matter, rather than a property of light. We will discuss the molecular basis of optical activity later (on page 351).

The Poincaré sphere.   Consider the most general case, elliptically polarized light. The shape of the ellipse may vary widely, from circular to linear, the orientation may vary, and the sense of rotation of the **E** vector may vary also. To describe the general state of polarization, therefore, we need three parameters: (1) the *ellipticity,* that is, the ratio of the lengths of the half-axes of the ellipse, $\epsilon = \tan^{-1}(b/a)$ (Figure 3.4-5). If the ellipticity is small, the ellipse is highly elongated, and in the limit when $\epsilon = 0$, the polarization is linear. If $\epsilon = 45°$, the polarization is circular. Furthermore, we need (2) the orientation, or *azimuth*, α, of the major axis of the ellipse with respect to the z axis, and (3) the direction of rotation, or *handedness*, of the **E** vector, clockwise or counterclockwise (looking toward the source).

All the possible states of polarization can be uniquely represented by a point P on the surface of a sphere called the *Poincaré sphere** (Figure 3.4-6). Latitude and longitude of P have the values 2ϵ and 2α, respectively. All linear polarizations lie on the equator. Circular polarizations correspond to the north pole and south pole. Right-elliptical oscillations are in the northern hemisphere and left-elliptical oscillations in the southern hemisphere. Points along a given parallel represent ellipses of the same ellipticity but of different azimuth.

Orthogonal polarizations are at diametrically opposite points. This means that two bundles of polarized light are orthogonal if (a) the light is

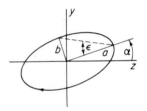

Figure 3.4-5 Projection of elliptically polarized light. Note definitions of azimuth α and ellipticity ϵ. In the example shown the azimuth is 17°, the ellipticity 25°, and the handedness clockwise.

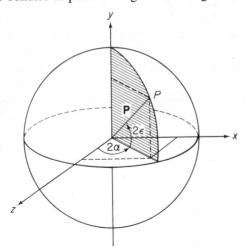

Figure 3.4-6 Poincaré sphere.

*Named after Jules Henri Poincaré (1854–1912), French mathematician. A child prodigy in mathematics while still in high school, Poincaré studied at the École Polytechnique and the École des Mines, and became professor of mathematical physics at the Sorbonne, the University of Paris. Poincaré's most important contributions were to the theory of differential equations and their applications to dynamics, especially to planetary motion and the three-body problem. He was a brilliant lecturer, much sought after also for popular presentations, but had few students as his research assistants. He liked to work alone; he published some 500 articles and several books, among them *Théorie Mathématique de la Lumière* (Paris: Georges Carré, 1892). In volume 2, chapter 12, there is a description of the Poincaré sphere.

linearly polarized in planes that differ by 90°, or (b) one polarization is right-circular and the other left-circular, or (c) the ellipticities are the same but the azimuths differ by 90° and the handednesses are opposite.

The Poincaré sphere (as well as the matrix formalism of polarization which I will discuss on page 344) has a very practical purpose. It provides us with a simple and elegant means of describing any state of polarization of light. Moreover, any changes in the state of polarization, as they may be caused by crystals or wave retardation plates, can be predicted by a suitable rotation of the Poincaré sphere.

Production of Polarized Light

Light can be polarized in various ways. We distinguish the following.

Polarization by scattering.   Consider again a bundle of unpolarized light that passes through an assembly of small particles in suspension (Figure 3.4-7). From a distance, at right angles to the direction of propagation, an observer sees the light as *linearly polarized*. The direction of oscillation of the scattered light will be *normal* (perpendicular) to the plane defined by the direction of propagation and the direction of observation.* (Clearly, the light scattered toward S cannot be horizontally polarized because that would require the incident light to oscillate back and forth along the axis, "longitudinally," and there is no such light.)

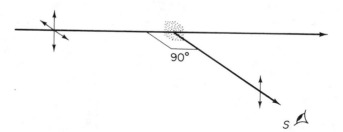

Figure 3.4-7   Light scattered at right angles is linearly polarized.

---

*Polarization by scattering occurs in the light we receive from the sky. Look at the sky in a direction 90° away from the sun, hold a polarizing filter in front of your eyes, and rotate the filter: The sky will periodically appear lighter and darker. Polarization by scattering is important to bees, ants, and other insects that use it for navigation, in contrast to man, whose eyes cannot discriminate between different states of polarization. The Vikings, though, some 1000 years ago, may have found their way across the oceans and back to port by polarization: the "sunstones" mentioned in old sagas were probably cordierite, a polarizing crystal; these pebbles are found on the coast of Norway, and can be used even when the sky is overcast. See R. Wehner, "Polarized-Light Navigation by Insects," *Scientific American* **235** (July 1976), 106–15.

Polarization by reflection. When light is reflected at the surface of matter, it penetrates a short distance into the surface and induces molecular oscillations. As in polarization by scattering, the reradiated light can contain only light oscillating normal to the plane of incidence (the plane defined by the incident light and the surface normal), and no light oscillating parallel to it. Therefore, the reflected light is linearly polarized. Maximum polarization occurs when the reflected light and the refracted light subtend an angle of 90° (Figure 3.4-8). In that case,

$$\alpha_2 + 90° + \beta = 180°$$

$$\alpha + \beta = 90°$$

and

$$\beta = 90° - \alpha$$

Since

$$\sin(90° - \alpha) = \cos \alpha$$

Snell's law, $\sin \alpha / \sin \beta = n_2/n_1$, becomes

$$\frac{\sin \alpha}{\cos \alpha} = \frac{n_2}{n_1}$$

Thus

$$\boxed{\alpha = \tan^{-1}\left(\frac{n_2}{n_1}\right)}$$

[3.4-1]

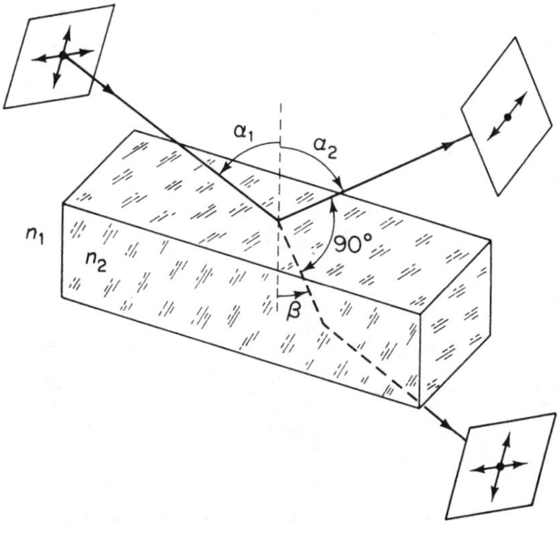

Figure 3.4-8 Polarization by reflection.

which is *Brewster's angle of maximum polarization on reflection.* * (We have seen on page 304 that Brewster's angle can also be derived from Fresnel's first equation.)

Polarization by reflection is seen on wet pavement, water, and similar surfaces, causing *glare*. A pair of polarizing glasses, with their planes of transmission vertical, will suppress such glare.

Polarization by transmission.

Each time light is reflected at a surface at Brewster's angle, the reflected component is linearly polarized normal to the plane of incidence. The transmitted light thus contains a greater proportion of light polarized parallel to the plane of incidence. If the process is repeated often enough, the transmitted light becomes more and more purely linearly polarized (Figure 3.4-9). This is how *polarization multilayers* work. (These are used in the infrared where thin films of selenium take the place of the glass.)

Polarization by selective absorption.

Polarization by selective absorption is an important case which can be best understood by the example of the *wire-grid polarizer*.

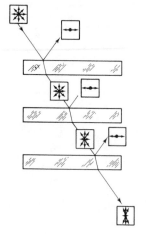

Figure 3.4-9 Polarization by transmission.

A wire-grid polarizer is a grid of parallel wires. In 1888, Heinrich Hertz used such grids as polarizers to test the properties of radiowaves he had discovered the year before. The electric field impinging on the grid may be resolved into two orthogonal components, one oscillating parallel to the wires and the other normal to them. The parallel component drives the conduction electrons in the wires along their lengths, the current encounters resistance, heats up the wire, and thereby loses energy. The normal component, however, has no electrons to drive very far and hence passes through without much loss. In short, it is the *normal* component that goes through (contrary to what one might think).

What the wire-grid polarizer does for longer waves can be accomplished for visible light by a *dichroic crystal*. A good example is *tourmaline,* an aluminoborosilicate containing Al_2O_3, B_2O_3, and SiO_2. On passing through the crystal, the light is split into two components, the

---

*Named after Sir David Brewster (1781–1868), Scottish physicist, professor of physics at St. Andrews College. Initially a minister in the Church of Scotland, Brewster became interested in optics, found the angle named after him, contributed also to dichroism, absorption spectra, and stereophotography, invented the kaleidoscope, and wrote a book about it. A prolific writer, Brewster edited several journals, published a 526-page *Treatise on Optics,* the *Life of Sir Isaac Newton,* several more books, and some 315 journal articles. Brewster's law, in his own words, states that "when a ray of light is polarised by reflexion, the reflected ray forms a right angle with the refracted ray." D. Brewster, "On the laws which regulate the polarisation of light by reflexion from transparent bodies," *Phil. Trans. Roy. Soc. London* **105** (1815), 125–59.

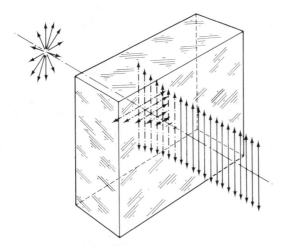

Figure 3.4-10   Polarization by selective absorption.

horizontal component in Figure 3.4-10 being attenuated, the vertical component going through. The light emerging from the crystal, therefore, is linearly polarized. (The transmitted light, because of the natural color of tourmaline, is green; in orthogonally polarized light the crystal appears black, hence the term "dichroic" meaning "two colors.")

The same dichroic effect is seen with crystals of quinine sulfate periodide, called *herapathite*.* These crystals are very small, and cannot be used for polarization. However, after a method was found to align the crystals, an efficient and economical polarizing material was produced, now known as *Polaroid*.[†]

---

*Named after William Bird Herapath (1820–1868), physician and surgeon at Queen Elizabeth's Hospital in Bristol, England. Herapath worked on a variety of subjects, from chlorophyll and the identification of bloodstains by microspectroscopy to home sanitation. His best known discovery came when his pupil, a Mr. Phelps, dropped iodine into the urine of a dog that had been fed quinine and saw that tiny emerald-green crystals formed there. Herapath, examining them under a microscope, noticed that the crystals were in some places light where they overlapped and in some places they were dark, an observation which he published under the slightly laborious title "On the Optical Properties of a newly-discovered Salt of Quinine, which crystalline substance possesses the power of polarizing a ray of Light, like Tourmaline, and at certain angles of Rotation of depolarizing it, like Selenite," *Phil. Mag.* (4) **3** (1852), 161–73.

[†]Invented by Edwin Herbert Land (1909–), American physicist, inventor, and businessman. While an undergraduate at Harvard College, Land set out to develop a better way of producing polarized light, hoping to replace the single crystals that were so difficult to grow. Clearly, he reasoned, a great many small crystals would do the same if only they could be aligned properly. He took a suspension of herapathite and placed it in a strong magnetic field. It worked: he had the first man-made polarizer for light. Later he used an electric field to align the crystals. He also suspended them in a thermoplastic matrix, extruding the plastic, while soft, through a narrow slit. The result was "J sheet" polarizer;

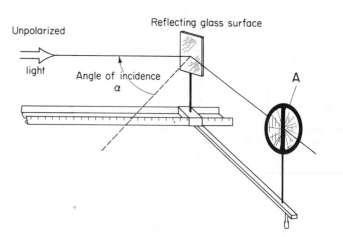

Figure 3.4-11 Axis finder (*A*) placed in path of light linearly polarized by reflection.

The most common type of a polarizer used today is *H sheet*. It has a matrix of parallel chains of the synthetic material polyvinyl alcohol (PVA, —CH$_2$—CHOH—), impregnated with iodine. The conduction electrons associated with the iodine move along the chains, like the electrons in a wire-grid polarizer, and thus absorb the electric field oscillating parallel to them; the plane of transmission is normal. *K sheet* is a polyvinylene polarizer optically similar to H sheet but more resistant to high temperatures and humidity.

Polaroid can also be made with the molecules arranged in circular symmetry. Such a filter, called an *axis finder,* is useful for finding the direction of oscillation of (linearly) polarized light. But how are the molecules oriented, circumferentially (tangentially) or radially? The answer is found by placing an axis finder in a beam of light reflected off a glass surface, at Brewster's angle, as illustrated in Figure 3.4-11. The axis finder will show a *vertical* black band. *Explanation:* We know that the reflected light is linearly polarized, oscillating normal to the plane of incidence (which, in this case, means oscillating up and down). We also know that attenuation occurs for oscillations parallel to the orientation of the molecules; therefore, this orientation is *radial*.

Polarization by double refraction.   Take a pen and make a dot on a sheet of paper. Place a crystal of calcite or "Iceland spar," calcium carbonate, CaCO$_3$, over the dot and look through. You will see two dots instead of one. Rotate the crystal and observe the dots. One of

it contains a great many herapathite crystals all aligned parallel to one another. The first patent on a synthetic sheet polarizer was issued to E. H. Land and J. S. Friedman, *Polarizing refracting bodies,* U.S. Patent 1,918,848, July 18, 1933. For more details, see E. H. Land, "Some Aspects of the Development of Sheet Polarizers," *J. Opt. Soc. Am.* **41** (1951), 957–63.

them will remain in place while the other moves around it. This was how Bartholin* discovered *double refraction,* or *birefringence.* Crystals such as calcite, quartz, mica, and ice which show this effect are called *aniso-tropic.* In contrast, most noncrystalline solids such as unstressed glass are *isotropic.*

Calcite is a good example of a system of anisotropic crystals called *hexagonal.* In order to visualize the shape of a hexagonal crystal, think first of an orthogonal (right-angled) block of material. Two diagonally opposite corners may be called *A* and *B,* as in Figure 3.4-12. Slightly compress the block along the line *A–B:* The faces that before were rectangular now become rhombic and corners *A* and *B* become blunt; that is, the angles adjacent to them grow larger, the angles away from them acute.

The line *A–B* is called the *crystal axis,* not to be confused with the optic axis of the system. In the direction specified by this axis the crystal is isotropic and its refractive index, the *ordinary* refractive index, ω, is constant (1.6584 for calcite and sodium light). In other directions, the index changes with the angle subtended with the crystal axis. In a direction 90° from the axis, the index is least (1.4864). This is the *extraordinary* refractive index, ϵ.

If unpolarized light is incident on the crystal, in a direction different from that of its axis, the light is split into two components, the *ordinary* or *o ray,* and the *extraordinary* or *e ray.* The *o* ray advances at the same velocity in all directions. If we could set up a point source of light within the crystal, Huygens' wavefronts would advance outward in the form of *spheres* (Figure 3.4-13).

The extraordinary ray, however, has different velocities in different directions and the wavefronts advance in the form of *ellipsoids of revolu-*

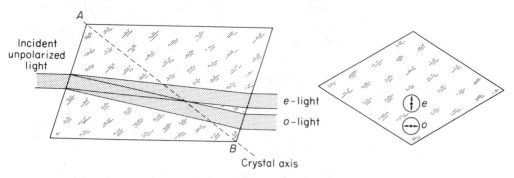

Figure 3.4-12   Double refraction and polarization of light in calcite.

---

*Erasmus Bartholin (1625–1698), Danish physician and professor of mathematics at the University of Copenhagen. He described his discovery in a 60-page publication, *Experimenta crystalli Islandici disdiaclastici quibus mira et insolita refractio delegitur* (Hafniae, 1670). Bartholin, however, was not aware of what we now call polarization; he thought his observation to be an unusual case of refraction, *double refraction.*

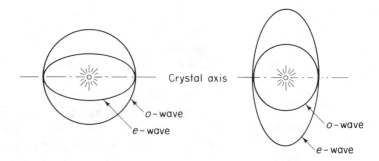

Figure 3.4-13 Huygens' wavefronts in a positive uniaxial crystal (*left*), negative uniaxial crystal (*right*).

tion. In some crystals, such as quartz, the ellipsoidal wavefronts lie within the spherical wavefronts because the velocity of the *e* ray is less, and the ϵ index is higher, than that of the *o* ray. Such crystals are called *positive* uniaxial. In others, such as calcite, it is the other way around; these are called *negative* uniaxial. The two bundles emerging from the crystal are both linearly polarized, at right angles to each other. Hence, such crystals make good polarizers if we could only eliminate one of the bundles. This is done in the *Nicol prism*.

Nicol prism.   The Nicol prism* is made out of a crystal of Iceland spar, cut into two and cemented together with a thin layer of Canada balsam in between (Figure 3.4-14). The two endfaces are polished down from 71° to 68° as shown. The *o* light is refracted more (downward) and is incident on the Canada balsam at an angle slightly larger than the *e* light. But since the refractive index of Canada balsam (1.526) lies be-

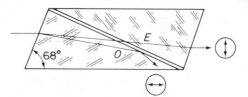

Figure 3.4-14 Nicol prism.

---

*Named after William Nicol (1768–1851), Scottish geologist and physicist. A lecturer at the University of Edinburgh, Nicol published his first paper at age 58. His interests were primarily in the fields of crystallography, mineralogy, and paleontology. In 1828 he invented his prism, and described it in an article "On a Method of So Far Increasing the Divergency of the Two Rays in Calcareous Spar That Only One Image May Be Seen at a Time," *Edinburgh New Phil. J.* **6** (1829), 83–4.

tween ω and ϵ for calcite (1.6584 and 1.4864, respectively), the *e* light will pass through but the *o* light will not. The *o* light undergoes total internal reflection and is absorbed in black paint on the sides of the prism, and only the *e* light will emerge from it.

Nicol prisms are good polarizers, but they are expensive and have a limited field of view (28°). This is because, if the light is not nearly parallel, either the *o* light will be incident on the balsam at an angle less than the critical angle and go through, or the *e* light will be reflected out of the prism too and be lost.

There are other prisms besides the Nicol type which are useful for special applications. The *Glan-Thompson prism* has a wider angular aperture (40°), but it is wasteful of calcite and hence even more expensive. The *Glan–Foucault prism* has no cement (but a narrow field) and thus is less likely to be damaged at high power densities. In the *Rochon prism* and the *Wollaston prism* the *o* and the *e* rays are transmitted side by side.

Types of birefringence.

The numerical difference between the ϵ and the ω index of refraction provides us with one definition of birefringence. But since $\epsilon - \omega$, in effect, is a measure of *path difference,* Γ, this path difference *per unit path length, L,* gives us another definition; thus

$$\text{birefringence} \equiv \epsilon - \omega = \frac{\Gamma}{L} \qquad [3.4\text{-}2]$$

Crystalline birefringence occurs if a crystal has an asymmetric structure. Crystalline birefringence is independent of the surrounding medium. Examples are all anisotropic crystals and many biologic materials, such as collagen, cellulose, and muscle fibers, whose molecular structure has a certain preferential orientation and which are so densely packed that liquid cannot penetrate them. In *form birefringence,* on the other hand, elemental units, called *micelles,* of a given refractive index are distributed throughout a matrix of a different index. If such a material is immersed in a liquid of a different index, the liquid will penetrate between the micelles, replace the original matrix, and the form birefringence will change.

If a transparent isotropic substance such as plastic or glass is subject to mechanical stress, it becomes temporarily birefringent. Such birefringence is the basis of *photoelastic stress analysis,* a method widely used in the engineering world. For testing purposes, either a transparent replica of the structure is made (and examined in transmitted light) or a plastic coating is applied to the real structure (and examined in reflected light). In either case, the birefringence reveals the strains induced in the structure.

Malus' law. Consider a beam of linearly polarized light, coming toward you. The electric vector of the light may oscillate in the vertical direction, as shown in Figure 3.4-15, left. The light may then be incident on a polarizer whose *plane of transmission* subtends an angle θ with the direction of oscillation (center). How will the amplitude of the light change?

Draw a normal from the tip of vector **A** to the plane of transmission (right). If $\theta = 0$, all the light will go through and $\mathbf{A} = \mathbf{A}'$. If $\theta = 90°$, none will go through and $\mathbf{A}' = 0$. In order to find the amount of light transmitted at intermediate angles, note that $\mathbf{A}'/\mathbf{A} = \cos \theta$. Therefore, since the energy of light is proportional to the square of its amplitude, the energy transmitted, Q', relates to the energy incident, Q, as

$$\boxed{Q' = Q \cos^2 \theta}$$ [3.4-3]

where θ is the angle through which one of the polarizers has been turned with respect to the other. This relationship is known as *Malus' law.* *

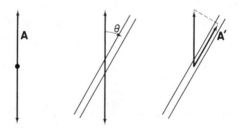

Figure 3.4-15 Malus' law.

---

*Étienne Louis Malus (1775–1812), French army officer and engineer. One evening in 1808 while standing near a window in his home in the Rue d'Enfer in Paris, Malus was looking through a crystal of Iceland spar at the setting sun reflected in the windows of the Palais Luxembourg across the street. As he turned the crystal about the line of sight, the two images of the sun seen through the crystal became alternately darker and brighter, changing every 90° of rotation. After this accidental observation Malus followed it up quickly by more solid experimental work, described his observation in "Sur une propriété de la lumière réfléchie," *Mémoires phys. et chim. Soc. d'Arcueil, Paris* **2** (1809), 143–58. He concluded that the light, by reflection on the glass, became *polarized* but, by chance, he defined the "plane of polarization" of the reflected light as being the same as the plane of incidence. Today we know that the **E** vector oscillates *normal* to the plane of incidence; thus Malus' "plane of polarization" is perpendicular to the plane of oscillation.

The Analysis of Light of Unknown Polarization

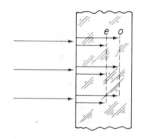

Figure 3.4-16 Retardation of light in a (positive) birefringent crystal such as quartz.

If light is completely linearly polarized and incident on a good polarizer which is rotated about the optic axis, the amount of light transmitted will vary between maximum and zero. But if some unpolarized light were added to the light (= partially polarized light), at no setting would the light be completely extinguished. The same would happen if circularly polarized light were added.

How, then, can we distinguish between the two? This is possible by using a relatively simple device known as a *quarter-wave plate*. Such a plate is made out of birefringent material which, as we have seen, makes light of orthogonal polarizations traverse the material at different velocities. One of the wavefronts will be ahead of the other by a distance that depends on the thickness of the plate (Figure 3.4-16). If one component is ahead by 90°, we have a *quarter*-wave plate. If the thickness is such that the path difference is 180°, we have a *half*-wave plate.

> **Example.** *Quarter-wave plates are often made of thin sheets of mica. Although mica is biaxial and has three refractive indices, we are justified in using 1.5936 for one index and 1.5977 for the other. How thick must the mica be to provide $\frac{1}{4}\lambda$ retardation, using sodium light, 589 nm?*

Solution. From Equation [3.4-2] we find the *path* difference,

$$\Gamma = L(\epsilon - \omega)$$

and from Equation [2.2-38], the *phase* difference,

$$\delta = \left(\frac{2\pi}{\lambda}\right)\Gamma = \left(\frac{2\pi}{\lambda}\right)L(\epsilon - \omega)$$

Then, for the phase difference to be $\delta = \frac{1}{4}\lambda = \frac{1}{2}\pi$,

$$\frac{1}{2}\pi = \left(\frac{2\pi}{\lambda}\right)L(\epsilon - \omega)$$

$$\frac{1}{2}\lambda = 2L(\epsilon - \omega)$$

$$L = \frac{\lambda}{4(\epsilon - \omega)}$$

and thus

$$L = \frac{589 \times 10^{-9}}{(4)(1.5977 - 1.5936)} = \boxed{0.036 \text{ mm}}$$

Let me illustrate by graphical construction how a quarter-wave plate works. A sinusoidal wave, A_1 in Fig. 3.4-17, left, rises up from zero at a

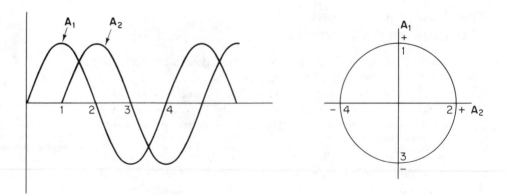

Figure 3.4-17   Adding two linear polarizations of $\lambda/4$ phase difference gives circular polarization.

time $t = 0$. A second wave of the same amplitude and frequency, A_2, is delayed by one-fourth of a wavelength, 90°. We consider four points evenly spaced in time, 1–2–3–4, and plot the amplitudes of these two waves, at these times, on coordinates labeled A_1 and A_2 (right). At point 1, the A_1 wave has reached a positive maximum and A_2 is zero; at point 2, $A_1 = 0$ and A_2 has reached a positive maximum; at point 3, A_1 is at a negative maximum and $A_2 = 0$; and at point 4, $A_1 = 0$ and A_2 is at a negative maximum. The plot is that of a *circle*, one example of many possible *Lissajous figures*.

This means that the superposition of two linear orthogonal polarizations of equal amplitude, which have a 90° phase difference between them, gives circularly polarized light. For different amplitudes, or for phase differences other than 90°, elliptically polarized light results. Conversely, linearly polarized light is the resultant of two circular polarizations of the same amplitude but of opposite sense of rotation, one right-handed and the other left-handed.

Such interaction of light, clearly, is the result of *interference*. Polarized light, like other light, comes to interference but only if certain conditions are met. These are set forth in *Fresnel–Arago's laws:* *

1. Two bundles of light, polarized in the same direction, do interfere.

---

*Dominique François Jean Arago (1786–1853), French physicist. Working at the Paris observatory, Arago's most noteworthy contributions were to the wave theory of light and to the magnetic effect of electric current. By running a current through a copper wire, Arago in 1820 disproved the then held theory that iron was necessary to produce electromagnetism. When Napoleon made himself emperor in 1852, Arago refused an oath of allegiance and resigned; the new emperor did not accept his resignation and let Arago stay on anyway. A. Fresnel and F. Arago, "Sur l'Action que les rayons de lumière polarisés exercent les uns sur les autres," *Ann. Chim. et Phys.* (2) **10** (1819), 288–305. (For Fresnel footnote, see page 224.)

Table 3.4-1 ANALYSIS OF LIGHT OF UNKNOWN POLARIZATION

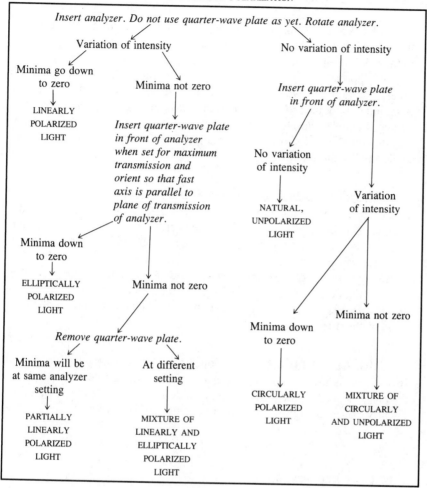

2. Two bundles, linearly polarized in orthogonal directions, do not interfere.

3. Two bundles, derived from orthogonal components of unpolarized light and subsequently brought into the same plane of oscillation, do not interfere.

4. Two bundles, derived from coherent portions of light and subsequently brought into orthogonal planes, do interfere. In this case, however, varying degrees of ellipticity result, as we have seen.

Returning to our initial question, by using a polarizer and a quarter-wave plate we can analyze almost any type of light of unknown polarization. The proper steps are listed in Table 3.4-1.

Matrix Formalism of Polarization

Matrix algebra, like Poincaré's sphere, greatly facilitates the description of polarized light. It also helps describe the interaction of polarized light with various types of polarizers and phase retardation plates. There are two approaches which in part overlap but otherwise complement each other. The *Mueller calculus* is based on a phenomenological description; it includes unpolarized light but requires larger (4 × 4) matrices. The *Jones calculus* is derived from the mathematical description of a monochromatic polarized wave; it has smaller (2 × 2) matrices but some contain complex numbers.

Mueller calculus.   In the Mueller calculus, any form of light including partially and elliptically polarized light is described by four parameters called *Stokes' vectors*. These vectors, **I, M, C, S**, in this order, define "intensity," preference for horizontal polarization, preference for +45° polarization, and preference for right-circular polarization. They are written in the form of a column matrix, below one another. The characteristics of the polarizer or the retardation plate are condensed in the 4 × 4 *Mueller matrix*. The matrix, then, contains 16 elements, but fortunately most of them are zero.*

Jones calculus.   The Jones calculus is used in either of two forms, depending on whether we need to know the phase of the light. If the phase is needed, the light requires *full Jones vectors*.† For example, for linearly polarized light oscillating in the horizontal direction,

$$\begin{bmatrix} A_z \ e^{i\phi_z} \\ 0 \end{bmatrix}$$

If no phase information is needed,

$$\begin{bmatrix} 1 \\ 0 \end{bmatrix}$$

---

*For details, see M. J. Walker, "Matrix Calculus and the Stokes Parameters of Polarized Radiation," *Am. J. Physics* **22** (1954), 170–74.

†Named after Robert Clark Jones (1916–   ), senior physicist and Research Fellow in Physics at the Polaroid Corporation, Fellow of the Optical Society of America and recipient of its Lomb and Ives medals. Jones has also made noteworthy contributions to radiation detection, hydrodynamics, acoustics, photography, and isotope separation, and as a hobby likes to ride railroad locomotives all over the world. R. Clark Jones, "A New Calculus for the Treatment of Optical Systems," *J. Opt. Soc. Am.* **31** (1941), 488–503.

Similarly,

$$\begin{bmatrix} 0 \\ 1 \end{bmatrix}$$

represents light linearly polarized in the vertical direction and

$$\begin{bmatrix} \frac{1}{2} \\ \frac{1}{2} \end{bmatrix} = \begin{bmatrix} 1 \\ 1 \end{bmatrix}$$

light linearly polarized at an azimuth of 45°. Circular polarization is represented by

$$\begin{bmatrix} 1 \\ -i \end{bmatrix} \quad \text{for right-} \quad \text{and} \quad \begin{bmatrix} 1 \\ i \end{bmatrix} \quad \text{for left-handed rotation}$$

Then, if right- and left-handed circularly polarized light (of equal amplitude) is combined,

$$\begin{bmatrix} 1 \\ -i \end{bmatrix} + \begin{bmatrix} 1 \\ i \end{bmatrix} = \begin{bmatrix} 1+1 \\ -i+i \end{bmatrix} = \begin{bmatrix} 2 \\ 0 \end{bmatrix} = 2\begin{bmatrix} 1 \\ 0 \end{bmatrix} \qquad [3.4\text{-}4]$$

which means that the result is linearly polarized light of twice the amplitude of either of the initial contributions.

Let the matrix $\begin{bmatrix} A \\ B \end{bmatrix}$ represent the incident light. From Table 3.4-2 we take the matrix of the element inserted in the path and write it to the left of the *AB* matrix,

Table 3.4-2 SIMPLIFIED JONES MATRICES REPRESENTING
VARIOUS POLARIZING ELEMENTS

Linear polarizer, transmission axis horizontal	⟷		$\begin{bmatrix} 1 & 0 \\ 0 & 0 \end{bmatrix}$
transmission axis vertical	↕		$\begin{bmatrix} 0 & 0 \\ 0 & 1 \end{bmatrix}$
transmission axis at +45°	↗	$\frac{1}{2}$	$\begin{bmatrix} 1 & 1 \\ 1 & 1 \end{bmatrix}$
Circular polarizer, right-handed	↻	$\frac{1}{2}$	$\begin{bmatrix} 1 & i \\ -i & 1 \end{bmatrix}$
left-handed	↺	$\frac{1}{2}$	$\begin{bmatrix} 1 & -i \\ i & 1 \end{bmatrix}$
Quarter-wave plate, fast axis horizontal			$\begin{bmatrix} 1 & 0 \\ 0 & -i \end{bmatrix}$
fast axis vertical			$\begin{bmatrix} 1 & 0 \\ 0 & i \end{bmatrix}$
Half-wave plate, fast axis at +45°			$\begin{bmatrix} 0 & 1 \\ 1 & 0 \end{bmatrix}$

$$\begin{bmatrix} b & a \\ d & c \end{bmatrix} \begin{bmatrix} A \\ B \end{bmatrix} = \begin{bmatrix} A' \\ B' \end{bmatrix} \qquad [3.4\text{-}5]$$

The resulting matrix, $\begin{bmatrix} A' \\ B' \end{bmatrix}$, describes the emerging light.

Example 1. *Linearly polarized light oscillating in the horizontal direction is incident on a half-wave plate oriented at +45°. What is the state of polarization of the emerging light?*

Solution.

$$\begin{bmatrix} 0 & 1 \\ 1 & 0 \end{bmatrix} \begin{bmatrix} 1 \\ 0 \end{bmatrix} = \begin{bmatrix} 0 \\ 1 \end{bmatrix}$$

The emerging light, therefore, is linearly polarized in the vertical direction.

Example 2. *A quarter-wave plate, with its fast axis horizontal, is inserted into a beam of light linearly polarized and oscillating at 45°. Determine the polarization of the emerging light.*

Solution.

$$\begin{bmatrix} 1 & 0 \\ 0 & -i \end{bmatrix} \begin{bmatrix} 1 \\ 1 \end{bmatrix} = \begin{bmatrix} 1 \\ -i \end{bmatrix}$$

The emerging light is right-circularly polarized.

Polarization Microscopy and Crystal Optics

A brief description of the polarizing microscope will help us understand many aspects of polarized light. At the same time it will introduce the field of *crystal optics*.

The polarizing microscope.
The addition of two polarizing filters and of a rotatable stage converts a laboratory microscope into an elementary polarizing microscope. One polarizer is located below the condenser, the other above the objective; thus the specimen is placed *between* the polarizers. The second polarizer, called an *analyzer,* is often set on top of the eyepiece.

There are two modes of operation, *orthoscopic* and *conoscopic*. In orthoscopic observation, the microscope is used like any other microscope; that is, it is focused on the specimen. In the conoscopic mode, an auxiliary lens, called the *Amici–Bertrand lens,* * is inserted about halfway between

---

*Giovan Battista Amici (1786–1863), Italian physicist and astronomer. Educated as an architect and engineer, Amici became professor of mathematics at the University of Modena, later director of the Astronomical Observatory of the Royal Museum of Physics and Natural History in Florence. He invented the roof prism, and built a large astronomical

Table 3.4-3 OUTLINE OF STEPS FOR IDENTIFYING A CRYSTAL

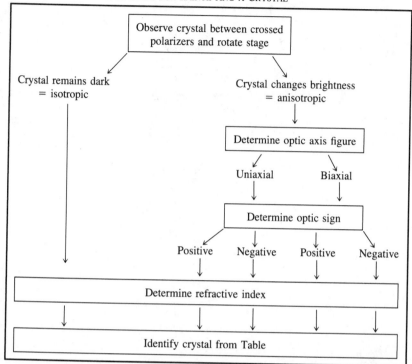

objective and eyepiece. This turns the microscope into a low-power tele-scope so that the observer, instead of looking *at* the specimen, looks *through* it (Figure 3.4-18).

First we determine whether the sample is *isotropic* or *anisotropic*. If a piece of strain-free glass is placed on the stage between crossed polar-izers, the glass remains dark no matter how the stage is turned; such a material is isotropic. Anisotropic matter, in contrast, goes through four positions of maximum extinction, 90° apart.

If the material is isotropic, its refractive index is determined next (Table 3.4-3, left-hand column). Only a few inorganic substances belong to this group: all amorphous materials and the crystals of the isometric system. A good way of determining the refractive index is by the *Becke line method*. Immerse the crystal in a liquid of known refractive index* and observe its contour while slowly raising or lowering the tube of the micro-

---

doublet of 5.33 m focal length, 28.5 cm in diameter. He designed microscope objectives with the front element in the form of a hemisphere; this gave the needed short focal length and high aperture, an ingenious idea still in use today.

*Such liquids can be obtained from R. P. Cargille Laboratories, Inc., 55 Commerce Road, Cedar Grove, NJ 07009.

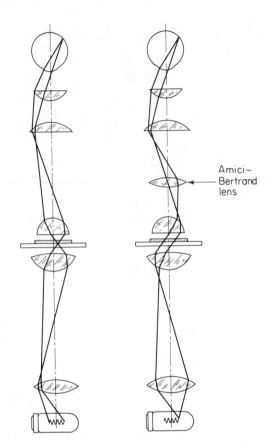

Amici–
Bertrand
lens

Figure 3.4-18   Ray diagrams for orthoscopic (*left*) and conoscopic (*right*) use of the polarizing microscope. Note that in the left-hand example the retina of the eye is conjugate to the specimen; in the right-hand example it is conjugate to the light source.

scope. When the tube is *raised* and the Becke line is seen to *contract* and to move *into* the crystal, the crystal has a refractive index *higher* than that of the liquid. If the effect is opposite, the liquid has the higher index.

Crystal optics.   On the other hand, if the crystal is *anisotropic*, we look first for the *optic axis figure*. Focus on a group of crystals. Change to the conoscopic configuration and rotate the analyzer to maximum background darkness. Using a high-aperture objective, slowly search the specimen until an optic axis figure is found. This figure may be either of the *uniaxial* or the *biaxial* type (Figure 3.4-19).

Uniaxial crystals have two principal indices of refraction, ω and ε. Biaxial crystals have three, α, β, and γ (see Tables 3.4-4 and 3.4-5).

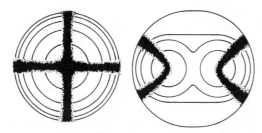

Figure 3.4-19 Optic axis figures of uniaxial and biaxial crystals are distinctly different. Figure on left is that of uniaxial crystal (such as calcite); on the right, biaxial crystal (such as mica). Concentric rings on left are *isochromatic curves*, representing equal path differences, 1λ, 2λ, . . . Dark cross is formed by *isogyres*, representing directions of oscillation.

Since birefringence can be defined as the numerical difference between the extraordinary and the ordinary refractive index, any crystal may have either one of two *optic signs*, positive or negative. If, for a uniaxial crystal, $\varepsilon > \omega$, the crystal is positive. If $\omega > \varepsilon$, it is negative. If, for a biaxial crystal, β is closer to α than it is to γ, the crystal is positive; if β is closer to γ, it is negative. These signs can also be determined by observation, using a half-wave gypsum plate.* Use conoscopy and insert the plate. If the crystal is uniaxial positive, the two quadrants that lie in direction of the length (the fast axis) of the $\lambda/2$ plate are yellow. The two other quadrants are blue (Figure 3.4-20). If the crystal is biaxial, the results are as shown in Figure 3.4-21.

Table 3.4-4 OPTICAL CHARACTERISTICS OF CRYSTALS

Type of material	Axes	Number of refractive indices	Notation	Crystal system
Isotropic		1	n	Isometric
Anisotropic	Uniaxial	2	ω, ε	Tetragonal Hexagonal
	Biaxial	3	α, β, γ	Orthorhombic Monoclinic Triclinic

---

*Such a plate is made from clear gypsum, selenite, cut to a thickness such as to produce a path difference of 550 nm. When the plate is placed between crossed polarizers, 550 nm light, which is yellow-green, does not pass through the analyzer and is missing from the incident white light; the transmitted light is *red*. This is why such plates are customarily referred to as *first-order red* plates.

Table 3.4-5 INDICES OF REFRACTION OF VARIOUS CRYSTALS
(FOR SODIUM D LIGHT, 589 nm)

Uniaxial crystals			
$\omega = 1.6584$	$\varepsilon = 1.4864$	Birefringence $= -0.172$	Calcite
1.5442	1.5533	$+0.009$	Quartz
1.5874	1.3361	-0.251	Sodium nitrate
Biaxial crystals			
$\alpha = 1.5601$	$\beta = 1.5936$	$\gamma = 1.5977$	Mica (muscovite)
1.3346	1.5056	1.5061	Potassium nitrate
1.4953	1.5353	1.6046	Tartaric acid

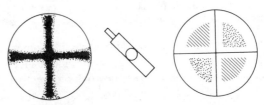

Figure 3.4-20 Determining the optic sign with a first-order red plate. Crystal is *uniaxial* positive. Hatched areas are yellow, stippled areas blue. If you like, color these areas.

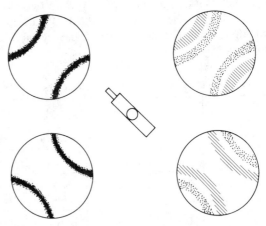

Figure 3.4-21 Crystal is *biaxial* positive. As before, hatched areas are yellow, stippled areas blue.

The birefringence of a crystal, together with its thickness, determines the *interference color* in which the crystal will appear, as we see from the following example.

Example. *A thin flake of mica, known to be 0.015 mm thick, is viewed (using orthoscopic observation). When oriented at 45° between crossed polarizers, what color is the crystal?*

Solution. While, from Equation [2.2-40], $\Gamma = L(n_2 - n_1)$, orientation at 45° refers to the indices α and γ,

$$\Gamma = L(\gamma - \alpha)$$

Thus

$$\Gamma = (0.015 \text{ mm})(1.5977 - 1.5601) = 564 \text{ nm}$$

which is green. This means that green light will exit from the crystal linearly polarized, and will not be transmitted through a crossed analyzer. But blue,

yellow, and red will; combined these colors add up to | purple |. If

polarizer and analyzer were parallel, the crystal would look green.

Optical Activity

When a bundle of linearly polarized light passes through a crystal of quartz, the light remains linearly polarized but its plane of oscillation rotates, that is, changes in orientation (Figure 3.4-22). This effect is not limited to crystals; in fact, it is found even in liquids such as turpentine and sugar solutions. Such substances are called *optically active,* and the effect termed *optical activity* or *optical rotation.* To an observer looking toward the source, the rotation may be clockwise (as shown), which is called positive or right-handed and the substance *dextrorotatory;* or it may be counterclockwise, which is called negative or left-handed and the substance *levorotatory.*

In 1825, Fresnel recognized that optical activity arises from circular double refraction. This means that a substance has an index of refraction for right-circularly polarized light that is different from the index for left-circularly polarized light. Since linearly polarized light is the resultant of two circular polarizations of the same amplitude but opposite sense of rotation, linearly polarized light passing through the substance will experience a rotation of its plane of oscillation. The degree of rotation is proportional to the difference between the two indices and to the length of path traversed.

Therefore, it is the *asymmetry* of the molecular structure that is responsible for the phenomenon of optical rotation. Even if two molecules contain the same atoms and bonds in the same sequential order, they may

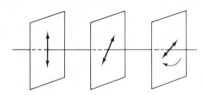

Figure 3.4-22   Optical rotation.

still have different orientations relative to the coordinate system. Such molecules are called *isomers*. If the isomers are exact mirror images of each other, they are *enantiomers*.

The degree of rotation, for a given wavelength, caused by a 1-mm-thick solid is called *specific rotation*. For liquids, specific rotation is defined as the rotation caused by a solution of 1 g of substrate in 1 ml of solvent through a path 10 cm long. This dependence on concentration is of practical interest, for example when testing syrups or urine for the amount of sugar present. At different wavelengths, the rotation is generally different, an effect called *rotatory dispersion*.

Electrooptic and Magnetooptic Effects

Certain materials, under the influence of an electric field, change their polarization characteristics; in particular, they change in birefringence. Such materials are called *electrooptic materials*. Other materials show similar changes when exposed to a magnetic field; these are called *magnetooptic materials*. Hundreds of solids and liquids exhibit these effects, but only a few of them to a degree that is of theoretical or practical interest. Depending whether the field is electric or magnetic, and depending whether the field is oriented normal (transverse) or parallel to the direction of the light, we distinguish the following.

Kerr effect.   A *Kerr cell** is a vessel filled with nitrobenzene. Light is passed through the cell, linearly polarized by a first polarizer. A second polarizer, at the end of the cell, is oriented at right angles, causing extinction. If an electric field of several kV cm^{-1} is applied, *normal* to the optic axis and at 45° to the direction of oscillation of the entering light, the light becomes elliptically polarized and part of it is transmitted.

Pockels effect.   In the *Pockels effect*, the field is *parallel* to the direction of the light. Often used are crystals of potassium dihydrogen phosphate, KH_2PO_4 (KDP), mounted between two crossed polarizers. Other electrooptic materials include lithium niobate, $LiNbO_3$, lithium tantalate, $LiTaO_3$, and barium sodium niobate, $Ba_2NaNb_5O_{15}$. When a high enough dc field, usually several hundred V mm^{-1}, is applied to the crystal, the refractive index (in one axis of the crystal) changes and linearly polarized light becomes elliptically polarized. Such electrooptic crystals make

---

*John Kerr (1824–1907), Scottish physicist. In 1875 Kerr used a block of glass (still on display at the Kelvin Museum of the University of Glasgow), with two wires from an induction coil attached to it and placed between crossed Nicol prisms. With the current on, the glass becomes positive uniaxial birefringent and light passes through.

compact and reliable high-speed shutters. They have replaced the toothed wheel in speed-of-light and electronic distance measurements and have numerous other applications as light switches and modulators.

Cotton–Mouton effect.

The *Cotton–Mouton effect* is the magnetic equivalent of the Kerr effect. When a liquid such as nitrobenzene or carbon disulfide is exposed to a transverse magnetic field, the liquid becomes birefringent.

Kerr magnetooptic effect.

When linearly polarized light is reflected off the polished pole piece of an electromagnet, it becomes elliptically polarized, with the major axis of the ellipse rotated with respect to the oscillation of the incident light. While ordinarily the *Kerr magneto-optic effect* amounts to a rotation through only a few arc min, it can be enhanced by applying a multilayer coating to the pole surface, thereby causing multiple reflections.

Faraday effect.

If a transparent medium is placed in a magnetic field, linearly polarized light passing through, parallel to the field, will show a rotation of its plane of oscillation. This phenomenon, historically the first experimental evidence of an interaction between light and an electromagnetic field, is called the *Faraday effect* or *Faraday rotation** (Figure 3.4-23). Again, the magnetic field changes the index of refraction; but it is the index for right-circularly polarized light that changes with

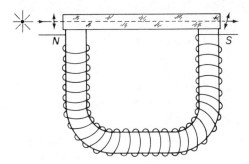

Figure 3.4-23   Faraday's experiment: Light is passed through slab of glass mounted on top of electromagnet.

---

*Named after Michael Faraday (1791–1867), British physicist. The son of a blacksmith, Faraday started out as a bookbinder's apprentice. Reading some of the books that passed through his hands, he became interested in chemistry and worked for a while as Sir Humphry Davy's lab assistant. As a young man, Faraday proclaimed that women were nothing in his life, and even published a poem ridiculing love. At age 30, he met, fell in love with, and married Sarah Barnhard, the 21-year-old daughter of a silversmith. Soon after Oerstedt's 1819 discovery that an electric current produced a magnetic field, Faraday, in

respect to the index for left-circularly polarized light. The net result is a rotation of the plane of oscillation.

Faraday knew that *optical* rotation (as caused by molecular asymmetry) is different from *magnetic* (magnetically induced) rotation: the direction of optical rotation depends on the direction of the light, while the direction of magnetic rotation depends on the direction of the field. Light passing through an optically active medium, and reflected back through it, shows no net rotation, because both rotations cancel. In magnetic rotation, the total rotation is cumulative; it is twice the one-way rotation. The angle of rotation ϕ in a given substance is proportional to the magnetic field strength $\mathbf{H}$ and to the distance L the light travels through the medium:

$$\phi = \mathbf{H}LV \qquad [3.4\text{-}6]$$

The term V is called *Verdet's constant*.

SUGGESTIONS FOR FURTHER READING

W. A. Shurcliff and S. S. Ballard, *Polarized Light* (Princeton, NJ: Van Nostrand Company, Inc., 1964).

W. A. Shurcliff, *Polarized Light, Production and Use* (Cambridge, MA: Harvard University Press, 1962).

W. Swindell, editor, *Polarized Light* (New York: Halsted Press, John Wiley & Sons, 1975).

J. M. Bennett and H. E. Bennett, "Polarization," in W. G. Driscoll and W. Vaughan, editors, *Handbook of Optics,* Sec. 10, pp. 1–164 (New York: McGraw-Hill Book Company, 1978).

P. S. Theocaris and E. E. Gdoutos, *Matrix Theory of Photoelasticity* (New York: Springer-Verlag, 1979).

E. E. Wahlstrom, *Optical Crystallography,* 5th edition (New York: John Wiley & Sons, Inc., 1979).

N. W. Jones and F. D. Bloss, *Laboratory Manual for Optical Mineralogy* (Minneapolis: Burgess Publishing Company, 1980).

E. Charney, *The Molecular Basis of Optical Activity: Optical Rotatory Dispersion and Circular Dichroism* (New York: John Wiley & Sons, Inc., 1979).

1821, placed a current-carrying wire near a magnetic pole and succeeded in making it rotate, opening the way to electric *motors*. Seeing this, he wondered whether the reverse might be true—whether a magnet could be made to produce a current. After casual experimentation over several years (during which time he liquefied gases, discovered benzene, and found a better way to make steel), he returned to his idea. For days he tried without success until, one day in 1831, in desperation he plunged a magnet down a coil: Suddenly there was an electric pulse. Why hadn't he seen that before? Clearly, it is the *change* of the magnetic field that did it. For this discovery, which opened the way to the electric *generator*, the whole scientific world honored him. In 1845, Faraday discovered the magnetic rotation of polarized light.

PROBLEMS

3.4-1. Consider a series of points that lie on the same *meridian* of the Poincaré sphere. What types of polarized light do these points represent?

3.4-2. What are the numerical limits of the ellipticity of polarized light? Describe the types of polarization to which these limits correspond, assuming that the azimuth is:
(a) Zero.
(b) +45°.

3.4-3. Unpolarized light is incident on a small volume of a scattering medium. What is the state of polarization of the light scattered at angles of 0°, 45°, and 135°, measured from the forward direction?

3.4-4. Circularly polarized light is passing through a colloidal suspension. Describe the state of polarization of the light scattered in directions of 0°, 45°, 90°, and 180° from the transmitted light?

3.4-5. What is the angle of incidence for complete polarization to occur on reflection at the boundary between water ($n = \frac{4}{3}$) and flint ($n = 1.72$) assuming that the light comes from the side of
(a) The water?
(b) The glass?

3.4-6. Light is reflected at a plane facet of a diamond ($n = 2.42$) immersed in oil ($n = 1.62$).
(a) What is the angle of incidence at which maximum polarization occurs?
(b) What is the angle of refraction?

3.4-7. Two Nicol prisms are set for maximum transmission. If one prism is rotated through 60°:
(a) What percentage of the amplitude is still going through?
(b) What percentage of the energy is still going through?

3.4-8. A bundle of unpolarized light passes through two polarizers whose planes of transmission are parallel. Through what angle must one of the polarizers be turned in order to reduce the amount of light to three-fourths of its initial value?

3.4-9. The sensitivity of photographic film is determined by exposing successive portions of it to light transmitted through two polarizers. Through what angles must one of the polarizers be turned in order to expose the film to light in ratios of 1.0 : 0.8 : 0.6 : 0.4 : 0.2 : 0?

3.4-10. Two polarizers are crossed at 90° so that no light is going through. If a third polarizer, oriented at 45°, is placed between the two polarizers, what percentage of light will emerge from the system? (Try it; there *is* light going through.)

3.4-11. Ice is a positive uniaxial crystal with indices of refraction of 1.309 and 1.310. How thick must the ice be to act as a quarter-wave plate for light of 600 nm wavelength?

3.4-12. A thin plate of calcite is cut with its axis parallel to the plane of the plate. What is the minimum thickness required to produce a $\lambda/4$ path difference for sodium light (589 nm)?

3.4-13 Following the example outlined in the text (page 342), plot the resultant polarization of two orthogonal, linearly polarized waves of equal amplitude which differ in phase by 30°.

3.4-14. A beam of unpolarized light passes through a linear polarizer and a quarter-wave plate, is reflected by a plane mirror, and passes again through the quarter-wave plate and the polarizer. What happens to the light?

3.4-15. In terms of the Jones calculus, describe a linear polarizer oriented with its transmission axis horizontal and:
(a) Linearly polarized light of horizontal orientation incident on it.
(b) The same light but now with vertical orientation incident on it.
What is the result?

3.4-16. Imagine that linearly polarized light of horizontal orientation is incident on a linear polarizer oriented at +45°.
(a) The light then enters a polarizer whose transmission axis is vertical. What happens? Describe in terms of the Jones calculus.
(b) What happens if the vertical polarizer is placed ahead of the 45° polarizer?

3.4-17. Two quarter-wave plates, their fast axes horizontal, are placed one behind the other. Determine the system matrix of the combination.

3.4-18. A quarter-wave plate, fast axis horizontal, is followed by a linear polarizer, transmission axis at $+45°$, and then by another quarter-wave plate, fast axis vertical. Describe the system matrix and the type of device it represents.

3.4-19. The specific rotation of quartz in light of 589 nm wavelength is $21.7°$/mm. What thickness of quartz, cut perpendicularly to its axis and inserted between *parallel* polarizers, will cause no light to be transmitted?

3.4-20. A glucose solution of unknown concentration is contained in a 12-cm-long tube and seen to rotate linearly polarized light by $2.5°$. Since the specific rotation of glucose is $52°$, what is the concentration?

3.4-21. If the Verdet constant for monobromonaphthalene, in yellow D light, is 0.104 arc min/A and if a magnetic field of 2.6×10^5 A m^{-1} is applied and the light traverses a path 10 cm long, through what angle will the plane of oscillation turn?

3.4-22. Continuing with Problem 3.4-21, assume that one of the two polarizers is rotated with respect to the other through the angle just found. Then, light proceeding through the system in one direction will *pass through,* but in the opposite direction (everything else being equal) *it will not.* How do you reconcile this fact with the assumption that the path of light is *reversible?*

3.5

Gradient-Index, Fiber, and Integrated Optics

THE TERM GRADIENT INDEX REFERS to local variations of the refractive index in an otherwise uniform transparent medium. Optical effects caused by such media have been known since ancient times. The *mirage,* seen when looking across an expanse of hot desert, is one example. It is only recently that gradient-index optics (GRIN optics) has become a prominent part of modern optics. GRIN optics includes lenses that, in contrast to conventional lenses, do not have a uniform refractive index throughout. For example, the index may be higher in the center than in the periphery. Such a lens, even if made in the form of a plane-parallel plate, will act as a positive, converging lens. If the plate in addition has curved surfaces, like an ordinary lens, it is equivalent to a combination of lenses and can be corrected for various aberrations, using no more than a single element. Other applications of GRIN materials extend to optical waveguides (*fiber optics*) and to miniaturized systems (*integrated optics*).

Atmospheric Refractive Index Gradients

Gradient-index phenomena are seen in everyday life. The atmosphere of Earth has a refractive index that decreases with height, because the density becomes less at higher altitudes. This causes light to proceed

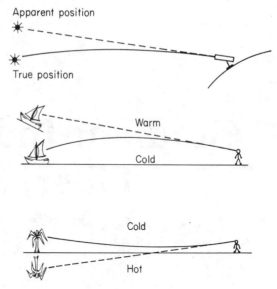

Figure 3.5-1   Regular atmospheric refraction (*top*), looming (*center*), and mirage (*bottom*). Dashed lines indicate apparent positions, full lines true positions.

in a *curved path,* an effect known as *regular atmospheric refraction* (Figure 3.5-1, top). A similar effect, called *looming,* is seen when looking across a snowfield or across a body of cold water. It makes objects on the surface appear to be lifted up (center). The opposite effect is the mirage or *Fata Morgana;* it occurs on hot sunny days when the air directly above the ground is hotter than the air higher up. Since the refractive index is lower near the ground, distant objects appear below the horizon as though reflected in a pool of water (bottom).* *Random atmospheric refraction* is due to turbulence. It causes the twinkling, or scintillation, of the stars. Refractive gradients were once called *schlieren,* from the German meaning inhomogeneities, and the study of such phenomena has been known as *schlieren optics.*[†]

Theory of Refractive Gradients

The quantitative relationship between the refractive gradient and the deflection of light is derived as follows. Consider a medium whose index

---

* For more details, see A. B. Fraser and W. H. Mach, "Mirages," *Scientific American* **234** (Jan. 1976), 102–11.

[†] The classic paper on schlieren optics is H. Schardin, "Die Schlierenverfahren und ihre Anwendungen," *Erg. exakt. Naturwiss.* **20** (1942), 303–439.

is lowest at the top and highest at the bottom. The refractive index *gradient*, therefore, is positive downward, along the $-y$ axis.

In Figure 3.5-2, the light is incident from the left. Outside the gradient field the light has a wavelength λ_0. At the top of the field where the index is n, the wavelength is λ. At the bottom, the index is n' and the wavelength λ'. The dashed lines represent wavefronts. If there are N wavefronts within the volume shown, the upper arc has a length

$$\Delta L = N\lambda = N\frac{\lambda_0}{n} \qquad [3.5\text{-}1]$$

and the lower arc

$$\Delta L' = N\lambda' = N\frac{\lambda_0}{n'} \qquad [3.5\text{-}2]$$

The arc lengths depend on the radius of curvature, R, of the rays and on the angle of deviation, δ, measured in radians:

$$\Delta L = R\delta \qquad [3.5\text{-}3]$$

and

$$\Delta L' = (R - \Delta y)\delta \qquad [3.5\text{-}4]$$

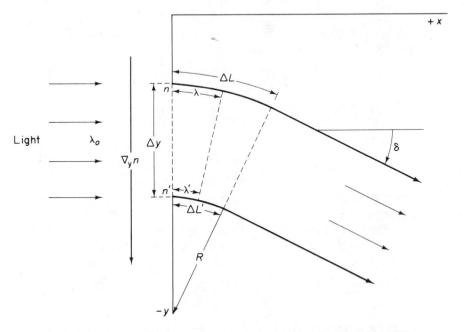

Figure 3.5-2 Deflection of light incident from the left and passing through the gradient-index field.

Subtracting Equation [3.5-4] from [3.5-3] gives

$$\Delta L - \Delta L' = \Delta y \delta \qquad [3.5\text{-}5]$$

Since

$$\Delta L - \Delta L' = N\frac{\lambda_0}{n} - N\frac{\lambda_0}{n'} = N\lambda_0\left(\frac{n'-n}{nn'}\right)$$

and since from Equation [3.5-1]

$$N\lambda_0 = n\Delta L$$

$$\Delta L - \Delta L' = n\Delta L\left(\frac{n'-n}{nn'}\right) \qquad [3.5\text{-}6]$$

Combining Equations [3.5-5] and [3.5-6] by eliminating $\Delta L - \Delta L'$, and solving for δ yields

$$\delta = \frac{1}{\Delta y}\left(1 - \frac{n}{n'}\right)\Delta L \qquad [3.5\text{-}7]$$

The gradient, Δn, need not extend along the y axis. In fact, Δn is the sum of three partial derivatives,

$$\Delta n = \Delta x\frac{\partial n}{\partial x} + \Delta y\frac{\partial n}{\partial y} + \Delta z\frac{\partial n}{\partial z} \qquad [3.5\text{-}8]$$

which, as on pages 185 and 286, can be written in the form of a vector operator,

$$\nabla n = \frac{\partial n}{\partial x}\hat{\mathbf{i}} + \frac{\partial n}{\partial y}\hat{\mathbf{j}} + \frac{\partial n}{\partial z}\hat{\mathbf{k}} \qquad [3.5\text{-}9]$$

The angle of deflection, δ, then becomes

$$\boxed{\delta = \frac{1}{n}\nabla n \cdot \Delta\mathbf{L}} \qquad [3.5\text{-}10]$$

This angle, hence, is a function of the gradient ∇n, of the index n, and of the length of path traversed, $\Delta\mathbf{L}$.

For light proceeding horizontally, along the $+x$ axis, it is sufficient to resolve δ into two components, δ_y and δ_z. The y component, for example, is given by

$$\delta_y = \frac{1}{n}\int\frac{\partial n}{\partial y}\,dL \qquad [3.5\text{-}11]$$

where the integral is taken over the length, L, traversed by the light. If this length is relatively short and the medium homogeneous in that direction,

$$\delta_y = \frac{1}{n}\frac{\partial n}{\partial y}L$$

[3.5-12]

Inserting this equation into Equation [3.5-3] gives

$$\Delta L = R\,\frac{1}{n}\frac{\partial n}{\partial y}L$$

[3.5-13]

and setting $\Delta L = L$ and solving for R,

$$R = \frac{n}{\partial n / \partial y}$$

[3.5-14]

Thus *the steeper the gradient, the shorter the radius of curvature through which the light is bent.*

> **Example.** *Assume that a cuvette is filled with a salt solution whose refractive index varies from $n_1 = 1.4$ at the top to $n_2 = 1.6$ at the bottom. If the distance between top and bottom is 5 cm and the path through the solution 1 cm:*
>
> *(a) Determine the angle of deflection of the light.*
> *(b) Determine the radius of curvature through which the light is bent.*
> *(c) Determine the apex angle of a prism (made of glass of $n = 1.5$) which would give the same deflection.*
> *(d) Check your result by considering the path difference between two rays, passing through top and bottom of the solution, and of the prism.*

Solution. (a) The angle of deflection is found from Equation [3.5-12], using for n the average refractive index, $(1.4 + 1.6)/2 = 1.5$:

$$\delta = \frac{1}{n}\frac{\partial n}{\partial y}L = \frac{1}{1.5}\frac{0.2}{0.05}(0.01) = 0.02667 \text{ rad} = \boxed{1.53°}$$

(b) The inverse of the angle, in radians, is the radius of curvature,

$$R = \frac{1}{0.02667} = \boxed{37.5 \text{ m}}$$

(c) Note that the 1.53° angle refers only to deflection within the solution. Outside, after passing through the rear surface of the cuvette, the angle becomes

$$\beta = \sin^{-1}(1.5)(\sin 1.53°) = 2.29°$$

The apex angle of a (thin) prism that gives the same deflection is found from Equation [1.2-13],

$$\sigma = \frac{\delta}{n-1} = \frac{2.29°}{1.5-1} = \boxed{4.58°}$$

(d) The path difference between top and bottom of the solution, from Equation [2.2-40] is

$$\Gamma = L(n_2 - n_1) = (1)(1.6 - 1.4) = 0.2 \text{ cm}$$

Using the same equation, now solved for L, we find the thickness of the base of the prism:

$$L = \frac{\Gamma}{n_2 - n_1} = \frac{0.2}{1.5 - 1} = 0.4 \text{ cm}$$

Since by construction

$$\tan \tfrac{1}{2}\sigma = \frac{0.4/2}{5}$$

the apex angle must be

$$\sigma = (2)\,\frac{\tan^{-1}(0.4/2)}{5} = \boxed{4.58°}$$

the same result as before.

Gradient-Index Lenses

In a conventional lens, the refractive index is uniform. Refraction takes place only at the surfaces of the lens. In a *gradient-index lens,* or GRIN lens, refraction takes place also within the lens. That has major advantages. A single GRIN lens can be corrected to specifications that could not be met before. Hard-to-grind aspherics can be replaced by (spherical) GRIN lenses. And in complex optical systems the number of elements can be reduced, without sacrificing performance, by replacing one or several conventional lenses by GRIN lenses.

GRIN lenses are made in three forms. One form has a *radial* gradient (cylindrical symmetry). The index of refraction varies as a function of distance *from* the optic axis.* The endfaces of such lenses are often plane

---

*GRIN lenses of radial symmetry were made as early as 1905 by Robert Williams Wood (1868–1955), professor of experimental physics at Johns Hopkins University. Wood used a mixture of glycerol and gelatin, and let it gel between two parallel plates, forming a cylinder a few centimeters in diameter. When the cylinder is soaked in water, the water slowly diffuses into the jelly, replacing (part of) the glycerol, and lowering the index. The cylinder is then cut into several slices, each slice representing a GRIN lens of radial symmetry—which forms an image just as a conventional (converging) lens does. Wood made notable contributions also to UV and IR spectroscopy, and was a superb lecturer with a flair for showmanship. For some of the spectroscopic experiments he did at his farm on Long Island, he used a grating at the end of a long section of sewer pipe; he cleaned the pipe of cobwebs by letting a cat run through it. Wood wrote two books, one of them a collection of nonsense poetry, *How to Tell the Birds from the Flowers* (New York: Dover Publications, Inc., 1959). When he sent a copy of it to President Theodore Roosevelt, the President asked for more of his writings, so Wood sent him a copy of his other book, *Physical Optics.* Even today, *Physical Optics* is worth reading for its wealth of experimental detail. (The GRIN

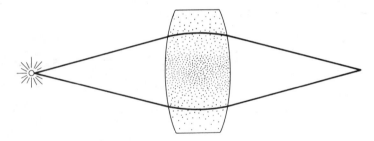

Figure 3.5-3   Converging GRIN lens with spherical endfaces. Note curved ray paths within the lens.

but they can be ground as curved surfaces to give additional power. A lens of this type is illustrated in Figure 3.5-3.

Another type of GRIN lens contains an *axial* gradient. The refractive index varies as a function of distance *along* the axis: the surfaces of constant index are plane and normal to the axis. Axial refractive gradients are particularly useful for the correction of spherical aberration, replacing aspheric surfaces.

In Figure 3.5-4, top, we have an example of spherical aberration, as seen earlier in Chapter 1.6. The essential point is that the marginal rays come to a focus *closer* to the lens than the paraxial rays. Now take a GRIN lens blank containing an axial gradient (center left). The refractive index is highest on the left (front) and from there decreases toward the right (back). If the front surface is ground spherical (center right), the refractive index is less at the edge of the GRIN lens than it is at the center and the marginal rays are bent less. If the gradient profile is properly chosen, rays passing through the lens at any height above the axis come together at a common focus, eliminating spherical aberration (Figure 3.5-4, bottom).

The third type is the *spherical* GRIN lens. Here the index of refraction varies symmetrically about a point: the surfaces of constant index are spheres. Such lenses were studied by Clerk Maxwell (Maxwell's fish eye lens) and later by Luneburg,* but they have never actually been made. The crystalline lens of the human eye (page 143) is of this type.

The design and use of GRIN lenses is in its infancy. Photographic singlets (single-element lenses) have been built; they take the place of more expensive, multielement lenses. GRIN lenses are also used in copying machines, as microscope objectives, for video disk readout, and as Schmidt corrector plates in large-aperture astronomical telescopes.

lenses Wood made are described on pp. 89–90 of the third edition, published in 1934 by The Macmillan Company, New York, and reprinted by Dover Publications, Inc., 1968.) See also W. Seabrook, *Doctor Wood—Modern Wizard of the Laboratory* (New York: Harcourt Brace & Co., 1941).

  *R. K. Luneburg, *Mathematical Theory of Optics*, pp. 164–67, 172–88. (Berkeley: University of California Press, 1964).

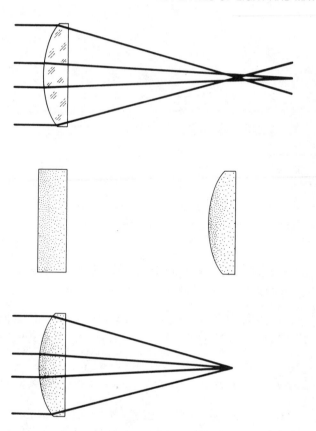

Figure 3.5-4 Eliminating spherical aberration by an axial GRIN lens. [For details, see D.T. Moore and D.P. Ryan, "Gradient index optical lenses," *The Physics Teacher* **15** (1977), 409–13.]

The main problem is *fabrication*. Most promising are *ion exchange* and *ion stuffing*. Ion exchange is explained as follows. Most glasses (page 30) are of the soda–lime type. The Na^+ ions are only loosely bound to the glass structure by weak interactions with the Si and O atoms. These Na^+ ions are exchanged by other ions known to increase the refractive index, most notably Ag^+, K^+, Li^+, Tl^+, Cu^{2+}, and Pb^{2+}. Salts containing any of these ions, most often AgCl and $AgNO_3$, are heated in a crucible and a solid glass cylinder is placed on top of the salt. At a temperature between 200 and 400°C, the salt melts but the glass remains solid and floats on top of the salt. The crucible is maintained at that temperature for several days. During that time, the silver (or other) ions will replace the sodium ions of the glass, gradually moving farther into the glass, raising the index.

Ion stuffing requires special glasses containing, for example, silicon dioxide, boron pentoxide, or rubidium oxide. These glass formers, when heated, separate into two phases. One phase is rich in silica and insoluble; the other phase is rich in alkali and soluble. This alkali phase can be leached out. That leaves the insoluble phase a microporous, spongy, and often opaque structure. The blank is then placed in a salt solution, such as $NaNO_3$ or KNO_3, and the solution sucked through by a vacuum applied to one side until the material is completely saturated. Next the blank is placed for a short time in distilled water. This leaches out some of the salt, more from the outer, less from the deeper layers. After that the sample is heated to 40 to 60°C. This process evaporates the solvent and causes the salt to precipitate. Then the temperature is raised to 900 to 1000°C and held there for 2 h. This closes the pores and makes the glass transparent again.*

Experimental Gradient-Index Optics

Töpler's method. In *Töpler's schlieren method*[†] light from a point source [(1) in Figure 3.5-5] is focused by two lenses (2 and 4) onto a knife edge (5). In addition, the gradient-index field (3) is imaged by a third lens (6), which is often the objective lens of a camera, onto a screen or photographic film (7). Lenses 2 and 3 must be large enough to cover the gradient field. They must also be of high quality and free of polishing defects; otherwise, such defects would show. Frequently, these lenses are replaced by concave first-surface mirrors.

If the sampling field is empty, homogeneous, and of uniform refractive index, the light passing through is blocked out by the knife edge and no light reaches the screen (solid lines in Figure 3.5-5). If there are

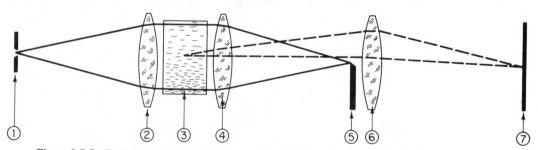

Figure 3.5-5 Töpler's method. 1, Light source; 2 + 4, schlieren head; 3, gradient-index field; 5, knife edge; 6, camera lens; 7, image plane.

---

*E. W. Deeg and D. A. Krohn, *Ophthalmic lens with locally variable index of refraction and method of making same,* U.S. Patent 4,073,579, Feb. 14, 1978.

[†]Robert Hooke in 1672 was the first to use schlieren techniques [according to J. Rienitz, "Schlieren experiment 300 years ago," *Nature, London* **254** (1975), 293–95]. Léon Foucault applied them to testing large astronomical objectives [L. Foucault, "Mémoire sur la construction des télescopes en verre argenté," *Ann. de l'Observatoire Imp. de Paris* **5** (1859), 197–237] and August Joseph Ignaz Töpler, German scientist (1836–1912), applied them to the study of optical inhomogeneities in gases [A. Töpler, *Beobachtungen nach einer neuen optischen Methode* (Bonn: Max. Cohen u. Sohn, 1864)].

inhomogeneities (refractive gradients) present, these gradients deflect part of the light, which bypasses the knife edge, and the gradients show on the screen as bright streaks, real images of the gradients (dashed lines).

Foucault's knife-edge test is used to evaluate lenses and mirrors. Light from a point source is brought to a focus by the lens or mirror, and falls on a screen placed some distance beyond the focus. A knife edge is positioned *at* the focus. If the knife edge is moved across the axis, and if the lens or mirror is of perfect quality, the disk of light seen on the screen changes rapidly and evenly from light to dark. If the knife edge is positioned a short distance *behind* the focus and moved across the axis, a shadow moves across the disk in the same direction; if the knife edge is placed *ahead* of the focus, the shadow moves in the opposite direction. As in Töpler's method, any imperfections show as either lighter or darker patches.

If a diaphragm with a matrix of holes is placed in front of the lens or mirror and if a screen is held a short distance ahead of or behind the focus, the deviation of the various rays passing through these holes is particularly easy to see. This is called the *Hartmann test.* *

Ronchi grid method.

This is one of numerous variations of Töpler's method. Both dimensions of a photograph are needed to show the $y-z$ position of any image detail. But it would be good to have another dimension available to show the magnitude of the gradients. This additional dimension could be some distortion in an otherwise regular pattern. To accomplish this, a *Ronchi grid* is placed halfway between the objective and the eyepiece of a microscope (see Figure 3.5-6 and page 251). No other changes are needed in the system.

The exact position of the grid is not critical. When the grid is close to the eyepiece, its shadow is sharpest but the sensitivity of the system is least. When the grid is near the objective, the opposite is true. When the slit (number 16 in Figure 3.5-6) is oriented parallel to the grid, the image seen in the microscope has a series of equidistant parallel lines superimposed on it. If there are no optical-thickness gradients present (optical thickness = thickness × refractive index), these lines are straight. But *with* such gradients present, the grid's shadow shows characteristic distortions, seen in Figure 3.5-7.

The slit below the condenser may be replaced by a second Ronchi grid. This allows more light to pass through and thus makes the method useful at high magnifications. The two grids need not be in register and they need not be conjugate; they need only be parallel.

---

*J. Hartmann, "Objektivuntersuchungen," *Zschr. f. Instrumentenkunde* **24** (1904), 1–21, 33–47, 97–117.

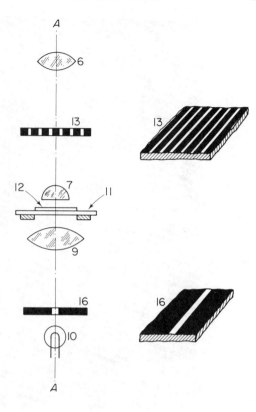

Figure 3.5-6 Ronchi-grid gradient-index microscope. *A–A*, Optic axis, 6, eye-piece; 7, objective lens; 9, condenser; 10, light source; 11 + 12, specimen; 13, Ronchi grid; 16, rotatable slit diaphragm. [From J.R. Meyer-Arendt, *Optical System for Microscopes or Similar Instruments*, U.S. Patent 2,977,847; Apr. 4, 1961.]

Color schlieren methods.

The magnitude of the gradients can also be represented by *color*. Early color schlieren systems were built in the form of prism spectrographs, with the light passing first through the prism and then through the gradient field. A second slit takes the place of the knife edge in Töpler's method. However, a slit significantly reduces both the amount of light and the resolution. Much better than a prism is a *wedge-type interference filter* and to use no slit at all (Figure 3.5-8).

Three processes take place:

1. *Without* index gradients, the filter merely casts a shadow on the film in the camera. This shadow provides a "background spectrum."

2. *With* index gradients, the two lenses 2 and 3 form a real image of the gradients, as in Töpler's method.

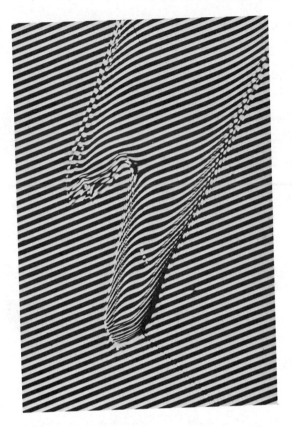

Figure 3.5-7   Photomicrograph of a thin layer of colorless cement spilled across a glass slide. Thickness gradients near the boundary of the layer cause characteristic distortions of the line pattern.

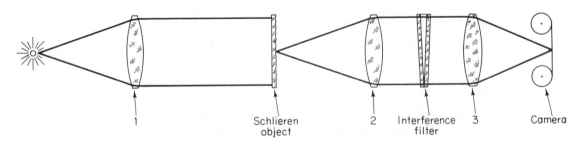

Figure 3.5-8   Color schlieren system using wedge-type interference filter. 1, Collimating lens; 2 + 3, schlieren head.

3. *In addition,* the gradients, which in effect are prisms, deflect the light through different parts of the filter. The result is that these gradients take on different colors, highly saturated and pleasing to look at.*

Fiber Optics

Optical fibers or, as they are called today, *fiber guides,* are of two types, *step-index fibers* and *GRIN fibers*. Step-index fibers were developed first.[†] They consist of a transparent core of glass or plastic of a given refractive index surrounded by a cladding layer of lower index. Most of the light travels inside the core and is contained there by total internal reflection. But if the fiber is bent too sharply, the angle of incidence at the core–cladding interface becomes less than the minimum angle and part of the light escapes, causing "cross-talk."

Assume that n_0 in Figure 3.5-9 is the refractive index of the outside medium, n_1 the index of the core of the fiber, and n_2 the index of the cladding. To keep the light inside the core, the angle of incidence at the core–cladding interface must not be less than

$$\sin \gamma = \frac{n_2}{n_1} = \frac{c}{a} \qquad [3.5\text{-}15]$$

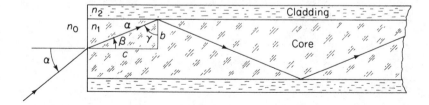

Figure 3.5-9 Propagation of light in a step-index fiber.

---

* See, for example, J. R. Meyer-Arendt, "Microscopy as a Spatial Filtering Process," in R. Barer and V. E. Cosslett, editors, *Advances in Optical and Electron Microscopy,* Vol. 8, p. 14 and facing page (London: Academic Press, Inc., 1982).

[†] At night, when we look at a water fountain illuminated from below, the light seems to follow the curved streams of water. Actually, the light follows a zigzag path, because of total internal reflection. The first to show this effect, using a stream of water flowing from a tank, was John Tyndall, "On Some Phenomena Connected with the Motion of Liquids," *Proc. Roy. Inst.* **1** (1854), 446–48. J. L. Baird extended the principle to fibers to convey an image, *An Improved Method of and Means for Producing Optical Images,* Brit. Patent 285,738, Feb. 15, 1928.

Since

$$a^2 = b^2 + c^2$$

$$1 = \frac{b^2}{a^2} + \frac{c^2}{a^2}$$

$$\left(\frac{b}{a}\right)^2 = 1 - \left(\frac{c}{a}\right)^2 = 1 - \left(\frac{n_2}{n_1}\right)^2$$

$$\frac{b}{a} = \sqrt{1 - \left(\frac{n_2}{n_1}\right)^2} = \sin \beta$$

Then, from Snell's law,

$$n_0 \sin \alpha = n_1 \sin \beta = n_1 \sqrt{1 - \left(\frac{n_2}{n_1}\right)^2} = \sqrt{n_1^2 - n_2^2} \qquad [3.5\text{-}16]$$

which is the numerical aperture of the fiber.

In a *GRIN fiber*, the refractive index varies radially, decreasing from the axis outward. The light travels inside along the axis; if it deviates from the axis, it is returned by the gradient.

Assume, for simplicity, that the GRIN material is divided into individual strata (layers) and that n_1 in Figure 3.5-10 is the material of the highest index. A ray is trying to break away, making an angle of incidence α_{12} at the 1–2 boundary.

From Snell's law,

$$n_1 \sin \alpha_{12} = n_2 \sin \beta_{12} = n_2 \sin \alpha_{23} \qquad [3.5\text{-}17]$$

At the 2–3 boundary,

$$n_2 \sin \alpha_{23} = n_3 \sin \alpha_{34}$$

and so on. At the nth boundary, a distance r away from the axis,

$$n_1 \sin \alpha_{12} = n(r)\sin \alpha(r)$$

Thus

$$n(r)\sin \alpha(r) = \text{constant} \qquad [3.5\text{-}18]$$

which means that, if n decreases, α increases and the light is turned back toward the axis: the light is confined within the fiber. The ray never even touches the wall. This is completely different from a step-index fiber, which relies on total internal reflection. The advantage of GRIN fibers is that all rays traveling through have the same optical path length; in step-index fibers a pulse of light injected at one end becomes considerably wider when it emerges at the other end.

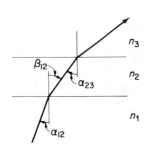

Figure 3.5-10 Refraction of light in a stratified medium, with the refractive index highest in n_1. Light incident from below.

Applications.   The most elementary application of fiber optics is the *transmission of light,* either to illuminate hard-to-reach places or to conduct light out of such places, perhaps to a photocell located at a more accessible site. In punch-tape and card readers, brush contacts and arrays of numerous tiny light bulbs have given way to *light pipes* and a single light bulb that can easily be cooled and replaced when needed.

More interesting is the *transmission of images.* The best known example is the *flexible fiberscope.* As shown in Figure 3.5-11, some of the fibers conduct light into the cavity to be examined while the others carry the image back to the observer. The image-conducting fibers, up to 140 000 of them, are by necessity very thin, often no more than 10 μm in diameter each, and the entire fiber bundle may be no more than a few millimeters thick. Fiberscopes are used extensively in medicine and engineering; they make it possible to inspect just about any cavity in the human body, from the respiratory to the digestive tract, and to look inside the heart while it beats.

Of increasing interest is the use of fiber guides for *communications.* Compared to electrical conductors, optical fibers are lighter in weight, less expensive, equally flexible, not subject to electrical interference, and more secure to interception. Most important, fibers can now be made which have losses as low as one decibel per kilometer, 1 dB km^{-1}. This is a remarkable achievement considering that only a decade ago the best fibers had losses in excess of 1000 dB km^{-1} and 20 dB km^{-1} was thought to be the limit.

Example. *Both the* bel* *and the* decibel *are comparative units. One bel means that the power in one channel, or at one time, is 10 times that in another channel, or at another time; 2 bel means 100×, 3 bel 1000×, and*

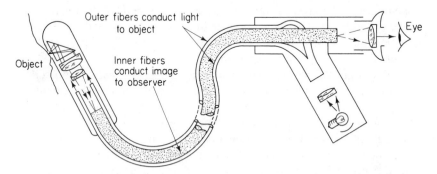

Figure 3.5-11   Flexible fiberscope.

* Named after Alexander Graham Bell (1847–1922), American scientist. Working at the Boston School for the Deaf, Bell reasoned that if sound could be made into fluctuating electric currents, these currents could be carried through wires over a distance and converted back into sound. When he accidentally spilled battery acid on his pants and, with his audio

so on. For practical use the unit bel is too large, hence the decibel, dB, 1 bel = 10 dB. The difference, in dB, between two powers, ϕ, therefore, is $10 \log_{10}(\phi_2/\phi_1)$. If one-half the initial power in a fiber is lost, $\phi_2 = \frac{1}{2}\phi_1$:

$$10 \log_{10} \frac{\frac{1}{2}\phi_1}{\phi_1} = 10 \log_{10} 0.5 = \boxed{-3 \text{ dB}}$$

the minus sign indicating the loss. Over a distance of 1 km, this is not much.

Integrated Optics

In recent years, fiber optics has developed into complex systems of miniature dimensions. This is the field of *integrated optics*. There are two types of integrated waveguides: *planar* guides, which are relatively wide, and *strip* guides, which are more narrow, more similar to fibers. Both types are exceedingly thin, of the order of a wavelength of light, and either type can be made in the form of step-index guides or as GRIN guides, like fibers.

The light travels (mostly) inside the guide. But how does the light enter? Shining light head-on into the narrow side of a thin film would cause so much scatter that very little light would enter the film. Instead, *beam couplers* are used, which are either small prisms placed on top of the film, or gratings made interferometrically in a layer of photoresist.

A variety of (passive) elements can be made by changing the thickness of the film. In a thinner film the effective velocity of the light is higher (because its zigzag path is stretched out longer); in a thicker film it is less. A *thin-film prism* is made by adding, through a mask with a triangular opening, another layer of transparent material. When the light reaches the base of the triangular-shaped thicker film, it is delayed and, as in a conventional prism, it deviates from its initial direction. The same result can be obtained by a refractive gradient.

Thin-film lenses are of two varieties, *Luneburg lenses* and *geodesic lenses*. A Luneburg lens looks like a flat circular mound; it combines a thicker film and the GRIN feature, the index being highest in the center of the mound and decreasing toward the periphery (Figure 3.5-12, left). Light incident on the lens from *any* direction within the plane of the film comes to a focus, interestingly enough without aberrations except curvature of field. A geodesic lens has the form of a dome or depression in a step-index film of uniform thickness (right).

---

equipment nearby, called out to his assistant, near the other end of the line on another floor, "Watson, please come here. I want you," these famous words became the first telephone communication.

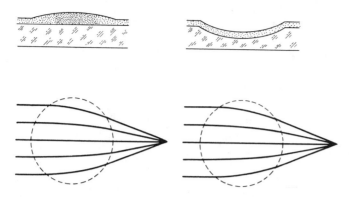

Figure 3.5-12   Thin-film Luneburg lens (*left*) and geodesic lens (*right*); ray trace for collimated light shown below.

Active elements in integrated optics include light modulators, light switches, beam deflectors, and scanners. Modulators can be made to operate on the amplitude, phase, frequency, or state of polarization of the light. Beam deflectors and scanners change the direction of the light in response to an electric signal.

Manufacturing such elements is a rapidly evolving art. Earlier planar guides were simply films such as tantalum oxide or lithium niobate, coated on a substrate by vacuum deposition. Modern strip guides are made by diffusion techniques, ion implantation, proton bombardment, or material removal by electron or laser beam writing. Particularly interesting is the fabrication of *monolithic integrated optics* where all functions, from the generation of the light to guiding to modulation to detection, are performed in a single crystal, most commonly gallium arsenide, GaAs.

Gallium arsenide is a material well suited also for the construction of lasers (see Chapter 5.4, page 509). Such lasers serve as tiny, highly efficient light sources that together with other passive and active elements can be incorporated into small chips, often no larger than a few centimeters, or even millimeters, in size. Figure 3.5-13 gives an example.

SUGGESTIONS FOR FURTHER READING

Ch. K. Kao, *Optical Fiber Systems: Technology, Design, and Applications* (New York: McGraw-Hill Book Company, 1982).

S. E. Miller and A. G. Chynoweth, *Optical Fiber Telecommunications* (New York: Academic Press, Inc., 1979).

E. W. Marchand, *Gradient Index Optics* (New York: Academic Press, Inc., 1978).

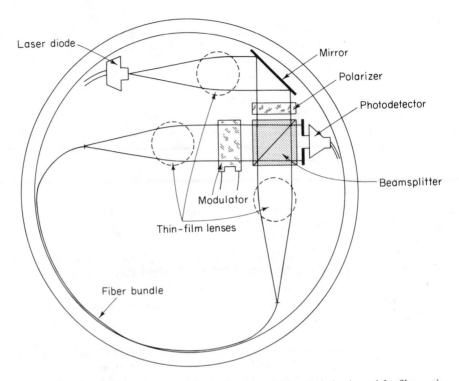

Figure 3.5-13 Schematic representation of integrated circuit used for fiber-optics gyroscope. For details on gyros, see Chapter 6.1, page 539.

D. T. Moore, "Gradient-index optics: a review," *Applied Optics* **19** (1980), 1035–38.

D. Marcuse, *Light Transmission Optics*, 2nd edition (New York: Van Nostrand Reinhold, 1982).

L. Levi, *Applied Optics, a Guide to Optical System Design*, Vol. 2, Chapter 13, "Integrated and Fiber Optics," pp. 147–241 (New York: John Wiley & Sons, Inc., 1980).

T. Tamir, editor, *Integrated Optics*, 2nd edition (New York: Springer-Verlag, 1979).

PROBLEMS

3.5-1. A slab of transparent material measures 3 mm × 10 mm × 60 mm. If the refractive index of the material varies linearly in the lengthwise direction from 1.51 to 1.55, how much will a beam of collimated light, incident normally on the widest face, be deflected at a distance of 2 m from the slab?

3.5-2. A palm tree is seen across 10 km of desert that has become so hot that directly above the ground the refractive index of the air is 1.000 290, whereas 5 m higher up it is 1.000 292. By how much does the tree appear to be displaced?

3.5-3. A plane GRIN plate of radial symmetry is 40 mm in diameter and 10 mm thick. Its refractive index varies from 1.6 in the center to 1.5 in the periphery and its rear face is ground so that no refraction occurs there. What is the focal length?

3.5-4. A plane plate, 50 mm in diameter and 10 mm thick, is made to have a peripheral refractive index $n_2 = 1.50$ and a focal length of 120 cm. What should be the refractive index in the center?

3.5-5. Imagine a solid sphere of gradient-index material whose index is highest in the center and 1.4 at the periphery.
(a) What is the *least* index in the center required to keep light injected into the sphere traveling around inside without touching the surface?
(b) What must be the *diameter* of the sphere?

3.5-6. Some GRIN material is made in the form of a doughnut. The inside diameter (the diameter of the hole in the doughnut) is 34 cm and the outside diameter 46 cm; the cross section of the ring, therefore, is 6 cm in diameter.
(a) If the average index is 1.6, what radial gradient is needed to keep a beam of light traveling along exactly in the center of the ring?
(b) What should be the highest, and the lowest, refractive index of the material?

3.5-7. Schlieren systems can be *calibrated* by placing a small, low-power lens in the schlieren field and replacing the knife edge by a Ronchi grid. What will the image look like and how can it be used for calibration?

3.5-8. Continue with Problem 3.5-7 and describe how the image will change if the grid is replaced by a wedge-type color filter.

3.5-9. If a step-index fiber has a core of index 1.55 and a cladding of 1.53:
(a) What is its numerical aperture?
(b) What is the maximum angle (of incidence) at which light can enter the fiber?

3.5-10. Parallel light enters a fiber of diameter $D = 0.1$ mm (Figure 3.5-14). Determine the least radius R through which the fiber may be bent, assuming that the core has an index of 1.54, the cladding 1.52.

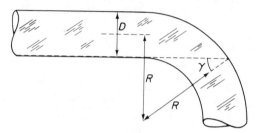

Figure 3.5-14

3.5-11. Design a diverging (negative) lens of the Luneburg type.

3.5-12. Three individual fibers enter an integrated circuit, and three other fibers leave it. A switching element can be designed such that any incoming fiber is connected to any outgoing fiber, but no incoming fiber is connected to more than one fiber, nor is any fiber *not* connected to any other fiber. How many permutations are possible?

3.6

Radiometry

RADIOMETRY IS THE PROCESS OF MEASURING radiant power and other quantities derived from it. Radiometry refers to such quantities anywhere in the electromagnetic spectrum. A subdivision of radiometry is *photometry*, which refers only to quantities perceived by the human eye as the sensation of *light*.

All radiometric quantities have the adjective *radiant* and their symbols carry the subscript *e*, for "electromagnetic." All photometric quantities have the adjective *luminous* and their symbols carry the subscript *v*, for "visual."

Radiometry and photometry have long been confounded by an overabundance of terms and units. At times, different terms are used for the same quantities. At other times, a single term (such as "intensity") is used for too many quantities. Terms like "candlepower" are ill conceived. Still others, such as nox, phot, glim, skot and scot, bril and brill, helios, lumerg, pharos, stilb, and blondel, merely delight the historian.

In recent years much progress toward clarity has been made, especially since international agreement has been reached to adopt simple, logical, and easily convertible units, based on the International System of Units, SI. These terms and units are defined and used in this chapter.

Terms and Units

Energy.   We distinguish *radiant energy,* which refers to electro-magnetic radiation in general, and *luminous energy,* which refers to light.

Radiant energy.   The unit of radiant energy, Q_e, the same as the unit of energy in general, is the *joule,* J.

Luminous energy.   The unit of luminous energy, Q_v, is the *lumen-second* or talbot.

Power.   *Power, ϕ, is energy, Q, generated, transmitted, or received, per unit of time, t:*

$$\phi = \frac{Q}{t}$$ [3.6-1]

Power is the crucial term. It is the *term central to all of radiometry and photometry.* From it all other terms are derived. The expression used earlier was "flux," a very descriptive term because it refers to the *flow* of energy. But since elsewhere in physics energy per unit of time is called *power,* this term is recommended for use in radiometry and photometry.

Radiant power.   *Radiant power, ϕ_e,* is the total amount of electro-magnetic energy transmitted per unit of time. The unit of radiant power is the same as that of power in general, *watt,* W:

$$1 \text{ watt} = \frac{1 \text{ joule}}{1 \text{ second}}; \qquad W = J\ s^{-1}$$

Luminous power.   *Luminous power, ϕ_v,* by inference, is that part of radiant power that appears as light. Its unit is the *lumen,* lm.

Radiant-luminous conversion.   To convert radiometric quantities into photometric quantities, or vice versa, we take into account the relative visibility, or *luminous efficiency,* of the light at each particular wavelength. A blue lamp, for example, may emit the same amount of radiant energy as a green lamp, but the latter appears brighter because the eye is more sensitive to green light than to blue. For daylight or *photopic vision,* the peak sensitivity of the human eye is in the bright yellow, at 555 nm (Figure 3.6-1). Photopic vision is a function of the *cones* of the retina. If the eye is adapted to night or *scotopic vision,* which is a function of the *rods,* the peak efficiency shifts to 510 nm, as shown. Luminous efficiency has no units; it is given as a percentage.

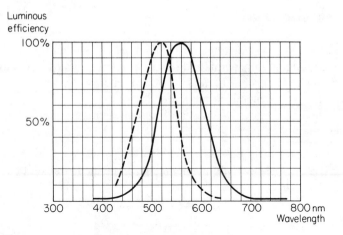

Figure 3.6-1 Sensitivity of the human eye. Photopic vision shown by solid line, scotopic vision by dashed line. Both curves are normalized to their own maxima.

At the peak wavelength of photopic vision, 555 nm, 1 watt is set equal to 673 lumens. (Conversely, 1 lumen is equivalent to 1.486 milliwatts, mW.) To use this conversion anywhere else in the spectrum, we need to know the luminous efficiency at that wavelength. For example, at a wavelength of 600 nm the efficiency is 63.1 percent; thus 1 watt of monochromatic light of 600 nm wavelength is equal to (673)(0.631) = 425 lumens, or 1 lumen = 1/(673 × 0.631) = 2.35 mW. Factors for other wavelengths are given in Table 3.6-1.

The term *efficacy* applies to light sources only. It is the ratio of power emitted to the (electric or other) power consumed; that is, efficacy is the *ratio of output* (in lumens) *to input* (in watts). Its unit is lm W^{-1}. (Electric power is the product of voltage and current: 1 W = 1 V × 1 A.)

Example. *A light bulb rated at 200 watts has an output of 3500 lumens. What is:*
(a) The luminous efficiency of the light bulb?
(b) The efficacy of the light bulb?

Solution. (a) Divide the output by the equivalence, 1 watt = 673 lumens. Then

$$\phi = \frac{3500 \text{ lm}}{673 \text{ lm/W}} = 5.2 \text{ "light-watts"}$$

(The term "light-watt" refers to the fraction of the total power that is perceived by the eye as luminous power.) Therefore, only

$$\frac{5.2}{200} = \boxed{2.6 \text{ percent}}$$

Table 3.6-1 SPECTRAL DISTRIBUTION OF PHOTOPIC
LUMINOUS EFFICIENCY

Wavelength	Luminous efficiency	Wavelength	Luminous effiiciency
410	0.001	570	0.952
420	0.004	580	0.870
430	0.012	590	0.757
440	0.023	600	0.631
450	0.038	610	0.503
460	0.060	620	0.381
470	0.091	630	0.265
480	0.139	640	0.175
490	0.208	650	0.107
500	0.323	660	0.061
510	0.503	670	0.032
520	0.710	680	0.017
530	0.862	690	0.008
540	0.954	700	0.004
550	0.995	710	0.002
560	0.995	720	0.001

reaches the eye in the form of light. These 2.6 percent represent the
luminous efficiency of the lamp.

(b) The *efficacy* of the same lamp is

$$\frac{3500 \text{ lm}}{200 \text{ W}} = \boxed{17.5 \text{ lm}/\text{W}}$$

Radiation geometry.   Terms such as energy and power, both
radiant and luminous, are general; they do not take into account the specific
geometry of the source, the light, or the receiver. To include these
configurations, we need another set of terms: *areance, sterance,* and
pointance (Figure 3.6-2). These terms and the associated symbols and
units are summarized in Table 3.6-2.

Areance I.   *Areance* (exitance), M, is the amount of power, ϕ,
emitted per unit area, A:

$$\text{areance} = \frac{\text{power}}{\text{area}}$$

$$\boxed{M = \frac{\phi}{A}} \qquad\qquad [3.6\text{-}2]$$

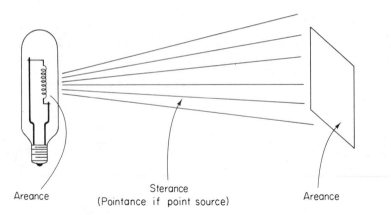

Areance   Sterance
(Pointance if point source)   Areance

Figure 3.6-2   Terms used in radiation geometry.

Table 3.6-2 RADIOMETRIC AND PHOTOMETRIC TERMS,
SYMBOLS, AND UNITS

Term	Symbol	Unit	
Energy	Q		
Radiant energy	Q_e	joule	J
Luminous energy	Q_v	lumen-second	lm s
Power (flux)	ϕ		
Radiant power	ϕ_e	watt	W
Luminous power	ϕ_v	lumen	lm
Areance (exitance)	M		
Radiant areance	M_e	watt/meter^2	W m^{-2}
Luminous areance	M_v	lumen/meter^2	lm m^{-2}
Pointance (intensity)	I		
Radiant pointance	I_e	watt/steradian	W sr^{-1}
Luminous pointance	I_v	lumen/steradian	lm sr^{-1}
		(candela)	(cd)
Sterance	L		
Radiant sterance	L_e	watt/(meter^2 steradian)	W m^{-2} sr^{-1}
(radiance)			
Luminous sterance	L_v	lumen/(meter^2 steradian)	lm m^{-2} sr^{-1}
(luminance)		(candela/meter^2)	(cd m^{-2})
Areance (irradiance/	E		
illuminance)			
Radiant areance	E_e	watt/meter^2	W m^{-2}
Luminous areance	E_v	lumen/meter^2	lm m^{-2}
		(lux)	(lx)

The term areance (exitance) applies to all light sources, including a piece of ground glass through which light is shining, or a mirror or a sheet of paper from which light is reflected.

    Radiant areance.   *Radiant areance* (radiant exitance), M_e, refers to the emission of electromagnetic power in general. Its unit is watts per square meter, W m^{-2}.

    Luminous areance.   The photometric equivalent is *luminous areance* (luminous exitance), M_v, with the unit lumens per square meter, lm m^{-2}.

Pointance.   *Pointance*, as the name implies, refers to *point* light sources. In fact, pointance is *restricted* to point sources. (The term "intensity" is deprecated.) Pointance, *I*, is the amount of power, ϕ, originating at a point source and proceeding within a cone of a given solid angle, ω:

$$\text{pointance} = \frac{\text{power}}{\text{solid angle}}$$

$$\boxed{I = \frac{\phi}{\omega}} \qquad\qquad [3.6\text{-}3]$$

Since, from Equation [3.1-8], the total solid angle about a point is $\omega = 4\pi$ (steradians), a point source that emits uniformly in all directions has a total pointance of

$$I = \frac{\phi}{4\pi} \qquad\qquad [3.6\text{-}4]$$

Other solid-angle relationships worth noting: If σ is the apex angle of a right-circular cone, the solid angle contained within the cone is

$$\omega = 2\pi\left[1 - \cos\left(\frac{\sigma}{2}\right) \right] = 4\pi \sin^2\left(\frac{\sigma}{4}\right)$$

For small angles,

$$\cos\alpha \approx 1 - \frac{\alpha^2}{2}$$

and hence,

$$\omega \approx \pi\left(\frac{\sigma}{2}\right)^2 \qquad\qquad [3.6\text{-}5]$$

where σ is measured in radians.

Radiant pointance. *Radiant pointance* (formerly radiant intensity), I_e, is radiant power per solid angle. Its unit is watts per steradian, W sr^{-1}.

Luminous pointance. In *luminous pointance* (formerly luminous intensity), I_v, the unit is lumens per steradian, lm sr^{-1}, a unit called the *candela*.

The international standard of luminous pointance is a blackbody radiator surface 1 cm^2 in size, at the temperature of solidification of platinum, 2045 K or 1772°C, and at normal atmospheric pressure, 101.325 kPa, kilopascal (1 Pa = 1 Newton/meter^2, N m^{-2}). To this source is arbitrarily assigned the luminous pointance of 60 cd.* In other words, a blackbody surface of an area of 1/60 of 1 cm^2, or approximately 1.7 mm^2, produces 1 cd of luminous pointance.

Example 1. *A point source emits uniformly into all space a total power of 220 lm. What is the luminous pointance?*

Solution. From Equation [3.6-4] we note that a 1-cd point source emits a total of 4π lumens. Thus

$$I = \frac{\phi}{\omega} = \frac{220 \text{ lm}}{4\pi \text{ sr}} = \boxed{17.5 \text{ cd}}$$

Example 2. *A 70-cd light source is placed 20 cm from a screen with a hole 4 cm in diameter. How much light goes through the hole?*

Solution. The area of the hole is

$$A = \pi R^2$$

and the solid angle subtended at the source by the circumference of the hole

$$\omega = \frac{A}{d^2} = \frac{\pi R^2}{d^2}$$

Therefore,

$$\omega = \pi\left(\frac{2}{20}\right)^2 = \pi\left(\frac{1}{100}\right) \text{ sr}$$

from which we find that

$$\phi = I\omega = (70 \text{ lm})\left[\pi\left(\frac{1}{100}\right) \text{ sr}\right] = \boxed{2.2 \text{ lm}}$$

Example 3. *A searchlight projects 0.012 lm on a billboard 1 km away, forming a disk of light 3.2 m in diameter. Find the pointance.*

Solution. Since the beam illuminates an area

---

*Resolution adopted by the 13th General Conference on Weights and Measures, Paris, France, October 1967.

$$A = \pi R^2 = \pi \left(\frac{3.2}{2}\right)^2 = 8 \text{ m}^2$$

in size, it subtends a solid angle of

$$\omega = \frac{A}{R^2} = \frac{8 \text{ m}^2}{(1000 \text{ m})^2} = 8 \times 10^{-6} \text{ sr}$$

Thus, the pointance is

$$I = \frac{\phi}{\omega} = \frac{0.012 \text{ lm}}{8 \times 10^{-6} \text{ sr}} = \boxed{1500 \text{ cd}}$$

Sterance.   The term *sterance* is a hybrid of *ster*adian (solid angle) and are*ance* (exitance). It refers to light that comes from an *extended* source, rather than from a point source. Sterance, L, is the amount of power emitted from, transmitted through, or reflected off a surface per unit area of that surface per solid angle. In short, sterance is pointance per area. More precisely, sterance is pointance per *projected* area, $A \cos \theta$,

$$\text{sterance} = \frac{\text{power}}{\text{area} \times \text{solid angle}} = \frac{\text{pointance}}{\text{area}}$$

$$\boxed{L = \frac{\phi}{(A \cos \theta)(\omega)} = \frac{I}{A \cos \theta}} \qquad [3.6\text{-}6]$$

where θ is the angle subtended by the normal to the (emitting) surface and the direction of the light. Thus the amount of light that leaves a surface element ΔA in a given direction is proportional to $\cos \theta$:

$$I_\theta = I_\perp \cos \theta \qquad [3.6\text{-}7]$$

which is *Lambert's cosine law.* Substituting Equation [3.6-7] in [3.6-6] gives

$$L = \frac{I_\perp \cos \theta}{A \cos \theta} = \text{constant} \qquad [3.6\text{-}8]$$

which shows that the subscript θ can be dropped: *The sterance of a Lambertian surface is the same in all directions.*

Example. *This definition of a Lambertian surface makes more sense when we consider a flat metal disk with a rough surface, heated to incandescence. If the disk has an area of 1 cm^2 and a sterance of 1 W cm^{-2}sr^{-1}, it radiates 1 W sr^{-1} in a direction normal to its surface. In a direction 45° to the normal, it radiates only (1 W sr^{-1})(cos 45°) = 0.707 W sr^{-1}. However, when seen from the 45° direction, the disk appears more narrow, com-*

pressed into an ellipse, and hence its projected area is less also. Both reductions cancel and the sterance remains the same.

Radiant sterance. The unit of *radiant sterance* (radiance), L_e, is watts per square meter per steradian, W m^{-2}sr^{-1}.

Luminous sterance. The unit of *luminous sterance* (luminance), L_v, is lumens per square meter per steradian. But since lumen per steradian is a candela, a more convenient unit is candelas per square meter, cd m^{-2}.

Several other units of luminous sterance (luminance) are not recommended for use:

Stilb, sb, $= 1$ cd/cm$^2 = 10^4$cd m^{-2}.
International apostilb, asb, or *blondel*, $= 1/\pi$ cd/m^2.
Skot $= 10^{-3}$ asb.
Lambert (or *centimeter-lambert*) $= 1/\pi$ cd/cm$^2 = 10^4/\pi$ cd m$^{-2} = 1/\pi$ sb.
Meter-lambert $= 1/\pi$ cd/m^2.
Foot-lambert $= 1/\pi$ cd/ft$^2 = 3.426$ cd m^{-2}. One foot-lambert is the luminous areance of a uniform diffuser that emits one lumen per square foot.

From these relationships we see that a Lambertian surface of 1 lm/m^2 areance (exitance) produces 1 apostilb ($1/\pi$ cd/m^2) sterance (luminance):

$$M_v = \text{lm}/\text{m}^2 \rightarrow L_v = 1/\pi \text{ cd}/\text{m}^2 \qquad [3.6\text{-}9]$$

Similarly, a Lambertian surface of 1 lm/cm^2 is a source of 1 lambert ($1/\pi$ cd/cm^2), and a Lambertian surface of 1 lm/ft^2 is a source of 1 foot-lambert ($1/\pi$ cd/ft^2). More conversion factors are found in Table 3.6-3, examples in Table 3.6-4.

Table 3.6-3 CONVERSION FACTORS FOR UNITS OF LUMINOUS STERANCE (LUMINANCE)

One	Equals		
Nit	1 cd/m^2 *	$\pi \times 10^{-4}$ lambert	0.2919 foot-lambert
Stilb	1 cd/cm^2	π lamberts	2.919×10^3 foot-lambert
Lambert	$1/\pi$ cd/cm^2	0.318 stilb	9.29×10^2 foot-lambert
Foot-lambert	$1/\pi$ cd/ft^2	3.43×10^{-4} stilb	1.1×10^{-3} lambert

*Recommended unit.

Table 3.6-4 LUMINOUS STERANCE (LUMINANCE)
OF VARIOUS OBJECTS

Object	Luminous sterance $(cd\ m^{-2})$
Atomic bomb	$2\ \times 10^{12}$
Sun	2.3×10^{9}
Xenon arc	$1\ \times 10^{9}$
60-W light bulb, inside frosted	1.2×10^{5}
Snow in bright sunlight	$4\ \times 10^{4}$
Average landscape in summer	8000
Average landscape under overcast sky	2000
Sterance recommended for comfortable viewing	1400

Areance II. Now let the light be *incident on a surface*. The terms formerly used were "irradiance" or "illuminance." The new term is again *areance* but now areance II, with the symbol E, power incident per unit area:

$$E = \frac{\phi}{A} \qquad\qquad [3.6\text{-}10]$$

the same definition as power *emitted* by a surface. This is easily justified: Many surfaces receive light from a source and reflect, scatter, or otherwise reemit part of this light, acting as secondary sources. The definitions of both types of areance thus are the same, *power per unit area*.

Since the total power emitted from a point source into all space is $\phi = 4\pi I$ and since the surface area of a sphere is $A = 4\pi R^2$, a source of pointance I produces a total areance of

$$E = \frac{4\pi I}{4\pi R^2}$$

Canceling 4π and changing R into d, for distance, gives

$$E = \frac{I}{d^2} \qquad\qquad [3.6\text{-}11]$$

which is the *inverse-square law*. It shows that the areance, at distance d, is directly proportional to the pointance of the source and inversely proportional to the square of the distance (Figure 3.6-3).

There is an interesting similarity worth noting that exists between areance and the magnetic induction, **B,** produced by a steady current. Consider again

Figure 3.1-3 used to derive Biot–Savart's law and Maxwell's second equation. The contribution dB made by each current element, $i\ dW$, was found to be

$$dB = \mu_0 \frac{1}{4\pi} \frac{1}{R^2} i\ dW \sin \theta \qquad [3.6\text{-}12]$$

This shows that dB is inversely proportional to the *square* of the distance.

But we may have an infinitely long wire. The total induction is found by integration,

$$B = \mu_0 \frac{1}{4\pi} i \int_{-\infty}^{+\infty} \frac{1}{R^2}\ dW\ dR \qquad [3.6\text{-}13]$$

Now, instead of integrating over the length of the wire, W, from $-\infty$ to $+\infty$, it is more convenient to change variables and replace W by the angle θ subtended by the wire and the direction to the test point (see again Figure 3.1-3). This means that the integration is performed from $\theta = 0$ (at infinity on one end) to $\theta = 180° = \pi$ rad (at infinity on the other end). The total induction then becomes

$$B = \mu_0 \frac{1}{4\pi} i \int_0^\pi \frac{1}{R^2} \sin \theta\ d\theta$$

$$= \mu_0 \frac{1}{4\pi} i \frac{1}{d} \int_0^\pi \cos \theta\ d\theta \qquad [3.6\text{-}14]$$

where d is the distance from the test point to the wire, rather than to the wire segment.

Solving the integral yields

$$B = \mu_0 \frac{1}{4\pi} i \frac{1}{d} (-\cos \pi + \cos 0) \qquad [3.6\text{-}15]$$

and thus

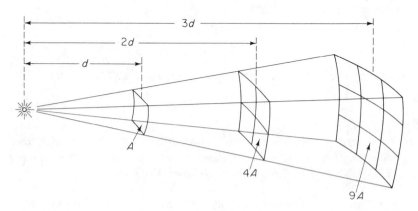

Figure 3.6-3 Inverse-square law.

$$B = \mu_0 \frac{1}{2\pi} i \frac{1}{d} \qquad [3.6\text{-}16]$$

which shows that instead of the inverse-*square* relationship we now have $1/d$.

This is exactly the same change that occurs when we proceed from a point light source to an extended source. In reality, of course, there are no true point sources, nor are there infinitely large sources. For any realistic light source the exponent in Equation [3.6-11] must be somewhere between 1 and 2, never reaching either limit.

Example. *Consider an illuminated slit 7 cm long and 25 cm away from a (small) photodetector. Determine the exponent that should replace the 2 in the inverse-square law.*

Solution. If we examine Equations [3.6-15] and [3.6-16] we notice that they differ by a factor $(\cos \theta_1 - \cos \theta_2)/2$. To find to which power the distance d needs to be raised, we set

$$\frac{1}{d^x} = \frac{1}{d} \left(\frac{\cos \theta_1 - \cos \theta_2}{2} \right)$$

$$d^x = d \left(\frac{2}{\cos \theta_1 - \cos \theta_2} \right)$$

and, since by definition $\theta_2 = 180° - \theta_1$,

$$d^x = d \left(\frac{1}{\cos \theta} \right) \qquad [3.6\text{-}17]$$

Inserting the actual figures $d = 25$ cm, $\theta_1 = \tan^{-1} [25/(7/2)] = 82°$ and $\theta_2 = 180° - 82° = 98°$, and using Equation [3.6-17] yields

$$25^x = 25 \left(\frac{1}{\cos 82°} \right) = 180$$

$$x = \frac{\log 180}{\log 25} = \boxed{1.613}$$

Indeed, with an extended light source, the exponent in the inverse-"square" law will always be *less than* 2.

So far we have assumed that the light falls on the surface at normal incidence, $\alpha = 0$. If it does not, the areance, as in Figure 3.4-15, is *less* by a cosine factor,

$$E = \left(\frac{I}{d^2} \right) \cos \alpha \qquad [3.6\text{-}18]$$

In practice, the finite size of the source is often neglected, provided that its diameter is less than one-tenth of the distance to the source. If the source is significantly larger than that, its radius r must be included and Equation [3.6-18] becomes $E = I \cos \alpha /(d^2 + r^2)$.

If both the surface normal at the source and the surface normal at the receiver subtend angles other than zero with the direction of the light, as shown in Figure 3.6-4, the cosines of both must be included. Furthermore, if we replace I, following Equation [3.6-6], by the product of sterance L and area A, then

$$E = \frac{LA}{d^2} \cos \theta \cos \alpha \qquad [3.6\text{-}19]$$

But A/d^2 is the solid angle, ω, which the source subtends at the receiver and therefore, setting both θ and α equal to 90°,

$$E = L\omega \qquad [3.6\text{-}20]$$

This shows that the areance (irradiance or illuminance) produced by light from an extended source depends only on the sterance (radiance or luminance) of the source and on the solid angle; it does *not* depend on the distance between source and receiver.

Radiant areance.   *Radiant areance* (irradiance), E_e, refers to the total power incident on a surface. Its unit is watts per square meter, W m^{-2}, the same unit as that of radiant exitance.

Luminous areance.   The unit of *luminous areance* (illuminance), E_v, is lumens per square meter, lm m^{-2}, a unit called *lux*, lx. In Figure 3.6-5 we have a source (A) that has a pointance of 1 cd. Since the candela is defined as lumen per steradian, and since 1 steradian is the solid angle subtended by an area 1 m^2 in size at a distance of 1 m, area B has a luminous areance (illuminance) of 1 lx.

Remember that a sphere of 1 m radius has a surface area $A = 4\pi R^2 = 12.57$ m^2. Thus a source of 1 cd pointance, from Equation [3.6-4],

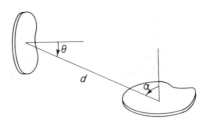

Figure 3.6-4   Source and receiver normals subtending acute angles with direction of the light.

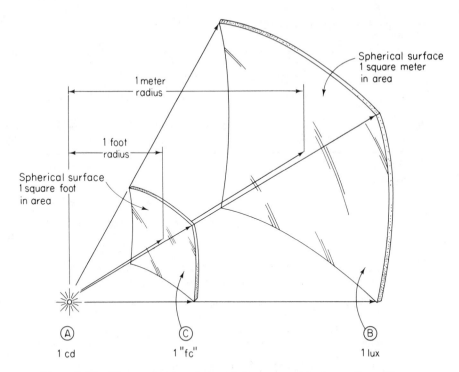

Figure 3.6-5 Diagram showing relationship between candela, footcandle, and lux. The power contained within the cone is 1 lumen.

has a *total* output, in all directions, of $\phi = 4\pi I = 12.57$ lm, but only 12.57 lm$/12.57$ m$^2 = 1$ lx reaches the area marked B in Figure 3.6-5.

The following units are obsolete and their use deprecated:
Phot $= 1$ lm/cm^2.

Footcandle, fc. This is an ill-chosen term which suggests that the luminous pointance (intensity) in "candles" ("candle power") is to be multiplied by a distance. Actually, it means lumen per square foot, lm/ft^2. If the area in Figure 3.6-5 were 1 ft^2 in size and 1 ft away from the source, it would receive a luminous areance (illuminance) of 1 "footcandle," fc (C). Table 3.6-5 lists some conversion factors.

Example 1. *A 75-W light bulb emits 1170 lm, distributed uniformly over a hemisphere. Find the area illuminated, and the luminous areance (illuminance), at a distance of 2.3 m from the bulb.*

Solution. A hemisphere of radius 2.3 m has an area

$$A = 2\pi R^2 = (2\pi)(2.3 \text{ m})^2 = \boxed{33.238 \text{ m}^2}$$

Table 3.6-5 CONVERSION FACTORS FOR UNITS OF LUMINOUS AREANCE (ILLUMINANCE)

One	Equals		
Lux *	1 lm/m² *	0.0001 phot	0.093 fc
Phot	1 lm/cm²	10 000 lx	929 fc
Footcandle	1 lm/ft²	10.76 lx	

*Recommended unit.

The luminous areance at that distance, from Equation [3.6-10], is

$$E = \frac{\phi}{A} = \frac{1170 \text{ lm}}{33.238 \text{ m}^2} = \boxed{35.2 \text{ lx}}$$

Example 2. *A light meter calibrated in footcandles (fc) held 2 ft from a light source, reads 14 fc. Find the pointance of the source.*

Solution. From Equation [3.6-11],

$$I = Ed^2 = (14 \text{ fc})(2 \text{ ft})^2 = \boxed{56 \text{ "cp"}}$$

[The term "cp" stands for "candle power," a now obsolete unit of luminous pointance (intensity). One "cp" produces 1 "fc" of luminous areance (illuminance).]

Example 3. *The top of a drafting table, inclined 30° with the horizontal, is illuminated by a single 200-cd lamp suspended 1.2 m directly above. Determine the luminous areance (illuminance), both in SI units and in "footcandles."*

Solution. From Equation [3.6-18],

$$E = \left(\frac{I}{d^2}\right) \cos \alpha = \left(\frac{200}{1.2^2}\right) \cos 30° = \boxed{120 \text{ lx}}$$

which, from Table 3.6-5, is equal to

$$\boxed{11.2 \text{ "fc"}}$$

Brightness is not the same as "intensity" or "illuminance." Brightness is a function of areance and reflectivity. *Illumination* is the process of exposing an object to light.

Throughput. In our discussion of magnification (page 55) we saw that the *Lagrange invariant, nhγ,* is constant throughout any optical system. If we extend the (two-dimensional) invariant to three dimensions, we arrive at a product, $n^2A\omega$, that has been given the name *throughput, T*. Throughput is a geometric measure of *how much light can be transmitted* through a system of index n and cross section A and that accepts a solid angle ω:

$$T = n^2A\omega \qquad\qquad [3.6\text{-}21]$$

Since A and ω occur in the definition of sterance, $L = \phi/(A\omega)$, solving for $A\omega$, substituting in Equation [3.6-21], and setting $n = 1$ shows that throughput is also the ratio of power to sterance, $T = \phi/L$. From this we conclude that the throughput of a given system *cannot be increased*, neither by focusing nor by "forcing" the light through a "funnel." In fact, the throughput within a given system becomes gradually *less*, due to inevitable losses on surfaces and in matter. The unit of throughput* is square meter steradian, m^2sr.

Radiometers and Photometers

Comparison photometers.   Systems for measuring radiant quantities are called *radiometers*. Systems for measuring visible light are called *photometers*. The classical example of a comparison photometer is the *Bunsen grease-spot photometer*. It consists of two mirrors arranged so that the two faces of a sheet of white paper can be viewed at the same time. When the two faces of the paper receive the same amount of light, a grease spot in the center of the paper is seen at least contrast. If d_1 and d_2 are the distances of the two sources from the paper, then at the equilibrium position and from the inverse-square law,

$$E_1 = \frac{I_1}{d_1^2} = E_2 = \frac{I_2}{d_2^2}$$

and thus

$$\frac{I_1}{I_2} = \left(\frac{d_1}{d_2}\right)^2 \qquad\qquad [3.6\text{-}22]$$

The *Lummer–Brodhun photometer* is similar but contains a series of prisms instead of two mirrors (Figure 3.6-6). As before, readings are taken at the equilibrium position.

---

*There is as yet no agreement on whether *throughput* is the best term. The French call it étendue, from which comes "extent," either *geometric extent, $A\omega$*, or *optical extent*, $n^2A\omega$. And there is more opposition. W. H. Steel of the Australian National Standards Laboratory, for one, asks rhetorically: "If the input is doubled, what happens to the through-put?" *Ans.* Nothing. Input and output are actual quantities of light entering and leaving a system; throughput describes the *capacity* of the system to convey light, rather than the quantity being conveyed.

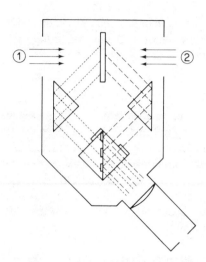

Figure 3.6-6 Lummer–Brodhun photometer.

Photoelectric radiometers.
A radiometer measures radiant areance (irradiance), although it may be calibrated instead in terms of power, pointance, or sterance. A *photometer* must have a photodetector whose spectral response, with or without any filters, matches that of the eye. Various quantities can be determined as follows.

Power. The total flux emitted by a light source is found by placing the source inside a hollow sphere with a diffusely reflecting inner surface, called an *integrating sphere*. The photodetector is located near a small window in the sphere's surface.

Areance II. Measuring areance (irradiance or illuminance) is simple. All that is needed is a photodetector. The surface of the detector must be completely filled with light, and the surface must be oriented normal to the direction of the light.

Pointance and sterance. Sterance, because it includes pointance, is the more general case. Measuring sterance is of much practical interest, but requires certain precautions: (1) all of the light to be measured must reach the detector and (2) extraneous light must be excluded. This calls for some auxiliary optics such as a lens, a suitably shaped stop, and sometimes a telescope or a microscope.

The simplest sterance meters are the small light meters customary on photographic cameras. Aim the meter at a large, uniformly illuminated wall: You will get a certain reading. Then move back. The reading will be the same—as long as the solid angle subtended at the meter by the size of

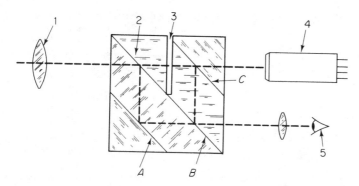

Figure 3.6-7 Schematic diagram of compact sterance meter (redrawn from D.A. Schreuder, "Optical System for a Universal Luminance Meter," *Applied Optics* **5** (1966), 1965–66, by permission). Light from the target proceeds to lens 1 and glass cube 2, which contains a slot 3 for a field-limiting mask, to photodetector 4. A series of partially coated interfaces *A*, *B*, and *C* direct light to observer 5, who can see the field both inside and outside the area limited by the mask.

the wall does not become less than the meter's acceptance angle. It is the sterance, not the distance between source and receiver, that matters.

If the target is small, and has an odd shape, the meter should look only at the sampling field. This is accomplished best by a see-through mask, such as that illustrated in Figure 3.6-7.

SUGGESTIONS FOR FURTHER READING

F. Grum and others, *Optical Radiation Measurements* (New York: Academic Press, Inc., 1983).

A. Stimson, *Photometry and Radiometry for Engineers* (New York: Wiley-Interscience, 1974).

C. L. Wyatt, *Radiometric Calibration: Theory and Methods* (New York: Academic Press, Inc., *1978).*

W. L. Wolfe, "Radiometry," in R. R. Shannon and J. C. Wyant, editors, *Applied Optics and Optical Engineering,* Vol. **8,** pp. 117–170 (New York: Academic Press, Inc., 1980).

E. Helbig, *Grundlagen der Lichtmesstechnik* (Leipzig: Akademische Verlagsgesellschaft Geest & Portig K.-G., 1977).

IES Lighting Handbook (New York: Illuminating Engineering Society, 1981).

PROBLEMS

3.6-1. A white screen receives 1050 lumens of blue light of 460 nm wavelength. How much radiant power is incident on the screen?

3.6-2. If a 60-cd light source consumes 0.68 A at 110 V and if it emits light uniformly in all directions, what is its luminous efficacy?

3.6-3. What is the solid angle subtended at the apex of a cone of light that emerges from a point source and at a distance of 14 cm fills an aperture 50 mm in diameter? Use:
(a) The area of the aperture and its distance.
(b) The plane angle subtended at the source by diametrically opposite marginal rays.

3.6-4. What is the radiant pointance of a source that projects 50.3 mW into a hemisphere?

3.6-5. The total power emitted by the sun is estimated to be 3.9×10^{26} J s^{-1}. What is the power incident per square meter just outside Earth's atmosphere, assuming a distance of 150 million km?

3.6-6. Light originating at a point source enters a lens 5 cm in diameter. If the light at the source subtends an angle of 0.17 sr, what is its vergence on entering the lens?

3.6-7. Light from a 600-cd street lamp falls on the pavement 5 m below. Determine the areance (illuminance).

3.6-8. A 64-cd source is placed in the focus of a mirror of 100 cm focal length and 10 cm diameter. How much light is incident on the mirror?

3.6-9. If a book is held 3 m from a high-intensity lamp and then is moved to 60 cm from the lamp, by how much has the areance (illuminance) increased?

3.6-10. A light bulb is suspended 4 m above the floor. To what distance should it be lowered in order to increase the areance (illuminance) to 1.77 of its initial value?

3.6-11. If a photographic print can be made in 10 s when the printing frame is held 20 cm from a point light source, what is the correct exposure time when the frame is held 30 cm away?

3.6-12. A 200-"candle power" incandescent lamp is mounted 5 ft above a drafting table, tilted 60° with respect to its horizontal position. What is the illuminance, in "footcandles"?

3.6-13. A 16-cd lamp is placed 50 cm from a screen on which it produces the same areance (illuminance) as another lamp, of unknown pointance, 75 cm away. Find the pointance of the unknown lamp.

3.6-14. Two high-intensity lamps of equal efficacy, one rated 25 W and the other 100 W, are mounted 90 cm apart from each other. How far from the 25-W lamp must the screen of a Bunsen grease-spot photometer be placed in order to have equal areance (illuminance) on both sides?

3.6-15. If the luminous sterance of an infinitely large, flat, diffusely reflecting surface has a certain value, how does the sterance change when determined:
(a) From an angle of 60°?
(b) From a distance one-half that used before?

3.6-16. The sterance of a field 10 cm in diameter is measured from a distance of 40 cm, using a calibrated photometer that has a (total) acceptance angle of 20°. By what factor must the reading of the photometer be corrected?

3.6-17. The flat center of a street receives light from two lamps of 2000 cd each, mounted on poles 14 m tall and 60 m apart from each other. Determine the areance (illuminance).

3.6-18. A screen is located 2.5 m from a point light source. When a sheet of smoked glass is placed between the source and the screen, the source must be moved 50 cm closer to the screen in order to produce the same areance (illuminance). What percentage of the light is lost in the glass?

3.6-19. A small light bulb is suspended 60 cm above the center of a table 1.2 m in diameter. If the power that reaches the table is 276 lm, what is:
(a) The pointance of the bulb?
(b) The areance (illuminance) near the edge of the table?

3.6-20. A point light source, as shown in Figure 3.6-8, is located directly above a point M on a horizontal plane. At which distance, h, above the plane must the light source be placed in order to produce maximum areance (illuminance) at a point P on the plane located 20 cm away from M?

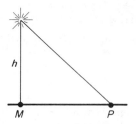

Figure 3.6-8

Part 4

Optics
of Transformations

TRANSFORMATION MEANS TO CHANGE from a given figure, or set of expressions, to an equivalent figure, or set of expressions. To change from the configuration of an object to the diffraction pattern caused by that object, or to change from a set of interference fringes, an "interferogram," to the corresponding "spectrogram," are examples of optical transformations. Knowledge of such transformations has rapidly advanced in recent years, inspired by work in electric network theory.

Transformations include many fundamental processes that occur in optics. In fact, transforms are the basis of all image formation. They are essential for optical data processing and for holography. Pattern recognition, to name another example, is best appreciated if we are familiar with the principles of the *optics of transformations*.

4.1

Fourier Transform
Spectroscopy

AN EXAMPLE OF TRANSFORMATION THAT CONNECTS with a topic we have discussed is the formation of spectra. Spectra can be obtained in two ways: (1) By dispersion, either by a prism or by a diffraction grating, and (2) by interferometry and subsequent transformation. To be sure, diffraction gratings also make use of interference; thus the principal difference is whether the spectrum is obtained *directly* or *indirectly*.

Fourier transform spectroscopy, with little exaggeration, is the one major advance in the technique of spectroscopy since Bunsen invented the spectroscope more than a hundred years ago. Fourier spectroscopy may not be the most convenient way to obtain a spectrum but it has an undisputed advantage over any prism or grating instrument: It is less wasteful of light.

Fringe Contrast Variations

We return for a moment to coherence and double-slit interference. If the light were perfectly monochromatic, the fringes would extend from the center out to infinity on both sides. But since there is no such light, the contrast farther away from the axis becomes less (Figure 4.1-1a). To scan across these fringes, we move a slit with a photodetector behind it (b), which results in the trace shown in (c).

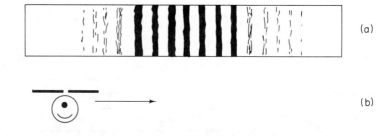

Figure 4.1-1 Interference fringes becoming less distinct farther away from the center (a); scanning across fringes (b) gives tracing (c).

If the light had been *more* monochromatic, the fringes would have extended farther, and an envelope drawn across the peaks of the trace would have been wider. (If the light had been "perfectly" monochromatic, the envelope would be a straight line, extending from infinity on the left to infinity on the right.) If the light had been *less* monochromatic, the envelope would have been more narrow. Obviously, there exists some relationship between the degree of monochromaticity of the light and the shape of the envelope.

Contrast, as defined on page 209, is the ratio of the difference between maximum areance, E_{max}, and minimum areance, E_{min}, to the sum of such areances:

$$\gamma \equiv \frac{E_{max} - E_{min}}{E_{max} + E_{min}} \qquad [4.1\text{-}1]$$

This definition should be compared to the definition of contrast preferred in the visual sciences for nonperiodic targets such as dark symbols on a light background. If L_1 is the luminous sterance of the symbol and L_2 the luminous sterance of the surrounding field, then the contrast

$$C = \frac{L_2 - L_1}{L_2} = 1 - \frac{L_1}{L_2} \qquad [4.1\text{-}2]$$

If the object is darker than the background, the contrast is positive and less than 1. If the object is lighter than the background, C is negative and can have any value.

Now comes a slight complication. Assume that we use sodium D light and observe the fringe pattern produced by it in a Michelson inter-

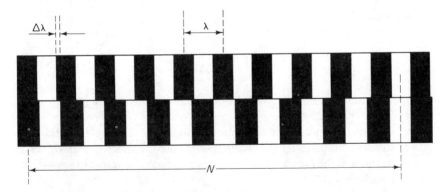

Figure 4.1-2 Vernier scale shows coincidences of spacing N, comparable to *beats* in acoustics and to the interference of doublet lines.

ferometer. When one of the mirrors is moved through a considerable distance, we notice that at some times the fringes are more visible than at other times. In fact, the fringes go through periodic variations of contrast, reaching maximum contrast every 980th fringe. Fizeau, as early as 1862, followed such variations through 52 cycles; he then concluded that yellow sodium light has two components of about equal intensity, that it is a *doublet* (i.e., two lines) with a separation of 1/980 of the average wavelength.* How did he arrive at this conclusion?

Think of two waves of slightly different wavelengths, λ_1 and λ_2. One wave is represented by the upper bar pattern in Figure 4.1-2, the other wave by the lower bar pattern. The difference between them is

$$\lambda_2 - \lambda_1 = \Delta\lambda \qquad [4.1\text{-}3]$$

On the left in the diagram the two waves coincide; on the right they coincide again. But obviously there is one more wave in the upper row. Then, if N is the number of waves between any two coincidences,

$$(N + 1)\lambda_1 = N\lambda_2$$

$$N\lambda_1 + \lambda_1 = N\lambda_2$$

$$\lambda_1 + \frac{\lambda_1}{N} = \lambda_2$$

$$\lambda_2 - \lambda_1 = \frac{\lambda_1}{N}$$

---

*H. Fizeau, "Recherches sur les modifications que subit la vitesse de la lumière dans le verre et plusieurs autres corps solides sous l'influence de la chaleur," *Ann. Chim. et Phys.* (3) **66** (1862), 429–82.

and, since $\lambda_1 \approx \lambda_2$,

$$\boxed{N = \frac{\lambda}{\Delta\lambda}} \qquad [4.1\text{-}4]$$

the same equation that defines the *resolvance* of a diffraction grating, in the first order (page 251).

If the light were perfectly monochromatic, such light would have wavetrains of infinite length, with an unlimited number, N, of waves in each train, and the *bandwidth*, $\Delta\lambda$, would be zero. Needless to say—there is no such light.

The term *coherence length*, Δs, was introduced earlier, on page 208, as $\Delta s = N\lambda$. If we substitute for N Equation [4.1-4],

$$\Delta s = \frac{\lambda^2}{\Delta\lambda} \qquad [4.1\text{-}5]$$

which connects coherence length, wavelength, and bandwidth.

Continuing with Fizeau's observation on sodium light, the wavelength difference between the two lines, from Equation [4.1-4], is

$$\Delta\lambda = \frac{\lambda}{N} = \frac{589 \text{ nm}}{980} = 0.6 \text{ nm} \qquad [4.1\text{-}6]$$

This agrees with the facts known today: The sodium D_2 line has a wavelength of 588.9953 nm and the D_1 line 589.5923 nm.

In theory, then, if we had two infinitesimally narrow spectrum lines, we would have periodic fluctuations out to infinity. With a real source, whose lines must by necessity be of finite width, the fluctuations lose contrast farther away from the axis and gradually fade out completely. From the variations of contrast we can deduce the profile of the spectrum lines that cause the interference pattern.

Example. *A Michelson interferometer is used for measuring the separation of a doublet known to have an average wavelength of 677 nm. Periodic variations of the fringe contrast are seen whenever one of the mirrors is moved through 0.44 mm. What are the two wavelengths?*

Solution. From the Michelson interferometer equation, page 199,

$$m = \frac{2d}{\lambda} = \frac{(2)(4.4 \times 10^{-4})}{677 \times 10^{-9}} = 1300$$

But $m = N$ and hence, from Equation [4.1-6],

$$\Delta\lambda = \frac{\lambda}{N} = \frac{677}{1300} = 0.52 \text{ nm}$$

If both lines have the same intensity, then $\lambda = 677 \mp 0.26$ nm, and thus

$$\boxed{\lambda_1 = 676.74 \text{ nm}, \qquad \lambda_2 = 677.26 \text{ nm}}$$

Fizeau's observation is of historic interest. It marks the first time that interferometry was used to reveal the doublet structure of a spectral line which had been believed to be monochromatic. From these beginnings, this technique has developed into a field of its own. Stated simply, we project light into an interferometer, obtain an interferogram, and convert the interferogram into a spectrogram, using Fourier transformation. This is the principle of *Fourier transform spectroscopy*.

Michelson Interferometer Spectroscopy

We now turn to the practical side of Fourier transform spectroscopy. Take a Michelson interferometer and let light from a source S be incident on the beamsplitter (Figure 4.1-3). The light proceeds to the two mirrors. One of the mirrors is stationary while the other is made to oscillate sinusoidally. If the incident light were monochromatic, the center of the field would alternately become bright and dark, and the photodetector would produce a sinusoidal alternating-current (ac).

The frequency of the signal depends both on the frequency of oscillation of the mirror and on the wavelength. Not only is it essential how fast the mirror moves but also how many wavelengths fit into the distance of motion. The shorter the wavelength, the higher the signal frequency: Light of one-half the wavelength gives twice the frequency. If the incident light contains several wavelengths, the output of the detector is no longer

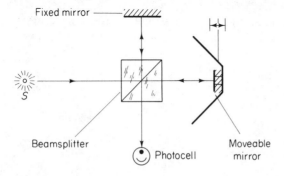

Figure 4.1-3 Michelson interferometer as used for Fourier transform spectroscopy.

Figure 4.1-4 Compact Michelson-type interferometer spectrometer.

sinusoidal; the ac signals that correspond to the individual wavelengths will superimpose.

The motion of the mirror, or *modulation,* can be realized in several ways. Mechanically driven mirrors, tuning forks with a mirror cemented to one of the tines, and piezoelectric crystals of various shapes have all been used for this purpose. In the prototype model shown in Figure 4.1-4, the beamsplitter is an *Abbe cube* (a partially reflecting 45° boundary inside a solid glass cube). Light enters from the left. To the right of the cube is a loudspeaker, with a small first-surface mirror glued to its membrane. To the rear of the cube, not visible in the illustration, is a larger, stationary first-surface mirror. The recombined beams emerge from the side of the cube facing the reader and are picked up there through a fiber guide, which in turn leads to a photodetector (not shown).

Data reduction.   Call x the path difference introduced by the motion of the mirror. If the light were monochromatic, the *amplitude* distribution in the interferogram would follow Equation [2.1-37],

$$y = \mathbf{A} \sin\left(2\pi\frac{x}{\lambda}\right)$$

Therefore, if the mirror is moving at a velocity $v = x/t$, the *intensity* distribution is

$$I = I_0 \sin^2\left(2\pi\frac{vt}{\lambda}\right) \qquad\qquad [4.1\text{-}7]$$

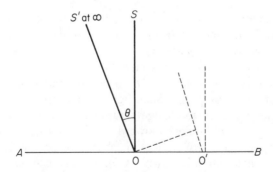

Figure 4.1-5 Deriving the concept of Fourier transformation. [Reprinted with permission from R.C. Jennison, *Fourier Transforms and Convolutions for the Experimentalist* (Elmsford, NY: Pergamon Press, Inc., 1961).]

Using the identity $\sin^2 \alpha = \frac{1}{2} - \frac{1}{2}\cos 2\alpha$, this is equal to

$$I = \tfrac{1}{2}I_0 \left[1 - \cos\left(4\pi \frac{vt}{\lambda} \right) \right] \qquad \text{[4.1-8]}$$

Thus the signal obtained from the detector is modulated sinusoidally at a frequency $2v/\lambda$. In practice, it is best to let the mirror oscillate at a rate that causes the signal frequency to fall into the audio range—for which amplifiers and wave analyzers are readily available. If, from the variation of the photocurrent, we recover the power distribution (of the light), we have the spectrum of the source. To do this, we use *Fourier transformation*.* Its principle is explained as follows.

A point source S is located in space at infinity, as illustrated in Figure 4.1-5. Light coming from S is incident at right angles on a line AB. An observer, standing at O anywhere along AB, will find that the areance (illuminance) is the same at all points. Furthermore, if the phase of the light at points O and O' were measured, the observer would find that the phase of the light intersecting AB is the same at both points.

---

*Jean Baptiste Joseph Baron de Fourier (1768–1830), French physicist and mathematician. Because of his mathematical skills, Fourier became an artillery officer, served as Napoleon's science adviser to Egypt, later became prefect of Isère, near Grenoble, Secretary of the French Academy of Sciences, and head of the École Polytechnique. After his return from Egypt, Fourier wrote a 21-volume treatise on Egyptology. His *Analytical Theory of Heat* became the foundation of thermodynamics. Fourier's first paper on series summations, submitted to the French Academy, was rejected because Lagrange did not believe that the series would converge; today this paper is considered one of the classics of mathematical physics. See also J. Herivel, *Joseph Fourier: The Man and the Physicist* (New York: Oxford University Press, 1975).

But then let the source be at S', off the normal constructed on AB. Now the observer, moving from O to O', will find that there is a continual change of phase. This phase change has a certain periodicity, a function of both the angle θ and the wavelength.

Conversely, assume line AB to be an emitter of light of uniform sterance and uniform phase. Such light will proceed toward S. But if the phase along AB were to change sinusoidally, with a certain space frequency, the light will proceed toward S'. A *reciprocal relationship* exists between the spatial distribution of the phase along a line (in terms of x) and the angular distribution (in terms of θ); and one can be *transformed* into the other:

$$f(x) \rightleftharpoons F(\theta) \qquad\qquad [4.1\text{-}9]$$

The transformation, however, need not be confined to *spatial* distribution versus *angular* distribution. Frequently, we have two spatial distributions, especially in the case of two-dimensional transforms (to be discussed in Chapters 4.3 and 4.4). Thus, to be as general as possible, we replace the variable θ by another variable, ξ, and write

$$f(x) \rightleftharpoons F(\xi) \qquad\qquad [4.1\text{-}10]$$

where we follow the convention that the capital F represents the Fourier transform of the function designated by the lower case f.

Fourier transformation.   Recall the superposition of waves from Chapter 2.2: If we have two or more sine waves of equal frequency, the result is another sine wave. If we have waves of slightly different frequencies, the result is a complex wave. Now we examine how a complex wave is *decomposed* into its components.

This is exactly the same problem as in acoustics, where sounds produced by most musical instruments are composed of a fundamental and higher harmonics, of different amplitudes and varying decay times. The purpose of *acoustical spectral analysis* is to find the frequencies that make up these composite sounds.

According to *Fourier's theorem*, any periodic function can be represented by a sum of sine and cosine terms of the form

$$y = \mathbf{A}_0 + \mathbf{A}_1 \sin \omega t + \mathbf{A}_2 \sin 2\omega t + \mathbf{A}_3 \sin 3\omega t + \cdots$$
$$+ \mathbf{B}_1 \cos \omega t + \mathbf{B}_2 \cos 2\omega t + \mathbf{B}_3 \cos 3\omega t + \cdots \quad [4.1\text{-}11]$$

This equation is an extension of Equation [2.1-1]: y is the displacement of the resultant wave at a time t, $\mathbf{A}_0$ is a constant, the $\mathbf{A}$'s and $\mathbf{B}$'s are the amplitudes of the component waves, and the ω's are their angular velocities. The resultant wave, therefore, is built up of waves whose frequencies vary as $1:2:3: \ldots$.

Equation [4.1-11] is a *Fourier series;* it represents the composite periodic wave, extending from infinity to infinity. By including a theoretically infinite number of terms, Fourier's theorem makes it possible to synthesize any waveform, even a "square" wave.

But Fourier analysis is not limited to periodic events, or to infinitely long waves. A wave*train*, that is, a group of waves of limited coherence length, can, and must, be represented by a *Fourier integral*. A property of these integrals is that the component waves differ only by infinitesimal increments of wavelength, not by multiples of wavelengths.

Figure 4.1-5 shows the relationship between spatial and angular distribution. Figure 4.1-6 uses a different approach to derive the relationship between two spatial distributions. As before, the light comes from infinity and proceeds along the axis; thus we have plane wavefronts incident on the aperture (left). A lens placed close to the aperture focuses the light into the F_2 plane (right).

For simplicity, consider only one dimension, normal to the direction of propagation and lying *in* the plane of the diagram. The height of the object is called x (not shown), the conjugate height of the image x', and the height in the transform ξ. We now determine the amplitude of the light that is *diffracted* toward an arbitrary point, P, in the image. This amplitude is the sum of the contributions from all elements of width $d\xi$ in the aperture. Inside the aperture we construct a line on which the optical path length is constant from any point to P. But P is off-axis; hence at any given point on the line there must be a path *difference,* Γ, between the incident light and the diffracted light. If the path difference is measured in wavelengths, then from Equation [2.2-38], the *phase* difference is

$$\delta = 2\pi\Gamma \qquad\qquad [4.1\text{-}12]$$

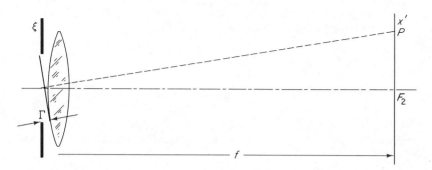

Figure 4.1-6 Deriving reciprocal relationship between light distribution within aperture (*left*) and light distribution in the image plane (*right*).

Each element of width $d\xi$ thus contributes an amplitude at P that may be represented by

$$dA = d\xi \cos 2\pi\Gamma \qquad [4.1\text{-}13]$$

Now, since the cosine term represents the real part of a complex function, we have, in analogy with Equation [2.1-21],

$$dA = d\xi \, e^{2\pi i\Gamma} \qquad [4.1\text{-}14]$$

From similar triangles in Figure 4.1-6,

$$\frac{\Gamma}{\xi} = \frac{x'}{f} \qquad [4.1\text{-}15]$$

We solve this equation for Γ, substitute it in Equation [4.1-12], and for convenience set $f = 1$. This gives

$$\delta = 2\pi x'\xi \qquad [4.1\text{-}16]$$

The total amplitude distribution in the image, $f(x')$, is then found by integration:

$$f(x') = \int_{-\infty}^{+\infty} F(\xi)e^{2\pi i x'\xi} \, d\xi \qquad [4.1\text{-}17]$$

where $F(\xi)$ is the amplitude distribution inside the aperture.

In the absence of absorption, the propagation of light is reversible. Therefore, we can reverse the process; from the amplitude in the image, we can find the amplitude in the aperture. Therefore, the *Fourier transform*, $F(\xi)$, of a function $f(x)$ is

$$\boxed{F(\xi) = \int_{-\infty}^{+\infty} f(x)e^{-2\pi i x\xi} \, dx} \qquad [4.1\text{-}18]$$

and the *inverse Fourier transform*, $f(x)$, of $F(\xi)$ is

$$\boxed{f(x) = \int_{-\infty}^{+\infty} F(\xi)e^{2\pi i x\xi} \, d\xi} \qquad [4.1\text{-}19]$$

The two functions $f(x)$ and $F(\xi)$ are called a *Fourier transform pair*. The exponential terms, as in Euler's formula, page 180, represent trigonometric functions,

$$e^{\pm 2\pi i x\xi} = \cos(2\pi x\xi) \pm i \sin(2\pi x\xi) \qquad [4.1\text{-}20]$$

There are several, slightly different ways of writing these integrals. For example, we may reverse the signs in the exponents of Equations [4.1-18] and [4.1-19] but there must always be $+i$ in one equation and $-i$ in the other. Furthermore, we could take the 2π terms outside the integral:

$$F(\xi) = \frac{1}{2\pi} \int_{-\infty}^{+\infty} f(x)e^{-ix\xi}\,dx \qquad [4.1-21]$$

but then

$$f(x) = \int_{-\infty}^{+\infty} F(\xi)e^{ix\xi}\,d\xi \qquad [4.1-22]$$

Or we could split the $1/(2\pi)$ term between the two functions:

$$F(\xi) = \frac{1}{\sqrt{2\pi}} \int \cdots \quad \text{and} \quad f(x) = \frac{1}{\sqrt{2\pi}} \int \cdots \quad [4.1-23]$$

Note the symmetry in all these expressions. Except for the sign in the exponent, x and ξ can be interchanged at will to convert one of the equations into the other. This illustrates the purpose of transformations in optics: We have seen a spectrum produce an interferogram, and now we see an interferogram converted into a spectrogram.

Point sampling and results.

Now that we have a recording of the interference fringes, the irradiance distribution across the fringes must be processed further. Michelson merely used the visibility curve, that is, the envelope over the fringes. Today we sample the irradiance at preselected, equally spaced *data points* (irrespective of whether they are located in a maximum or between maxima). The irradiance received at each datum point need not be recorded; it is simply received by a photo-detector and fed directly into a computer, which is programmed to perform the transformation.

Sometimes as many as 1000 data points are used (which increases the resolution), but usually 40 to 100 points are sufficient. Again, if the light were monochromatic, the irradiance distribution would follow Equation [4.1-8]. But if the light contains more than one frequency, each frequency contributes its own set of fringes and integration is needed to add these contributions. The integral can be evaluated, that is, the transformation can be carried out, in real time. This means that as soon as the interferogram is received, its transform, the spectrogram, is ready for inspection.

Figure 4.1-7 shows several typical examples. As we have seen, if the fringes extend out to infinity, the light would be perfectly monochromatic

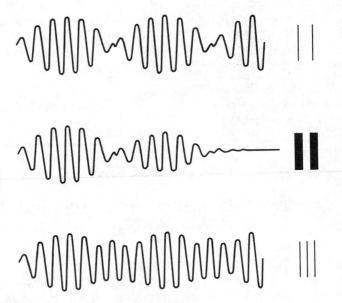

Figure 4.1-7   Interferograms obtained by scanning (*left*) and resulting spectrograms obtained by Fourier transformation (*right*).

and the spectrum line causing the fringes would be of infinitesimal width (neither of which is possible). If there are simple harmonic variations of contrast, out to infinity, we would have two lines, close together and of infinitesimal width (top), which also is impossible. In reality we may have such variations and a gradual loss of contrast farther away from the axis. This then means two lines, each of finite width, like the sodium D example referred to earlier (center). Three lines result from an even more complex interferogram (bottom).

Advantages of Fourier Transform Spectroscopy

1. Most important, a Fourier transform spectrometer can have a *wide aperture*. A grating or prism spectrometer must have a narrow slit. Expressed another way, a grating or prism scanning spectrometer looks at one wavelength at a time, rejecting all others. A Fourier transform instrument looks at all wavelengths all of the time. This has the advantage that spectra can be obtained even of very faint light sources or within a very *short interval of time*. High-resolution spectra often require no more than several seconds.

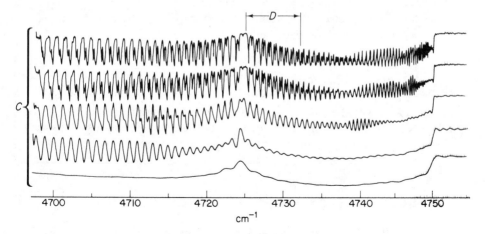

Figure 4.1-8   Resolution of a near-IR N_2O absorption band as a function of path difference. Successive curves correspond to double the travel distance. (From P. Connes and G. Michel, "Real Time Computer for Fourier Spectroscopy," Aspen Conf. on Fourier Spectroscopy, Mar. 1970. Courtesy J. Connes, Laboratoire Aimé Cotton, Orsay, France. Reproduced by permission.)

2. The resolvance (resolving power) of a Fourier transform spectrometer is almost as good as that of a grating spectrometer. At a wavelength of 500 nm, for example, a typical diffraction grating resolves 0.01 Å, a Fourier-type instrument about 0.025 Å.

3. In theory, the resolvance of a Fourier spectrometer is unlimited if the path difference is also unlimited. In practice, though, the mirror in a Michelson interferometer can move through a limited distance only, limiting the resolution (Figure 4.1-8). In various commercial Fourier instruments, most of them intended for use in the IR, the travel distances range from 0.25 mm to 5 cm. In analytical terms, the interferogram cannot be integrated from $-\infty$ to $+\infty$; the integral is *truncated*.

4. In any spectrometer, throughput and resolvance are inversely proportional. Thus some dimension of the spectroscope, for example the slit width, must be made smaller to gain higher resolvance. But the product of throughput × resolvance is not the same in every case: For a grating spectrometer it is slightly higher than for a prism, and for a Michelson-type Fourier spectrometer it is at least two orders of magnitude higher than for a grating.

5. If there are N spectral elements to be resolved during a total observation time t, then if the spectrum is scanned, each element is seen for a time t/N. The signal-to-noise ratio is then proportional to $\sqrt{t/N}$. But if we observe each element for the whole time t (as with a prism or grating spectrometer), the signal-to-noise ratio is proportional to $\sqrt{t}$. A Fourier

transform spectrometer, therefore, performs better by a factor of $\sqrt{N}$, a factor called *Fellgett's advantage*.*

> Fourier transform spectroscopy is the superior method. Even more important, Fourier spectroscopy is not simply the application of another little invention; rather, it marks a turning point in philosophy, *away* from high-precision delicate optics, *toward* a simple, rugged sensor coupled with sophisticated *electronic data processing*.

SUGGESTIONS FOR FURTHER READING

R. J. Bell, *Introductory Fourier Transform Spectroscopy* (New York: Academic Press, Inc., 1972).

L. Mertz, *Transformations in Optics* (New York: John Wiley & Sons, Inc., 1965).

R. N. Bracewell, *The Fourier Transform and Its Applications,* 2nd edition. (New York: McGraw-Hill Book Company, 1978).

D. Baker, A. Steed, and A. T. Stair, Jr., "Development of infrared interferometry for upper atmospheric emission studies," *Applied Optics* **20** (1981), 1734–46.

D. R. Matthys and F. L. Pedrotti, "Fourier transforms and the use of a microcomputer in the advanced undergraduate laboratory," *Am. J. Physics* **50** (1982), 990–95.

H. Stark, editor, *Applications of Optical Fourier Transforms* (New York: Academic Press, Inc., 1982).

PROBLEMS

4.1-1. A target consists of dark bars on a bright background. If the bright parts emit 5 arbitrary units of light and the dark parts 3 units, what is the contrast as defined:
(a) By Equation [4.1-1]?
(b) In the visual sciences?

4.1-2. At which numerical values of luminance will both types of contrast (see Problem 4.1-1) yield the same result?

4.1-3. If we assume that the wavetrains from the 546-nm green mercury line are 11 mm long, what is the hypothetical bandwidth? (Hypothetical because of *Doppler broadening,* to be discussed in a later chapter.)

4.1-4. What is the bandwidth of the orange krypton line, 606 nm, assuming that the coherence length is 80 cm?

4.1-5. If the mirror in a Michelson–type Fourier spectrometer moves at a velocity of 4 mm s^{-1} and the light comes from a helium–neon laser (633 nm):
(a) What is the frequency of modulation of the

---

*Named after Peter Berners Fellgett, who established this principle in his Ph.D. thesis, Cambridge University, 1951. P. Fellgett, "A propos de la théorie du spectromètre interférentiel multiplex," *J. Physique Radium* **19** (1958), 187–91.

photocurrent?

(b) Is this within the audio range?

4.1-6. If a Michelson spectrometer is used with sodium light (589 nm), what should be the velocity of the mirror to modulate the photocurrent at a frequency of 5 kHz?

4.1-7. If two tuning forks are vibrating side by side at 255 and 257 Hz, respectively, what is the beat frequency?

4.1-8. The spectrum of calcium contains two closely spaced lines centered around 395.1 nm.

If interference fringes produced by these lines reach maximum contrast at every 116th fringe, what are the two wavelengths?

4.1-9. Write the equation of a complex wave that results from the superposition of two simple harmonic waves which have amplitudes in the ratio of $2:1$, frequencies in the ratio of $1:3$, and which are in phase, meaning that at time $t = 0$ their displacement is $y = 0$.

4.1-10. Continue with Problem 4.1-9 and draw plots of the two contributions and of the resultant wave.

4.2

Transfer Functions

Transfer functions are a measure of performance. With their help we can predict *theoretically*, as well as confirm or disprove *experimentally*, the performance of individual lenses and of complete systems. Transfer functions also apply to peripheral components such as photographic film, television, the human eye, and even the atmosphere through which the light passes.

Spread Functions

When an object point (a *pixel*) is focused through a lens, the conjugate image, because of diffraction and aberrations, is not a point but a diffuse patch of light. When this patch is scanned using a pinhole aperture, the distribution of light in the image represents the *point spread function*, $S(y, z)$, of the lens.

Instead of a pinhole aperture, it is more convenient to use a *line* image and a *slit* aperture. This leads to the *line spread function, $S(z)$* (Figure 4.2-1). The line spread function is simply the point spread function integrated over the length of the slit, y,

$$S(z) = \int_{-\infty}^{+\infty} S(y, z)\, dy \qquad [4.2\text{-}1]$$

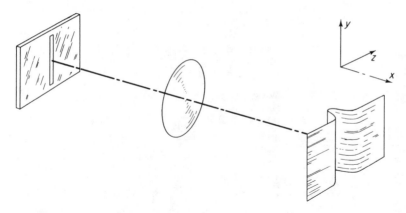

Figure 4.2-1   The line spread function. Object left, image right. The distribution of light in the image is shown as a cylindrical mound.

Now consider an extended object, composed of many pixels and assume that the light is incoherent. Each pixel forms its own patch of light, the point spread function of the lens; the composite of these patches is the sum of the spread functions, integrated over two-dimensional space. The result is the *areance* (irradiance) *distribution, E(y, z)*. In one dimension,

$$E(z) = \int_{-\infty}^{+\infty} I(z)S(z) \; dz \qquad\qquad [4.2\text{-}2]$$

which shows that the areance distribution is found from the integral of the spread function.

The converse holds true as well. The spread function is the derivative of the areance distribution,

$$\frac{dE(z)}{dz} = I(z)S(z) \qquad\qquad [4.2\text{-}3]$$

This means that the slope of the scanning trace, $dE(z)/dz$, at a given point, z_0, in the image is proportional to the line spread function at that point. Therefore, if the trace is found experimentally, a plot of its slope versus position in the image is the spread function of the lens.

Modulation

Definition.   In earlier times lenses were often tested by forming an image of a target, such as that shown in Figure 4.2-2. Such targets have high contrast: the lines are either black on a white background, or white on a black background.

But realistic objects are rarely black and white. Most often they come

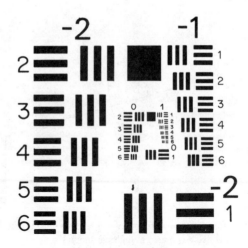

Figure 4.2-2   USAF test chart.

in shades of gray. According to Rayleigh's criterion (which says that resolution depends on wavelength and on the diameter of the lens), the result should be the same. But clearly, two lines of *low* contrast more easily fuse into one line than do two lines of the same spacing and *high* contrast. In addition, other factors besides wavelength and aperture affect, and generally reduce, the resolution. Think of atmospheric haze, and of lenses poorly made, badly scratched, or covered with dust or fingerprints. All these factors affect the contrast. Any determination of resolution must take into account the *contrast*.

Contrast, as we have seen (on pages 209 and 398), is defined as

$$\gamma \equiv \frac{E_{\text{max}} - E_{\text{min}}}{E_{\text{max}} + E_{\text{min}}} \qquad [4.2\text{-}4]$$

Consider, for example, an object of sinusoidal areance distribution (Figure 4.2-3). Scanning across the object, we find an alternating sequence of higher and lower areances. Instead of recording the maxima, E_{max}, and

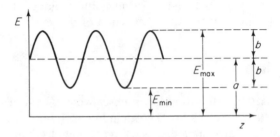

Figure 4.2-3   Sinusoidal areance distribution as a function of space (or time).

minima, E_{min}, it is more practical to introduce two new parameters, the mean areance, $a = (E_{max} - E_{min})/2 + E_{min}$, and the variation of the areance around the mean, $b = E_{max} - a$. The ratio of both is called the *modulation*, **M**,

$$\mathbf{M} = \frac{b}{a} \qquad [4.2\text{-}5]$$

This *normalized* variation of the areance is independent of the absolute areance.

Modulation transfer.

Whenever we test for performance, we are not as interested in modulation as in *how much modulation can be transferred* from object to image. It is the *modulation transfer* that counts. The ratio of modulation in the image to modulation in the object is called the *modulation transfer factor*, **T**,

$$\mathbf{T} = \frac{\mathbf{M}_{image}}{\mathbf{M}_{object}} \qquad [4.2\text{-}6]$$

In Figure 4.2-4, the image modulation is one-half the object modulation, and hence $\mathbf{T} = \frac{1}{2}$.

Is the image contrast always less than the object contrast? Not necessarily. For example, a portrait photograph taken on high-contrast film used for technical reproductions would show harsh, unnatural shadows, of a contrast *higher* than that of the object.

Contrary to what Equation [4.2-6] suggests, the transfer factor **T** of a given lens does not have a single, unique value. Instead, *it varies as a function of spatial frequency, R*:

$$\mathbf{T} = f(R) \qquad [4.2\text{-}7]$$

This dependence of **T** on R is of great practical interest. The term

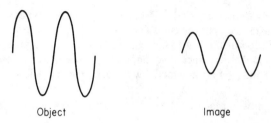

Object   Image

Figure 4.2-4   Scanning across lines shows high contrast in the object (*left*) and low contrast in the image (*right*).

spatial frequency refers to the number of lines, or other detail, within a given length. A technical line drawing, for example, may have four lines, and four intervals adjoining these lines, per centimeter. The spatial frequency, then, is 4 lines/cm, but since spatial frequency is customarily written in units of inverse *millimeters*, we have $R = 0.4$ mm^{-1}. If 1000 lines, and the intervals between them, fit into 1 mm, $R = 1000$ mm^{-1}. If one bar alone is 1 cm wide, $R = 0.05$ mm^{-1}. The higher the spatial frequency that a given optical system can resolve, the better the system.

In the propagation of light, in wave motion in general, and in electric circuits, frequency usually means *time frequency*. In conjunction with transfer functions and in optical data processing, it means *space frequency*.

According to Equation [4.2-7] the transfer factor is a function of spatial frequency. Therefore, instead of a single-number transfer *factor*, **T**, we have a transfer *function*, **T**(R), called *modulation transfer function*, **MTF**.

Convolution.   Assume that the areance (exitance), M, of a test object varies sinusoidally across z:

$$M(z) = a + b \cos 2\pi R z \qquad [4.2\text{-}8]$$

where a and b have been introduced before (see Figure 4.2-3).

Now, whenever an image is formed in incoherent light, each pixel in the object is subject to the spread function of the lens. Thus the image is the sum of the individual spread functions. But some image details may have changed in position, due to aberrations; hence, we use in the image another coordinate, ζ, the Greek equivalent of z. Then the function $M(z)$ refers to the exitance (in the object), and $E(\zeta)$ refers to the conjugate irradiance (in the image). To find $E(\zeta)$, we multiply $M(z)$ by the spread function of the lens, $S(z)$, and integrate over space, using the process of *convolution*.

Here we note a close relationship to electronic data processing. The exitance function, $M(z)$, in optics is equivalent to the *input* in electrical engineering; the spread function of the lens, $S(z)$, corresponds to the "impulse response" of the network; and the irradiance function, $E(\zeta)$, corresponds to the *output*. In either case, $M(z)$ and $S(z)$ are convolved with each other to yield $E(\zeta)$.

We now replace $S(z)$ by $S(\zeta - z)$ and instead of Equation [4.2-2] write

$$E(\zeta) = \int_{-\infty}^{+\infty} M(z)S(\zeta - z)\, dz \qquad [4.2\text{-}9]$$

which is the *convolution integral*. Substituting Equation [4.2-8] in [4.2-9] gives

$$E(\zeta) = a \int_{-\infty}^{+\infty} S(z) \, dz + b \int_{-\infty}^{+\infty} S(z)\cos 2\pi R(\zeta - z) \, dz \qquad [4.2\text{-}10]$$

Solving these integrals is easier if we separate the two variables z and ζ. For simplicity we also assume that the area under the first integral, the spread function, is unity. Then the irradiance distribution becomes

$$E(\zeta) = 1 + bS_{\cos} \cos 2\pi R\zeta + bS_{\sin} \sin 2\pi R\zeta \qquad [4.2\text{-}11]$$

where

$$S_{\cos}(R) = \int_{-\infty}^{+\infty} S(z)\cos 2\pi Rz \, dz \qquad [4.2\text{-}12]$$

and

$$S_{\sin}(R) = \int_{-\infty}^{+\infty} S(z)\sin 2\pi Rz \, dz \qquad [4.2\text{-}13]$$

These are the cosine Fourier transform and the sine Fourier transform, respectively, of the line spread function, $S(z)$. Since these transforms are functions of the same variable, R, they can be combined into an exponential function,

$$\boxed{\text{MTF} = \int_{-\infty}^{+\infty} S(z)e^{-2\pi iRz} \, dz} \qquad [4.2\text{-}14]$$

which is again the *modulation transfer function*. Note the similarity between Equations [4.2-14] and [4.1-18]; the *modulation transfer function of a lens is the Fourier transform of the spread function of that lens.*

Phase transfer.   I have mentioned that due to aberrations certain image pixels may become dislocated with respect to their nominal ("correct") positions (Figure 4.2-5). Coma and distortion are typical examples.

The extent of such dislocation is measured by the *phase* (shift) ϕ. The image coordinate ζ, therefore, must be replaced by $\zeta - \phi$. But ϕ, like T, is a function of spatial frequency, $\phi(R)$. Thus, in addition to the modulation transfer (function), we have a *phase transfer*. Both are connected, in exponential notation, in the complex *optical transfer function*, **OTF**:

$$\boxed{\text{OTF} = \text{MTF} \cdot e^{i\phi(R)}} \qquad [4.2\text{-}15]$$

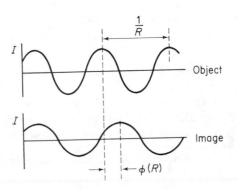

Figure 4.2-5 Dislocation of image (*bottom*) with respect to object (*top*). *R*, spatial frequency; ϕ, phase shift. Also note lower contrast in image compared to object.

The optical transfer function is a space-frequency-dependent complex quantity whose modulus, appropriately, is the modulation transfer function and whose phase is the phase transfer function. The optical transfer function, therefore, is the *ratio of the Fourier transform of the light distribution in the image to the Fourier transform of the light distribution in the object*:

$$\mathbf{OTF} = \frac{\text{Fourier transform of light distribution in image}^*}{\text{Fourier transform of light distribution in object}} \qquad [4.2\text{-}16]$$

In the most general terms, the optical transfer function describes the degradation of an image of different space frequencies. A perfect lens should have a modulation transfer factor **T** of unity, and a phase transfer factor ϕ of zero, *at all frequencies*. In reality, this is not possible, because of diffraction and aberrations. At low space frequencies, the **MTF** of a good lens may come close to unity. At higher frequencies the $\mathbf{T}(R)$ curve gradually declines, approaching $\mathbf{T} = 0$ somewhere between 100 and 1000 mm^{-1}. The $\phi(R)$ curve, at low frequencies, will be close to zero. At higher frequencies, it increases. In a way, the $\phi(R)$ curve is a mirror image of the $\mathbf{T}(R)$ curve, as shown in Figure 4.2-6.

---

*This important relationship was first formulated by Pierre-Michel Duffieux (1891–1976), French physicist, erstwhile assistant to Charles Fabry, and later professor at the universities of Rennes and Besançon. Duffieux, like Fourier before him, started out by working on the seasonal transfer of heat through Earth's soil, went on to Fourier transform spectroscopy, and wrote a highly original book, *L'Intégrale de Fourier et ses Applications à l'Optique* (Paris: Masson et Cie., 1970), in which he laid the foundation for transfer functions that have now replaced the earlier concept of resolution. P. M. Duffieux, *The Fourier Transform and its Applications to Optics*, 2nd edition (New York: Wiley-Interscience, 1983).

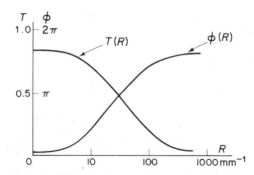

Figure 4.2-6   Modulation transfer factor, $\mathbf{T}(R)$, and phase transfer factor, $\phi(R)$, both as functions of spatial frequency.

There are some exceptions to this typical shape of the $\mathbf{T}(R)$ curve. Catadioptric systems of the Cassegrain type, for example, tend to show a hump before the curve finally reaches zero. Certainly, such behavior cannot be predicted from Rayleigh's criterion.

The Experimental Determination of Transfer Functions

The essential point in determining the optical transfer function of a lens or lens system is to form and scan an image of a test grid. Often used for this purpose are *Foucault grids* (also called Sayce targets); these contain a series of black-and-white, square-wave or sinusoidal, parallel bars of varying space frequency, widely spaced at one end and close together at the other. The grid can be used as the object, with a scanning slit in the conjugate image plane; or the slit can be the object, with the grid placed in the image plane. Sometimes the grid is made into a drum, as shown in Figure 4.2-7.

At the low space frequencies (widely spaced slots), the photocell faithfully registers all bars and intervals. At the high frequencies, depending on the quality of the lens, the individual signals fuse together, the lines are no longer "resolved," and the envelope across the trace goes *down*.

If a lens *system* is tested, the total **MTF** is the *product* of the **MTF**s of the individual lenses. (The individual transfer curves are multiplied, ordinate by ordinate, at each point on the abscissa; see, for example, Problem 4.2-5.) But that holds only if any aberrations introduced by one lens are not counteracted by aberrations of another.

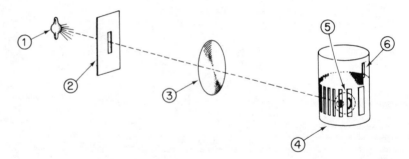

Figure 4.2-7 Experimental arrangement for measuring transfer functions. 1, Light source; 2, slit; 3, lens under test; 4, rotating drum; 5, photodetector; 6, reference mark for determining phase transfer.

Example 1. *Photographs are taken from a high-altitude aircraft of a cruise ship, painted white. Assume that the MTF of a typical camera lens is that shown in Figure 4.2-6. What focal length is necessary to obtain enough detail to identify the ship?*

Solution. The ship may have a brightness of 5 arbitrary units, and the surrounding ocean may have 2 units. Then its contrast is

$$\gamma = \frac{5 - 2}{5 + 2} = 0.43$$

Initially, the focal length of the system may be chosen so that the ship's image is 0.5 mm wide. This means a spatial frequency of $R = 1$ mm^{-1}. From Figure 4.2-6 we find that at this frequency the contrast transfer factor **T** is 0.8. At $R = 10$, **T** is 0.7 and at $R = 100$, it is 0.2.

　　Knowing that the object is of limited contrast (0.43), we now find that for different space frequencies the contrast, transferred to the image, changes as follows:

Object contrast		Transfer Space frequency R Transfer factor **T**		Image contrast
0.43	$\longrightarrow$	$R = 1$ **T** = 0.8	$\longrightarrow$	0.34
		$R = 10$ **T** = 0.7	$\longrightarrow$	0.3
		$R = 100$ **T** = 0.2	$\longrightarrow$	0.086

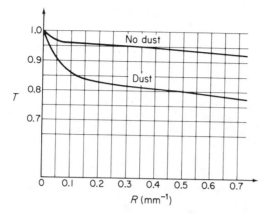

Figure 4.2-8 Modulation transfer function of a clean lens (*top curve*) and of the same lens covered with dust (*bottom*). [From E. Diederichs, "Optische Übertragungs-theorie angewandt auf die Oberflächengüte," Jos. Schneider & Co., Optische Werke, Kreuznach, *Hausmitteilungen* **13** (1961), 7/8, 82. Reproduced by permission.]

A contrast of 34 percent is satisfactory for visual observation. That means the ship is easy to see. But fine details, at $R = 100$, cannot be seen. At this frequency the image contrast is so low, less than 9 percent, that a lens of longer focal length is needed to obtain a larger image. The larger image contains *lower* frequencies, which in turn permit more contrast to be transferred.

Example 2. *A given lens is (a) well cleaned and (b) covered with dust. How will this affect the modulation transfer?*

Solution. Experimental observation shows, as illustrated in Figure 4.2-8, that the **MTF** of the dust-covered lens is fairly evenly depressed across all frequencies. However, aside from this general loss of contrast, even high frequencies come through surprisingly well. This is easy to understand. High space frequencies (fine details) in the object correspond to large Fourier elements (as we will see in Figure 4.3-8). These elements are appreciably larger than the dust particles covering the lens; hence, the Fourier elements are not much affected and the image retains a reasonable degree of contrast.

Example 3. *What effect does atmospheric turbulence have on aerial reconnaissance?*

Solution. From Figure 4.2-9 we conclude that if the details on the ground have high contrast, details of a given space frequency can be resolved well both through calm air and through turbulent air. Turbulence merely causes the contrast to become less (arrow *A*) but since the details have high contrast, they remain well visible. On the other hand, if the details have low contrast, for example if the target is camouflaged, only large details can be identified (arrow *B*), a result that surely we have anticipated by intuition.

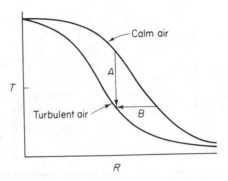

Figure 4.2-9 Modulation transfer through calm air and turbulent air.

SUGGESTIONS FOR FURTHER READING

F. H. Perrin, "Methods of Appraising Photographic Systems. Part I—Historical Review," *J. Soc. Motion Pict. Telev. Eng.* **69** (1960), 151–56; "Part II—Manipulation and Significance of the Sine-Wave Response Function," *ibid.* **69** (1960), 239–49.

W. B. Wetherell, "The Calculation of Image Quality," in R. R. Shannon and J. C. Wyant, editors, *Applied Optics and Optical Engineering,* Vol. VIII, pp. 171–315 (New York: Academic Press, Inc., 1980).

F. D. Smith, "Optical Image Evaluation and the Transfer Function," *Applied Optics* **2** (1963), 335–50.

K. R. Barnes, *The Optical Transfer Function* (New York: American Elsevier Publishing Company, Inc., 1971).

PROBLEMS

4.2-1. A camera lens of 28.6 mm focal length resolves details that in the image plane have a spatial frequency of 10 mm^{-1}. What is the angular resolution, in degrees?

4.2-2. What is the spatial frequency, in mm^{-1}, of the bar pattern in the USAF test chart shown in Figure 4.2-2:
(a) Near the left-hand lower corner, number 5?
(b) At the right-hand margin, number 4?

4.2-3. The curve in Figure 4.2-10 shows the relationship between luminance and the resulting absorbance of a photographic emulsion, known as the *Hurter–Driffield curve.*
(a) Which part of the curve will give the highest contrast?

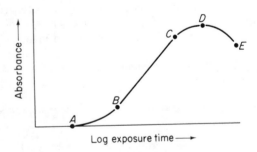

Figure 4.2-10

(b) How will the result change in section $D-E$ (severe overexposure), a case known as *solarization?*

4.2-4. Continuing with Problem 4.2-3, plot the

magnitude of the transfer factor as a function of the logarithm of the exposure time.

4.2-5. A test object of 76 percent modulation is photographed through a camera lens that for 4 lines mm^{-1} has a transfer factor $\mathbf{T} = 0.85$. The image is formed on film whose transfer factor, at the same frequency, is 0.62. What is the total modulation?

4.2-6. Assume that the primary mirror of a reflecting telescope is:
(a) Of perfect paraboloidal shape.
(b) Distorted and not truly paraboloidal.
(c) Of the correct shape but merely ground to a silky finish, rather than polished.
Plot the **MTF**s that correspond to these three conditions.

4.2-7. An extended object is projected by a converging lens into a real image, at about unit magnification. Assume that a sheet of distorted window glass is placed:
(a) Directly over the object.
(b) Immediately in front of, or closely behind, the lens.
(c) Directly in front of the image.
Neglecting the need for refocusing, how will the image change?

4.2-8. Continue with Problem 4.2-7 and assume now that the sheet of glass is precisely plane-parallel but only finely ground, rather than polished. How does this change the image?

4.3

Optical
Data Processing

EVEN AN ORDINARY LENS is an optical data processor. It transforms two-dimensional information, the object, into a two-dimensional display, the image, and, in contrast to electronic data processing, it does so without scanning.

In this chapter the emphasis is on practical applications. The description is phenomenological. The groundwork has been laid in the two preceding chapters, in our discussion of Fourier transformation. But there are important differences. Transfer functions are one-dimensional and essentially incoherent. Optical data processing is two-dimensional, and can either be coherent or incoherent. It all began with *Abbe's theory of microscopic image formation*.

Diffraction Theory of Image Formation

According to Abbe's theory,* the image formed by a microscope is produced in two steps. As illustrated in Figure 4.3-1, light from a point

---

*Ernst Abbe (1840–1905), German physicist, professor at the University of Jena, chief designer and partner of Carl Zeiss at the Carl Zeiss Optical Company in Jena. Abbe is known also for the theory of pupils, the sine condition, refractometer, and blood counting

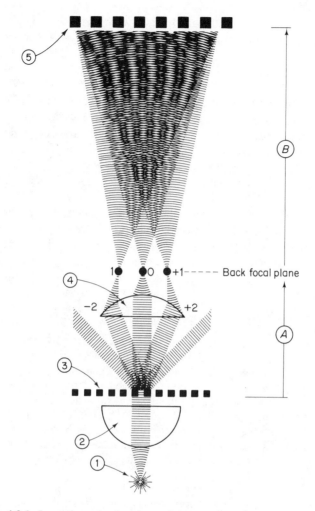

1 ● ●0 ●+1 - - - - - Back focal plane

−2   +2

Figure 4.3-1 Image formation in coherent light. 1, light source; 2, condenser; 3, object; 4, objective lens; 5, image; *A*, *B*, Fourier transformations.

source (bottom) enters a collimating lens (the condenser) and is incident on the object. Some of the light passes through the object and forms the zeroth-order maximum. Other light is diffracted by the object structure; it forms the higher orders. But because any lens is necessarily limited in

chamber, all named after him. In the course of his work, Abbe discovered that microscope objectives of low numerical aperture, however well they were made, produced images inferior to those produced by simpler lenses of high aperture, and on this basis formulated his theory of image formation. E. Abbe, "Beiträge zur Theorie des Mikroskops und der mikroskopischen Wahrnehmung," *Arch. mikr. Anat.* **9** (1873), 413–68.

diameter, only the zeroth and the first few higher orders enter the objective and contribute to image formation. All other higher orders are lost.

In the back focal plane of the objective the diffraction maxima are most distinct; they are *Fraunhofer diffraction* maxima or, in more modern terminology, they are the *Fourier transform* of the light distribution across the object. A subsequent transformation converts the first Fourier transform into the (intermediate) image, which is seen through the eyepiece.

According to Abbe, if more higher-order maxima are admitted, the image will contain more information, more detail. This is why a microscope objective of high numerical aperture yields a "better" image. It is the *information content* that is essential; "magnification" does not account for much.

By the same argument (the limited diameter of any lens), an image always contains *less* information than the object. This is most obvious if the object has details that are very small or very close together. If the details are so small, or the angular aperture of the objective so narrow, that *no* higher-order maxima enter the front lens, then the object structure cannot be resolved, no matter what the magnification.

At first, Abbe's theory was thought to apply only to illuminated targets, or imagery by coherent light. But the same form of argument applies also to self-luminous objects (incoherent imagery).

Lab Experiment Illustrating Abbe's Theory

1. Use as the object a grid that has about 50 lines to the millimeter. Place the grid on an optical bench and illuminate it, from the left, with collimated white light. Orient the grid so that its lines are horizontal.

2. Assemble a compound microscope, using a +10-diopter lens as the objective and a 10× Huygens eyepiece placed 30 cm to the right of the objective. Looking through the eyepiece, focus on the grid by slightly moving the objective back and forth.

3. Place a sheet of ground glass 10 cm to the right of the objective and slowly move the ground glass back and forth. Looking at the ground glass from the right, toward the light source, you will notice that in one position the light forms a distinct pattern (Figure 4.3-2, left). This is the Fourier transform; it appears in the back focal plane of the objective.

The bright spot in the center is the zeroth-order maximum. The less intense spots above and below are the higher-order maxima. The zeroth order is white. The higher orders are colored, red on the outside, blue on the inside.

4. Without moving anything else, replace the ground glass with an adjustable slit. Adjust the slit with the transform pattern. Block out all maxima except the zeroth order (Figure 4.3-2, center). A small piece of paper, held closely behind the slit, will help. Look through the eyepiece. Sufficient light passes through, but *no image is seen:* the zeroth-order maximum alone does not produce an image.

5. Open the upper and lower bracket of the slit so that the two first-order maxima are admitted also (Figure 4.3-2, right). The lines in the grid become

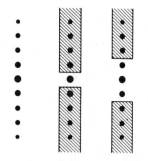

Figure 4.3-2 Fourier transform of a grid object, using point light source (*left*). Same transform, but the zeroth-order maximum only is admitted to the image space (*center*). Zeroth- and two first-order maxima admitted (*right*).

visible. Fully open the slit, admitting all maxima. The grid is seen even better. Higher-order maxima are needed for seeing more detail.

Conclusion: To get a satisfactory image we need two contributions: luminous *energy,* most of which is contained in the zeroth order, and *information,* which is contained in the higher orders.

Data Processing

Abbe's theory and the experiments just described have laid the foundation for optical data processing. But why should data be processed at all? To free the human mind from the routine chores of filing, storing, searching for, and retrieving data, in order to get—as efficiently as possible—at the information itself.

Certain data are binary, black or white, yes or no, 0 or 1. Such data are handled very well by electronic, *one*-dimensional processing. But human beings, since time immemorial, have become accustomed to *two*-dimensional data, from hieroglyphics to letters of the alphabet to pictorial representations. Such data can be processed one-dimensionally by scanning; but clearly, two-dimensional transformation is preferable because it is more direct. This two-dimensional approach is the domain of *optical data processing*.

Look again at Abbe's experiment and consider the general case of Fourier transformation, not limited to a microscope. Figure 4.3-3 shows a sequence of three lenses, lined up coaxially on an optical bench. Lens 1 collimates the light that comes from a point source. The light then reaches the object, located in the first (left-hand) focal plane of lens 2. Lenses 2 and 3 together form a real image on the right. The fundamental question is: What happens in the second (right-hand) focal plane of lens 2, the first

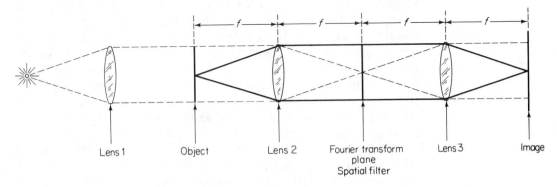

Figure 4.3-3 General arrangement of lenses in optical data processing; *f*, focal length.

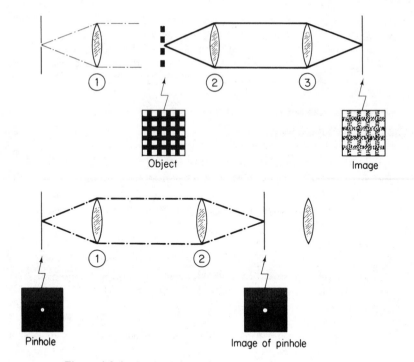

Figure 4.3-4   Spatial filtering with cross-grating as the object.

(left-hand) focal plane of lens 3? *This is where the light distribution across the object has undergone Fourier transformation.*

The object may be a fine wire mesh or *cross-grating* (Figure 4.3-4, top). Lenses 2 and 3 together form an image of the cross-grating. In addition, lenses 1 and 2 together form an image of the pinhole (bottom left). Note that the three lenses have two functions. They project, simultaneously but independently, two objects into two images: the cross-grating (top) is made into an image, and the pinhole (bottom) is made into an image also, two processes that remind us of the conjugate relationships between object and image, and between entrance pupil and exit pupil.

Now insert a *mask* in the plane of the pinhole image. In the top sequence in Figure 4.3-5 this mask is another *pinhole,* which admits only the undiffracted light, the zeroth order. The result is that the image plane *receives light but no information:* The screen is uniformly bright. In the center sequence the mask is a vertical *slit.* The result is that only the (vertically displayed) diffraction maxima caused by the horizontal lines in the cross-grating are transmitted. The image, therefore, shows only these horizontal lines. In the bottom sequence the mask is wide open, or there

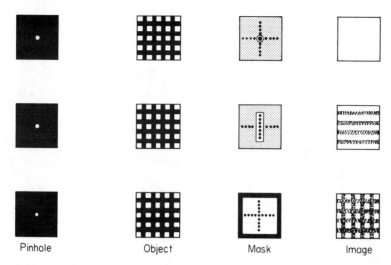

Pinhole   Object   Mask   Image

Figure 4.3-5 Results of inserting a variety of spatial filters in the optical data processor.

is *no mask*. Now the full structure of the cross-grating, with all its lines, is seen. This is the basis of *spatial filtering*.*

Spatial filtering is the process of enhancing or suppressing certain details in an image, making use of Fourier transformation. In particular, spatial filtering facilitates the detection of a *signal,* in the presence of unwanted *noise.* Therefore, the design of the (spatial) filter must be based on prior knowledge of preselected characteristics of the signal and the noise, as we see from the next example.

Some more examples of spatial filtering.
We have a wire mesh contaminated with dust (Figure 4.3-6, left). The Fourier transform generated by the grid alone is a regular, two-dimensional array of maxima. If a spatial filter containing a matrix of holes is placed in the transform plane such that the holes exactly line up with the maxima, then only light diffracted by the grid, *not light diffracted by the dust,* will pass through (center). The image, therefore, shows the grid, without the dust (right). If the maxima were blocked out, the dust would show, instead of the grid.

Precision alignment of the mask with the maxima is difficult. This alignment is made easier by using a sheet of *photochromic glass.* Such

---

* A series of spatial-filtering experiments, well suited for the more advanced optics lab, has been described by A. Eisenkraft, "A closer look at diffraction: experiments in spatial filtering," *The Physics Teacher* **15** (1977), 199–211.

Figure 4.3-6 Wire mesh contaminated with dust (*left*). Maxima diffracted by wires, and mask with holes admitting these maxima (*center*). Image shows grid without dust (*right*).

glass, as we will see in more detail later (on page 490), ordinarily is transparent, like any other glass. Under the influence of light, however, photochromic glass becomes dark. When a sheet of photochromic glass is placed in the Fourier transform plane, the more intense zeroth order causes the glass to darken more than the higher orders will. The higher orders pass

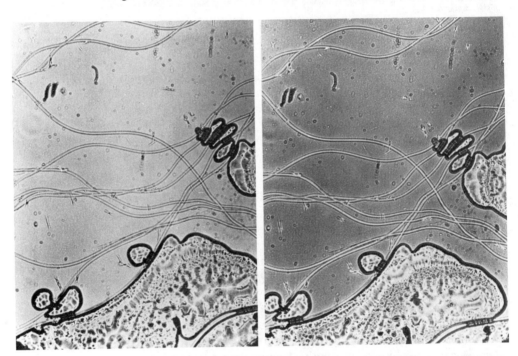

Figure 4.3-7 Nylon fibers and air bubbles in clear plastic, seen at naturally low contrast (*left*) and after insertion of photochromic glass in the Fourier transform plane (*right*). Photographic processing was identical. For more details see W. J. Baldwin and J.R. Meyer-Arendt, *Optical Contrast Enhancement System*, U.S. Patent 3,598,471, Aug. 10, 1971.

through the glass without much attenuation and, since *they* carry the information, the contrast increases. Figure 4.3-7 gives an example.

The relationship between object and transform can be expressed in a more quantitative way. When we compare the three rectangular apertures and their transforms in Figure 4.3-8, we note that an object of higher spatial frequency (short slit!) gives maxima that are farther apart. An object of lower spatial frequency (long slit!) gives maxima that are closer together. This is the same inverse relationship that we found earlier in Fraunhofer diffraction. Indeed, the *Fourier transform of an object is the Fraunhofer diffraction pattern of that object*. Following Equation [2.3-3],

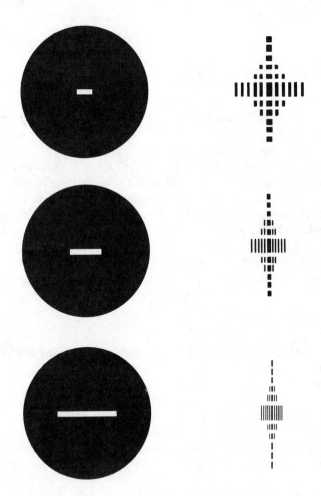

Figure 4.3-8   Narrow slits of various lengths (*left*) and their Fourier transforms (*right*).

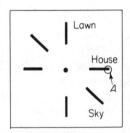

Figure 4.3-9 Theta modulation. Landscape scene in half-tone representation (*left*), theta-modulated (*center*), and Fourier transformed (*right*).

$$s \propto \frac{1}{\sin\theta} \qquad [4.3\text{-}1]$$

which shows that the size, *s*, of the details in the object structure is reciprocal to the angular size, *θ*, of its transform.

Theta modulation.

*Theta modulation,** another form of spatial filtering, also derives from Abbe's theory. Consider a scene that contains a house, a lawn, and sky (Figure 4.3-9, left). These elements are made out of small pieces of plastic diffraction grating oriented at different angles theta, *θ*—hence the term "theta modulation" (center).

The object is projected by a lens onto a screen. Focus the image. Cover the lens with a sheet of aluminum foil, the dull side facing the light source. On the foil you will see three pairs of spectra that correspond to the elements in the object scene, red being diffracted most, blue least. Cut out part of the foil (such as hole *A* in Figure 4.3-9, right); let only the red pass through in the two spectra that belong to the house, green for the lawn, blue for the sky. The mask thus made transmits only part of the light and the image appears in these colors.

Phase contrast.

Spatial filtering is not limited to manipulating the amplitude of the light, that is, blocking out or attenuating certain maxima. We can as well operate on the *phase*.

Abbe had assumed the object to be an *amplitude* grating. The diffraction maxima produced by such a grating have all the same phase. But the object may be a *phase* grating (page 252). Such gratings introduce a phase difference between the zeroth order and the higher orders. In addition, the zeroth-order maximum produced by a phase object is much more intense

*J. D. Armitage and A. W. Lohmann, "Theta Modulation in Optics," *Applied Optics* **4** (1965), 399–403.

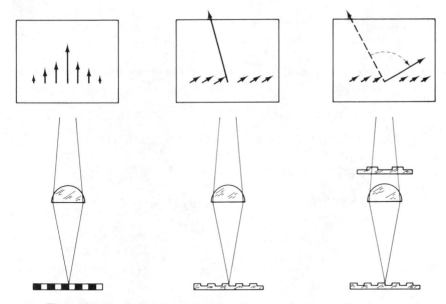

Figure 4.3-10 Vectorial representation of amplitude object (*left*), phase object (*center*), phase contrast (*right*).

than that of an amplitude object. In short, the transform of an amplitude grating is distinctly different from that of a phase grating.

It then occurred to Zernike* that, if the transform were modified properly, a phase object could be made to look like an amplitude object. That would be particularly useful in biology and mineralogy, where many microscopic specimens are phase objects that are sometimes hard to see without staining. Zernike called his process the "phase-strip method for observing phase objects in good contrast" or *phase contrast,* for short.

Figure 4.3-10 shows various transforms in vector representation. An amplitude grating, left, has a strong zeroth-order maximum and less intense higher orders, all of the same phase. With a phase grating, the zeroth-order maximum is even more intense (longer vector!) and the higher orders differ in phase by $\pi/2$ radian or 90° (center). All that is needed, then, to *make a phase grating look like an amplitude grating* is to attenuate the amplitude and delay the phase. This is done by inserting in the transform plane a $\pi/2$ phase-shifting element, conjugate to an annular aperture in the first focal plane of the condenser, and by making the phase-shifting

---

* Frits Zernike (1888–1966), Dutch physicist, professor of physics at the University of Groningen, received the Nobel prize in physics in 1953 for the discovery of phase contrast. F. Zernike, "Beugungstheorie des Schneidenverfahrens und seiner verbesserten Form, der Phasenkontrastmethode," *Physica* **1** (1934), 689–704; and "How I Discovered Phase Contrast," *Science* **121** (1955), 345–49.

coating slightly absorbent (right).

Phase contrast, however, and *variable phase contrast* are not nearly as promising for data processing as they once appeared because of the invention and rapid refinement of *holography,* the topic discussed in detail in the next chapter.

SUGGESTIONS FOR FURTHER READING

A. R. Shulman, *Optical Data Processing* (New York: John Wiley & Sons, Inc., 1970).

K. Preston, Jr., *Coherent Optical Computers* (New York: McGraw-Hill Book Company, 1972).

G. L. Rogers, *Noncoherent Optical Processing* (New York: John Wiley & Sons, Inc., 1977).

S. H. Lee, editor, *Optical Information Processing. Fundamentals* (New York: Springer-Verlag, 1981).

W. K. Pratt, *Digital Image Processing* (New York: John Wiley & Sons, Inc., 1978).

G. Harburn, C. A. Taylor, and T. R. Welberry, *Atlas of Optical Transforms* (Ithaca, NY: Cornell University Press, 1975).

PROBLEMS

4.3-1. (a) If the apertures in Figure 4.3-8 were moved up by a distance twice the width of the slit, how would the diffraction pattern change?
(b) If the slits were rotated clockwise, how would the patterns rotate?
(c) If the slits were made narrower, what would you see?

4.3-2. If the maxima in a transform pattern form a straight line extending between the numbers 8 and 2 on a clock, how are the lines in a grating oriented that cause these maxima?

4.3-3. The object illustrated in Figure 4.3-11 is known as an *Abbe diffraction plate.*
(a) If such an object is placed in a microscope of high numerical aperture, what do you see in the Fourier transform plane?
(b) How would the transform, and the image, change if the objective were of low numerical aperture?

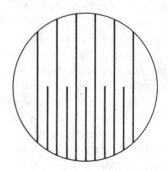

Figure 4.3-11

4.3-4. If an optical data processor is to be built so that integration is performed in the vertical dimension only, which of the lenses in Figure 4.3-3 should be replaced by cylinder lenses, and how should these lenses be oriented?

4.3-5. A collage is made up of a number of

pieces, each containing a series of closely spaced vertical lines.

(a) What kind of a spatial filter is needed to remove the lines while having the least effect on the composition?

(b) Where should the filter be placed?

4.3-6. The scene illustrated in Figure 4.3-12 is

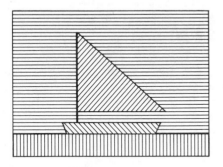

Figure 4.3-12

theta-encoded. How will the spectra, seen with white light, be oriented? What should a space filter look like that is to make the sky blue, the water green, the boat white, and the sail red?

4.3-7. A wire mesh containing 25 wires per centimeter each way serves as the object in an optical data processor. If two lenses, one on each side of the mesh, have focal lengths of 60 cm each and if the light has a wavelength of 600 nm, what is the distance between adjacent maxima in the transform plane?

4.3-8. A faint star at twilight is barely visible. When a rotating reticle composed of alternating opaque and clear sectors is placed in the image plane of a telescope and a photodetector is mounted behind it, how will the photocurrent change if:

(a) The star were brighter?

(b) The sky in general were darker?

(c) The sky in general were lighter?

(Problem suggested by R. Bruce Herrick.)

4.3-9. (a) What should be the shape (or size) of an aperture that acts as a *low-pass filter*, transmitting the low, and rejecting the high, space frequencies?

(b) Conversely, what should be the shape of an aperture that acts as a *high-pass filter*, transmitting the high, and rejecting the low, space frequencies?

4.3-10. Imagine that you are looking at the skyline of a large city. How much of the high-rise buildings, its contours, windows, scattered clouds, birds, and other detail do you see if a spatial filter is placed in the transform plane so as to transmit:

(a) Only the low space frequencies?

(b) All frequencies?

(c) Only the high frequencies?

4.4

Holography

Discoveries are often made because the time is right. Some discoveries have been made almost simultaneously yet independently. This is not so with *holography*. Holography is the discovery of one man, Dennis Gabor, at that time, 1948, working in a British industrial research laboratory. His discovery came at a time when the technology was not ready to take full advantage of it. It was several years later, with the invention of the laser, that Gabor's discovery reached its full potential.

Like other transforms, holography is a process of two-dimensional transformation. We deal with holography in a separate chapter for expediency, mainly because of the wealth of material generated by widespread interest.

Introductory Example

Let us discuss the basic concept of holography using a simple example. Light, as shown in Figure 4.4-1, top left, is incident on a point object and is diffracted by it, forming a series of concentric rings, a pattern of maxima and minima. The pattern is recorded photographically and made into a transparency. *In a second step* (top right), light is incident on the ring pattern and focused by it into a point, as if focused by a zone plate.

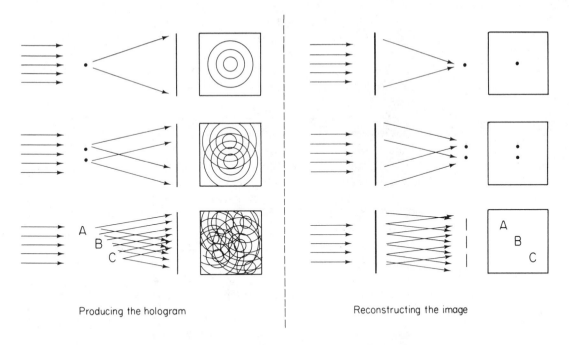

Figure 4.4-1 The principle of holography. *Top row*: Point object forming concentric diffraction rings as in zone plate; reconstruction of zone plate gives point image (*top right*). *Center row*: The same for two points; *bottom row*: for a more complex object.

Next (center row), imagine that the object consists of two points (pixels). The diffraction pattern then is two sets of concentric rings. In the second step each of these two sets focus, and thus the image consists of two points. Finally (bottom row), the object is an aggregate of many pixels, such as the letters ABC. The intermediate recording is an unrecognizable multiplicity of lines and rings. But since each set represents one pixel in the object, it focuses back again into one point (in the image), and the sum of all of these points is the ABC.

The essence of holography is that *the process of image formation is being interrupted and split into two:* In a first step the object is transformed into a photographic record, called the *hologram,* and in a second step, called *reconstruction,* the hologram is transformed into the image. No lens is needed in either step.*

_____

* Gabor had originally conceived of holography, or "wavefront reconstruction" as it was called then, as a means of correcting for the spherical aberration of the electron

Coherent background.   So far, it is easy to see that *hologra-phy is equivalent to two successive Fourier transformations,* and that it is very similar to Abbe's theory of microscopic image formation (page 425). The important difference is that in conventional image formation the two steps follow each other immediately, whereas in holography one is *de-layed.* But this poses a serious difficulty. A photographic emulsion, or any other detector, can record energies (intensities) but it cannot directly record the *phase* of the light. (Indirectly, the distortion of fringes gives us a clue as to phase changes.)

The solution can be found by returning briefly to the **ABC** example. In the first step of holography, each pixel in the object forms its own set of fringes. True, *within* each set the light interferes, but *between* sets there is no fixed phase relationship and hence no interference. To make the different *signals* compatible in phase, another wave, called the "coherent background" or *reference,* is added. If the coherent background is made significantly stronger than the signal, the resultant (phase) of both signal and reference becomes similar to that of the reference alone (Figure 4.4-2), and the contributions from different pixels interfere.

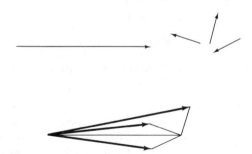

Figure 4.4-2   Adding strong background (*top left*) to random-phase signals (*top right*) gives similar resultant vectors (*bottom*).

---

microscope. Ironically, that goal has still not been reached. The term *holography* comes from the Greek $\overset{\varsigma'}{o}\lambda o\varsigma$ = all, whole and $\gamma\rho\alpha\varphi\epsilon\widehat{\iota}\nu$ = to write.

Dennis Gabor (1900–1979), Hungarian-born physicist, inventor, and philosopher. At age 14 Gabor read advanced physics texts, at 17 he wondered what happens to light as it travels from object to image. Gabor studied at the universities of Budapest and Berlin, became research engineer at the British Thomson-Houston Co. Ltd. in Bristol, professor of applied electron physics at the University of London, and later staff scientist at the CBS Laboratories in Stamford, Connecticut. His first publication on the subject was "A New Microscopic Principle," *Nature, London* **161** (1948), 777–78. Shortly thereafter he wrote "Microscopy by reconstructed wave-fronts," *Proc. Roy. Soc. London* **A197** (1949), 454–87, and "Microscopy by Reconstructed Wave Fronts: II," *Proc. Physical Soc.* **B64** (1951), 449–69, which both became classics in holography. In 1971, Gabor received the Nobel prize in physics for his discovery.

Example. *Using the size of the amplitude vectors drawn in Figure 4.4-2, what is:*
(a) The range of energies (intensities) possible?
(b) The contrast resulting from these energies?

Solution. Measuring the lengths of the vectors we find that the ratio of signal versus reference is $1:4.36$.

(a) The least possible amplitude (when signal and reference are out of phase, pointing in opposite directions) is $4.36 - 1 = 3.36$. The highest possible amplitude (when signal and reference are in phase, pointing in the same direction) is $4.36 + 1 = 5.36$. The amplitude ratio, $3.36/5.36$, is $1:1.6$ and the energy ratio is the square of that,

$$\left(\frac{3.36}{5.36}\right)^2 = \boxed{1:2.5}$$

(b) The contrast is

$$\frac{5.36^2 - 3.36^2}{5.36^2 + 3.36^2} = \boxed{0.44}$$

which is high enough to make the reconstruction visible.

Producing the Hologram

Based on these principles, holography developed further in several steps. Gabor had added the coherent background *coaxially*. This is easily done with objects that have enough open spaces between them, such as a wire mesh or opaque letters on a clear background. Signal and reference, in other words, travel in about the same direction.

In Leith and Upatnieks' *off-axis technique,* the reference beam is added at an offset angle. This made possible holography of *solid, three-dimensional objects.* * As we see from Figure 4.4-3, part of the light goes to a mirror and from there is reflected toward the film. This is the reference beam. The signal beam is scattered (diffracted) off the object and from there reaches the film. Both contributions combine and form fringes. Ordinarily, these fringes are very closely spaced and cannot be seen by the unaided eye, hence the typical hologram appears to be uniformly gray. Under the microscope, however, a hologram is seen to consist of a myriad

---

* See E. N. Leith and J. Upatnieks, "Wavefront Reconstruction with Continuous-Tone Objects," *J. Opt. Soc. Am.* **53** (1963), 1377–81; "Wavefront Reconstruction with Diffused Illumination and Three-Dimensional Objects," *J. Opt. Soc. Am.* **54** (1964), 1295–1301; and *Holograms,* U.S. Patent 3,894,787, July 15, 1975.

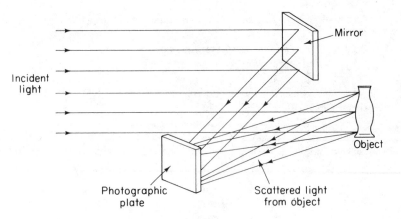

Figure 4.4-3 Recording the hologram.

of tiny "cells," each cell containing a series of fringes of various lengths and spacing.

At this point I emphasize again the fundamental difference between a hologram and a conventional photograph. In a photograph, the information is stored in an *orderly* fashion: each point in the object relates to a *conjugate* point in the image. In a hologram there is no such relationship; light from every object point goes to the entire hologram. This has two consequences. If the hologram were shattered, or cut into small pieces, each fragment would still reconstruct the whole scene, not just part of the scene (although the resolution, that is, the information content, would be less). Furthermore, since the hologram film is often larger in size than the object, it receives light not only from the side facing the film but also from the adjoining sides (which could not be seen in conventional photography). Thus, as the observer's head moves sideways in viewing the hologram, a truly *three-dimensional image* is seen.

A laser is not really needed for holography; it is merely the use of solid, three-dimensional objects that calls for light whose coherence length exceeds the path differences due to the unevenness of such objects.

There are many good descriptions of how to make, and view, your own hologram, useful both in the optics lab and suitable even as a lecture hall demonstration. See for example T. H. Jeong's article, listed in the Suggestions for Further Reading at the end of this chapter.

Reconstruction

Holography is a matter of *interference*. The interference occurs between the amplitude of the signal, A_1, scattered by the object, and the amplitude of the reference, A_2, reaching the emulsion directly (Figure

4.4-4). But photographic emulsions respond only to *energies;* thus instead of $A_1 + A_2$, we have

$$Q(y, z) = (A_1 + A_2)^2$$
$$= (A_1 + A_2)(A_1 + A_2)*$$
$$= |A_1|^2 + |A_2|^2 + A_1A_2^* + A_1^*A_2 \qquad [4.4\text{-}1]$$

where the asterisk * indicates the complex conjugate.

In the reconstruction process, light of amplitude A_3 is used to illuminate the hologram. The hologram, that is, the emulsion with the interference pattern recorded on it, has a transmittance function $T(y, z)$ and, since the hologram has been produced by the amplitudes A_1 and A_2,

$$T(y, z) = A_1A_2^* + A_1^*A_2 \qquad [4.4\text{-}2]$$

As shown in Figure 4.4-5, the A_3 light is diffracted (*modulated*) by the hologram

$$A_4 = A_3T(y, z) \qquad [4.4\text{-}3]$$

Note that in Equation [4.4-1] it is only the two right-hand terms that carry the information because only *they* are the result of any interference between signal and reference. Hence, if we consider merely these terms, then

$$A_4 \propto A_3A_1A_2^* + A_3A_1^*A_2 \qquad [4.4\text{-}4]$$

If the reconstruction amplitude A_3 is equal, or at least proportional, to the reference amplitude A_2, the amplitudes A_2 and A_3 in Equation [4.4-4] cancel and

$$A_4 \propto A_1 \qquad [4.4\text{-}5]$$

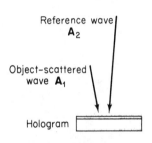

Figure 4.4-4 Two wave contributions forming hologram.

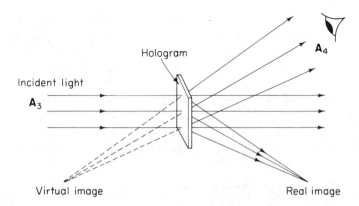

Figure 4.4-5 Reconstruction. The hologram diffracts the incident light A_3, producing a real and a virtual image.

This means that the amplitude diffracted by the hologram is proportional to the amplitude initially diffracted by the object: *The image is a reconstruction of the object*.

If A_2 and A_3 are plane waves, the virtual and the real image have the same magnification. The two images are mirror images, with the hologram in the plane of symmetry. This is an example of a *Fraunhofer* or *Fourier transform* hologram. On the other hand, if the hologram has been recorded with divergent light, the real image is magnified and the virtual image reduced in size. This is an example of a *Fresnel* hologram. A Fresnel hologram has focusing properties, as shown in Figure 4.4-5. Such a hologram produces without an additional lens a real and a virtual image, the same as a Fresnel zone plate which has both positive and negative focal lengths.

In either case, the observer looking into the A_4 beam will find this beam hard to distinguish from the A_1 beam. Indeed, the image seen in holography looks startlingly similar to the object, including its three-dimensionality and the parallax that is typical of the real world. It is this realism that has made holography so fascinating to both scientists and laypersons.

Practical considerations.   Holograms must be recorded on photographic plates or film of high resolvance.* Look again at Figure 4.4-3 and note that the reference, the light reflected by the mirror, and the signal, the light scattered by the object, subtend a certain angle at the photographic emulsion. If this angle is too large, more than a few degrees, the fringes formed between signal and reference are so closely spaced that even the best emulsion cannot resolve them. The experimental setup must also be stable and free from vibrations to prevent the fringes from smearing during exposure.

Reconstruction always yields a photographic positive. If a contact print were made of a hologram and, hence, the blacks and whites reversed, the image would still look the same, just as the "negative" of a simple diffraction grating would show the same spectrum lines as the original "positive."

Holograms can cover a full 360°. The film, as illustrated in Figure 4.4-6, completely surrounds the object. As before, part of the light falls on the object and is scattered from there toward the film. The reference part falls on the film directly (or it reaches the film by way of a convex mirror

---

* Photographic emulsions suitable for holography are Kodak Holographic Film type EK 120, SO-173, SO-253, and SO-343, and Agfa-Gevaert Holotest plates, 8E75HD or 10E75.

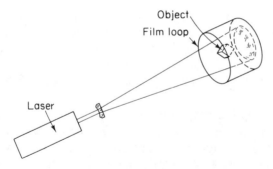

Figure 4.4-6   360° holography.

inside, near the end of the tube). When looking at the film loop, after processing, the image can be seen from any direction.*

In an emulsion that is sufficiently *thick,* wavefronts traveling in one direction can be made to interfere with wavefronts traveling in the opposite direction. The wavefronts then form a three-dimensional standing-wave pattern, rather than the two-dimensional pattern of a conventional hologram. If viewed in white light, such *volume holograms* give reconstructions in full color.

Applications of Holography

There are many aspects to holography. Its influence on photography, microscopy, astronomy, pattern recognition, and even art has only begun to bear fruit. Most important of all is that holography has provided us with *conclusive proof that any image formation is a two-step process.* Whereas in conventional image formation these two steps follow each other without delay, in holography they can be separated, conceptually as well as experimentally.

Holographic interferometry.   Testing for stresses, strains, and surface deformations is one of the most useful practical applications so far. In the *double-exposure technique* two exposures are made of the object, one before loading and the other after. The original object and the object after deformation are recorded holographically on the same plate. When the hologram is reconstructed by illuminating it with the reference wave, both images are viewed simultaneously and, since they are slightly different, the two images interfere. Thus any distortions of the object will

---

*T. H. Jeong, "Cylindrical Holography and Some Proposed Applications," *J. Opt. Soc. Am.* **57** (1967), 1396–98.

show in the form of fringes. In the *continuous-exposure* or *real-time technique* the object keeps moving during exposure (such as the vibrating body of a violin). This produces interference by light coming from different positions of the object during motion and, likewise, generates fringes. These fringes represent contours of constant amplitude of the vibration.

Holographic microscopy.

In contrast to a conventional high-power microscope, a *holographic microscope* has an appreciable depth of field; in fact, it need not be focused at all. As before, the light is split into two (Figure 4.4-7). One beam is passing through the specimen and through the microscope; the other beam is led around it. The two beams recombine, producing the hologram.

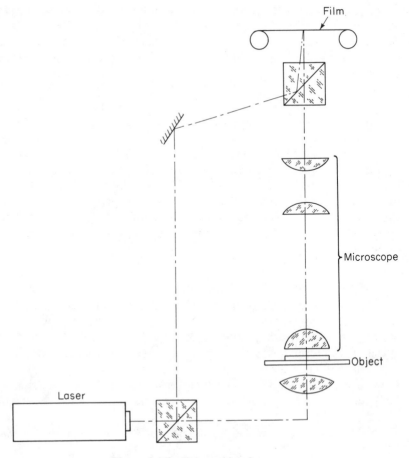

Figure 4.4-7 Holographic microscope.

The reconstructed image can be viewed in any cross section desired. The observer merely looks at the cross section he or she wishes to see, moving back and forth throughout the depth of the image, without the object being present anymore at all.

Information storage.
Information can be stored and retrieved more efficiently in the form of holograms than in the form of real images. Perhaps this is how information, then called *engrams,* is stored in the brain; at least it would help explain why attempts to locate certain "centers" in the brain have never met with much success and why major brain injury often does not lead to predictable, circumscribed defects.

Similar considerations apply to technical information storage. Storing images on microfilm is subject to errors because of misfocusing and contamination. Information storage by holography, either on photographic film or in single crystals of lithium niobate, $LiNbO_3$, does not have these problems.

Acoustic holography.
In our daily life there is no evidence for images formed by sound. But actually, such images should exist and, indeed, holography has made them possible because it requires no lenses.

There are different ways to realize acoustic holography.* In one of them, two sound generators, as in Figure 4.4-8, emit the reference and the signal, respectively. On a calm surface of water, these two contributions produce ripples. The ripple pattern *is* the hologram. The pattern may be

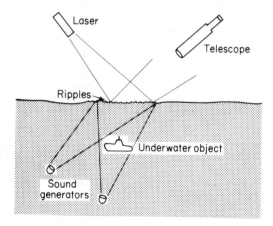

Figure 4.4-8   Acoustic holography.

---

*See P. Greguss, "Acoustical holography," *Physics Today* **27** (Oct. 1974), 42–49.

photographed and then reconstructed, or reconstruction may be made in real time by coherent light reflected off the surface as shown.

Pattern recognition.
One of the most exciting applications of holography is *pattern recognition,* also called *character recognition* in reference to alphanumeric characters. Early pattern recognition systems were based on geometric optics. Consider, for example, that we want to read the letter A (Figure 4.4-9). A set of characters, A, B, C, . . . , are printed as negatives on a strip of film and this film is moved through the image plane. If the character to be read matches the character on the film, the output from a photodetector is zero, triggering a printer. But in reality this does not work: the character and the negative must be aligned perfectly, both in position and size, an unrealistic requirement.

Instead, modern pattern recognition systems are based on holography. In place of a mask containing a real image of the letter A, we now use the *Fourier transform,* the hologram, of the A.

As in holography in general, the Fourier transform hologram of the A is the result of a superposition of two sets of wavefronts, the signal and the reference. The signal is diffracted by an original A and the reference is a beam of collimated light. Subsequently, when the hologram of the A is illuminated by collimated light, an A is reconstructed. When the transform is illuminated with light from another A, plane wavefronts result that can be focused *into a bright spot* (Figure 4.4-10, top). The spot is easily

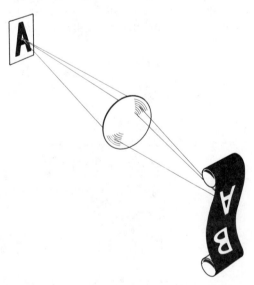

Figure 4.4-9 Pattern recognition based on geometric optics.

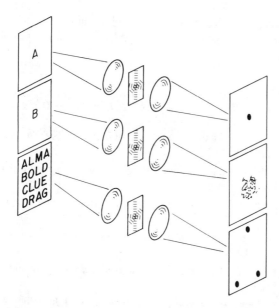

Figure 4.4-10 Pattern recognition by holographic *Vander Lugt filter* (the filters are seen between the lenses).

recognized by the eye, by photography, or photoelectrically. On the other hand, if the wavefronts are coming from a B, or from some other character, they do *not* transform into perfectly plane wavefronts and do *not* produce a focused spot but rather a diffuse patch of light (center). Hence, we can search a given matrix of characters and determine whether a particular character is present (bottom).

The holograms shown in Figure 4.4-10 appear to be amplitude filters. But because they are generated by interference between signal and reference, they represent in fact both the amplitude and the phase of the light. They are called "complex," "matched," or *Vander Lugt filters.* *

Plenty of difficulties still lie ahead before true *reading machines* can become a reality. Some letters and words are "inside" others. For example, F is inside E, P is inside R and B, T and L have the same horizontal and vertical lines, and *arc* is inside *search*. Clearly, the more alike the two characters, the less the power of discrimination. A major problem, also, will be to teach the machine to recognize the "meaning" of a letter set in

---

*The term "complex" is used, not because of the complexity of making them, but because of the analogy to complex numbers, which also carry both amplitude and phase information. The term "matched" comes from radar technology but now refers also to light coming from an identical, "matched" character. Filters of this type were first made by A. Vander Lugt, "Signal Detection by Complex Spatial Filtering," *IEEE Trans. Info. Theory* **IT-10** (1964), 139–45.

different typeface; the letter A, for example, can be printed A, **A**, 𝔄, 𝒜, a, **a**, ɑ, α, and an almost infinite number of variations is possible when it comes to *handwriting*.

SUGGESTIONS FOR FURTHER READING

T. H. Jeong, "The one-beam transmission hologram," *The Physics Teacher* **19** (1981), 129–33.

E. N. Leith, "White-Light Holograms," *Scientific American* **235** (Oct. 1976), 80–95.

H. M. Smith, *Principles of Holography,* 2nd edition (New York: John Wiley & Sons, Inc., 1975).

H. J. Caulfield, editor, *Handbook of Optical Holography* (New York: Academic Press, Inc., 1979).

N. Abramson, *The Making and Evaluation of Holograms* (London: Academic Press, Inc., 1981).

C. M. Vest, *Holographic Interferometry* (New York: John Wiley & Sons, Inc., 1979).

Y. I. Ostrovsky, M. M. Butusov, G. V. Ostrovskaya, *Interferometry by Holography* (New York: Springer-Verlag, 1980).

B. P. Hildebrand and B. B. Brendon, *An Introduction to Acoustical Holography* (New York: Plenum Publishing Corp., 1974).

PROBLEMS

4.4-1. If the amplitude vectors representing the signal and the reference, when they combine to form the hologram, have a ratio of 1:4, what is the greatest possible phase difference between them? Express as a fraction of wavelength.

4.4-2. Following an example by Gabor, assume that the amplitudes of the signal and the reference are related as 1:10. Since the two beams when they combine may be completely *in* phase, or completely *out of* phase, what is the maximum ratio of their energies (intensities)?

4.4-3. If the angle subtended at the hologram by the signal and the reference is 15°, what is the spacing of the fringes provided the wavelength is 492 nm?

4.4-4. Holograms are taken with light of 632.8 nm wavelength. What is the limiting angle between the signal and the reference if the space frequency in the hologram is not to exceed 200 lines/mm?

4.4-5. Reconstruction is made of a hologram that is part of a zone plate, *not* including its center. If the hologram is viewed in collimated light, show where the real and the virtual image will be formed.

4.4-6. A projectile is fired through a wind tunnel placed in one arm of a Mach–Zehnder interferometer. If the fringe pattern were "reconstructed," what would it show?

4.4-7. Consider a three-dimensional object and image and discuss the essential difference between *stereoscopic* and *holographic* image formation.

4.4-8. The theoretical storage capacity of a hologram is one bit of information per λ^3 of volume and the storage *density* n^3/λ^3. Therefore, how many bits can be stored in 1 mm³ if the index is $n = 1.483$ and the wavelength $\lambda = 546$ nm?

4.4-9. A pattern recognition system is built as a reading-aid for the blind, using a hand-held probe as the reading head. How critical is it to follow printed material exactly line by line?

4.4-10. Continuing with Problem 4.4-9, what happens when the reading head is correctly aimed at the character to be read and then is slightly *twisted* (turned around its own axis)?

Part 5

Quantum Optics

ALMOST THREE CENTURIES AGO, Isaac Newton thought that a light source sent out a stream of minute particles and that diffraction and interference occurred because of attractive and repulsive forces exerted on them by an obstacle. Wave theory laid these ideas to rest, at least for a while. Today we know that electromagnetic energy is indeed generated, emitted, transmitted, and received in the form of discrete units of energy, called *quanta*.

No introduction to optics can be complete without reference to the quantum aspect of light. The laser cannot be explained without it. What is even more important is the completeness and beauty of the system of optics which we discover through understanding phenomena such as atomic spectra, absorption, fluorescence, and lasers, all of which are based on the concepts of *quantum optics*.

5.1

Light as a Quantum Phenomenon

A BEAM OF INTENSE LIGHT, it seems, should cause more of any effect than a beam of dim light. But this is not always so. The wavelength of the light plays an important role. In fact, there are certain limits beyond which there are no effects at all, no matter what the intensity. An example is the *photoelectric effect*.

The Photoelectric Effect

Experimental observations. Consider the circuit shown in Figure 5.1-1. Light falls on an evacuated tube and is incident on a plate M. This plate is made out of a certain material which, when irradiated, releases electrons, then called *photoelectrons*. Opposite M is another plate, the collector plate C. If C is made *positive* with respect to M, the photo-electrons released by M are attracted by, and travel to, C. As the potential V, read on an high-impedance voltmeter, is increased, the current, i, read on an ammeter, increases too, but only up to a given *saturation level*, because then all of the electrons emitted by M are collected by C.

On the other hand, if C is made *negative*, some photocurrent will still exist, provided that the electrons ejected from M have enough kinetic energy to overcome the repulsive field at C. But as C is made more

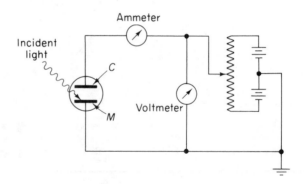

Figure 5.1-1 Photoelectric effect.

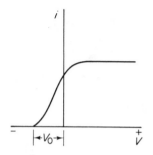

Figure 5.1-2 Dependence of photocurrent on potential applied to collector plate.

negative, a point is reached where *no* electrons reach *C* and the current drops to zero. This occurs at the *stopping potential,* V_0 (Figure 5.1-2). In short: A significant amount of photocurrent is present only if the collector, *C,* is made positive, that is, if it is the *anode.* The light-sensitive surface *M,* then, is the photo*cathode.*

From the classical point of view we expect that:

1. The maximum kinetic energy of the photoelectrons will depend on the amount of light falling on the photocathode; that is, it will depend on the *areance* (irradiance). Experimental observation shows that this is not so. The maximum kinetic energy of the electrons and, since KE $= \frac{1}{2}mv^2$, their maximum velocity, are independent of the areance. True, at higher areance *more* electrons will be ejected but their energies and velocities remain the same, no matter how much light reaches the photocathode.

2. Instead, the maximum energy of the photoelectrons turns out to be proportional to the *frequency:* blue light produces more energetic photoelectrons than red light.

3. There seems to be *no time lag,* at least not more than about 10^{-8} s, between the onset of irradiation and the resulting photocurrent.

These facts are hard to reconcile. Conceivably, the energy imparted to the electrons could come from either of two sources: It could come from the light (but that would mean higher energies with higher areance and presumably some time delay for very low areance). Or it could come from heat energy stored in the material, the light merely acting as a trigger (but we do not find higher energies as the surface is heated).

Planck's constant. The answer was found by Einstein, following a suggestion by Planck.* Planck reasoned that the atoms or mole-

---

*Max Karl Ernst Ludwig von Planck (1858–1947), German physicist, professor of theoretical physics at the University of Berlin, and president of the then Kaiser Wilhelm

cules that make up a surface act as electromagnetic dipole oscillators. These oscillators then emit energy, not in the form of a smooth continuous flow, but as a stream of discrete units, each called a *"quantum of energy"* or *quantum,* for short, or, in the case of light, a *photon.**

According to Planck, the least amount of energy, E, that can be emitted is given by

$$E = h\nu$$ [5.1-1]

where h is a factor of proportionality, since called *Planck's constant,* and ν is the frequency of oscillation. The numerical value of Planck's constant, to the best of our current knowledge, is

$$h = (6.626\,176 \pm 0.000\,036) \times 10^{-34} \text{ J Hz}^{-1}$$ [5.1-2]

where the unit, joule/hertz, can also be written Js, joule-second. Planck's constant must be determined experimentally; it cannot be predicted from theory. It has been measured in a variety of ways, and all of these measurements agree.

Example. *What is the energy of a quantum of ultraviolet radiation of 340 nm wavelength?*

Solution. Solving $v = \lambda\nu$, Equation [1.1-7], for ν, replacing v by c, and substituting in Equation [5.1-1] gives

$$E = h\frac{c}{\lambda}$$ [5.1-3]

and therefore

$$E = (6.63 \times 10^{-34} \text{ Js})\frac{3 \times 10^8 \text{ m s}^{-1}}{340 \times 10^{-9} \text{ m}} = \boxed{5.85 \times 10^{-19} \text{ J}}$$

Gesellschaft, now renamed Max Planck Gesellschaft. In 1900, Planck introduced the concept of discrete oscillators to account for the shape of the spectrum of radiation from an incandescent source. Acceptance of this idea was slow at first; Planck himself resisted it for several years. "My futile attempts to fit the elementary quantum of action somehow into the classical theory continued for a number of years," he wrote, "and they cost me a great deal of effort." M. Planck, "Ueber irreversible Strahlungsvorgänge," *Ann. Physik* (4) **1** (1900), 69–122. In 1918 Planck received the Nobel prize for physics.

In 1905, Albert Einstein applied Planck's theory to the photoelectric effect, postulating that light is not only *emitted* but also *received* in the form of quanta of energy. A. Einstein, "Über einen die Erzeugung und Verwandlung des Lichtes betreffenden heuristischen Gesichtspunkt," *Ann. Physik* (4) **17** (1905), 132–48. This paper by Einstein on the photoelectric effect is much more radical than the one he wrote in the same year on the theory of relativity. In fact, it was this paper that earned him the Nobel prize. For Einstein biography, see chapter 6.1, page 525.

*The term quantum comes from the Latin *how much.* The plural is *quanta.*

Work function.   It is often convenient to measure energies on an atomic scale not in joule but in a unit called *electron volt*, eV. By definition, 1 electron volt is the amount of (potential) energy acquired, or lost, by an electron moving through a potential difference of 1 volt:

$$1 \text{ eV} = (1e)(1V)$$

Since the charge on the electron, e, is very nearly 1.60×10^{-19} C,

$$1 \text{ eV} = 1.60 \times 10^{-19} \text{ J} \qquad [5.1\text{-}4]$$

which applies to energies in any form, including kinetic energy.

The more intense the light, the greater the number of quanta. However, from Equation [5.1-1], light of low frequency (red light) has low-energy quanta. Light of high frequency (blue light) has high-energy quanta, even though, at low intensities, there may be only a few quanta per unit of time. Some examples are shown in Table 5.1-1.

Solving Equation [5.1-3] for λ and substituting Equation [5.1-4] leads to

$$\lambda = \frac{hc}{E} = \frac{(6.63 \times 10^{-34} \text{ Js})(3 \times 10^8 \text{ m/s})}{1.6 \times 10^{-19} \text{ J/eV}}$$

$$= \frac{1240 \text{ nm eV}}{E} \qquad [5.1\text{-}5]$$

where the wavelength is given in nanometers and the energy in electron volts.

We return to the photoelectric effect. Part of the energy contained in an incident quantum, $h\nu$, is required to free the electron and let it escape from the surface. This part is called the *work function*, $\mathcal{W}$. It is only the excess energy, beyond what is used up as work function, that appears as kinetic energy (of the electron). The maximum kinetic energy with which the photoelectron can escape, therefore, is

$$\boxed{KE_{max} = h\nu - \mathcal{W}} \qquad [5.1\text{-}6]$$

which is *Einstein's photoelectric effect equation.*

Table 5.1-1 Quantum Energy (in eV) in Various Regions of the Electromagnetic Spectrum

Infrared	1 eV
Visible	2.2 eV
Ultraviolet	5 eV
X rays	10^4 eV
γ rays	10^7 eV (10 MeV)

From Einstein's equation we conclude that no electrons can escape unless $h\nu$ is greater than $\mathcal{W}$. The kinetic energy, and hence the velocity, of the electrons depends not only on the frequency of the light but also on the work function of the surface.

Example 1. *(a) If the threshold wavelength for rubidium is 582 nm, what is its work function, in eV?*
(b) With blue light of 475 nm wavelength incident, what is the maximum kinetic energy of the photoelectrons, also in eV?
(c) What is the stopping potential?

Solution. (a) At the threshold, no photoelectrons are released and Equation [5.1-6] becomes zero:

$$\mathrm{KE}_{\mathrm{max}} = h\nu - \mathcal{W} = 0$$

$$\mathcal{W} = h\nu = h\frac{c}{\lambda}$$

Converting into eV, using Equations [5.1-1] and [5.1-5], gives

$$\mathcal{W} = \frac{1240 \text{ nm eV}}{\lambda} = \frac{1240}{582} = \boxed{2.13 \text{ eV}}$$

(b) Since the energy of a photoelectron of 475 nm wavelength, also from Equation [5.1-5], is

$$E = \frac{1240 \text{ nm eV}}{475} = 2.61 \text{ eV}$$

the maximum kinetic energy, from Equation [5.1-6], is

$$\mathrm{KE}_{\mathrm{max}} = h\nu - \mathcal{W} = 2.61 - 2.13 = \boxed{0.48 \text{ eV}}$$

(c) The stopping potential, then, is

$$V_0 = \boxed{0.48 \text{ V}}$$

The term eV, as we have seen, is a unit of energy. Hence, it can be set equal to $h\nu$, and

$$\frac{V_0}{\nu} = \frac{h}{e} \qquad\qquad [5.1\text{-}7]$$

If now the stopping potential, V_0, is plotted versus the frequency, ν, as in Figure 5.1-3, we obtain a straight line which has the slope h/e. The line need not go through zero; any displacement, in fact, is due to the work function of the particular surface. For different work functions, we obtain different lines but the slope of these lines is the same. This suggests that h/e is a fundamental property of light, as indeed it is. The intercept of a

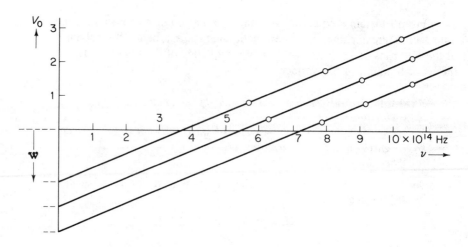

Figure 5.1-3 Plots of stopping potential, V_0, versus frequency, ν, for three surfaces with different work functions.

given line with the y axis tells us the (negative value of the) work function of the surface, and the intercept with the x axis, where $V_0 = 0$, the *threshold frequency* of the surface.

Example 2. *Assume that radiation of 250 nm wavelength is incident on a surface of unknown work function and that at that wavelength the stopping potential is 2.60 V. At 375 nm wavelength the stopping potential is 0.94 V. Based on these figures, what is Planck's constant?*

Solution. First we convert the wavelengths into frequencies:

$$\nu_1 = \frac{c}{\lambda} = \frac{3 \times 10^8}{250 \times 10^{-9}} = 1.2 \times 10^{15} \text{ Hz}$$

$$\nu_2 = \frac{3 \times 10^8}{375 \times 10^{-9}} = 0.8 \times 10^{15} \text{ Hz}$$

Then we replace V_0 in Equation [5.1-7] by the difference of the two stopping potentials and ν by the difference of the two frequencies,

$$\frac{V_1 - V_2}{\nu_1 - \nu_2} = \frac{h}{e} \qquad\qquad [5.1\text{-}8]$$

so that

$$h = \frac{(V_1 - V_2)(e)}{\nu_1 - \nu_2} = \frac{(2.60 - 0.94)(1.6 \times 10^{-19})}{(1.2 - 0.8) \times 10^{15}}$$

$$= \boxed{6.64 \times 10^{-34} \text{ J/Hz}}$$

Applications of the photoelectric effect.   There are two types of photo-detectors, *thermal detectors* and *quantum detectors*. Thermal detectors are based on absorption, rather than on the photoelectric effect, and do not have the wave-length dependency typical of the latter. As radiant energy is absorbed by the detector, it raises the temperature of the material, which in turn changes its electric resistivity, produces a current (*thermocouples*), or alters its capacitance.

Quantum detectors respond to quanta, rather than to quantum energies (heat): an incident quantum transfers its energy directly to an electron, without much heating. Consequently, the response of a quantum detector is generally faster than that of a thermal detector (although for maximum efficiency a quantum detector must be *cooled*). Examples are emission photodetectors such as *diode phototubes* and *photomultipliers*; they usually contain cesium or potassium cath-odes, which have low work functions. In spectral regions where their quantum efficiency is high (below 550 nm) photoemitters are very nearly ideal; their sensi-tivity is high enough to count individual photons.

Photoconductive detectors are thin-film devices where each incident quan-tum releases an electron–hole pair, thereby increasing the electric conductivity. They are most useful above 1 μm, where efficient photoemitters are not available.

Photovoltaic detectors or *photodiodes* are *p-n* junction semiconductors (see Figure 5.4-12). They respond to radiation from the near UV, through the visible, into the far IR; their response time can be as short as a few nanoseconds, ns. Within limits, the photocurrent produced, i, is linearly proportional to the power of the incident light, ϕ, the ratio of both being called the *photoelectric sensitivity,* γ:

$$\gamma = \frac{i}{\phi} \qquad\qquad [5.1\text{-}9]$$

But not every quantum incident causes a photoelectron to be released, even if the quantum energy is above the threshold. This ratio is the *quantum efficiency,* η:

$$\eta = \frac{\text{number of photoelectrons released}}{\text{number of light quanta received}} \qquad [5.1\text{-}10]$$

The most widely used photodiodes contain a thin transparent layer of cadmium oxide deposited on a layer of selenium. These detectors require nothing more than a current indicator; they are used in photographic exposure meters.

De Broglie Waves

The next step after the introduction of quanta was made by Louis de Broglie.* De Broglie thought that if light is acting at some times as waves and at others as particles, perhaps "real" particles would show some wave behavior too.

---

*Louis Victor Pierre Raymond Duc de Broglie (1892-), French physicist. After graduating from the Sorbonne with a degree in medieval history, de Broglie, pronounced "d' Bro'lie," turned to science. He wrote a thesis on quantum theory, became professor at the Sorbonne, and in 1929 won the Nobel prize in physics. L. de Broglie, "A Tentative Theory of Light Quanta," *Phil. Mag.* (6) **47** (1924), 446–58.

We have seen that $E = h\nu$. We now take Einstein's *mass–energy relation*,

$$E = mc^2 \qquad\qquad [5.1\text{-}11]$$

(which will be derived in Chapter 6.1, page 540) and combine both equations. This gives

$$h\nu = mc^2$$

Replacing ν by v/λ and c by v leads to

$$h\frac{v}{\lambda} = mv^2$$

and thus

$$\lambda = \frac{h}{mv} \qquad\qquad [5.1\text{-}12]$$

The waves themselves are called *de Broglie waves* and the wavelength *de Broglie wavelength*. The product, mv, is the *momentum*, p, of the particles. Together with Equation [5.1-1] we then have two postulates,

$$\boxed{E = h\nu} \qquad \text{and} \qquad \boxed{p = \frac{h}{\lambda}} \qquad [5.1\text{-}13]$$

known as the *Einstein–de Broglie relations*. Note that ν and λ on the right-hand sides are typical wave parameters, while the left-hand terms, E and p, are typical of particles.

De Broglie's hypothesis has been extensively tested and completely verified. It applies to all particles, of any size, even to a 10-ton truck, although *its* diffraction pattern can probably never be observed. But electrons, neutrons, and similarly small particles, when incident on a crystal of the proper interplanar spacing, are diffracted much the same as X rays are diffracted—which shows that such "real" particles have wave properties as well.

The Uncertainty Principle

To the casual investigator, it might appear that measurements could be made so delicate that they would help us decide whether light consists of quanta or waves. This is not so. There is an inherent interaction between the process of measuring and the phenomenon being measured which frustrates any such attempt. This is the essence of *Heisenberg's uncertainty principle.* *

Consider again the double-slit experiment discussed in detail in Chapter 2.2, page 194. As shown in Figure 5.1-4, the light comes from a source on the left, passes through two slits, *A* and *B*, and falls on a photographic film on the right. After development, the maxima on the film negative will appear *black,* the minima *white.* Using the wave theory of light, we have no difficulty explaining such fringes. Similarly, a sequence of maxima and minima results with electrons; we only need to substitute in the double-slit equation the wavelength of electrons as derived from de Broglie's relationship.

But now we come to a puzzling dilemma. Point *C* in Figure 5.1-4 is at a minimum, that is, at a location where, under the rules of double-slit interference, no electrons could ever be. However, if either *A* or *B* is covered so that the passage through that slit is blocked, the situation changes and some electrons *do reach C,* because the central maximum in what is now *single-*slit diffraction is fairly broad and extends beyond the first-order minima in *double*-slit interference.

Perhaps, as the electrons pass through the double slits they influence each other and form the typical double-slit pattern (nothing at *C*). But even if we reduce the electron flux, or the flux of light, to a level so low that only *one* electron—or *one* quantum—at a time passes through, the result

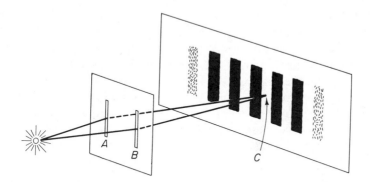

Figure 5.1-4   Double-slit interference. Location *C* is that of a *minimum.* Lateral distances exaggerated.

---

*Werner Curt Heisenberg (1901–1976), German physicist, teacher, and philosopher. Heisenberg studied under Sommerfeld, Born, and Bohr; he formulated quantum mechanics, quantum field theory, and the uncertainty principle named after him. He became professor of physics at Leipzig, director of the Max Planck Institutes in Berlin, Göttingen, and Munich, and received numerous honors, including the Nobel prize in physics in 1932.

is still the same: *No* quantum will ever reach *C* if *both A* and *B* are open. But a quantum *can* reach *C* if *either A or B* is open. In short, a quantum that has passed through *B* can reach *C* if *A* is closed, but it cannot reach *C* if *A* is open.

How can the quantum, passing through *B,* know whether *A* is closed or open? If we place a detector near *A,* the double-slit fringes disappear: The mere presence of the detector, even if it does not block the slit, changes the result from double-slit interference to single-slit diffraction. Many physicists, including Einstein, have tried to think of an experiment to show which slit a quantum has gone through, without destroying the interference—without success. The current view is that the quanta are represented by *fields,* entities that travel through space in the form of waves but carry energy, mass, charge, and other parameters ordinarily attributed to particles.

Imagine a sequence of quanta, all of energy $h\nu$, moving in the $+x$ direction. To isolate a single quantum, we try using a fast shutter that opens for only a very short interval of time, Δt. It seems that we can thus isolate a wavetrain of length $c\,\Delta t$, traveling at the speed of light, c (Figure 5.1-5).

Suppose that we detect the light with a photocell. But there is only one "point in time" at which the quantum makes the detector click. How can we reconcile the *finite length* of time Δt and the *point* in time? We have to assume an *uncertainty in position* of the quantum

$$\Delta x = c\,\Delta t \qquad\qquad [5.1\text{-}14]$$

Next consider that the light is passed through a spectroscope in order to determine its frequency. If we open the shutter instantaneously, without time delay, the leading edge of the light would be square and would abruptly rise from zero to maximum (power). But a square edge is equivalent to the sum of an infinite number of Fourier elements. This requires additional frequencies to be present and, therefore, requires an *uncertainty in frequency,* $\Delta\nu$, also.

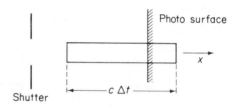

Figure 5.1-5   Position uncertainty.

From Equation [1.1-7] we know that

$$\frac{v}{\nu} = \lambda$$

But now, replacing v by c and substituting the uncertainties of both ν and x, we have

$$\frac{c}{\Delta\nu} = \Delta x \qquad\qquad [5.1\text{-}15]$$

Combining this equation with Equation [5.1-14] gives

$$\boxed{\Delta\nu\,\Delta t = 1} \qquad\qquad [5.1\text{-}16]$$

which is the *uncertainty principle*. Replacing ν in Equation [5.1-15] by c/λ and substituting, from Equation [5.1-13], h/p for λ then leads to

$$\boxed{\Delta x\,\Delta p = h} \qquad\qquad [5.1\text{-}17]$$

This shows that whereas we can determine *either* the position *or* the momentum of a quantum, we cannot determine *both* at the same time.

Phase space. Extending Equation [5.1-17] to three dimensions, we find that

$$\Delta x\,\Delta p_x = h; \qquad \Delta y\,\Delta p_y = h; \qquad \Delta z\,\Delta p_z = h$$

Thus, by multiplication,

$$(\Delta x\,\Delta y\,\Delta z)(\Delta p_x\,\Delta p_y\,\Delta p_z) = h^3 \qquad\qquad [5.1\text{-}18]$$

Based on classical physics, we may be tempted to describe the properties of a particle by a *point* in six-dimensional space, using three coordinates for space and three for momentum. But the uncertainty principle does not permit such description. Instead, a quantum can be represented only by a *six-dimensional hypercube*, called a *phase cell,* of volume h^3. Naturally, it is hard to visualize six-dimensional space; Figure 5.1-6, at least, illustrates the idea for two dimensions.

The plot is that of one space coordinate versus one momentum coordinate. If the momentum is known, as in the photoelectric effect, the position cannot be (phase cell 1). If the position is known, the momentum cannot be (2). Position and momentum together can be determined only to a certain degree (3). Thus a phase cell has no definite shape; it has only a certain (six-dimensional) volume.

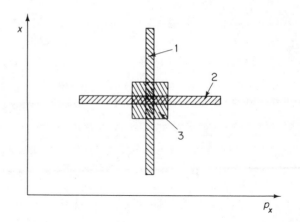

Figure 5.1-6   Two-dimensional projection of a phase cell. Actually, cells 1 and 2 should be infinitely long and infinitesimally narrow.

The principle of complementarity.   The "dualistic nature of light" is a term that has intrigued many writers and implies some kind of dichotomy. I take a more *unitary* view and think of light as something *unique*. Light is neither a true wave (its oscillations come in finite wavetrains, like no other wave) nor is light made up of conventional particles (quanta can only move at the speed of light, while "real" particles can have any speed less than c). Light's wave nature and quantum nature are not mutually exclusive; they are, as Bohr put it, *complementary*.

Light is both, quanta and waves. The quanta contain the energy. The waves guide them. At shorter wavelengths, and in the emission and absorption of light, the quantum aspect becomes predominant. At longer wavelengths, and in diffraction and interference, the wave aspect becomes predominant. But wavelength is not the divisive factor: Light of the same wavelength can act as waves in one experiment and as quanta in another.

SUGGESTIONS FOR FURTHER READING

A. P. French and E. F. Taylor, *An Introduction to Quantum Physics* (New York: W. W. Norton & Company, Inc., 1978).

F. J. Bockhoff, *Elements of Quantum Theory,* 2nd edition (Reading, MA: Addison-Wesley Publishing Company, Inc., 1976).

W. Heitler, *The Quantum Theory of Radiation,* 3rd edition (London: Oxford University Press, 1954).

R. H. Kingston, *Detection of Optical and Infrared Radiation* (New York: Springer-Verlag, 1978).

L. Levi, *Applied Optics, A Guide to Optical System Design,* Volume **2,** Chapter 16, "Photoelectric and Thermal Detectors," pp. 441–609 (New York: John Wiley & Sons, Inc., 1980).

PROBLEMS

5.1-1. Determine the wavelength of a quantum whose energy is 2.48 eV.

5.1-2. Express Planck's constant in terms of electron volts.

5.1-3. What is the energy, in eV, of a quantum of:
(a) Infrared of 2.0 μm wavelength?
(b) Ultraviolet of 250 nm wavelength?

5.1-4. If a ruby laser ($\lambda = 694$ nm) delivers a pulse of 0.29 J energy, how many photons are contained in that pulse?

5.1-5. If cesium has a work function of 1.9 eV, what is the longest wavelength (cutoff wavelength) that could produce any photoelectrons?

5.1-6. What is the work function, in eV, of a surface from which 540 nm light could eject photoelectrons of 0.2 eV maximum kinetic energy?

5.1-7. A photon of 332 nm wavelength strikes a potassium surface ($\mathcal{W} = 2.25$ eV). What is the maximum velocity of the photoelectrons that come off the surface? (Mass of electron: 9.1×10^{-31} kg.)

5.1-8. Light of 400 nm wavelength hits a surface that has a work function of 2.48 eV. What is:
(a) The highest kinetic energy, in joules, of the electrons ejected from the surface?
(b) The cutoff wavelength of the light?

5.1-9. A certain metal is irradiated with light of 400 nm wavelength. The photoelectrons emitted are stopped by a potential of -0.625 V. What is the maximum velocity of the photoelectrons?

5.1-10. Continue with Problem 5.1-9 and determine the work function of the metal.

5.1-11. When the clean surface of sodium metal (in a vacuum) is irradiated at various wavelengths, the following stopping potentials are found:

Wavelength (nm)	283	329	476
Stopping potential (V)	2.11	1.49	0.33

(a) Plot the data so as to give a straight line.
(b) Determine Planck's constant.
(c) Determine the work function of sodium.

5.1-12. What are the stopping potentials for a surface of $\mathcal{W} = 1.5$ eV and the same wavelengths as in Problem 5.1-11?

5.1-13. What is the de Broglie wavelength of electrons traveling at one-tenth the speed of light?

5.1-14. A proton ($m = 1.67 \times 10^{-27}$ kg) is being accelerated to a de Broglie wavelength of 0.99 Å. What velocity is necessary?

5.1-15. Neutron beams show diffraction effects just as more conventional waves do. If 1.67×10^{-27} kg is the mass of a neutron, what size aperture will cause neutrons of 20 keV energy to give first-order minima at 0.19° angular distance from the axis?

5.1-16. Rocksalt crystals have lattice planes spaced 2.8 Å apart. What velocity must neutrons have to be diffracted through an angle of 28°?

5.1-17. An electron is being accelerated to a velocity of 200 m s^{-1}, which can be measured to an accuracy of 0.1 percent. To what accuracy can we determine the *position* of the electron?

5.1-18. Continuing with Problem 5.1-17, determine the uncertainty in *frequency* for the same electron.

5.2

Atomic Spectra

"IT IS IMPOSSIBLE EVER TO KNOW the chemical composition of the sun!" said French philosopher Auguste Comte, before Bunsen and Kirchhoff established *emission spectroscopy*. Although it was known before, they wrote, that "certain substances when heated in a flame impart definite bright lines to their spectra, . . . differences in temperature have no effect upon the position of these lines . . . which are characteristic of each substance. . . . From these facts it appears certain that the appearance of the bright lines may be regarded as . . . proof of the presence of the particular substance." In one of their early experiments, Bunsen and Kirchhoff were able to detect less than 1/3 000 000 mg of sodium chlorate, convincing proof of the sensitivity of their method.

Continuous Emission Spectra

Blackbody radiation. When solid materials, such as the filament in a light bulb, or gases under high pressure are heated to *incandescence,* they emit *continuous* spectra.

The maximum amount of radiation that can be emitted by a hot solid is that produced by a *blackbody radiator*. A blackbody radiator can be approximated by a hollow block of cast iron with a small hole in it. Radiant energy entering the hole becomes trapped inside; hence, the cavity is a near perfect absorber. On the other hand, if the cavity is heated, like an oven, it radiates energy out of the hole. The spectral distribution of the radiation emitted by a blackbody is a function of temperature alone; the material plays no role.

The spectrum emitted by a blackbody can be described by *Planck's radiation law*. This law is often written in terms of radiant power emitted per unit surface area, that is, in terms of areance, M, as a function of wavelength, λ:*

$$M_\lambda = \frac{C_1}{\lambda^5} \frac{1}{e^{C_2/\lambda T} - 1} \qquad [5.2\text{-}1]$$

where λ is the wavelength, T the absolute temperature, in degrees Kelvin, K, and C_1 and C_2 are two *radiation constants*. The first radiation constant, C_1, is defined as, and has a numerical value of,

$$C_1 = 2\pi hc^2 = 3.742 \ldots \times 10^{-16} \text{ W m}^2 \qquad [5.2\text{-}2]$$

and the second radiation constant, C_2:

$$C_2 = \frac{hc}{k} = 1.4388 \ldots \times 10^{-2} \text{ m K} \qquad [5.2\text{-}3]$$

where h is Planck's constant, c the velocity of light, and k *Boltzmann's constant*, $1.3806 \ldots \times 10^{-23}$ J K^{-1}.

Inserting the two radiation constants in Equation [5.2-1] gives

$$M_\lambda = \frac{2\pi hc^2}{\lambda^5} \frac{1}{e^{hc/\lambda kT} - 1}$$

and, replacing c/λ by ν,

$$M_\lambda = \frac{2\pi h\nu^5}{c^3} \frac{1}{e^{h\nu/kT} - 1} \qquad [5.2\text{-}4]$$

If Planck's radiation law is written in terms of *energy density*, u, per wavelength interval, rather than in terms of areance, M_λ in Equation [5.2-1] changes to u_λ and, since energy density is related to areance as

$$u = \frac{4}{c} M$$

the first radiation constant becomes

$$C_1 = 8\pi hc \qquad [5.2\text{-}5]$$

---

*M. Planck, "Ueber eine Verbesserung der Wien'schen Spectralgleichung," *Verh. Dtsch. Physik. Ges.* **2** (1900), 202–4.

Rayleigh–Jeans' law.
There are several approximations to Planck's law. If we substitute hc/k for C_2 in Planck's law and expand the exponential, then

$$e^{hc/(k\lambda T)} = 1 + \frac{hc}{k\lambda T} + \frac{1}{2!}\left(\frac{hc}{k\lambda T}\right)^2 + \cdots \qquad [5.2\text{-}6]$$

If $hc/(k\lambda T) \ll 1$, we may neglect terms of the second and higher orders and write

$$M_\lambda \approx \frac{2\pi ckT}{\lambda^4} \qquad [5.2\text{-}7]$$

This is the *Rayleigh–Jeans law,* * which holds for longer wavelengths, as we will see in the example on page 471.

Wien's displacement law.
If we plot Planck's law for different temperatures, we find that with increasing temperature (a) *more energy* is emitted and (b) the *peak emission shifts toward the shorter wavelengths*. This behavior is illustrated in Figure 5.2-1. The peak wavelength, λ_{max}, is given by

$$\lambda_{max} = \frac{2.8978 \times 10^{-3}}{T} \text{ m K} \qquad [5.2\text{-}8]$$

This is *Wien's displacement law,* † where T is again the absolute temperature ($0°C = 273.16$ K).

This agrees with common experience. At room temperature and somewhat above, a material body emits merely infrared. As the body is heated and the temperature raised, the body begins to glow dark red. As

---

*Named after Lord Rayleigh (see footnote page 319) and James Hopwood Jeans (1877–1946), British scientist, writer, and musician; professor of mathematics at Cambridge University. As a youth, Jeans was fascinated by clocks; he could tell time at age 3, at 9 wrote a short piece on how to build a clock from scraps of tin and other materials. Among the many books Jeans wrote were *The Dynamical Theory of Gases, Theoretical Mechanics,* and *Mathematical Theory of Electricity and Magnetism,* and his *Report on Radiation and the Quantum Theory,* where he expanded on Rayleigh's theory. He wrote several books on astronomy and cosmology, also a series of highly successful popular books, some of them on the science of music. As a musician (piano and organ) he stirred up controversy by insisting that a note played on a piano would sound no different with the key struck by an umbrella than by a finger of Paderewski.

†Wilhelm Carl Werner Otto Fritz Franz Wien (1864–1928), German physicist. Wien studied under Helmholtz, became his assistant, later successor to Röntgen at both the universities of Würzburg and Munich. Wien contributed to thermodynamics and electric discharges but is best known for his displacement law. He received the Nobel prize for physics in 1911. W. Wien, "Temperatur und Entropie der Strahlung," *Ann. Physik* (Neue Folge) **52** (1894), 132–65.

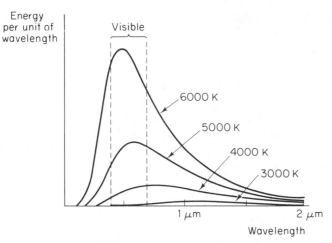

Figure 5.2-1   Distribution of radiant energy as a function of wavelength. At higher temperatures the peak emission shifts to shorter wavelengths.

the temperature is raised further, the body becomes bright red ("red hot"), then yellowish, and finally "white hot."

The *color temperature* of a light source refers to the temperature of a blackbody that emits radiation of the same subjective color. Color temperature is of interest in the visual sciences and in photography. For example, the color temperature of the sun, measured at sea level, is 5600 K. The color temperature of a 200-W tungsten filament lamp is about 3200 K, that of a photoflood lamp about 3400 K. Accordingly, daylight color film is balanced for 5600 K, indoor color film for about 3200 K.

Stefan–Boltzmann's law.   The *total* radiation output of a blackbody is found by integrating the areance, M, over all wavelengths:

$$M = \int_0^\infty M_\lambda \, d\lambda = \sigma T^4 \qquad [5.2\text{-}9]$$

This is *Stefan–Boltzmann's law*. * It shows that the areance increases as the

---

*Josef Stefan (1835–1893), Austrian, professor of physics at the University of Vienna, found the relationship by experiment; Ludwig Edward Boltzmann (1844–1906), professor of physics at the same university up to his death by suicide, derived it independently from theory. Their discoveries were published in J. Stefan, "Über die Beziehung zwischen der Wärmestrahlung und der Temperatur," *Sitzungsber. Math.-Naturwiss. Classe kais. Akad. Wiss., Wien* **79,** II. Abtheilung (1879), 391–428; L. Boltzmann, "Ableitung des Stefan'schen Gesetzes, betreffend die Abhängigkeit der Wärmestrahlung von der Temperatur aus der electromagnetischen Lichttheorie," *Ann. Physik* (Neue Folge) **22** (1884), 291–94.

fourth power of the absolute temperature. The constant of proportionality, σ, is called *Stefan's constant;* it has a numerical value of 5.67 . . . $\times 10^{-8}$ W m^{-2} K^{-4}.

Kirchhoff's law.
Finally, we note again that a blackbody is a perfect radiator and absorber. Transparent and reflective materials are poor radiators and absorbers. An object that is a good radiator at a given wavelength is also a good absorber at the same wavelength. The ratio between areance (exitance), M, and absorptivity, α, at a given temperature, is constant:

$$\frac{M}{\alpha} = \text{constant} \qquad [5.2\text{-}10]$$

which is *Kirchhoff's law.* [*]

It was the study of blackbody radiation that led Planck to the formulation of *quantum theory.* The classical explanation of the frequency distribution of blackbody radiation was based on the idea that the radiation forms *standing waves.* Evidently, the number of standing waves that fit into a cavity depends on the wavelength and on the dimensions of the cavity. For long wavelengths only a few modes are possible; as the wavelength decreases, the number of modes rapidly increases. But if this were true also on the atomic scale, the amount of energy emitted at the short-wavelength end of the spectrum would become infinitely large. This prediction of classical theory is called the *ultraviolet catastrophe,* "ultraviolet" because the difficulty arises at high frequencies and "catastrophe" because the classical prediction disagrees both with experiment and with common sense.

Example 1. *If the peak emission from the sun is at 475 nm, what is its surface temperature, in degrees Celsius?*

Solution. From *Wien's displacement law,*

$$T = \frac{2.8978 \times 10^{-3}}{475 \times 10^{-9}} = 6100 \text{ K}$$

$$6100 - 273 = \boxed{5827^\circ\text{C}}$$

Comment. The sun, however, is not a perfect blackbody. If it were, we could apply *Stefan–Boltzmann's law;* this would result in 5750 K. The discrepancy is due to absorption at shorter wavelengths in the outer layers

---

[*] G. Kirchhoff, "Ueber das Verhältniss zwischen dem Emissionsvermögen und dem Absorptionsvermögen der Körper für Wärme und Licht," *Ann. Physik* (4) **19** (1860), 275–301.

of the photosphere of the sun, which does not affect the position of the 475-nm maximum.

Example 2. *The Rayleigh–Jeans law is an approximation to Planck's law, at least in the IR. Consider, for instance, a temperature of 6000 K and a wavelength of 120 μm and compare the results.*

Solution. From Planck's law, Equation [5.2-1], we find that

$$M = \frac{3.74 \times 10^{-16}}{(0.12 \times 10^{-3})^5} \frac{1}{\exp[(1.44 \times 10^{-2})/(0.12 \times 10^{-3})(6 \times 10^3)] - 1}$$

$$= 1.5 \times 10^4 \frac{1}{2.718^{0.02} - 1}$$

$$= \boxed{7.4 \times 10^5 \text{ W m}^{-2}}$$

From the Rayleigh–Jeans law, Equation [5.2-7],

$$M = \frac{(2)(\pi)(3 \times 10^8)(1.38 \times 10^{-23})(6 \times 10^3)}{(0.12 \times 10^{-3})^4}$$

$$= \frac{156 \times 10^{-12}}{0.0002 \times 10^{-12}}$$

$$= \boxed{7.8 \times 10^5 \text{ W m}^{-2}}$$

Note that all of these radiation laws apply to blackbody radiation only. If the body is not black, the amount of energy radiated is *less* by a factor ϵ, called the *emissivity* of the surface. This factor usually lies between 0.2 and 0.9, but for highly polished metals it may be as low as 0.01.

Light sources. Both Wien's law and Stefan–Boltzmann's law are fundamental to the design of modern light sources. Early electric lights had *carbon* filaments. Then came metal filaments, in particular those that have a high melting point such as *tungsten*.

Tungsten melts at 3410°C, but before that happens it slowly evaporates and forms deposits on the inner surface of the glass envelope, reducing the output. Some halogens, in particular iodine, retard this process. On the inside of the hot (quartz) envelope, the iodine combines with the tungsten deposits to form tungsten iodide. Near the even hotter filament the molecule dissociates again, adding tungsten to the filament and letting the iodine reenter the cycle. The filament can then be heated close to the melting point, which accounts for the high efficacy of such *quartz iodine lamps*.

Fluorescent lamps, in contrast to incandescent lamps, do not produce much heat. They are often filled with argon, neon, and/or mercury. The inside of such lamps is coated with a variety of phosphors, often $3Ca_3(PO_4) \cdot Ca(FC1)_2$, and other additives such as manganese to enhance the reds. *High-pressure mercury lamps may contain traces of dysprosium, holmium, gallium, or indium iodide, which tend to broaden the Hg lines and generate new ones. High-pressure xenon lamps can be built for power outputs in excess of 120 kW and power densities higher than that of the sun.*

Line Emission Spectra

Introduction. Now we come to *line* emission spectra and to the specific realm of Bunsen and Kirchhoff's *emission spectroscopy.* [*] In contrast to continuous spectra, line emission spectra are produced by vapors and gases at *low* pressure, where the atoms or molecules are far apart and do not significantly interact.

Each chemical element has characteristic spectrum lines that permit a positive identification. Some elements, such as hydrogen and neon, have relatively few lines; others, such as iron, have many thousand. [†]

The comparatively simple spectrum of (atomic) hydrogen is illustrated in Figure 5.2-2. It is readily apparent that the lines follow each other in an orderly fashion. A group of such lines is called a *series*. The most prominent, brightest line (of atomic hydrogen) lies to the right; it has the longest wavelength, and is designated α. The next line to the left is called β, the third line γ, and so on. At the short-wavelength end, on the far left, the lines crowd together and form the *series limit*.

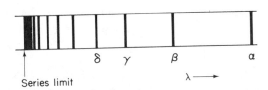

Figure 5.2-2   Series of lines in the hydrogen spectrum.

---

[*]Robert Wilhelm Bunsen (1811–1899), German chemist and professor of chemistry at the University of Heidelberg. Bunsen invented the gas burner and the grease-spot photometer, both named after him, the carbon zinc battery, and the prism spectroscope, at that time containing a hollow prism filled with carbon disulfide. Bunsen, a careful and patient experimentalist, and Gustav Robert Kirchhoff (1824–1887), a theoretical physicist and professor of physics at the same university, proved to be an ideal combination of talents. Their fundamental paper is G. Kirchhoff and R. Bunsen, "Chemische Analyse durch Spectralbeobachtungen," *Ann. Physik* (4) **20** (1860), 161–89.

[†]The spectrum *lines* are actually images of the entrance *slit* of the spectrograph. With a triangular aperture, the "line" spectrum would be a series of triangles.

The Rydberg equation.

The position of the lines (in the spectrum of hydrogen) follows a simple relationship. This relationship fits all known lines (of hydrogen) and has led to the prediction of new ones. It can be expressed in the form of *Rydberg's equation*,*

$$\frac{1}{\lambda} = \mathcal{R}\left(\frac{1}{m^2} - \frac{1}{n^2}\right)$$

[5.2-11]

where λ is the wavelength, in nm; $\mathcal{R}$ the *Rydberg constant*[†] for hydrogen, approximately 0.011 nm^{-1}; m an integer, 1, 2, 3, . . . , which is constant within any one series, and n the quantum number or "running figure," another integer which runs as a whole number 2, 3, 4, . . . through the series.

The first or α line in any one series is found when we set

$$n = m + 1$$

[5.2-12]

the second, β, when $n = m + 2$; the third, γ, when $n = m + 3$; and so on. At the series limit $n \to \infty$.

After Rydberg's discovery, Lyman applied Rydberg's equation using $m = 1$, and found a series of lines in the ultraviolet. Paschen then set $m = 3$ and found another series, in the infrared. As time went on, and more sensitive detectors became available, especially in the IR, more and more series were discovered. A summary of what is known today is shown in Table 5.2-1.

A particularly strong hydrogen line is the Lyman α line, $\lambda = 121.6$ nm. It is by far the most powerful line emitted by the sun. The Lyman α line is completely absorbed by air; it is also absorbed by deoxyribonucleic acid, DNA, the building block of all living matter on Earth. On other

---

*Named after Johannes Robert Rydberg (1854–1919), Swedish mathematician and physicist. Rydberg, unlike most others, used wavenumbers, $\sigma = 1/\lambda$, in trying to predict the position of the lines. When he heard that Johann Jakob Balmer (1825–1898), a Swiss secondary-school teacher, had found an empirical formula, $\lambda = 364.56 \, m^2/(m^2 - 4)$ nm; $m = 3, 4, 5, \ldots$, he changed Balmer's formula to wavenumbers, recognized that Balmer's is a special case and that $\mathcal{R}$ is a universal constant. Being interested all his life in the periodic system, Rydberg maintained that between hydrogen and helium there must be two more elements, which he called "nebulium" and "coronium," their alleged lines later found to be due to ionized oxygen/nitrogen and iron. Balmer's contribution is found in "Notiz über die Spectrallinien des Wasserstoffs," *Ann. Physik* (Neue Folge) **25** (1885), 80–87; Rydberg's in "Recherches sur la constitution des spectres d'émission des éléments chimiques," *Kungliga Svenska vetenskaps-akademiens handlingar* (4) **23** (1890), No. 11, 1–155.

[†]The most precise figure found to date is 0.010 973 731 476 (32) nm^{-1}. See J. E. M. Goldsmith, E. W. Weber, and T. W. Hänsch, "New Measurement of the Rydberg Constant Using Polarization Spectroscopy of Balmer-α," *J. Opt. Soc. Am.* **68** (1978), 1411.

Table 5.2-1 SUMMARY OF HYDROGEN SERIES

Series	Year of discovery	Integer, m	Quantum number, n	Spectral region
Lyman	1904	1	2, 3, 4, . . . , ∞	UV
Balmer	1885	2	3, 4, 5, . . . , ∞	Visible
Paschen	1908	3	4, 5, 6, . . . , ∞	IR
Brackett	1922	4	5, 6, 7, . . . , ∞	IR
Pfund	1924	5	6, 7, 8, . . . , ∞	IR
Humphreys	1953	6	7, 8, 9, . . . , ∞	IR
Hansen and Strong	1973	7	8, 9, 10, . . . , ∞	IR

planets which have no protective atmosphere, therefore, life either cannot exist or must be based on genetic material other than DNA.

Rutherford's model of the atom.

To understand why spectrum lines are arranged in such an orderly fashion, we need to review the *atomic structure of matter*. According to Rutherford,* an atom consists of a massive, positively charged nucleus and of one or several negatively charged electrons revolving around the nucleus.

For simplicity, assume that a single electron revolves in a circular orbit of radius R. Following the laws of classical mechanics, the centripetal force, F, needed to keep a body of mass m and velocity v in a circular orbit is

$$F = \frac{mv^2}{R} \qquad [5.2\text{-}13]$$

In the case of an atom, the centripetal force is provided by electrostatic attraction which, from Coulomb's law, is

$$F = \frac{1}{4\pi\varepsilon_0} \frac{q_1 q_2}{R^2} \qquad [3.1\text{-}1]$$

Combining both equations by eliminating F gives

$$\frac{mv^2}{R} = \frac{1}{4\pi\varepsilon_0} \frac{e^2}{R^2} \qquad [5.2\text{-}14]$$

where m is the mass of the electron, approximately 9.1×10^{-31} kg, e the charge on the electron, approximately 1.6×10^{-19} C, and ε_0 the electric permittivity of free space, approximately 8.85×10^{-12} C^2 N^{-1} m^{-2}.

---

*E. Rutherford, "The Scattering of α and β Particles by Matter and the Structure of the Atom," *Phil. Mag.* (6) **21** (1911), 669–88.

The kinetic energy of the electron is

$$\text{KE} = \frac{1}{2}\,mv^2 = \frac{1}{2}\frac{1}{4\pi\varepsilon_0}\frac{e^2}{R} = \frac{e^2}{8\pi\varepsilon_0 R} \qquad [5.2\text{-}15]$$

and the potential energy, by integrating Equation [3.1-1], is

$$\text{PE} = -\frac{e^2}{4\pi\varepsilon_0 R} \qquad [5.2\text{-}16]$$

Thus the total energy, Q, of the electron is the sum of its kinetic and potential energies,

$$Q = \frac{e^2}{8\pi\varepsilon_0 R} - \frac{e^2}{4\pi\varepsilon_0 R} = -\frac{e^2}{8\pi\varepsilon_0 R} \qquad [5.2\text{-}17]$$

The angular momentum, L, again from classical mechanics, is

$$L = mvR \qquad [5.2\text{-}18]$$

Such a system is mechanically stable, because the Coulomb force provides the centripetal force necessary to keep the electron in orbit. However, Rutherford's model does not account for the discrete frequencies that occur in the emission process.

Bohr's model of the atom.

Two years after Rutherford, Bohr *postulated* (assumed) that electrons move only in certain orbits, called *stationary states,* and that the angular momentum of an electron has only certain values which are integral multiples of $h/(2\pi)$.* Thus instead of Equation [5.2-18] we now have

$$mvR = n\,\frac{h}{2\pi} \qquad [5.2\text{-}19]$$

---

*Niels Henrik David Bohr (1885–1962), Danish physicist, professor of theoretical physics at the University of Copenhagen, founder of the Institute for Theoretical Physics. Bohr's scientific work extended through 57 years, a period of fundamental changes. He began when the structure of the atom was still unknown; he ended when atomic physics had reached maturity. Bohr established the quantum orbits of the hydrogen atom, discovered element 72, hafnium (after the Latin name for Copenhagen), laid the groundwork for quantum field theory, and in 1922 received the Nobel prize for physics. During World War II he escaped from Denmark aboard a small fishing vessel and came to the United States. Introduced at the Los Alamos Laboratories as a fictitious "Mr. Nicholas Baker," he helped see the Manhattan Project through to completion. Later in his life, Bohr's writings became more philosophical, expounding over and over again the concept of complementarity, which, he thought, "would afford people the guidance they need." N. Bohr, "On the Constitution of Atoms and Molecules," *Phil. Mag.* (6) **26** (1913), 1–25, 476–502, 857–75.

where n is the (principal) *quantum number,* an integer, $n = 1, 2, 3, \ldots$, and h is Planck's constant.

Eliminating v between Equations [5.2-14] and [5.2-19] then gives for the radii of the *quantized* orbits

$$R = n^2 \frac{h^2 \varepsilon_0}{\pi e^2 m} \qquad [5.2\text{-}20]$$

For example, for hydrogen, which has one electron,

$$R = (1)^2 \frac{(6.63 \times 10^{-34})^2 (8.85 \times 10^{-12})}{(\pi)(1.6 \times 10^{-19})^2 (9.1 \times 10^{-31})} = 5.3 \times 10^{-11} \text{m} \qquad [5.2\text{-}21]$$

which shows that the hydrogen atom in its lowest energy state ($n = 1$) has a diameter of approximately 1 Å.*

To find the total (quantized) energies in the various states, we substitute Equation [5.2-20] in [5.2-17] and obtain

$$Q = -\frac{e^4 m}{8 \varepsilon_0^2 n^2 h^2} \qquad [5.2\text{-}22]$$

Again for hydrogen,

$$Q = \frac{(1.6 \times 10^{-19})^4 (9.1 \times 10^{-31})}{(8)(8.85 \times 10^{-12})^2 (1^2)(6.63 \times 10^{-34})^2}$$

$$= 2.17 \times 10^{-18} \text{ J}$$

$$= \frac{21.7 \times 10^{-19}}{1.6 \times 10^{-19}} = 13.6 \text{ eV} \qquad [5.2\text{-}23]$$

a figure that will soon come up again.

Atomic transitions.

It is easier to represent the discrete energy states of an atom in the form of *energy levels,* rather than in the form of orbits. An energy-level diagram is shown in Figure 5.2-3, using again an atom of hydrogen. At a temperature of absolute zero, the atom is in its *ground state* and has an *ionization energy* of -13.6 eV, as derived. At higher temperatures, more and more atoms in the ensemble are in a higher, *excited state.* For a quantum number $n = 2$, following Equation [5.2-18], the electron has an energy of -3.4 eV, for $n = 3$, -1.5 eV, and so on;

---

*Here I have made several simplifying assumptions, neglecting the facts that the orbits are elliptical, rather than circular, that the proton is not stationary (which would require it to have infinite mass), and that the mass of the electron changes relativistically as a function of velocity.

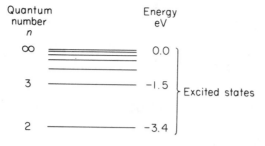

Figure 5.2-3 Energy-level diagram of hydrogen. In the ground state, $n = 1$, the electron has an energy of -13.6 eV, in the $n = 2$ state -3.4 eV, in the $n = 3$ state -1.5 eV, and so on.

but quantum theory does not allow the electron to have energies between -13.6 and -3.4 eV, or between -3.4 and -1.5 eV.

The *transition* from one energy state to another is accomplished by a transfer of energy. If energy, in whatever form, is supplied to the system, the system is raised from a lower energy state, E_1, to an higher, excited state, E_2. Such a transition, $E_1 \rightarrow E_2$, is called *absorption* (see Chapter 5.3). The reverse process, the downward transition $E_2 \rightarrow E_1$, is called *emission*. During the emission process, as the atom *returns from a higher energy state, E_2, to a lower energy state, E_1, it emits a quantum of energy.* The amount of energy being emitted is the difference between the energies in the two states,

$$h\nu = E_2 - E_1 \qquad\qquad [5.2\text{-}24]$$

which is *Bohr's frequency condition.*

Solving Equation [1.1-7], $v = \lambda\nu$, for λ and setting, from Equation [5.2-23], $h\nu = 13.6$ eV, it follows that the wavelength of the radiation

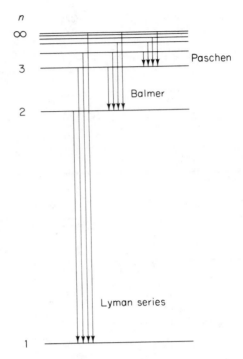

Figure 5.2-4   Energy-level diagram for hydrogen showing transitions in the Lyman, Balmer, and Paschen series.

emitted during a transition of hydrogen from the highest to the ground state is

$$\lambda = \frac{c}{\nu} = \frac{hc}{h\nu} = \frac{(6.63 \times 10^{-34} \text{ J s})(3 \times 10^{8} \text{ m/s})}{(13.6 \text{ eV})(1.6 \times 10^{-19} \text{ J/eV})} = 91.4 \text{ nm} \quad [5.2\text{-}25]$$

This is precisely the limit of the Lyman series. For the other series of hydrogen we use the same reasoning, noting that in Equation [5.2-22] the *square* of the quantum number occurs in the denominator. An energy-level diagram for the first three series of hydrogen is shown in Figure 5.2-4.

Finally, if the quantum number, n, is used to describe the higher energy state and another constant, m, is introduced to describe the lower state, then

$$\nu = \frac{me^{4}}{8\varepsilon_{0}^{2}h^{3}}\left(\frac{1}{m^{2}} - \frac{1}{n^{2}}\right) \quad [5.2\text{-}26]$$

which is again the *Rydberg equation*. In short, we have arrived at the same result, once by experimental observation and again by theoretical reasoning.

Table 5.2-2   WAVELENGTHS (IN nm) OF
SELECTED SPECTRUM LINES OF VARIOUS ELEMENTS,
LISTED IN ORDER OF INCREASING ATOMIC NUMBER

Helium	$_2$He	587.5618
Neon	$_{10}$Ne	470.440 470.885 471.534 488.492
Sodium	$_{11}$Na	588.9953 589.5923
Potassium	$_{19}$K	766.4907 769.8979
Krypton	$_{36}$Kr	587.09158 605.78021 760.15465 810.43660 811.29023
Cesium	$_{55}$Cs	852.110
Mercury	$_{80}$Hg	435.835 546.0740

Other line spectra.   The spectrum of *helium* can be described by an equation very similar to Rydberg's equation for hydrogen. The helium spectrum has four series, a first and second Lyman series ($m = 1$ and $m = 2$), both in the vacuum ultraviolet, the Fowler series ($m = 3$), and the Pickering series ($m = 4$).

The spectra of other atoms are more complicated. Only a few are understood as thoroughly as the spectra of hydrogen and helium; these include the spectra of the alkali metals, lithium, sodium, potassium, rubidium, and cesium. The spectra of most other elements are still more complex and a discussion of them is not needed for an understanding of atomic spectra in general. The most prominent lines of certain elements are listed in Table 5.2-2.

SUGGESTIONS FOR FURTHER READING

H. G. Kuhn, *Atomic Spectra*, 2nd edition (New York: Academic Press, Inc., 1969).

R. Chang, *Basic Principles of Spectroscopy* (New York: McGraw-Hill Book Company, 1971).

R. D. Cowan, *The Theory of Atomic Structure and Spectra* (Berkeley: University of California Press, 1981).

E. L. Grove, editor, *Applied Atomic Spectroscopy* (New York: Plenum Press, 1978).

PROBLEMS

5.2-1. Will the *color temperature* of light change and, if so, how will it change, when a blue filter is placed in front of the light source?

5.2-2. When, at a given temperature and wavelength, a material has an emissivity of 0.6, to which parameter(s) does this figure refer?

5.2-3. A carbon filament, which may be considered a blackbody, is heated successively to 500°C, 1000°C, and 2000°C. What are the peak wavelengths emitted at these temperatures?

5.2-4. Can daylight (peak wavelength 555 nm) be obtained from a tungsten filament lamp without using an additional filter?

5.2-5. What is the peak wavelength emitted by the human body (normal temperature = 37°C)?

5.2-6. If a certain material is heated to 1000°C and then emits infrared energy of 2.3 μm wavelength, what must be its temperature to emit yellow light of 575 nm?

5.2-7. Verify the numerical values and the units of the two radiation constants that are part of Planck's radiation law.

5.2-8. Express the Rydberg constant in terms of *frequency,* using the necessary parameters to three significant figures.

5.2-9. Using the Rydberg equation, find the wavelength and color of the Balmer α line in the spectrum of hydrogen.

5.2-10. What is the difference, in nm, between the wavelengths of the tenth line and the series limit of the Balmer series of hydrogen?

5.2-11. What is the series number, the quantum number, and the wavelength of:
(a) The Lyman β line?
(b) The Balmer γ line?
(c) The Paschen α line?

5.2-12. With the ionization energy of hydrogen being 13.6 eV, what is its energy in the $n = 4$ state?

5.2-13. Calculate the velocity of the electron in the first Bohr orbit of the hydrogen atom.

5.2-14. Determine the magnitude of the Coulomb attraction between the nucleus and the electron in the ground state of the hydrogen atom. Use the centripetal force and the electrostatic force equations. Take the velocity of the electron from Problem 5.2-13.

5.3

Absorption

WHEN A TRANSITION OCCURS from a higher energy level to a lower energy level, a quantum of light is emitted. Conversely, when a quantum of light is incident on an atom (and if the quantum has the same frequency as it would have when emitted), the quantum will *raise* the atom from the lower state to an excited state. This process is called *absorption*. If the quantum does not have the right frequency, it will pass through the medium without absorption or it will be scattered.

Sometimes, the terms absorption and extinction are used synonymously. This is not correct. Extinction is the broader term; it includes attenuation by both absorption and scattering.

Absorption Spectra

Emission spectra versus absorption spectra.
As in emission, *absorption spectra* can be either continuous spectra or line spectra. Continuous absorption spectra occur with dyes and other colored liquids and solids. Line absorption spectra result when (white) light passes through a gas at low pressure and at a temperature lower than that of the source. These lines have the same wavelengths as the lines which the gas

Table 5.3-1 PROMINENT FRAUNHOFER LINES*

Line	Position in spectrum	Wavelength (nm)	Origin
C	Red	656.3	H
D_1	Yellow	589.6	Na
D_2	Yellow	589.0	Na
b_1	Green	518.4	Mg
b_2	Green	517.3	Mg
F	Blue	486.1	H
g	Violet	434.0	H
h	Violet	410.2	H
H	Violet	396.8	Ca
K	Violet	393.4	Ca
L	Violet	382.0	Fe

would emit if it were hot; they identify an element just as emission spectra do. Absorption lines, superimposed on the continuous spectrum of the sun, are known as *Fraunhofer lines*; they are due to absorption in the outer layers of the sun's photosphere. The most prominent Fraunhofer lines are listed in Table 5.3-1.*

The exponential law.
We follow our earlier discussion of light scattering (page 321) and let a collimated beam of light pass through a thin slice of a medium of thickness Δx. We call ϕ_0 the power of the light incident on the medium and ϕ' the power that emerges from it (Figure 5.3-1).

On passing through the medium, a certain fraction of the light, $\Delta\phi$, will be lost:

$$\phi_0 - \phi' = \Delta\phi \qquad [5.3\text{-}1]$$

The magnitude of this loss is proportional to ϕ_0, to the thickness Δx, and to a constant of proportionality called *absorptivity, α:*

$$\phi' - \phi_0 = -\Delta\phi = \phi_0\alpha\,\Delta x \qquad [5.3\text{-}2]$$

an equation that is virtually the same as the earlier Equation [3.3-5]. The absorptivity (formerly called "absorption coefficient") is characteristic of the medium; it is also a function of wavelength.

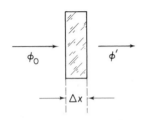

Figure 5.3-1 Loss of light in a medium of thickness Δx.

---

*By his own count, Fraunhofer saw 574 dark lines in the spectrum of the sun but, as we know today, some of these are due to absorption in the atmosphere of Earth, rather than the photosphere of the sun. J. Fraunhofer, "Bestimmung des Brechungs- und des Farbenzerstreuungs-Vermögens verschiedener Glasarten, in Bezug auf die Vervollkommnung achromatischer Fernröhre," *Ann. Physik* (2) **26** (1817), 264–313.

Assume that the medium is made into infinitesimally thin slices, each of thickness dx. Again, in each slice a constant fraction of light, $d\phi$, is lost and Equation [5.3-2] becomes

$$\frac{d\phi}{\phi_0} = -\alpha\,dx \qquad [5.3\text{-}3]$$

In order to find the total loss within the medium (of thickness x), we integrate Equation [5.3-3] between the limits of ϕ and x, respectively:

$$\int_{\phi_0}^{\phi'} \frac{d\phi}{\phi} = -\alpha \int_0^x dx \qquad [5.3\text{-}4]$$

Since

$$\int_a^b \frac{d}{dx} f(x)\,dx = f(b) - f(a) \qquad \text{and} \qquad \int \frac{dx}{x} = \ln x$$

$$\ln\left(\frac{\phi'}{\phi_0}\right) = -\alpha x \qquad [5.3\text{-}5]$$

and

$$\frac{\phi'}{\phi_0} = e^{-\alpha x} \qquad [5.3\text{-}6]$$

If the absorbing medium is a solution, the concentration of the solution, c, usually given in grams or moles per liter, must be included also and Equation [5.3-6] becomes

$$\phi' = \phi_0\, e^{-\alpha x c} \qquad [5.3\text{-}7]$$

which is the *exponential law of absorption*.* Note that in deriving this law, integration leads to *natural*, Napierian logarithms (to base e). In contrast,

---

*The exponential law as presented has evolved through many efforts. Most notable among its originators were the following: Pierre Bouguer (1698–1758), French oceanographer, was the first to find a relationship similar to the exponential law. He published it in his book *Essai d'optique sur la gradation de la lumière* (Paris: Jombert, 1729). Johann Heinrich Lambert (1728–1777), a largely self-taught German physicist and mathematician, contributed to thermodynamics, photometry (cosine law), absorption, planetary motion, and cosmology, liked to reduce his thoughts on logic and philosophy to mathematical models. He described the law in *Lamberts Photometrie (Photometria, sive de mensura et gradibus luminis, colorum et umbrae)* (Augsburg, 1760), the same year in which a second, greatly enlarged edition of Bouguer's book, renamed *Traité d'optique sur la gradation de la lumière* (Paris: H. L. Guerin & L. F. Delatour, 1760) was published (posthumously) that contained virtually the same relationship. August Beer (1825–1863), German physicist, professor of mathematics at the University of Bonn, recognized the role of the concentration, as he reported in his book *Einleitung in die höhere Optik* (Braunschweig: F. Vieweg und Sohn, 1853).

Table 5.3-2 TRANSMITTANCE
AND ABSORBANCE

Percent Transmittance, T	Absorbance, A
100	0.0000
90	0.0458
80	0.0969
70	0.1549
60	0.2218
50	0.3010
40	0.3979
30	0.5229
20	0.6990
10	1.0000
1	2.0000
0	∞ (opaque)

as we will see shortly when it comes to practical applications, we use *common* logarithms (to base 10).

Transmittance, T, is the ratio of the radiant power transmitted through the sample to the power incident on it (or transmitted through a blank or pure solvent), both measured at the same wavelength:

$$T = \frac{\phi'}{\phi_0} \qquad [5.3\text{-}8]$$

Absorbance, A, is the logarithm to base 10 of the inverse of the transmittance:

$$A = \log_{10}\left(\frac{1}{T}\right) = -\log_{10} T \qquad [5.3\text{-}9]$$

(The term "optical density" is deprecated because of possible confusion with "density," which is mass per unit volume. The term "absorption" is a process rather than a property.) Representative figures that show the relationship between transmittance and absorbance are listed in Table 5.3-2.

Absorbance is an important term. In principle we could measure the *transmittance* of a solution at different concentrations and plot the curve thus obtained. But if we use *absorbances,* rather than transmittances, plotting is easier because then the relationship is linear and only a few points are needed to establish a straight line. Furthermore, absorbances are additive, whereas transmittances are multiplicative. Figure 5.3-2 illustrates the connection.

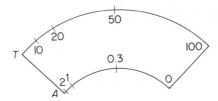

Figure 5.3-2 Scale of photometer showing logarithmic relationship between transmittance, T, and absorbance, A.

Absorptivity, α, as we saw in Equation [5.3-7], appears in the exponential law as a *natural* logarithm:

$$\ln\left(\frac{\phi'}{\phi_0}\right) = -\alpha x c \qquad [5.3\text{-}10]$$

But absorbance, A, is based on *common* logarithms:

$$A = \log_{10}\left(\frac{\phi_0}{\phi'}\right) \qquad [5.3\text{-}11]$$

In order to convert one into the other, we use the identities $\ln x = 2.3026\ldots\log_{10}x$ or $\log_{10}x = 0.4343\ldots\ln x$, as needed. This gives us a set of equations that we will use frequently:

$$
\boxed{
\begin{aligned}
A &= \log_{10}\left(\frac{1}{T}\right) \\[1em]
A &= 0.4343\,\alpha x c \\[1em]
T &= 10^{-A} = \frac{1}{10^A} \\[1em]
\alpha &= 2.3026\,\frac{A}{xc}
\end{aligned}
}
\qquad [5.3\text{-}12]
$$

From the second equation in [5.3-12] we see that the absorbance, A, of a given sample is directly proportional to its thickness, x:

$$A = (k)(x)$$

If we have two samples of the same material (same α and same c), we divide one equation by the other:

$$\frac{A_1}{A_2} = \frac{(k)(x_1)}{(k)(x_2)}$$

cancel the constant, k, and rearrange:

$$A_2 = A_1 \frac{x_2}{x_1} \qquad [5.3\text{-}13]$$

Multiplying by -1 and taking the logarithm of 10 to the Ath power gives

$$\log 10^{-A_2} = \log 10^{-A_1(x_2/x_1)}$$

and, substituting twice the third equation in [5.3-12],

$$\boxed{T_2 = T_1^{x_2/x_1}} \qquad [5.3\text{-}14]$$

This is a very useful equation which we will need whenever we compare the transmittance of a sample of thickness x_1 with the transmittance of another sample, of the same material but of a different thickness, x_2.

Example 1. *A neutral density filter of a given thickness absorbs 10 percent of the light incident on it. How much light will a filter absorb, made of the same material but 12 times as thick?*

Solution. If 10 percent is being absorbed by the first filter, 90 percent is transmitted. Hence the transmittance is

$$T_1 = 90\% = 0.9$$

The transmittance of the second (thicker) filter, from Equation [5.3-14], is

$$T_2 = T_1^{x_2/x_1} = 0.9^{12} = 0.28$$

Therefore, if the second filter transmits 28 percent of the light, it absorbs

$$100 - 28 = \boxed{72 \text{ percent}}$$

This problem could also be solved, much more laboriously, by converting the transmittance of the first filter, T_1, into its absorbance:

$$A_1 = \log_{10}\left(\frac{1}{T_1}\right) = -\log_{10} T_1 = -\log_{10} 0.9 = 0.04576$$

Then for the second filter,

$$A_2 = (12)(A_1) = (12)(0.04576) = 0.5491 = -\log_{10} T_2$$

Hence,

$$\log_{10} T_2 = -0.5491$$

and

$$T_2 = \text{inv log (antilog) } -0.5491 = 0.28$$

from where, as before,

$$100 - 28 = \boxed{72 \text{ percent}}$$

Example 2. *Another filter, 4 mm thick, absorbs 30 percent of the light. How thick a filter is needed to absorb 60 percent?*

Solution. Converting these percentages into transmittances, we have $T_1 = 70$ percent and $T_2 = 40$ percent. Also, $x_1 = 4$ mm, while x_2 is to be found. Then again from Equation [5.3-14],

$$0.4 = 0.7^{x_2/4}$$

$$0.4^4 = 0.7^{x_2}$$

$$4 \log 0.4 = x_2 \log 0.7$$

(For this last step, either common or natural logarithms may be used.) The thickness of the second filter, then, is

$$x_2 = \frac{4 \log 0.4}{\log 0.7} = \boxed{10.3 \text{ mm}}$$

In general, and in accordance with the exponential law, a solution of a given thickness (path length) and concentration will absorb the same fraction of light as the same solution twice as thick but of half the concentration. There are situations, however, where α varies as the concentration is changed.

Selective absorption.   Some materials, such as metals, have high absorptivity from the ultraviolet through the infrared; even a thin metal membrane is virtually opaque to light. Other materials such as dyes block out only part of the spectrum. This is called *selective absorption*. In general, there are no substances that have no absorption at least somewhere in the electromagnetic spectrum: glass is opaque to UV, water absorbs below 190 nm and in the IR, and absorption by air extends from the region of soft X rays to and including the vacuum ultraviolet.

This makes somewhat obsolete the earlier distinction between *general absorption* and *selective absorption*. In general absorption, the absorptivity of the material is very nearly the same over a wide range of wavelengths. In selective absorption, attenuation occurs at certain wavelengths more than at others. Within the visible spectrum, this causes the sensation of *color*.* A sample of green glass, for instance, will absorb

*Early in 1672, Newton wrote to the Royal Society of London announcing that he wanted to report "a philosophical Discovery . . . in my judgment the oddest if not the most

mainly red, and since red is complementary to green, the transmitted light will lack red and the sample appear green.

Any color has three physical characteristics: *hue, lightness,* and *saturation.* Hue gives the color its name, such as blue, green, or red. Lightness refers to whether a color is pale (light) or dark (deep). Saturation is described by terms such as reddish white, moderate red, strong red, and vivid red. If more and more white is added to red paint, the dominant hue, red, stays the same but a series of tints will result, ending in a pale pink.

Experimental Methods

In order to determine the absorption characteristics of a sample as a function of wavelength, we need a *spectrophotometer,* that is, a photometer or radiometer connected to a device for wavelength selection. If the sample is a solid such as glass or plastic, the sample is made into a plane-parallel, polished plate. If it is a liquid or a substance in solution, it is placed in a cuvette and a comparison made between the solution and the pure solvent. Modern spectrophotometers plot the transmittance, or the absorbance, or both, as a function of wavelength.

Types of spectrophotometers.
Spectrophotometers can be built as one-beam-indicator, one-beam-substitution, and two-beam instruments. The latter may have either one or two detectors. In the *one-beam one-detector method,* the light proceeds in a single path through a monochromator and through the sample to the photodetector (Figure 5.3-3). The sample and the reference (blank) are alternately brought into the path and the transmittance through each is measured at each wavelength desired.

In the *one-beam substitution* (zero) *method,* not shown, sample and reference again are alternately brought into the path. While measuring the reference, however, a neutral density filter or other attenuator is brought into the beam to make the two powers equal. The advantage is that the response of the detector need not be linear.

In the *two-beam two-detector method,* the light coming from the monochromator is divided by a beamsplitter, passed through the sample

---

considerable detection which hath hitherto been made in the operations of Nature." He let sunlight fall on a "triangular glass-Prisme" and found that "Light it self is a Heterogeneous mixture of differently refrangible Rays." The colors obtained could not be divided further; they could be combined back into white. I. Newton, "A New Theory about Light and Colours," *Phil. Trans. Roy. Soc. London* **5,** No. 80 (Feb 19, 1672), 3075–87.

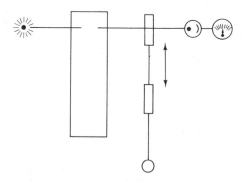

Figure 5.3-3   One-beam one-detector method.

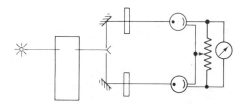

Figure 5.3-4   Two-beam two-detector method.

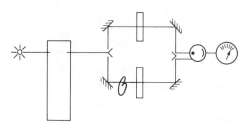

Figure 5.3-5   Two-beam one-detector method.

and the blank, respectively, and is incident on two detectors (Figure 5.3-4). The potentiometer is adjusted so that the galvanometer reads zero. Two-beam instruments are independent of voltage fluctuations.

In the *two-beam one-detector method* (Figure 5.3-5), the light is split, passes through the sample and the blank, is recombined by another beamsplitter, and falls on *one* detector; thus there can be no mismatch between two photosurfaces. The reference beam is attenuated (chopped), and the degree of attenuation directly read in terms of transmittance or absorbance.

Photochromic glass. Materials are called *photo-chromic* if, under the influence of light, they become darker or change color. When the light is removed, they return to their initial state. The process of darkening, or *activation,* takes about a minute, until the absorbance reaches a certain saturation level. Light between 320 and 400 nm wavelength is most effective. The return process, or clearing, depends on the ambient temperature (even at room temperature there is some *thermal fading*); clearing can be accelerated by light between 550 and 650 nm (*optical bleaching*). In the region between activation and bleaching, between 450 and 500 nm, light has no effect.

Photochromism is similar to photography. In both, there is a dissociation of silver halides:

$$AgCl \xrightleftharpoons{h\nu} Ag^+ + Cl^- \qquad\qquad [5.3\text{-}15]$$

But, in a photographic emulsion the halogen ion diffuses away while the silver remains in place. After development, the silver grains produce the dark regions in the photographic negative. In photochromic glass, the glass matrix is so tight that the halogen cannot diffuse away; it remains close to the silver and recombines with it when activation ceases. Hence, the process of photochromism is *reversible.* Samples have been exposed for half a million cycles of activation and bleaching, and have shown no fatigue. Typically, a small amount of copper, or other catalysts, is added to the glass; this increases the rate and depth of darkening by a factor of 100 or more.

Thermal fading, as the name implies, depends on the ambient temperature (on a hot day photochromic glass clears faster), but activation (darkening) does not. Thus the absorbance will reach a state of dynamic equilibrium that for a given areance (irradiance) decreases with increasing temperature. On a cold day, and with a given amount of light, photochromic glass becomes darker than on a hot day.

SUGGESTIONS FOR FURTHER READING

G. D. Christian and F. J. Feldman, *Atomic Absorption Spectroscopy* (New York: Interscience Publishers, John Wiley & Sons, Inc., 1970).

W. J. Price, *Analytical Atomic Absorption Spectroscopy* (London: Heyden & Son Ltd., 1972).

F. W. Billmeyer, Jr., and M. Saltzman, *Principles of Color Technology,* 2nd edition (New York: John Wiley & Sons, Inc., 1981).

D. B. Judd and G. Wyszecki, *Color in Business, Science and Industry,* 3rd edition (New York: John Wiley & Sons, Inc., 1975).

D. L. MacAdam, *Color Measurement: Theme and Variations* (New York: Springer-Verlag, 1981).

PROBLEMS

5.3-1. (a) If 38 percent of the incident light is *transmitted* through a sample, what is the absorbance (optical density)?
(b) What is the absorbance if 38 percent of the light were *absorbed?*

5.3-2. A metallic coating absorbs 88 percent of the incident light. A thin film has one-half the absorbance of the coating. What is the transmittance of the film?

5.3-3. If a filter absorbs 20 percent of the light incident on it, how much light would the filter absorb if it were twice as thick?

5.3-4. A thin layer of a certain material absorbs 40 percent of the radiant energy passing through. How much energy will a layer absorb that is nine times as thick?

5.3-5. A 2.7-mm-thick neutral density filter causes 35 percent absorption. If the filter were 5 mm thick, how much would it transmit?

5.3-6. Twelve percent of the light is being absorbed in a filter 3.2 mm thick. How much light would be lost if the filter were 5.6 mm thick?

5.3-7. A pipe 5 m long contains a gas at normal atmospheric pressure. If the gas has an absorptivity of $\alpha = 0.08$ m^{-1}, how much light, in percent, is being absorbed?

5.3-8. If 2.3-mm-thick sunglasses transmit 40 percent, how thick should the glasses be to transmit only 20 percent?

5.3-9. A cylinder, 25 cm long and filled with a liquid, absorbs 62 percent of the light of a given wavelength. What is the absorptivity of the liquid?

5.3-10. A certain type of protective glass has an absorptivity of $\alpha = 0.39$ mm^{-1}. What percentage of the light is transmitted through a plate 4 mm thick?

5.3-11. The absorptivity of seawater at 570 nm is 0.22 m^{-1}. How long a path will reduce the transmittance to 80 percent?

5.3-12. If the memory unit in an optical computer is made of photochromic glass, which color of light should be used for:
(a) Storing the information?
(b) Reading the information?
(c) Erasing the information?

5.4

Lasers

THE WORD LASER IS AN ACRONYM for light amplification by stimulated emission of radiation. In 1916, Einstein predicted that the existence of equilibrium between matter and electromagnetic radiation required that besides emission and absorption there must be a third process, now called *stimulated emission*. This prediction attracted little attention until 1954, when Townes and coworkers developed a microwave amplifier (maser) using ammonia, NH_3. In 1958, Schawlow and Townes showed that the maser principle could be extended into the visible region and in 1960, Maiman built the first laser using ruby as the active medium. From then on, laser development was nothing short of miraculous, giving optics new impetus and wide publicity.

Stimulated Emission of Radiation

Einstein's prediction.
So far we have discussed the transitions that occur in the processes of emission and absorption of radiation. But Einstein, in a much celebrated paper, actually a lecture given in 1916 and later published in 1917, postulated that there must be a third process.*

---

* A Einstein, "Zur Quantentheorie der Strahlung," *Mitt. Physikal. Ges. Zürich*, No. 18 (1916), and *Physikal. Zschr.* **XVIII** (1917), 121–28.

First, consider the number of atoms, per unit of volume, that exist in a given energy state. This number, called the *population, N,* is given by *Boltzmann's equation,*

$$N = e^{-E/kT} \qquad\qquad [5.4\text{-}1]$$

where E is the energy level of the system, k Boltzmann's constant, and T the absolute temperature.

In any process of absorption or emission, at least two energy states are involved. Thus the populations in the two states, a lower state, E_1, and a higher state, E_2, are compared. The ratio of the populations in these two states, N_2/N_1, is called *Boltzmann's ratio* or *relative population,*

$$\frac{N_2}{N_1} = \frac{e^{-E_2/kT}}{e^{-E_1/kT}}$$

from which it follows that

$$N_2 = N_1 e^{-(E_2-E_1)/kT} \qquad\qquad [5.4\text{-}2]$$

A plot of Equation [5.4-2] will yield an exponential curve, known as a *normal* or *Boltzmann distribution* of atomic populations. The term "normal" means that the system is in thermal equilibrium. If we substitute Bohr's frequency condition in Equation [5.4-2], we obtain

$$N_2 = N_1 e^{-h\nu/kT} \qquad\qquad [5.4\text{-}3]$$

Following Einstein, assume first that an ensemble of atoms is in thermal equilibrium and that it is *not subject to* any external radiation field. At room temperature, or higher, a certain number of atoms will be in an excited state. These atoms may lose energy, in the form of radiation in units of quanta $h\nu$, without any outside stimulus. This is called *spontaneous emission.*

Now consider the *rate* of this transition, that is, the number of atoms in the higher state that make the transition to the lower state, per second. Since N_2 is the number of atoms in the higher state, the rate of the spontaneous transitions, P_{21}, is

$$P_{21} = N_2 A_{21} \qquad\qquad [5.4\text{-}4]$$

where A_{21} is a constant of proportionality.

Assume next that the system *is subject to* some external radiation field. In that case, one of two processes may occur. Which of these will occur depends on the direction (the *phase*) of the radiation field with respect to the phase of the oscillator. If the phases of the two systems coincide, a quantum of the radiation field may cause the emission of another quantum. This process, anticipated by Einstein, is now called *stimulated emission.* Its rate is

$$P_{21} = N_2 B_{21} u_\nu \qquad\qquad [5.4\text{-}5]$$

where B_{21} is another constant of proportionality and u_ν is the energy density, as a function of frequency, ν.

On the other hand, if the phase of the radiation field is in a direction opposite to that of the oscillator, the impulse transferred counteracts the oscillation, energy is consumed, and the system is raised to a higher state. This is the case of *absorption*. Its rate is

$$P_{12} = N_1 B_{12} u_\nu \qquad\qquad [5.4\text{-}6]$$

where N_1 is the number of atoms in the lower state and B_{12} is still another constant of proportionality. The three constants, A_{21}, B_{12}, and B_{21}, are known as *Einstein's coefficients*.

Before proceeding further we summarize these findings in a schematic diagram. If a quantum of energy $h\nu$, of the right frequency, is incident on an atom, the transition is upward: the quantum will be *absorbed* (Figure 5.4-1, top). Conversely, if the transition is downward, a quantum is *emitted*. This emission may occur either *spontaneously* (center), or it may occur with the atom being *stimulated* to return to the lower state (bottom).

With the system in thermal equilibrium, the net rate of downward transitions must equal the net rate of upward transitions:

$$N_2 A_{21} + N_2 B_{21} u_\nu = N_1 B_{12} u_\nu \qquad\qquad [5.4\text{-}7]$$

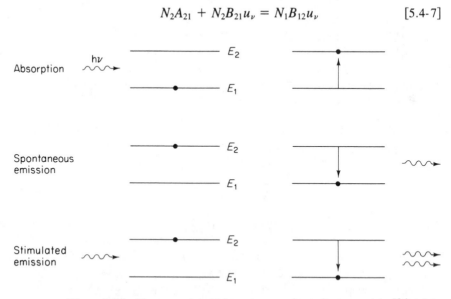

Figure 5.4-1 Three ways by which an atom can change its energy state: If the atom is in a lower state, it may absorb a quantum of energy (*top*). If the atom is in the higher state, it may release a quantum spontaneously (*center*) or it may be *stimulated* to release a quantum, triggered by another quantum passing through (*bottom*).

We divide both sides of Equation [5.4-7] by N_1. This yields

$$\frac{N_2 A_{21}}{N_1} + \frac{N_2 B_{21} u_\nu}{N_1} = B_{12} u_\nu$$

$$\frac{N_2}{N_1}(A_{21} + B_{21} u_\nu) = B_{12} u_\nu$$

$$\frac{N_2}{N_1} = \frac{B_{12} u_\nu}{A_{21} + B_{21} u_\nu} \qquad\qquad [5.4\text{-}8]$$

If we then substitute Equation [5.4-3], we obtain

$$\frac{B_{12} u_\nu}{A_{21} + B_{21} u_\nu} = e^{-h\nu/kT} \qquad\qquad [5.4\text{-}9]$$

Solving for u_ν gives

$$u_\nu = \frac{A_{21}}{B_{12}} \frac{1}{e^{h\nu/kT} - B_{21}/B_{12}} \qquad\qquad [5.4\text{-}10]$$

In order to maintain thermal equilibrium, the system must release energy, in the form of radiation. The spectral distribution of this radiation follows *Planck's radiation law*, Equation [5.2-1]. If we then substitute for C_1 Equation [5.2-5] and for C_2 Equation [5.2-3], we can write Planck's law in terms of energy density (as a function of wavelength):

$$u_\lambda = \frac{8\pi hc}{\lambda^5} \frac{1}{e^{-hc/\lambda kT} - 1} \qquad\qquad [5.4\text{-}11]$$

The energy density must be consistent with Planck's law for any value of T. This is possible only when

$$B_{21} = B_{12} \qquad\qquad [5.4\text{-}12]$$

and*

$$\frac{A_{21}}{B_{21}} = \frac{8\pi h\nu^3}{c^3} \qquad\qquad [5.4\text{-}13]$$

where h is Planck's constant, ν the frequency, and c the velocity of light.

These last two equations are called *Einstein's relations*. Ordinarily, when an atomic system is in equilibrium, absorption and spontaneous emission take place side by side, but because $N_2 < N_1$, absorption dominates: An incident quantum is more likely to be absorbed than to cause emission. But if we can find a material that could be *induced* to have a

---

* See D. C. O'Shea et al., *Introduction to Lasers and Their Applications*, page 58, listed in the Suggestions for Further Reading.

majority of atoms in the higher state than in the lower state, $N_2 > N_1$, then we have a condition called *population inversion* and the system will possibly lase.

Example. *A relative population, or Boltzmann ratio, of $1/e$ is often considered representative of the ratio of populations in two energy states at room temperature, $T = 300 \, K$. As an example, determine the wavelength of the radiation emitted at that temperature.*

Solution. If the Boltzmann ratio, Equation [5.4-2],

$$\frac{N_2}{N_1} = e^{-(E_2 - E_1)/kT}$$

is set equal to

$$\frac{N_2}{N_1} = \frac{1}{e} = e^{-1}$$

then

$$\frac{E_2 - E_1}{kT} = 1$$

so that

$$E_2 - E_1 = kT = h\nu$$

Solving for ν and, using $c = \lambda\nu$, converting ν into wavelength gives

$$\lambda = \frac{ch}{kT} = \frac{(3 \times 10^8)(6.63 \times 10^{-34})}{(1.38 \times 10^{-23})(300)} = 0.048 \times 10^{-3} \, \text{m}$$

$$= \boxed{48 \, \mu\text{m}}$$

which means that at room temperature mostly infrared radiation of around 48 μm wavelength will be emitted.

The concept of population inversion can be best illustrated if we consider a system that has *three energy states (a three-level system)*. These states may be designated E_1, E_2, and E_3. With the system in equilibrium, the uppermost state, E_3, is populated least, and the lowest state, E_1, is populated most (Figure 5.4-2).

The curve shown represents a Boltzmann distribution, Equation [5.4-2]. Since the population in the various states is such that $N_3 < N_2 < N_1$, the system is absorptive rather than emissive. But assume that the system is being excited by supplying extraneous energy. This may cause either N_3 or N_2, or both, to exceed N_1; thus the system becomes top-heavy and reaches population inversion (Figure 5.4-3).

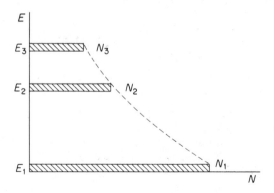

Figure 5.4-2 Population of three-level system under conditions of thermal equilibrium.

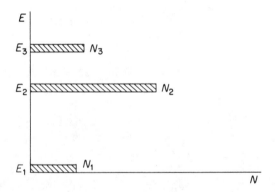

Figure 5.4-3 Three-level system showing population inversion of E_2 with respect to E_1.

General Properties of Lasers

Construction and components. A laser requires three components for operation. There is first an *energy source* that will raise the system to an excited state. Next there is an *active medium* that, when excited, achieves population inversion and subsequently lases. Third, in most cases, there is an *optical cavity* that provides the feedback necessary for laser oscillation. These components are shown schematically in Figure 5.4-4.

Energy is conveyed to the active medium, the material that is raised to the excited state and that ultimately lases, in various ways. The medium may be a solid, liquid, or gas, and it may be one of thousands of materials that have been found to lase. The process of raising the medium to the

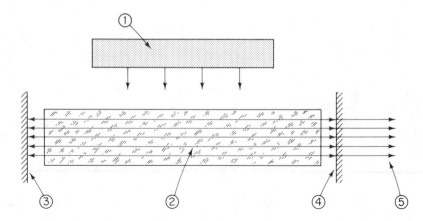

Figure 5.4-4 Basic components of a laser oscillator: Energy source (1) supplies energy to active medium (2). Medium is contained between two mirrors (3 and 4). Mirror 3 is fully reflective while mirror 4 is partially transparent. Laser radiation 5 emerges through partially transparent mirror.

excited state is called *pumping,* in analogy to pumping water up to a higher level of potential energy. Some lasers are built as laser *amplifiers;* they need no optical cavity. Most lasers, though, are laser *oscillators;* they require the medium to be enclosed in a resonant cavity, usually formed by two mirrors facing each other. One of the mirrors is coated to full reflectance; the other mirror is partially transparent to let some of the radiation pass through.

In both laser amplifiers and oscillators, the first few quanta will probably be emitted spontaneously and will trigger the in-phase release of more quanta. These quanta will become part of a continually growing wave, and *avalanche action* will ensue. In a laser *oscillator,* because the radiation is reflected back and forth between the mirrors and the path is so much longer, the avalanche action will be much more pronounced. Excitation, population inversion, and cavity resonance are the fundamental processes that have made lasers a reality and that we will now discuss in detail.*

---

*The laser is an example of the same idea having come to different people, in different parts of the world, at about the same time. The idea first occurred to Charles Hard Townes (1915–   ), then professor of physics at Columbia University, in 1951 while sitting on a park bench in Washington, DC. Independently, Alexandr Mikhailovich Prokhorov (1916–   ), head of the Oscillation Laboratory of the P. N. Lebedev Institute of Physics at the University of Moscow, and his assistant Nicolai Gennadievich Basov (1922–   ), now director of the Lebedev Institute, in 1952 presented a paper at an All-Union (SSSR) Conference on radio spectroscopy in which they discussed the possibility of building a "molecular generator." The pertinent publications are A. L. Schawlow and C.

Excitation. There are several ways of pumping a laser and producing the population inversion necessary for stimulated emission to occur. Most commonly used are the following:

Optical pumping
Electric discharge
Inelastic atom–atom collisions
Direct conversion
Chemical reactions

In *optical pumping,* a light source is used to supply luminous energy. Most often this energy comes in the form of short flashes of light, a method first used in Maiman's ruby laser and still widely employed today in solid-state lasers. The laser material is simply placed inside a helical xenon flashlamp, of the type customary in photography.

Another means of excitation is by direct electron excitation as it occurs in an *electric discharge*. This method is preferred in gaseous ion lasers, of which the argon laser is a good example. The electric field, typically several kV m^{-1}, causes electrons emitted by the cathode to be accelerated toward the anode. Some of these electrons will impact on the atoms of the active medium (*electron impact*), ionize the medium, and raise it to the excited state, thus producing the population inversion needed.

An electric discharge may occur under the influence of a direct current, dc, or an alternating current, ac, although radio frequencies, rf, of about 20 to 30 kHz may be used. In the dc and the ac mode, two electrodes are needed that protrude into the discharge tube as shown later in Figure 5.4-10. The ac mode is easiest to realize because not much more is needed than an ordinary high-voltage transformer.

In a particularly important class of lasers excitation by electric discharge still provides the initial excitation which raises one type of atoms to their excited state, or states. These atoms then *collide* inelastically with another type of atoms, and it is these latter atoms that provide the population inversion needed for laser emission. An example of this is the helium–neon laser.

A *direct conversion* of electric energy into radiation occurs in light-emitting diodes, LEDs, and in the type of semiconductor lasers that derive from them.

H. Townes, "Infrared and Optical Masers," *Phys. Rev.* **112** (1958), 1940–49; and A. M. Prokhorov, "Molecular Amplifier and Generator for Submillimeter Waves," *Soviet Physics, J.E.T.P.* **34** (1958), 1658–59. In 1964, Townes, Prokhorov, and Basov, and in 1981 Schawlow, were awarded the Nobel prize for physics.

In a *chemical laser,* the energy comes from a chemical reaction, without need for any other energy source. Hydrogen, for example, can combine with fluorine,

$$H_2 + F_2 \rightarrow 2HF \qquad\qquad [5.4\text{-}14]$$

a reaction that generates enough energy to pump a CO_2 laser.

Cavity configurations.   If a *single* passage of light through the active medium is sufficient to pump a laser to population inversion, the laser will amplify the light; this is called a laser *amplifier*. Most lasers, however, are designed so that the light makes *multiple* passes through the medium; the light is reflected back and forth within a *cavity*. This makes available many more atoms for the excitation process, for population inversion, and for avalanche action; such a system is called a laser *oscillator*.

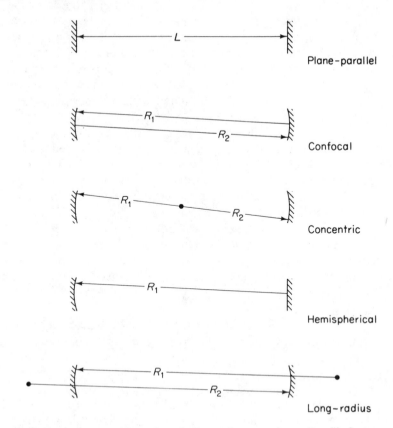

Figure 5.4-5   Cavity configurations. L, distance between mirrors; R, radii of curvature of the mirrors.

Optical cavities can have many configurations. In most the light is reflected between two mirrors that are facing each other, as in a Fabry–Perot interferometer. The simplest form of a resonant cavity is limited by two *plane-parallel mirrors* (Figure 5.4-5, top). These mirrors require very precise alignment. The plane-parallel configuration is rarely used today; it is mainly of historic interest.

A *confocal* cavity has two identical concave mirrors separated by a distance equal to their radius. These lasers are easier to align. A *concentric* (also called *spherical*) resonator contains two concave mirrors of the same radius, separated by a distance twice their radius, $L = 2R$. A *hemispherical* cavity consists of a concave spherical mirror at one side and a plane mirror at the other. The plane mirror is placed at the center of curvature of the spherical mirror, $L = R_1$. Such resonators are considerably easier to align than those mentioned so far but the power output is only about one-half of what it could be.

A *"long-radius"* cavity has two concave mirrors of equal, and fairly long, radius of curvature. This is often a good compromise between the plane-parallel and the confocal variety; it is the type used most often in today's commercial lasers.

Mode structure.   Assume for simplicity that the cavity is limited by two plane-parallel mirrors. Since the mirrors form a *closed* cavity, the radiation field inside the cavity generates a *standing-wave pattern* that has nodes, rather than antinodes, at both ends. If L is the length of the cavity, the longest wavelength that will produce a standing wave is $\lambda_0 = 2L$. The next shorter wavelength is $\lambda = L$, the next shorter wavelength $\lambda = \frac{2}{3}L$, and so on. Thus, in general terms,

$$\lambda = \frac{2}{q} L \qquad\qquad [5.4\text{-}15]$$

where q is an integer, 1, 2, 3, . . . , called the *axial mode number,* which describes the number of half-wavelengths or *axial modes* that fit into the cavity. For these wavelengths the cavity is *resonant.*

It is often preferable to write this equation in terms of frequency, ν. In order to do so, we replace the velocity, v, in $v = \lambda\nu$ by the velocity of light, c, solve for ν, and substitute Equation [5.4-15]. Furthermore, since a laser cavity must, by necessity, contain a medium (of index n), rather than free space, we replace the actual length of the cavity, L, by the optical path length, $S = Ln$. Thus, instead of Equation [5.4-15], we now have

$$\nu = q\,\frac{c}{2S} \qquad\qquad [5.4\text{-}16]$$

which is the *resonance condition for axial modes,* in terms of frequency.

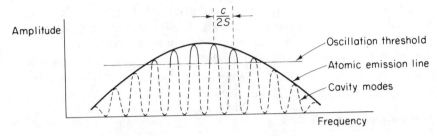

Figure 5.4-6   Output of laser consists of a series of resonance modes separated by $c/2S$ and often contained within a single emission line.

Since two consecutive modes differ by $q = 1$, they are separated by a *frequency difference, $\Delta \nu$,*

$$\Delta \nu = \frac{c}{2S} \qquad\qquad [5.4\text{-}17]$$

These slightly different frequencies are closely, and evenly, spaced. Often, several of them lie within the frequency range where excitation exceeds the threshold; that is, they lie within the width of a single emission line. The output of a laser, therefore, need not be exactly monochromatic; it consists of several resonant lines, separated from each other by $c/2S$, as shown in Figure 5.4-6.

In addition to the axial modes we also have *transverse modes. Their* standing-wave pattern is best described by the acronym TEM, which stands for **transverse electromagnetic mode**. The transverse modes, in

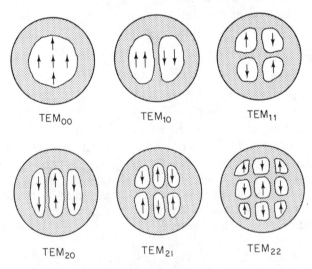

Figure 5.4-7   Schematic representation of various TEM modes.

contrast to the axial modes, are generally few in number, and they are easy to count. Just look at the (dark) intervals, the "nodal lines," between the patches of light that are seen on a distant screen, especially when the beam has been spread out by placing a negative lens in its path. Within each patch, the phase of the light is the same, but between patches the phase is reversed. Figure 5.4-7 shows several examples. Note that the TEM_{00} mode has no nodal lines; it has the same phase (uniphase) across the aperture.

The TEM_{00} mode is widely used for several reasons: there is no phase reversal across the beam, the spatial coherence is the highest possible, and the beam's divergence is least; thus the beam can be focused to the least spot size and reach the highest power density.

Gain. The *gain* of a laser depends on several factors. Foremost among them is the separation of the energy levels that provide laser transition. If the two levels are farther apart, the gain is higher because the laser transition energy is a larger fraction of the pump transition energy. If the two levels are closer together, the gain is lower.

In a way, gain is the opposite of absorption. Return for a moment to the exponential law, Equation [5.3-6],

$$\phi' = \phi_0 e^{-\alpha x} \qquad [5.4\text{-}18]$$

Evidently, considering the *loss* of light that occurs in absorption, the exponent (αx) in this equation is negative and the absorptivity α is positive. This is true for thermal equilibrium (normal Boltzmann distribution) where $N_2 < N_1$. Conversely, if population inversion occurs where $N_2 > N_1$, the exponent is positive and α is negative. Indeed, laser emission may be considered *negative absorption*.

It is often convenient to write Equation [5.4-18] in the form

$$\phi' = \phi_0 e^{\beta x} \qquad [5.4\text{-}19]$$

where β is the *gain coefficient,* the negative of the absorptivity ("absorption coefficient"), $\beta = -\alpha$.

As the wave is reflected back and forth between the mirrors, it will lose some of its energy, mainly because of the limited reflectivity of one of the mirrors. (This limited reflectivity is needed to let radiation pass through the mirror to be utilized.) If the two mirrors have reflectivities r_1 and r_2, then with each round trip the initial power in the cavity, ϕ_0, decreases by a factor $r_1 r_2$ and the resultant power becomes

$$\phi' = \phi_0 r_1 r_2 \qquad [5.4\text{-}20]$$

Eliminating ϕ'/ϕ_0 between Equations [5.4-18] and [5.4-20] and calling γ

the loss per round trip, we find that

$$r_1 r_2 = e^{-\gamma} \qquad [5.4\text{-}21]$$

and hence,

$$\gamma = -\ln{(r_1 r_2)} \qquad [5.4\text{-}22]$$

where γ, like α in absorption, is positive.

Clearly, in order to attain laser emission, the gain must at least equal the sum of the losses. Thus we extend Equation [5.4-18] to read

$$\phi' = \phi_0 e^{\beta - \alpha x} \qquad [5.4\text{-}23]$$

where β is the gain coefficient, α the absorptivity, and x the separation of the mirrors. If $\beta > \alpha x$, the radiation field inside the medium builds up and the system will lase. By contrast, if $\beta < \alpha x$, the oscillations will die. The equality

$$\beta = \alpha x \qquad [5.4\text{-}24]$$

hence, is the *threshold condition* necessary to sustain laser emission.

Types of Lasers

Solid-state lasers. The classical example of a three-level solid-state laser is the *ruby laser*.* Ruby is synthetic sapphire, aluminum oxide, Al_2O_3, with 0.05 percent per weight of chromium oxide, Cr_2O_3, added to it. The Cr^{3+} ions are the active ingredient; the aluminum and oxygen atoms are inert. The ruby crystal is made into a cylindrical rod, several centimeters long and several millimeters in diameter. The ends of the ruby are polished flat, normal to the axis, to act as the cavity mirrors. Pumping is by light from a xenon flash tube.

As shown in Figure 5.4-8, chromium-doped ruby has three energy levels, E_1, E_2, and E_3. The uppermost level, E_3, is fairly wide (to accept a wide range of wavelengths) and has a short lifetime: the excited Cr^{3+} ions rapidly relax and drop to the next lower state, E_2. This transition is non-radiative. The E_2 state is *metastable;* it has a lifetime of about 10^{-3} s, several orders of magnitude longer than that of E_3. Consequently, the Cr^{3+} ions remain that much longer in E_2 before they drop down to the ground state, E_1. The $E_2 \rightarrow E_1$ transition is radiative; it produces the spontaneous, incoherent red fluorescence typical of ruby, with a peak near 694 nm. But, as the pumping energy is increased beyond the threshold, population

---

*First described by T. H. Maiman, "Stimulated Optical Radiation in Ruby," *Nature, London* **187** (1960), 493–94, and "Stimulated Optical Emission in Fluorescent Solids. I. Theoretical Considerations," *Phys. Rev.* **123** (1961), 1145–50.

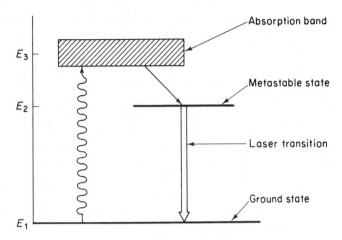

Figure 5.4-8 Three-level energy diagram representative of ruby laser. Pumping transition is shown by wavy line, nonradiative transition by simple arrow, stimulated transition by hollow arrow.

inversion occurs in E_2 with respect to E_1, the output becomes coherent, and the system lases, with a sharp peak at 694.3 nm.

A three-level system such as ruby requires high pumping powers, because the laser transition terminates at the ground state and more than one-half of the ground-state atoms must be pumped up to the higher state to achieve population inversion. By contrast, if we have a *four-level system,* much less power is needed for excitation. As shown in Figure 5.4-9, excitation, return to a metastable state, and stimulated emission are

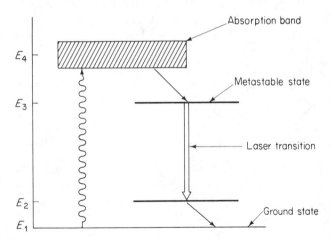

Figure 5.4-9 Four-level system typical of neodymium laser. Pumping transition is shown by wavy arrow, nonradiative transitions by simple arrows, and laser transition by hollow arrow.

all the same as in a three-level system except that the radiative transition does not terminate at the ground state. There is now an additional energy state above the ground state, split off from it by the crystal field.

As before, excitation raises the system from the ground state, E_1, to the highest of the four levels, E_4. From there the atoms decay to the metastable state, E_3. Since both E_4 and E_2 drain rapidly, laser action will commence as soon as the metastable state, E_3, has reached population inversion with respect to E_2. This inversion can be maintained with only moderate pumping, which allows such lasers to operate in the *continuous wave* (c.w.) mode, in contrast to most three-level lasers, which emit *pulses*.

The typical example of a four-level system is the *neodymium laser,* where trivalent neodymium ions, Nd^{3+}, are added to a matrix either of glass or of yttrium aluminum garnet, YAG. Emission from a neodymium–glass laser is at 1.060 μm, from the (more efficient) Nd:YAG laser at 1.064 μm.

Liquid lasers.
In contrast to solids, liquids do not crack or shatter, and can be made in sizes almost unlimited. Early liquid lasers contained chelates, metallo-organic compounds formed of a metal ion and an organic radical, such as europium benzoyl acetonate. Emission is at 613.1 nm.

Nonchelate *dye lasers* are often built around rhodamine or other fluorescent organic dyes. There exist so many fluorescent dyes that lasers can be made for almost any wavelength desired, from the UV to the IR. Moreover, dye lasers can be *tuned,* often through a considerable range, about 100 nm, either by varying the concentration of the dye or by adding, and turning, a diffraction grating in place of one of the cavity mirrors.

Fluorescence, together with phosphorescence, is part of a "cold" emission of light called *luminescence*. According to the type of energy supplied, we distinguish *photoluminescence,* where the energy is supplied in the form of light; *electroluminescence,* where it is supplied as electric energy; *chemo-* and *bioluminescence,* where it comes from chemical reactions usually involving an oxidation–reduction process; and *triboluminescence,* which is due to mechanical forces.

In fluorescence, the emission of the fluorescent light lasts no longer than about 10^{-8} s after excitation has ended. Some of the energy absorbed in the excitation process ($1 \rightarrow 2$ transition) goes into heating, and is lost for emission ($2 \rightarrow 1$ transition). Thus, compared with the primary (exciting) light, the energy of the secondary (fluorescent) light is less, $h\nu_{21} < h\nu_{12}$, and its wavelength longer, $\lambda_{21} > \lambda_{12}$. This is called *Stokes' law*. In phosphorescence, the emission persists for some time, up to several hours, after excitation has ended.

Gas lasers. Depending on the gas used, we distinguish atomic, metal–vapor, and molecular gas lasers. An example of an atomic gas laser is the *helium–neon laser,* probably the most widely used type of all.* A typical He–Ne laser consists of a Pyrex tube, about 35 cm in length and 2 mm in diameter, with electrodes on the side and fused silica windows set at Brewster's angle. The tube contains a mixture of 5 parts helium and 1 part neon, kept at a pressure of 133 Pa (approximately equal to 1 mm Hg). The electrodes are connected to a high-voltage source of about 4 kV dc. The mirrors are placed outside the tube as shown in Figure 5.4-10.

As mentioned, the end windows are set at Brewster's angle. This causes light, returning from the outside mirrors back to the cavity and oscillating *normal* to the plane of incidence, to be reflected away from the cavity. Hence, the component oscillating *parallel* to the plane of incidence becomes dominant and will sustain laser emission: the radiant energy that emerges from the laser will be *linearly polarized.*

Excitation of an He–Ne laser initially involves only the helium. Interestingly enough, two of the higher helium states have almost the same energy as two of the higher neon states. The excited helium atoms, when they collide with neon atoms still in the ground state, will then transfer energy to the neon. The advantage of this collision process is that the fairly light helium atoms can be easily pumped up to their excited states; the much heavier neon atoms could not be raised efficiently without them. The neon, after collision, is raised to two excited states, $3S$ and $2S$, while the helium returns to the ground state. These higher neon states are meta-stable and quickly reach population inversion relative to the lower $3P/2P$ states.

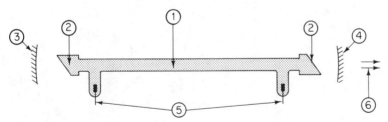

Figure 5.4-10 Helium–neon laser and its essential components: 1, Gas discharge tube; 2, end windows set at Brewster's angle; 3, fully reflective mirror; 4, partially reflective mirror; 5, electrodes connected to power supply; 6, laser output.

---

*First disclosed by A. Javan, W. R. Bennett, Jr., and D. R. Herriott in "Population Inversion and Continuous Optical Maser Oscillation in a Gas Discharge Containing a He–Ne Mixture," *Phys. Rev. Letters* **6** (1961), 106–10, who discovered emission at 1.15 μm. The familiar red laser light of 632.8 nm wavelength was obtained a year later by A. D. White and J. D. Rigden, "Continuous Gas Maser Operation in the Visible," *Proc. IRE* **50** (1962), 1697.

The neon returns to these lower states by various transitions, most notably by laser emission at 632.8 nm, in the c.w. mode (Figure 5.4-11).

An example of an *ion laser* is the argon laser. It generates light of 488.0 nm and 514.5 nm, in either pulsed or c.w. operation. Best known among *metal–vapor lasers* is the helium–cadmium laser. Like the He–Ne laser, it operates in the c.w. mode; it emits a brilliant blue of 441.6 nm wavelength.

A good example of a *molecular gas laser* is the carbon dioxide laser. This laser is virtually in a class by itself because in contrast to atomic gas lasers (such as He–Ne or He–Cd), it has a particularly high power output while retaining the high monochromaticity and spatial coherence characteristic of atomic lasers. Even in the c.w. mode a CO_2 laser can produce more than 200 kW, enough to cut through a 20-mm-thick steel plate in a matter of seconds. Excitation is by way of nitrogen, which takes the place of the helium in the He–Ne laser, and subsequent collision with CO_2. Emission is at 10.6 μm.

A special type is the *TEA CO$_2$ laser,* the acronym standing for transverse electric excitation at atmospheric pressure. At that pressure the gas has more molecules per unit volume, the laser tube need not be sealed hermetically, and the electrodes can be placed closer together, which means lower pumping voltages and overall higher efficiency.

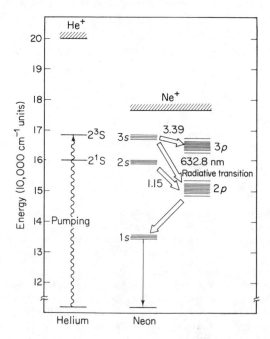

Figure 5.4-11 Energy-level diagram of helium–neon laser. Hollow arrows: radiative transitions.

Excimer lasers contain rare-gas halides such as XeF, KrF, ArF, or others. These molecules are unstable in the ground state but bound in the excited state. Laser emission is in the short ultraviolet.

Semiconductor lasers.
Still another class of lasers is derived from the *light-emitting diode,* LED. These diodes are small, comparable in size to a transistor (less than 1 mm in diameter), and convert electric energy directly into light, with no need for a separate energy unit as in optical pumping.

A semiconductor has energy levels that usually fuse into two major bands, a lower *valence band,* which is filled with electrons and thus carries a minus charge, and an upper *conduction band,* which is empty. Adding impurity atoms, donors, which supply electrons to the conduction band results in an n-type material; adding acceptors, which remove electrons from the valence band and leave holes in their place, results in a p-type material. If an electric field is applied to the p-n junction, and if the junction is forward-biased, that is, if the p material is connected to the positive terminal, the electrons flow from the n side to the p side, and the holes from p to n. As the electrons combine with the holes across the band gap, excess energy is released in the form of light.

The first LEDs were made from gallium arsenide, GaAs. They require very little current (about 10 mA at 1.7 V). But as more current is applied (and if, preferably, certain dopants are added to the GaAs), excited hole–electron pairs that have not had time to recombine spontaneously, are forced together causing *stimulated emission*. In the diode or injection laser, the light is trapped and reflected back and forth between the end faces of the crystal, much as in a conventional laser; the light leaves the junction as shown in Figure 5.4-12. Emission for GaAs is at 905 nm, but depending on the dopants, and on the semiconductor material or combination of

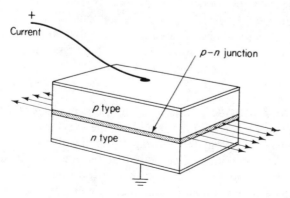

Figure 5.4-12   Schematic diagram of p-n junction laser.

Table 5.4-1   SUMMARY OF TYPICAL LASERS

Ion	Matrix	Principal wavelength
Cadmium	Helium	441.6 nm
Argon		514.5 nm
Neon	Helium	632.8 nm
Chromium	Aluminum oxide (ruby)	694.3 nm
Gallium arsenide		905 nm
Neodymium	YAG	1.064 μm
Carbon dioxide	Nitrogen	10.6 μm

materials, diode lasers can be made to emit almost anywhere in the spectrum, from the UV to the IR. Their efficiency is much higher than optical pumping (around 40 percent versus 3 percent), their c.w. output 10 mW, and their peak power up to 100 W. Their main application is in waveguide communication systems and integrated optics.

Since Maiman first used ruby, numerous other materials have been found to lase, emitting radiation from the extreme ultraviolet (109.8 nm, for a para-molecular hydrogen line) to millimeter waves (1.991 mm, for a methyl chloride transition). All gaseous elements and many molecules are known to lase, and the same is true of most metals. Any dye that fluoresces will lase if enough energy is supplied. The types of lasers most commonly used are listed in Table 5.4-1.

Applications

Compared to radiation from other sources, laser radiation stands out in several ways. It has a particular beam shape and divergence. It is highly coherent, both spatially and temporally. It is eminently suited to produce interference. Often it is polarized as soon as it leaves the laser cavity. It can be generated in the form of very short pulses, at high powers and, because of its high spatial coherence, at very high power densities. These are among the properties that make lasers the unique tool they are for a great many practical applications.

Beam shape.   A laser beam has a certain *profile*. Most of its energy is concentrated in the center, near the axis of propagation; toward the periphery, the energy falls off rapidly. A stable mode structure results whenever the amplitude distribution across the output mirror, after reflection back and forth between the mirrors, remains the same. The question then is what amplitude distribution will reproduce itself in the far

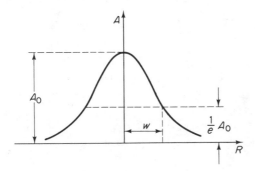

Figure 5.4-13 Amplitude distribution across laser beam oscillating in the TEM$_{00}$ mode.

field; that is, what function is its own Fourier transform? There are several solutions to this problem but the simplest is that of a *Gaussian distribution*.

The maximum amplitude, $\mathbf{A}_0$, exists in the center of the beam. With increasing distance, r, from the axis the amplitude drops off exponentially:

$$\mathbf{A}_r = \mathbf{A}_0 e^{-(r/w)^2} \qquad [5.4\text{-}25]$$

where w is the radius of the beam. The Gaussian function, $\exp\left[-(r/w)^2\right]$, falls to $1/e$,

$$\mathbf{A}_r = \frac{1}{e}\,\mathbf{A}_0 \qquad [5.4\text{-}26]$$

when the distance r is *equal* to the beam radius w, $r = w$ (Figure 5.4-13). Furthermore, since energy is proportional to the square of the amplitude, the beam radius, or *spot size*, w, is defined as that distance from the axis where the power has dropped to $1/e^2$ of its value in the center. Twice that distance, $2w$, is the *beam diameter*.

The beam radius, w, is a function of distance along the axis. If we call x the axial distance measured from the midpoint between the two (concave) mirrors, then the parameter w is given by*

$$w_x = w_0 \sqrt{1 + \left(\frac{\lambda x}{\pi w_0^2}\right)^2} \qquad [5.4\text{-}27]$$

where λ is the wavelength and w_0 is the minimum beam radius between the

*Following G. D. Boyd and J. P. Gordon, "Confocal Multimode Resonator for Millimeter through Optical Wavelength Masers," *Bell System Tech. J.* **40** (1961), 489–508.

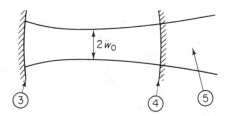

Figure 5.4-14   Confocal cavity causes beam to contract to minimum beam diameter $2w_0$. 3, fully reflective mirror; 4, semitransparent mirror; and 5, laser output.

mirrors. In the case of a confocal resonator, this simplifies to

$$w_0 = \sqrt{\frac{L\lambda}{\pi}} \qquad\qquad [5.4\text{-}28]$$

where L is the distance between the mirrors. Twice the radius w_0, that is, the minimum beam diameter halfway between the mirrors, is called the *beam waist*.

The diameter of the beam waist determines the *divergence* of the beam, simply because the shape of the beam outside of the cavity is an extension of its shape inside. This relationship is illustrated in Figure 5.4-14.

The *diffraction part* of the beam's divergence is given by Rayleigh's criterion, Equation [2.3-16],

$$\theta \approx 1.22 \frac{\lambda}{D} \qquad\qquad [5.4\text{-}29]$$

a limit that applies mainly to gas lasers. As a rule, the diffraction divergence is about twice as large as the beam-waist divergence.

We note from Equation [5.4-29] that the diffraction divergence can be reduced if the beam is passed through a *beam expander,* which is an inverted afocal telescope, either of the Galilean or the astronomical type. The Galilean type, shown schematically in Figure 5.4-15, is preferred, especially with high-power lasers because the rays are not brought to a focus, which might cause ionization and breakdown of the air.

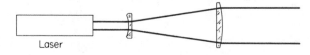

Figure 5.4-15   Beam expander. Type shown is inverted Galilean telescope.

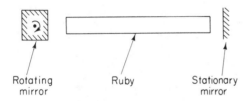

Figure 5.4-16   Q-switched laser.

Power.   Power is the amount of energy delivered per unit of time. A typical laser pulse contains about 10 J of energy. This does not seem to be very much, but if this energy is delivered within 0.5 ms, the power output is 20 000 W. Some lasers produce 300 J in 5 ns, which translates into 60 GW (gigawatt, 10^9 W). But even pulses as short as 30 femtoseconds (30×10^{-15} s) have been reached, and with other lasers energies of up to 600 kJ and powers as high as 2500 TW (terawatt, 10^{12} W). That is an incredible amount, about the same as the output of all power plants in the world combined. These plants, of course, generate power continually; a laser pulse may last for perhaps only a fraction of a nanosecond.

Such high powers are generally produced by *Q switching*, which means compressing the energy released into a very short period of time. One possibility of doing this is by using a rotating mirror at one end of the laser cavity (Figure 5.4-16). The semitransparent mirror at the other end is stationary, as usual. Then, even after the system has been pumped, oscillation cannot occur; it is delayed until one of the facets of the rotating mirror is exactly parallel to the stationary mirror: at that instant all of the accumulated energy is dumped into one single, very short "giant" pulse. Besides rotating mirrors there are various other schemes that accomplish the same purpose, Kerr cells, KDP crystals, and certain dyes that at first absorb all the light, preventing oscillation, but as the avalanche grows larger, the dye suddenly bleaches and oscillation occurs, with the same result as with the rotating mirror.

Industrial applications.   The high powers available from lasers have led to a great many industrial and medical applications, from more mundane tasks such as the *machining* of materials to the exotic, such as containing a plasma and trying to release energy from nuclear fusion.

Drilling by laser is commonplace. Diamond, about the hardest material known, has been drilled before, but the process is tedious and time consuming. A laser does it quickly, producing traces of (black) graphite which facilitates absorption. Holes have been drilled also into teeth, paper clips, even into single human hairs. Lasers are used also for *cutting* materials, from stacks of fabrics to ceramics to metals. A sheet of 2.5-mm-thick stainless steel, for example, can be cut at a rate of better than 25 cm min^{-1}, 25-mm-thick titanium at 75 cm min^{-1}.

Welding by laser meant earlier only welding on a microscopic scale, such as in integrated circuits, or in the retina of the eye in *retinal detachment*. Momentary exposure to the focused laser beam causes small burns that on healing keep the retina back in place by the formation of scar tissue. Lasers show promise also in the treatment of certain, easily accessible tumors such as *melanoblastoma,* a heavily pigmented, often completely black tumor of the skin and the choroid of the eye which sometimes erupts into highly malignant growth. But in recent years, as lasers have become more powerful, even heavy steel plates have been welded. For example, steel plates 5 cm thick have been welded together, using a 90-kW laser, at speeds higher than 2.5 m min^{-1}.

Because of their high directionality and high frequency, laser beams looked at first promising for *communications*. However, turbulence in the air severely limits the amount of information that can be transmitted, unless the beam is confined to a fiber waveguide. *Optical radar,* known as "lidar," for **l**ight **d**etection **a**nd **r**anging, makes it possible to obtain echoes from clouds, haze, and atmospheric pollutants too small to be detected by conventional (microwave) radar.

Very important among practical applications are *precision measurements,* including alignment and the measurement of distances, thicknesses, angles, and velocities. The distance to the moon, for example, has been determined to an accuracy *better than 15 cm.*

Nonlinear effects.
Ordinarily, the refractive index and the absorptivity of a material are properties of the material, independent of the light that passes through. With very intense light, this can be different. Some lasers have electric field strengths as high as 10^9 Vm^{-1}, more than enough to cause the breakdown of air (3×10^6 Vm^{-1}) and only a few orders of magnitude less than the electric fields holding a crystal together.

The Coulomb force between the nucleus and the electron in an hydrogen atom, from Problem 5.2-14, for example, is

$$\mathbf{F} \approx 8 \times 10^{-8} \text{ N}$$

If we convert this force into electric field strength, $\mathbf{E}$, then from Equation [3.1-4],

$$\mathbf{E} = \frac{\mathbf{F}}{q} = \frac{8 \times 10^{-8}}{1.6 \times 10^{-19}} = 5 \times 10^{11} \text{ Vm}^{-1} \qquad [5.4\text{-}30]$$

about the same as the local field inside a crystal.

This causes a number of phenomena long predicted by theory but observed only recently since the advent of high-power lasers. Light from such lasers can easily change the refractive index, causing "self-focusing," which means that a beam of light contracts into powerful, short-lived threads of light, and quickly shatters the material on which it is incident.

Frequency doubling is the generation of second harmonics. If red light from a ruby laser (694.3 nm) is passed through a KDP crystal, part of the light is converted into 347.15 nm UV, exactly one-half the wavelength of the incident light. By higher-order frequency conversion, still shorter wavelengths can be reached. For example, if we start out with the 1.064-μm emission from an Nd:YAG laser, we can, by successive frequency doubling, first in a crystal of KDP and then in potassium dideuterium phosphate, reach the fourth harmonic, 266 nm. Of this radi-

---

Laser speckle.   Laser light is very coherent and very intense. So it comes as a disappointment that this light, when projected on a screen, is rather mottled and uneven. Laser light has a definitely grainy structure, a phenomenon called *speckle*.

The theory of laser speckle is very complex. It is a matter of the statistics of random processes. With some simplification we can say that any surface has a certain roughness, with a vast number of individual facets and scatterers. The light reflected from such a surface consists of contributions from all of these facets and, since the light has high spatial coherence, these contributions form loci of interference, distributed at random in three-dimensional space in front of the surface. For example, look at a screen diffusely illuminated by laser light. While looking, slowly move your head from side to side as if saying "no." The path lengths traversed by these contributions will change and the speckle pattern will move, an effect aptly called a *"red snowstorm."*

The same effect is seen when, instead of moving the head, the screen is moved slowly transverse to the line of sight. If the head, or the screen, is moved slowly in one direction and the speckle grains are seen to move in the *opposite* direction, the observer is myopic (nearsighted). If the grains move in the *same* direction, the observer is hyperopic (farsighted). This can be explained by the position of the far point of the eye, which, as we have seen on page 144, in myopia is located in front of the eye, and in hyperopia behind it. This effect can be used to test for potential visual refractive deficiencies.

ation, the fifth, and even the seventh harmonic have been produced,* which means a wavelength *as short as 38 nm,* within the X-ray range.

Laser Safety

Any light, no matter where it comes from, can cause damage. But *laser radiation is particularly dangerous,* mainly because of its high spatial coherence (which means that it can be focused down to very high power densities). In addition, many lasers emit in a region of the spectrum (red) where the light just does not seem to be as powerful as if it were white.

Of all the parts of the human body, clearly the *eye* is the most vulnerable. Which part of the eye is subject to injury is a matter of wavelength. In the IR, above 10 μm, much of the energy is absorbed by water. Since water is the main constituent of most any biologic tissue, it is the *cornea* that is severely damaged first. Out of the 10.6-μm radiation produced by a CO_2 laser, for example, 90 percent is absorbed in a layer of water only 0.025 mm thick.

Light between 400 nm and 1.4 μm, on the other hand, will penetrate through the eye and be absorbed in the pigment epithelium behind the *retina*. The exact wavelength plays only a minor role. The major factor is whether the eye is relaxed or accommodating. If the eye is relaxed, the output of a 50-kW neodymium laser can be focused to 10^{16} Wm^{-2}, about 100 million times the power density on the surface of the sun. The pigment epithelium then virtually explodes, ejecting black granules into the vitreous, forming vapor bubbles, and causing severe hemorrhages surrounding the vaporized and destroyed tissue.

The *maximum permissible exposure,* MPE, when looking directly into a laser beam, is a function of the exposure *time.* For very short exposures, from 1 ns (10^{-9} s) to 2×10^{-5} s, and at wavelengths between 400 nm and 1.4 μm, the maximum permissible exposure is

$$\text{MPE} = 0.0005 \text{ mJ cm}^{-2} \qquad [5.4\text{-}31]$$

For exposure times from 2×10^{-5} to 10 s the limit is

$$\boxed{\text{MPE} = 1.8 \ t^{3/4} \text{ mJ cm}^{-2}} \qquad [5.4.\text{-}32]$$

---

*J. Reintjes, S. Chiao-Yao, and R. C. Eckardt, "Generation of Coherent Radiation in the XUV by Fifth- and Seventh-Order Frequency Conversion in Rare Gases," *IEEE J. Quantum Electron.* **QE-14** (1978), 581–96.

and for exposures longer than 10 s it is

$$\text{MPE} = 10 \text{ mJ cm}^{-2} \qquad\qquad [5.4\text{-}33]$$

The limit when looking at diffusely reflected laser light, for the same wavelengths and for exposure times from 1 ns to 10 s, is

$$\text{MPE} = 10 \, t^{1/3} \text{ J cm}^{-2} \text{ sr}^{-1} \qquad\qquad [5.4\text{-}34]$$

Needless to say, these limits are merely guidelines and should not be used to predict what is safe and what is not. They are based on the best information available today.

Example. *An individual is accidentally exposed to a 1 ms pulse from a ruby laser. If the beam and the pupil of the eye are both 6 mm in diameter, what is the maximum permissible power?*

Solution. A 6-mm-wide pupil has an area

$$A = \pi R^2 = (\pi)(0.3)^2 = 0.28 \text{ cm}^2$$

The maximum permissible power, from Equation [5.4-32], is then

$$\text{MPE } t^{-1} = (1.8)(10^{-3})^{3/4}(10^{-3}) \text{ J cm}^{-2} (0.28 \text{ cm}^2)(10^{-3} \text{ s})^{-1}$$

$$= (1.8)(10^{-9/4})(0.28) \text{ J s}^{-1} = \boxed{2.8 \text{ mW}}$$

For a 1-mW c.w. He–Ne laser the maximum permissible exposure time is $\frac{1}{10}$ of a second! That is a good figure to remember. It means that even for a simple laboratory laser *the human reaction time is too long to prevent possibly permanent damage.* EXTREME CAUTION MUST BE USED IN ANY WORK INVOLVING ANY TYPE OF A LASER!

SUGGESTIONS FOR FURTHER READING

D. C. O'Shea, W. R. Callen, and W. T. Rhodes, *Introduction to Lasers and Their Applications* (Reading, MA: Addison-Wesley Publishing Company, Inc., 1978).

B. A. Lengyel, *Lasers,* 2nd edition (New York: John Wiley & Sons, Inc., 1971).

G. R. Fowles, *Introduction to Modern Optics,* 2nd edition, (New York: Holt, Rinehart and Winston, Inc., 1975), pp. 264–281.

G. H. B. Thompson, *Physics of Semiconductor Laser Devices* (New York: Wiley-Interscience, 1980).

J. F. Ready, *Industrial Applications of Lasers* (New York: Academic Press, Inc., 1978).

W. Demtröder, *Laser Spectroscopy: Basic Concepts and Instrumentation* (Heidelberg: Springer-Verlag, 1981).

A. Mallow and L. Chabot, *Laser Safety Handbook* (New York: Van Nostrand-Reinhold Company, 1978).

Good accounts of the literature on lasers, their history, development, types, applications, laboratory experiments, and safety aspects, are found in B. A. Lengyel, "Evolution of Masers and Lasers," *Am. J. Physics* **34** (1966), 903–13; D. C. O'Shea and D. C. Peckham, "Resource Letter L-1: Lasers," *Am. J. Physics* **49** (1981), 915–25; P. F. Schewe, "Lasers," *The Physics Teacher* **19** (1981), 534–47; and in recent issues and volumes of *Scientific American*.

PROBLEMS

5.4-1. If laser action occurs by the transition from an excited state to the ground state, $E_1 = 0$, and if it produces light of 693 nm wavelength, what is the energy level of the excited state?

5.4-2. Transition occurs between a metastable state E_3 and an energy state E_2, just above the ground state. If emission is at 1.1 μm and if $E_2 = 0.4 \times 10^{-19}$ J, how much energy is contained in the E_3 state?

5.4-3. Certain lasers can be *optically* pumped by a chemical reaction. Discuss the probable construction of such lasers and their obvious advantages and disadvantages.

5.4-4. Describe the energy-level diagrams of the two most popular lasers that are pumped by atomic collisions.

5.4-5. Ruby has a refractive index of $n = 1.765$. If the crystal is 4 cm long, the wavelength 694.3 nm, and the laser operating in the TEM_{00} mode, what is the least number of standing waves that fit into the cavity?

5.4-6. Determine the frequency difference of a laser that is 1.5 m long and contains a gas of $n = 1.0204$.

5.4-7. Fluorescein in an alkaline solution has an absorption maximum near 495 nm and a fluorescence maximum at 525 nm. How much energy is lost in the process of absorption and reradiation of one quantum?

5.4-8. Fluorescence and phosphorescence can be distinguished by the time they last after excitation has ended. Using two rotating sector plates mounted on a common axis, how could you build a simple apparatus to distinguish the two? (Problem suggested by A. Szent-Györgyi.)

5.4-9. The index of refraction of gallium arsenide near 5 μm is 2.62. What is the reflectivity at the crystal–air interface?

5.4-10. Emission from a semiconductor laser occurs in the narrow junction between the two types of material. If the wavelength is 905 nm and the "slit width" 5.2 μm, what is the full-angle divergence of the output?

5.4-11. Where is the beam waist located in a laser cavity if the cavity is of:
(a) The confocal type?
(b) The concentric type?
(c) The hemispherical type?

5.4-12. If the divergence of a helium–neon laser is 1 arc min and determined only by diffraction at the exit aperture, how wide is the aperture?

5.4-13. A diffraction-limited ruby laser, outside the atmosphere, is aimed at a target 10 km away. If the beam is initially 11 mm wide, how wide is it when it hits the target?

5.4-14. The light from a He–Ne laser is 1.93 mm in diameter. If the beam is diffraction-limited, what is the full-angle divergence of the beam:

(a) As it leaves the aperture?
(b) After it has passed through a 5× telescope used as a beam expander?

5.4-15. If a laser delivers 1.48 mW in a beam 3.6 mm in diameter, what is the power density in a spot 1 μm in diameter, assuming a loss of 20 percent in the focusing system?

5.4-16. If a camera lens can focus collimated light to a spot 25 μm in diameter, what is the power density produced by a pulse from a Q-switched 100-MW laser?

5.4-17. In a laser accident, a He–Ne laser of only 1 mW power projects a beam of 6×10^{-3} rad divergence into the eye. Assume that the refractive power of the eye is +60 diopters, the refractive index inside the eye 1.34, and losses on passing through to the retina 10 percent. What is the power density on the retina?

5.4-18. Can working 8 hours a day under fluorescent lights have any adverse effects? With the safety limit set at 20 J cm^{-2} per day, how does this compare to the result of our example on page 517?

5.4-19. Looking at the speckle pattern on a diffuse screen, an uncorrected *myopic* observer turns his or her head slowly from left to right. (a) How do the loci of interference, generated in front of the screen, move relative to the screen?

(b) How do their conjugate images move on the retina?
(c) Therefore, what does the observer see?

5.4-20. Continue with Problem 5.4-19 and assume now that the observer is *hyperopic*.

5.4-21. What is the maximum power in a 10-ns pulse from a Q-switched ruby laser that can safely pass through a pupil 6.18 mm in diameter?

5.4-22. With the eye exposed for 8 s to a helium–cadmium laser, what is the maximum power that can be allowed to pass through a 5-mm-wide pupil?

5.4-23. What is the maximum energy density per steradian considered safe when looking at light from a 30-ns ruby laser pulse reflected off a diffuse white surface?

5.4-24. A hologram is viewed in the light from a 2.5-W He–Ne laser. If the hologram subtends at the eye an angle of $\frac{1}{2}$ sr, if 10 percent of the energy incident on the hologram is scattered into that angle, if the pupils are dark-adapted and 8.74 mm wide, and if the applicable equation holds true also outside the time limits given, how long may the observer look at the hologram without exceeding the safety limit?

Part **6**

Epilogue

WE HAVE ALMOST REACHED THE END of our introduction to classical and modern optics. And now comes a major constraint. Whenever the light source or the observer or any part of the system is moving, virtually all the laws of optics change. For example, when a point source is moving, the customary, uniform emission of light into space no longer holds and must be reformulated to include the velocity of motion of the source. The same applies to the laws of propagation, refraction, reflection, interference, electromagnetic theory, radiometry, spectroscopy, and others. This is the realm of *relativistic optics*.

6.1

Relativistic Optics

THE EXAMPLE OF RELATIVISTIC OPTICS cited most often is the Michelson–Morley experiment. Actually, the special theory of relativity, which is the foundation of most of relativistic optics, has not evolved through that experiment. Michelson and Morley's experiment stems from the preoccupation of nineteenth-century physics with the concept of the "aether," a concept that, while historically important, bears little relevance to our thinking today. Instead, we will go directly to a discussion of the facts of relativity, both Galileo's and Einstein's, and then go on to a series of examples that illustrate how relativity has changed the concepts and laws of conventional optics.

Transformations

Galileo's transformation. Assume that somewhere in three-dimensional space there occurs a single physical *event*, such as a collision of two particles. To describe the event fully we need four coordinates, three space coordinates, x, y, z, and a time coordinate, t. It is understood that the coordinate system is that of the laboratory including the observer and that it is an *inertial frame of reference*: it may be stationary or it may be in uniform motion but it is not subject to acceleration.

Now consider a given inertial frame, S, and another inertial frame, S', that is in uniform motion relative to S. For simplicity assume that the three sets of axes are parallel to each other and that the frame S' moves, at velocity v, along the x–x' axis, in the direction of $+x$ (Figure 6.1-1). The observer attached to S ascribes to the event the coordinates x, y, z, t. But to an observer moving along with S' the same event occurs at x', y', z', t'. Since the two frames did coincide at time $t = t' = 0$ and since $x = vt$,

$$\boxed{\begin{aligned} x' &= x - vt \\ y' &= y \\ z' &= z \end{aligned}}$$

[6.1-1]

which is *Galileo's space coordinate transformation*. The time at which the event is seen to occur, to both the observers in S and S', is the same; thus we add the statement that

$$\boxed{t' = t}$$

[6.1-2]

Instead of a point assume that we have an extended (one-dimensional) object such as a meter stick and that we wish to determine its length. The endpoints of the stick are called A and B and these are at rest in the S frame. If the stick is parallel to the x axis, the observer in S assigns to these points the coordinates x_A and x_B and the observer in S' the coordinates x_A' and x_B'. Using Galileo's transformation with respect to x, we find that

$$x_A' = x_A - vt_A \qquad \text{and} \qquad x_B' = x_B - vt_B$$

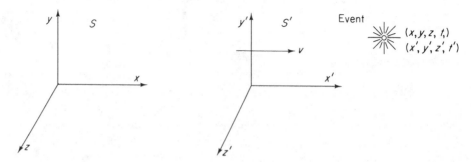

Figure 6.1-1   Two inertial frames: (*left*), S is stationary; (*right*), S' moves to the right at velocity v.

so that

$$x_B' - x_A' = x_B - x_A - v(t_B - t_A)$$

Since the two endpoints, A and B, are measured at the same time,

$$t_A = t_B$$

and

$$x_B' - x_A' = x_B - x_A \qquad\qquad [6.1\text{-}3]$$

The meter stick, therefore, has the same length in both the S frame and the S' frame.

Similar arguments can be made for the velocity, the acceleration, momentum, angular momentum, kinetic energy—in short, for Newton's laws and all other laws of mechanics that follow from them. These laws are the same in all inertial frames; they are *invariant* of the transformation. This is not so, however, for the laws of electrodynamics. Clerk Maxwell's equations, for example, are not preserved in form under Galilean transformation. Before experimental confirmation, Albert Einstein came forward with *two postulates* (assumptions) to resolve this dilemma.

Einstein's postulates.   These two postulates are the very foundation of the special theory of relativity. They are known as *Einstein's postulates.** They state that:

1. The laws of physics, including electrodynamics, are the same in all inertial frames of reference. There is no preferred frame. It is not possible to detect any absolute motion of bodies in space but only relative motions of one body with respect to another. This is the *principle of relativity*.

2. The velocity of light in free space is the same in all inertial frames

---

*Albert Einstein (1879–1955), German-born physicist, mathematician, philosopher, and humanitarian. From 1902 to 1909, Einstein worked as a patent examiner in the Swiss Federal Patent Office in Bern, then became professor at the universities of Zürich and Prague, the Swiss Polytechnic Institute, and the Prussian Academy of Sciences in Berlin. In 1933, Einstein came to the United States and joined the then newly organized Institute for Advanced Studies in Princeton, New Jersey, where he remained for the rest of his life.

The principal paper that laid the foundation for the special theory of relativity is A. Einstein, "Zur Elektrodynamik bewegter Körper," *Ann. Physik* (4) **17** (1905), 891–921, available in an English translation in A. Einstein et al., *The Principle of Relativity* (New York: Dover Publications, Inc., 1981). By his own account, Einstein began working on the special theory of relativity at the age of 16 and continued working on it, off and on, for 10 years.

and is independent of the motion of the source emitting the light. This is the *principle of the constancy of the speed of light.* *

The unique aspect of Einstein's work is that it was not based on prior experimental evidence. Rather, it led to a *prediction* of experiments whose results, over the years, have conclusively shown that Einstein was right.

Lorentz' transformation.

If we measure the length of an object *at rest*, in its own frame of reference, S, we measure its *proper length* ("rest length"). But the object may be moving, in a frame S', relative to S. This requires the two observers, in S and S', to determine the coordinates of the endpoints of the object simultaneously. If the endpoints are not determined at the same time, then, because of the finite speed of light, one point may be seen to have moved more, or less, than the other point and the length measured may not be the proper length.

Simultaneity is the crucial word. But how can we tell that two events are simultaneous or, in a more general sense, how can we *synchronize* two clocks? Within the same frame this is easy. But we have two frames, moving relative to each other. If it were possible to transmit a signal with infinite speed, there would be no problem. But we do not know any signals that could reach all points of the universe in zero time. The fastest signals known are those transmitted by light or other electromagnetic radiation, such as radio waves. No faster method of sending a signal has ever been found.

It is actually classical physics which makes the fictitious assumption of zero time (science fiction). Relativistic physics requires a finite, limiting speed. Nature itself shows that relativistic physics is *real* and not a philosophical concept.

Will the synchronized clocks of the observer in S also be in synchrony with the synchronized clocks of the observer in S'? Because of the finite speed of light, they will not. This means that simultaneity is not independent of the frame of reference. Time, and space, are both relative; to find a connection between the unprimed and the primed coordinates, we somehow have to tie the concept of time into the concept of space.

To do this we use a hypothetical *light clock*. This "clock" is illustrated in Figure 6.1-2 (and will come up again, somewhat modified, in the Michelson–Morley experiment, Figure 6.1-8). A source A emits a pulse of

---

*At first sight, this is hard to comprehend. If a bullet is fired from a moving truck, the velocity of the bullet is different depending on whether the gun is pointed forward or backward with respect to the motion of the truck. This is not so with light. Although experiments have been tried to prove the "ballistic theory," all of them failed, as shown, for example, by G. C. Babcock and T. G. Bergman, "Determination of the Constancy of the Speed of Light," *J. Opt. Soc. Am.* **54** (1964), 147–51.

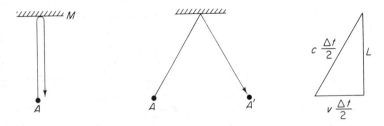

Figure 6.1-2 Light clock at rest (*left*) and moving at velocity v (*center*). (*Right*) Graphical construction.

light that travels through distance L to a mirror, M, and back again to A. The clock is attached to the S' frame. But to the observer who travels with it, the clock is at rest and the time interval for the pulse to go from A to M and back is

$$\Delta t' = \frac{2L}{c} \qquad\qquad [6.1\text{-}4]$$

Now let the clock be seen by the observer in S. To this observer the clock is moving and the path from A to the mirror and back to A' is longer. In order to describe Δt in terms of $\Delta t'$, consider the triangle in Figure 6.1-2, right. From Pythagoras' theorem we find that

$$\left(c\frac{\Delta t}{2}\right)^2 = L^2 + \left(v\frac{\Delta t}{2}\right)^2 \qquad\qquad [6.1\text{-}5]$$

$$c^2\,\Delta t^2 - v^2\,\Delta t^2 = 2L^2$$

and thus

$$\Delta t = \frac{2L}{\sqrt{c^2 - v^2}} = \frac{2L}{c}\frac{1}{\sqrt{1 - (v/c)^2}} \qquad\qquad [6.1\text{-}6]$$

Substituting $2L/c = \Delta t'$ yields

$$\Delta t = \Delta t'\frac{1}{\sqrt{1 - (v/c)^2}} \qquad\qquad [6.1\text{-}7]$$

which means that to the observer in S the (moving) clock in S' *runs slow*, a phenomenon called *time dilation*.

Consider again the two endpoints, A and B, of the object, and their coordinates, x_A and x_B. In the S frame, these coordinates determine the proper length, L_0, of the object. Then let the observer in S' move, at velocity v, through a distance equal to L_0:

$$L_0 = x_B - x_A = v\,\Delta t \qquad\qquad [6.1\text{-}8]$$

But to the observer in S', it takes time $\Delta t'$ for the object to move past. Its length, to that observer, is

$$L' = v \, \Delta t' \qquad\qquad [6.1\text{-}9]$$

Now we solve Equation [6.1-7] for $\Delta t'$, substitute it in Equation [6.1-9], and set $v \, \Delta t = L_0$. This gives

$$\boxed{L' = L_0 \sqrt{1 - \left(\frac{v}{c}\right)^2}} \qquad\qquad [6.1\text{-}10]$$

which is the *FitzGerald–Lorentz length contraction*.*

Evidently, an object has its greatest length when seen in its rest frame (where its velocity is zero). When seen from another frame, its length, measured in a direction parallel to the motion, becomes less by a factor $\sqrt{1 - (v/c)^2}$. In a direction normal to the direction of motion, the dimensions do not change; they are the same to both observers.

Does the object *really* become shorter? No, it retains its proper length for an observer flying along with it in the S' frame. But to an observer in S the object not only *appears* to be shorter, it really *is* shorter. The length contraction is not an illusion; it is real for an observer not moving along with the object.

If we apply the contraction factor to the x dimension only, and leave y and z unchanged, we obtain a new set of equations:

$$\boxed{\begin{aligned} x' &= \frac{x - vt}{\sqrt{1 - (v/c)^2}} \\ y' &= y \\ z' &= z \end{aligned}} \qquad\qquad [6.1\text{-}11]$$

called *Lorentz' space coordinate transformation*.

We now look at frame S from S' and reverse the direction of the velocity v. This gives

---

*Named after George Francis FitzGerald (1851–1901), Irish physicist and professor of natural philosophy at Trinity College in Dublin, and Hendrik Antoon Lorentz (1853–1928), Dutch physicist and professor of mathematical physics at the University of Leiden. FitzGerald and Lorentz, working at the time of intense search for the aether, used their contraction to explain the Michelson–Morley experiment (page 535). H. A. Lorentz, "De relatieve beweging van de aarde en den aether," *Versl. Zitt. Wis- en Natuurkundige Afdeeling Koninklijke Akad. Wetenschappen, Amsterdam* **1** (1892), 74–79.

$$x = \frac{x' + vt'}{\sqrt{1 - (v/c)^2}} \qquad [6.1\text{-}12]$$

Next we write the first equation in [6.1-11] in the form of

$$x - vt = x'\sqrt{1 - (v/c)^2}$$

and combine it with Equation [6.1-12] to eliminate x,

$$\frac{x' + vt'}{\sqrt{1 - (v/c)^2}} = vt + x'\sqrt{1 - (v/c)^2} \qquad [6.1\text{-}13]$$

Solving for t gives first

$$vt = \frac{x' + vt'}{\sqrt{1 - (v/c)^2}} - x'\sqrt{1 - (v/c)^2}$$

$$= \frac{x' + vt' - [x' - x'(v/c)^2]}{\sqrt{1 - (v/c)^2}}$$

and then

$$\boxed{t = \frac{t' + (vx'/c^2)}{\sqrt{1 - (v/c)^2}}} \qquad [6.1\text{-}14]$$

which is *Lorentz' time coordinate transformation,* quite different from Galileo's expression $t = t'$. Clearly, at speeds small compared to c, the Lorentzian equations reduce to the (approximate) Galilean equations.

Optics and the Special Theory of Relativity

Headlight effect.   Now we come to the specific applications of the theory of relativity to optics. When a firework explodes in the sky, it scatters sparks uniformly in all directions. But when the firework explodes while in rapid motion, it scatters most of its sparks in the forward direction.* It is much the same with a source of light except that, from Einstein's second postulate, the velocity of the "sparks" cannot exceed c.

Consider an isotropic point light source (which, by definition, emits light uniformly in all directions). Then imagine a screen with a hole in it that is placed over the source so that one half of the light goes out to the left of the screen, the other half to the right (Figure 6.1-3, left). With the source at rest, the screen is flat; it coincides with the y axis and the angle subtended by the screen and the $+x$ axis, θ, is 90°.

---

*As quoted from W. Rindler, *Essential Relativity: Special, General, and Cosmological,* 2nd edition, p. 260 (New York: Springer-Verlag, 1977).

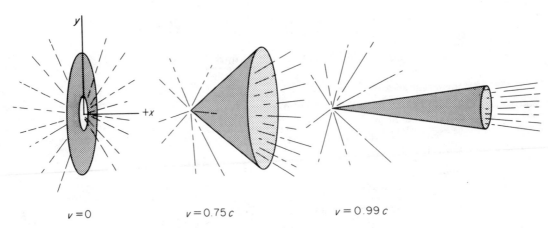

$$v=0 \qquad\qquad v=0.75\,c \qquad\qquad v=0.99\,c$$

Figure 6.1-3 Light emitted from moving source becomes concentrated in the forward direction.

Now let the source and the hypothetical screen travel to the right at a velocity v. Without stating proof, the screen folds into a *cone* (with $\theta < 90°$) but it still divides the light into equal parts, one half emitted *outside* of the cone and the other half *inside* of it. A photon traveling along the cone's surface covers, per unit time Δt, a distance $c\,\Delta t$. The projection of $c\,\Delta t$ on the $+x$ axis is Δx. But this is also the distance the *source* travels, $\Delta x = v\,\Delta t$. The cosine of the apex half-angle is the ratio of these two distances:

$$\cos\,\theta = \frac{v\,\Delta t}{c\,\Delta t} = \frac{v}{c} \qquad\qquad [6.1\text{-}15]$$

As the source moves faster (Figure 6.1-3, center and right), the cone becomes more narrow and the light more concentrated in the forward direction, a result that gave the phenomenon its name, *headlight effect.*

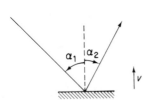

Figure 6.1-4 With the mirror moving toward the source, the angle of reflection becomes less, $\alpha_1 > \alpha_2$.

Reflection of light on a moving mirror.

When a mirror moves in a direction parallel (tangential) to its surface, nothing much happens. When the mirror moves normal (perpendicular) to its surface, several things change. First consider the *angle of reflection.* As the mirror moves *toward* the source, the angle becomes *less* (Figure 6.1-4). If, as before, α_1 is the angle of incidence, α_2 the angle of reflection, and v the velocity of motion of the mirror, then*

---

*Following W. Rindler, *Introduction to Special Relativity,* p. 54 (New York: Oxford University Press, 1982).

$$\boxed{\frac{\tan \frac{1}{2}\alpha_1}{\tan \frac{1}{2}\alpha_2} = \frac{c + v}{c - v}} \qquad\qquad [6.1\text{-}16]$$

If $v = 0$, clearly, $\alpha_1 = \alpha_2$. This is the conventional case, the first law of reflection in an inertial frame. As the mirror *recedes* from the source, the angle of reflection becomes *larger* than the angle of incidence and may even exceed 90°.

The same conclusions can be drawn from a construction of wavefronts. As a wavefront, inclined at a certain angle, approaches the mirror, one edge of the wavefront reaches the mirror earlier than the opposite edge does. But this makes the mirror *appear* (to the light) as if it were tilted with respect to the real mirror and reflection on the now "virtual" mirror yields the same ray trace as that obtained from the inequality $\alpha_1 \neq \alpha_2$ and the "real" mirror.

Example. *What velocity is needed to observe the effect, assuming the angle of incidence is 45°?*

Solution. The most sensitive goniometer, that is, angle-measuring instrument, is the Michelson stellar interferometer. It can resolve angles as small as 0.022 arc sec, or $0.022/3600 \approx 0.000\,006°$. Then, setting $\alpha_2 = 45° - 0.000\,006°$ and using Equation [6.1-16], we have

$$\frac{0.414\,213\,562}{0.414\,213\,501} = 1.000\,000\,147 = \frac{300\,000\,000 + v}{300\,000\,000 - v}$$

$$300\,000\,000 + v = 300\,000\,044 - 1.000\,000\,147\,v$$

so that

$$v \approx \frac{44}{2} = \boxed{22 \text{ m s}^{-1}}$$

a surprisingly low velocity.

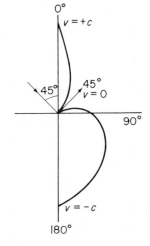

Figure 6.1-5 Relationship between mirror velocity and angle of reflection for $\alpha_1 = 45°$.

The relationship between the mirror velocity and the $\alpha_1 \neq \alpha_2$ inequality can be presented in the form of a three-dimensional plot. In Figure 6.1-5, for example, the angle of incidence is again 45° and, for $v = 0$, the angle of reflection is 45° also. All angles of reflection possible, from zero through 180°, are represented by points on the curve shown. The distance from the center of the coordinate system to a given point on the curve is a measure of the velocity of the mirror (*toward* the source in the upper half, *away* from it in the lower half) that produces such angles of reflection. Furthermore, if the angle of incidence is varied along a third coordinate, normal to the plane of the figure, these plots fuse into a *three-dimensional indented surface*.

There is another effect that occurs on a moving mirror, in addition to the change of the angle of reflection. Light reflected from a moving mirror changes in *wavelength*. This is easy to understand. As the mirror is moving toward the source and *against* the flow of photons incident on it, the momentum of the mirror *raises* the energy of the reflected photons, increasing $h\nu$, thus increasing the frequency and reducing the wavelength (Figure 6.1-6, left). As the mirror moves away from the source and *with* the flow of the photons, the opposite happens and the wavelength increases (right). The two wavelengths, λ_1 for the incident light and λ_2 for the reflected light, are related, in the first order, as

$$\frac{\lambda_1}{\lambda_2} = \frac{1 + (v/c) \cos \alpha_1}{1 - (v/c) \cos \alpha_2} \qquad [6.1\text{-}17]$$

If $v = 0$, then $\lambda_1 = \lambda_2$, which, again, is the conventional case.

Doppler effect.   Light reflected from a mirror can be considered as coming from a virtual source located behind the mirror. Thus, as the mirror moves, the virtual source moves as well (at twice the velocity of the mirror) and the light is subject to the *Doppler effect*.

Let us call c the velocity of the wave emitted by the source, not necessarily the velocity of light, and x_0 the distance from the source, real or virtual, through which the wave travels during time interval Δt:

$$x_0 = c \, \Delta t$$

With the source at rest, the distance x_0 contains N waves of length λ_0,

$$x_0 = N\lambda_0$$

Thus

$$\lambda_0 = \frac{x_0}{N} = \frac{c \, \Delta t}{N} \qquad [6.1\text{-}18]$$

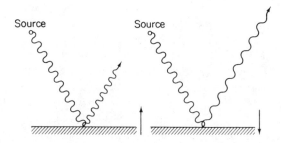

Figure 6.1-6   With the mirror moving toward the source, the wavelength of the reflected light becomes shorter (*left*); with the mirror moving away, it becomes longer (*right*).

Now let the source be moving at velocity v,

$$v = \frac{\Delta x}{\Delta t}$$

The distance x' contains the same number of waves but now of length λ':

$$x' = x_0 - \Delta x = N\lambda'$$

$$\lambda' = \frac{x_0 - \Delta x}{N} \qquad\qquad [6.1\text{-}19]$$

Dividing Equation [6.1-18] by [6.1-19] gives

$$\frac{\lambda_0}{\lambda'} = \frac{c\,\Delta t}{x_0 - \Delta x}$$

Substituting $x_0 = c\,\Delta t$ and $\Delta x = v\,\Delta t$, canceling Δt, and solving for λ' yields

$$\lambda' = \lambda_0\left(\frac{c - v}{c}\right) = \lambda_0\left(1 - \frac{v}{c}\right) \qquad\qquad [6.1\text{-}20]$$

and, setting $\lambda_0 - \lambda' = \Delta\lambda$,

$$\boxed{\Delta\lambda = \lambda_0\frac{v}{c}} \qquad\qquad [6.1\text{-}21]$$

which is the *classical, first-order Doppler effect.** With the source approaching, $\lambda' < \lambda_0$, which gives a "blue shift." Conversely, with the source receding, $\lambda' > \lambda_0$, which gives a "red shift."

Relativity has added a correction to the first-order Doppler effect, Equation [6.1-21]. The reason is that the source of any wave is by necessity an oscillator (a "clock") and, when moving, is subject to time dilation and *runs slow*. The distance $\Delta x = v\Delta t$ referred to before must then be modified according to Equation [6.1-7]:

$$x' = v\,\Delta t'\frac{1}{\sqrt{1 - (v/c)^2}}$$

---

*Named after Johann Christian Doppler (1803–1853), Austrian high school mathematics teacher, later professor of experimental physics at the University of Vienna. In his publication, "Ueber das farbige Licht der Doppelsterne und einiger anderer Gestirne des Himmels," *Abh. Königl. böhm. Gesellsch.* (Prag: Borrosch and André, 1842), p. 465, Doppler attributed the different colors of certain stars to their motion toward or away from Earth. He was wrong; the speed of light is so high that in order to change color, the stars would have to move at velocities too high even on an astronomical scale.

and Equation [6.1-21], for a receding source, becomes

$$\lambda' = \lambda_0 \frac{1 + v/c}{\sqrt{1 - (v/c)^2}} \qquad [6.1\text{-}22]$$

Expansion by the binomial theorem yields

$$\lambda' = \lambda_0\left[1 + \left(\frac{v}{c}\right) + \frac{1}{2}\left(\frac{v}{c}\right)^2 + \frac{1}{2}\left(\frac{v}{c}\right)^3 + \cdots\right] \qquad [6.1\text{-}23]$$

Note that the second-order term contains the square of v. Thus it is immaterial whether the source is approaching or receding. When the motion of the source is purely radial, along the line of sight, Equation [6.1-22] reduces to

$$\lambda' = \lambda_0 \sqrt{\frac{1 + v/c}{1 - v/c}} \qquad [6.1\text{-}24]$$

which describes the *longitudinal* (radial) *relativistic Doppler effect*.

In theory, the same relativistic correction applies also to the Doppler effect in *acoustics*. But sound emitters do not vibrate at as high velocities as oscillating atoms. In practice, therefore, the second-order term can be omitted.

If the direction of motion of the source subtends an angle θ with the direction of the light, a $\cos\theta$ factor must be added to the (v/c) term in the numerator of Equation [6.1-22]. But this means that even when the source moves at right angles to the line of sight; that is, when $\cos\theta = 0$ and Equation [6.1-22] becomes

$$\lambda' = \lambda_0 \frac{1}{\sqrt{1 - (v/c)^2}} \qquad [6.1\text{-}25]$$

there occurs a (red) shift of wavelengths. This is called the *transverse Doppler effect*, a purely relativistic phenomenon, due to time dilation of the moving source.

Example. *Certain double stars can be resolved by the Doppler effect even if they are too close together to be resolved otherwise. Spectra obtained from such* spectroscopic binaries *indicate their motion relative to Earth by characteristic blue and red shifts as illustrated in Figure 6.1-7.*
One of the largest red shifts on record is shown by the galaxy 3C123. There, lines normally found in the UV are shifted into the red, the actual figures indicating that this galaxy, the farthest out known, is moving away at a velocity of 0.637c, more than one-half the speed of light. Doppler broadening, of spectrum lines, is due to the random thermal motion of

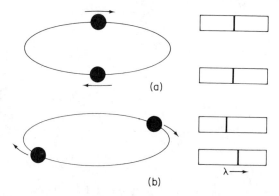

Figure 6.1-7 Schematic representation of the motion of a binary star system and of the Doppler shifts associated with it. (a) Sidewise motion, no first-order Doppler shift; (b) motion toward and away from observer produces blue and red shift.

molecules (in a gas). The result is a symmetric widening of emission lines, often up to several times their natural width.

The Michelson–Morley experiment.

Of all the interference experiments connected with relativistic optics the *Michelson–Morley experiment* is rightly the most famous.* A Michelson interferometer, as described on page 198, is set up so that one arm is parallel to the orbital motion of Earth and the other arm normal to it (Figure 6.1-8). The two arms are adjusted to equal length.

Now, while Earth moves, the beamsplitter and the two mirrors reach new positions, shown by the dashed lines. The time needed for the light to make a round trip in the upper (normal) arm, in analogy with Equation [6.1-6], is

$$t_\perp = \frac{2L}{\sqrt{c^2 - v^2}} \qquad [6.1\text{-}26]$$

In the parallel arm it is

$$t_\parallel = \frac{L}{c + v} + \frac{L}{c - v} \qquad [6.1\text{-}27]$$

---

*Albert A. Michelson, at the time of the experiment, was professor of physics at the Case School of Applied Science in Cleveland; Edward Williams Morley (1838–1923) was professor of chemistry at Western Reserve University, also in Cleveland. The pertinent publication is A. A. Michelson and E. W. Morley, "On the Relative Motion of the Earth and the Luminiferous Aether," *Phil. Mag.* (5) **24** (1887), 449–63. The Michelson–Morley experiment was later repeated with extraordinary care by Joos, with the same result. G. Joos, "Die Jenaer Wiederholung des Michelsonversuchs," *Ann. Physik* (5) **7** (1930), 385–407.

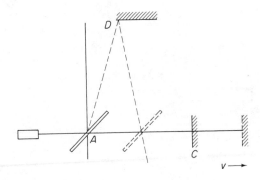

Figure 6.1-8 Michelson–Morley experiment. Note similarity to our "light clock" (Figure 6.1-2).

If we expand both Equations [6.1-26] and [6.1-27] by the binomial theorem and drop terms higher than $(v/c)^2$,

$$t_\perp = \frac{2L}{c}\left[1 + \frac{1}{2}\left(\frac{v}{c}\right)^2\right] \qquad [6.1\text{-}28]$$

and

$$t_\parallel = \frac{2L}{c}\left[1 + \left(\frac{v}{c}\right)^2\right] \qquad [6.1\text{-}29]$$

The interferometer is now turned through 90°. The total time difference, for the two positions, is then

$$\Delta t' - \Delta t = \frac{2L}{c}\left(\frac{v}{c}\right)^2 \qquad [6.1\text{-}30]$$

The quantity of interest is the optical path difference, Γ:

$$\Gamma = c\,\Delta t = 2L\left(\frac{v}{c}\right)^2 \qquad [6.1\text{-}31]$$

and since the number of fringes, m, is related to Γ as $m = \Gamma/\lambda$,

$$m = \frac{2L}{\lambda}\left(\frac{v}{c}\right)^2 \qquad [6.1\text{-}32]$$

In the actual experiment, L was 11 m, $\lambda = 590$ nm, and $v = 30$ km s^{-1}. This should have given a fringe shift of

$$m = \frac{(2)(11)}{590 \times 10^{-9}}\left(\frac{3 \times 10^4}{3 \times 10^8}\right)^2 = 0.37 \text{ fringe}$$

which would have been easy to see.

But *there was no fringe shift*. This result of the Michelson–Morley experiment, at that time, came as a complete surprise. Today we know that, according to the second postulate, the velocity of light is the same in all inertial frames and independent of the motion of the source.

Velocity of light in moving matter.

In 1818, Fresnel predicted* that light would be "dragged along" by a moving medium and that its new (phase) velocity, v, should be

$$v = \frac{c}{n} + w\left(1 - \frac{1}{n^2}\right) \qquad [6.1\text{-}33]$$

where c is the speed of light in free space, w the velocity of the medium, and n its refractive index. As we see from this equation, Fresnel assumed that the phase velocity of light, c/n, does not change by the full velocity w but only by a fraction thereof, $1 - 1/n^2$, called the *Fresnel drag coefficient*. The experiment was later performed by Fizeau[†] who confirmed Fresnel's prediction.

In Fizeau's experiment, shown in Figure 6.1-9, light from a source S is divided into two bundles by a beamsplitter B. One of the beams is traveling clockwise, deflected by three mirrors, and the other beam counterclockwise. The beams recombine at B and are observed through a telescope T, where interference fringes are seen.

The light passes through parts of a pipe filled with water. When the water is flowing as shown, one of the light beams is traveling in the direction of flow and the other opposite to the direction of flow. The fringes are then seen to *shift*.

Today we interpret Fizeau's experiment in terms of Lorentz' transformation. If we were to use Galileo's addition of velocities and simply add the two velocities, v' and w,

$$v = v' + w \qquad [6.1\text{-}34]$$

the numerical result would be too high. (This was, of course, why Fresnel needed his coefficient.) Instead we take Equation [6.1-12] and replace v in the denominator by the velocity of the water, w:

$$w = \frac{x' + wt'}{\sqrt{1 - (v/c)^2}} \qquad [6.1\text{-}35]$$

---

*A. Fresnel, "Sur l'influence du mouvement terrestre dans quelques phénomènes d'optique," *Ann. Chim. et Phys.* (2) **9** (1818), 57–66.

†H. Fizeau, "Sur les hypothèses relatives à l'éther lumineux, et sur une expérience qui parait démontrer que le mouvement des corps change la vitesse avec laquelle la lumière se propage dans leur intérieur," *Compt. rendu* **33** (1851), 349–55.

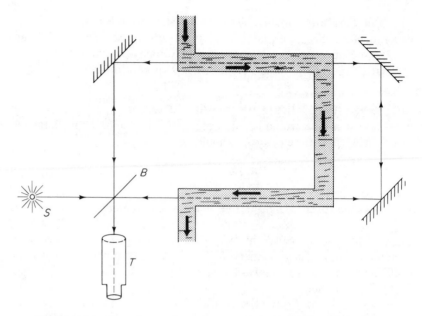

Figure 6.1-9 Fizeau's experiment for measuring the velocity of light in running water.

In Equation [6.1-14] we make the same change in notation:

$$t = \frac{t' + (wx'/c^2)}{\sqrt{1 - (v/c)^2}} \qquad [6.1\text{-}36]$$

Dividing Equation [6.1-35] by [6.1-36] gives

$$v = \frac{x}{t} = \frac{x' + wt'}{t' + \dfrac{wx'}{c^2}} = \frac{\dfrac{x'}{t'} + w}{1 + \dfrac{wx'/t'}{c^2}}$$

so that

$$v = \frac{v' + w}{1 + \dfrac{v'w}{c^2}} \qquad [6.1\text{-}37]$$

which is *Einstein's velocity addition theorem.*
    Expansion by the binomial theorem gives

$$v = (v' + w)\left(1 - \frac{v'}{c}\frac{w}{c} - \cdots\right) = v' + w - \frac{v'^2 w}{c^2} - \frac{w^2 v'}{c^2} - \cdots$$

The last term, a second-order quantity, may be dropped. Then we substitute $v' = c/n$, and $v'/c = 1/n$, and obtain

$$v = \frac{c}{n} + w\left(1 - \frac{1}{n^2}\right)$$

exactly the same as Fresnel's prediction, Equation [6.1-33], without assuming any "drag" or "aether."

Rotating systems.

A good example of a rotating system is *Sagnac's interferometer*. As in Fizeau's experiment, one of the beams travels clockwise and the other counterclockwise (Figure 6.1-10). The whole apparatus is mounted on a rigid support that can be rotated about a vertical axis. The rotation causes one beam to travel farther than the other so that its frequency falls. The frequency of the other beam correspondingly rises. The result, again, is a fringe shift.

For simplicity assume the path to be circular. We call v the tangential velocity, $v = R\omega$, where R is the radius of the path and ω the angular velocity. Then the time it takes the light to travel around in one direction is $t_1 = 2\pi R/(c + v)$, and $t_2 = 2\pi R/(c - v)$ in the other. The difference between the two times is

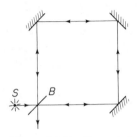

Figure 6.1-10 Sagnac's interferometer.

$$\Delta t = t_1 - t_2 = 2\pi R\left(\frac{1}{c + v} - \frac{1}{c - v}\right)$$

$$= \frac{4\pi R v}{c^2}\frac{1}{1 - (v/c)^2} \approx \frac{4\omega A}{c^2} \qquad [6.1\text{-}38]$$

where A is the area enclosed by the path. Since $c = m\lambda/\Delta t$, this time difference is seen as a fringe shift of

$$m \approx \frac{4\omega A}{c\lambda} \qquad [6.1\text{-}39]$$

which is *Sagnac's formula*. The shape of the loop and the index of the medium play no role. The same principle is the basis for modern *laser gyroscopes*.*

---

*Georges M. M. Sagnac (1869–1928), French physicist, professor of physics at the University of Lille, later at Paris. In one of his experiments, Sagnac used a square path about 1 m on each side. It took a velocity of 120 rpm to see the fringes shift but Sagnac, who retained a lifelong dislike for relativity, attributed the result to the "aether wind." G. Sagnac, "L'éther lumineux démontré par l'effet du vent relatif d'éther dans un interféromètre en rotation uniforme," *Compt. rendu* **157** (1913), 708–10.

Some time later, Michelson and Gale, in order to detect the lesser angular velocity of Earth, used a longer path, enclosing an area 339 × 613 m², near the present O'Hare airport in Chicago. A. A. Michelson and H. G. Gale, "The Effect of the Earth's Rotation

Miscellaneous other effects.   A special case of light coming from a moving source is that of *Vavilov–Cerenkov radiation*. The light is emitted by particles that move rapidly through a medium at a velocity higher than the phase velocity c/n in that medium. The result is a faint, airy blue light, easily seen in swimming-pool nuclear reactors.

Snell's law changes in a way similar to the reflection of light on a moving mirror. When a rare-to-dense boundary moves toward the source, the angle of refraction becomes smaller than predicted by the conventional form of Snell's law, and when the boundary moves away from the source, the angle becomes larger.

Clerk Maxwell's equations also must be rewritten, using four-dimensional tensor calculus, to accommodate the time component needed due to the limitations of simultaneity.*

The *mass* of a particle moving at relativistic speeds increases, inversely proportional to the Lorentz factor:

$$m' = m_0 \frac{1}{\sqrt{1 - (v/c)^2}} \qquad [6.1\text{-}40]$$

a fact that has been amply confirmed by a great many particle acceleration experiments. If we expand Equation [6.1-40] and drop terms higher than the second order, then

$$m' \approx m_0 + \frac{\frac{1}{2}m_0 v^2}{c^2} \qquad [6.1\text{-}41]$$

which shows that the inertial mass of a moving particle exceeds its proper mass (rest mass) by $1/c^2$ times its kinetic energy, $KE = \frac{1}{2}mv^2$. Consequently, energy appears to *add* to the mass by

$$E = mc^2 \qquad [6.1\text{-}42]$$

which is again Einstein's mass–energy relation that we used before, without proof, on page 460.

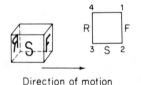

Direction of motion

Figure 6.1-11 Transparent cube with lettering R–S–F seen in perspectivic view (*left*) and from the top (*right*).

Terrell rotation.   Think of a cube, made out of transparent material and moving to the right as shown in Figure 6.1-11. The vertical edges of the cube are numbered 1, 2, 3, 4 and the front, side, and rear faces carry the letters F, S, and R, respectively. The width (length) of the cube

on the Velocity of Light," *Astrophys. J.* **61** (1925), 137–45. Today's gyroscopes are often fully integrated and have many turns of tightly wound optical fiber. See, for example, BMS, "Sensitive fiber-optics gyroscopes," *Physics Today* **34** (Oct. 1981), 20–22.

*Details are found in V. A. Ugarov, *Special Theory of Relativity*, pp. 180–257 (see the Suggestions for Further Reading).

is L_0 and its velocity v. We are looking at the cube from the side, at a right angle to its direction of motion, viewing the letter S.

If the cube is at rest, face S appears as an exact square (Figure 6.1-12, left). The rear face, R, between edges 3 and 4, cannot be seen. But then assume that the cube moves at high speed and that we look at it from the same direction as before or (which is easier to see) that we take a photograph of it using a camera with a high-speed shutter. This requires the photons that build up the image to arrive on the retina (or on the photographic film) *simultaneously*. But simultaneous *arrival* precludes simultaneous *emission* because edge 4 is farther away than edge 3. Thus light from edge 4 must have left edge 4 when it was still at 4′, to the left of 4. This makes the cube appear as if it had turned, an effect which I will call *Terrell rotation*.*

In addition, the side face, but not the front or rear face, is subject to Lorentz contraction (bottom right). Thus the total apparent width of the cube moving at

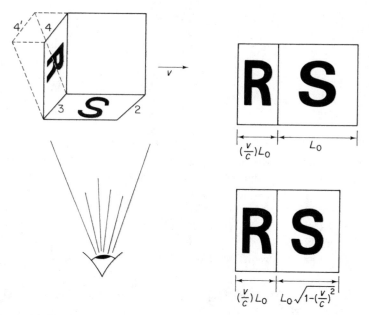

Figure 6.1-12 Cube moving to the right with velocity v and seen by observer looking at side face S.

___

*J. Terrell, "Invisibility of the Lorentz Contraction," *Phys. Rev.* **116** (1959), 1041–45. V. F. Weisskopf, "The Visual Appearance of Rapidly Moving Objects," *Physics Today* **13** (Sept. 1960), 24–27.

velocity v is

$$\Sigma L' = L_0 \frac{v}{c} + L_0 \sqrt{1 - \left(\frac{v}{c}\right)^2} \qquad [6.1\text{-}43]$$

At rest, $v = 0$, and at the velocity of light, $v = c$, the total apparent width is L_0. At intermediate velocities, the total width will be *greater* than L_0. In order to find the velocity at which maximum elongation occurs, we write Equation [6.1-43] in the form of

$$y = \left(\frac{L_0}{c}\right)v + \frac{L_0}{c}\sqrt{c^2 - v^2}$$

and differentiate with respect to v:

$$\frac{dy}{dv} = \frac{L_o}{c} + \frac{L_0(-2v)}{2c\sqrt{c^2 - v^2}} \qquad [6.1\text{-}44]$$

If we then set the right-hand term equal to zero, we find that

$$v = \sqrt{c^2 - v^2} = \frac{c}{\sqrt{2}} \approx 0.71c \qquad [6.1\text{-}45]$$

The second derivative,

$$\frac{d^2y}{dv^2} = -\frac{L_0}{c}\left[\frac{c^2}{(c^2 - v^2)^{3/2}}\right] \qquad [6.1\text{-}46]$$

is negative at $v = c/\sqrt{2}$, which shows that at that velocity there ,is indeed a maximum.

If the cube is seen from a direction other than from the side, the results are even more startling. At certain angles θ, subtended by the forward direction and the line of sight, and above certain threshold velocities, the front face becomes reverted and an F appears as ꟻ. Below the threshold the front face is seen as F, and the R is reverted, Я. At velocities higher than $v = c\cos\theta$, the front face is reverted, ꟻ, and the R is seen correctly. In summary, Terrell rotation and other distortions make an object look different than expected from Lorentz contraction alone.

Optics and the General Theory of Relativity

Newton's law of gravitation tells us that the force F causing acceleration is given by

$$F = G\frac{m_1 - m_2}{R^2} \qquad [6.1\text{-}47]$$

where G is the universal constant of gravitation, $6.672 \times 10^{-11}\,\text{N m}^2\,\text{kg}^{-2}$,

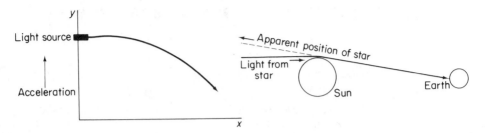

Figure 6.1-13   Bending of light under acceleration (*left*) and in a gravitational field (*right*).

and R is the distance between two masses, m_1 and m_2. But does a force between masses act instantaneously? If it did, this would violate the second postulate that no velocity can be higher than the speed of light. *Gravity waves* would solve the problem but, although predicted by theory, they have not been detected, as yet.

According to the *general theory of relativity*, the effects of acceleration cannot be distinguished from the effects of a gravitational field. This holds for light as well as for conventional masses.* Just as a projectile moves through a gravitational field in the form of a parabola, so does light, in an accelerated frame, follow a *curved path* rather than a straight line (Figure 6.1-13, left).

Likewise, light traveling through a gravitational field follows a curved path (right). In reality, though (if there is such a thing as "reality"), the path of the light may be straight and the space near the mass *warped*. For light passing by near the edge of the sun, Einstein predicted a deflection of 1.745 arc sec. Indeed, stars close to the solar disk were found to be displaced 1.70 ± 0.10 arc sec away from the sun, a brilliant confirmation of Einstein's theory.

> Sir Arthur Eddington, British astronomer, was once asked: "Is it true, Sir Arthur, that you are one of three men in the world who understands Einstein's theory of relativity?" The astronomer appeared reluctant to answer. "Forgive me," said the questioner, "I should have realized that a man of your modesty would find such a question embarrassing." "Not at all," said Eddington, "I was just trying to think who the third could be."

---

*A. Einstein, "Über den Einfluss der Schwerkraft auf die Ausbreitung des Lichtes," *Ann. Physik* (4) **35** (1911), 898–908.

SUGGESTIONS FOR FURTHER READING

A. Einstein, *The Meaning of Relativity,* 5th edition, (Princeton, NJ: Princeton University Press, 1955).

R. Resnick, *Introduction to Special Relativity* (New York: John Wiley & Sons, Inc., 1968).

W. Rindler, *Introduction to Special Relativity* (New York: Oxford University Press, Inc., 1982).

W. Rindler, *Essential Relativity: Special, General, and Cosmological,* 2nd edition (New York: Springer-Verlag, 1977).

V. A. Ugarov, *Special Theory of Relativity* (Moscow: Mir Publishers, 1979).

T. P. Gill, *The Doppler Effect* (New York: Academic Press, Inc., 1965).

PROBLEMS

6.1-1. A passenger walks forward in the aisle of a train at a brisk pace, 1.5 m s^{-1}. If the train moves along a straight track at 80 km/h, how fast is the passenger moving with respect to ground?

6.1-2. A gun with a muzzle velocity of 20 m s^{-1}, mounted on a tank, is aimed at an angle of $45°$ with the forward direction. If the tank moves at 40 km/h, what is the velocity of the projectile as it leaves the barrel?

6.1-3. If a 15.3-m-long space vehicle passes a planet at a velocity of 6×10^7 m s^{-1}, how long does the vehicle appear to an observer on that planet?

6.1-4. How fast must a meter stick travel for an observer to conclude that it is only one-half as long?

6.1-5. A linear object, subtending an angle of $45°$ with the direction of motion, travels at $0.8c$. What is the angle in the S frame?

6.1-6. If the mean half-life of pions in the laboratory frame is 1.78×10^{-8} s, what is their half-life when traveling at $0.9c$?

6.1-7. What is the (total) apex angle of a hypothetical cone equally dividing the light emitted by a point source traveling at 99 percent of the speed of light?

6.1-8. If a point light source moves at $0.9c$, what is the solid angle into which one-half the energy is emitted? (See page 381 for solid-angle conversion.)

6.1-9. Light is incident at an angle of $40°$ on a plane mirror moving toward the source at a velocity of $0.15c$. Find the angle of reflection.

6.1-10. A plane mirror is receding in a direction normal to its surface at a velocity of one-half the speed of light. Show how the light is reflected if it is incident on the mirror at an angle of:
(a) $30°$.
(b) $60°$.

6.1-11. If in Problem 6.1-9 the incident light has a wavelength of 500 nm, what is the wavelength of the reflected light?

6.1-12. If microwaves (radar) are reflected off an automobile traveling at 108 km/h, what is the relative change of frequency, in terms of (v/c), of the return signal?

6.1-13. A certain star shows a first-order Doppler shift of 1.4 Å involving the red (656 nm) hydrogen line. How fast does the star move relative to Earth?

6.1-14. If a distant star is receding from Earth at a velocity of 2.4×10^7 m s^{-1}, by how much would the 656 nm hydrogen line shift if we consider:
(a) Only the first-order Doppler effect?
(b) The total, relativistic Doppler effect?

6.1-15. The spectrum of a distant nebula shows the 434 nm hydrogen line to have shifted to 752 nm. How fast is the nebula moving away from Earth?

6.1-16. How fast would you have to go through a red traffic light ($\lambda = 630$ nm) so that it appears green (520 nm)?

6.1-17. A simple way of introducing the Michelson–Morley experiment is by the *row-boat analogy*. A man is rowing through a distance of 100 m, at a velocity of $c = 0.2$ m s^{-1}, across a river flowing at $v = 0.1$ m s^{-1} (Figure 6.1-14).
(a) What is the effective velocity of the boat?
(b) What is the time needed to complete a round trip?

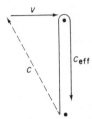

Figure 6.1-14

6.1-18. Continue with Problem 6.1-17 and now let the man row *parallel* to the river, downstream and upstream, covering the same distance as before. How much longer does this take as compared to rowing across and back?

6.1-19. Derive a general formula for the time delay, Δt, between the two beams of light in Fizeau's drag experiment, neglecting the length of pipe *not* traversed by the light and any small quadratic terms.

6.1-20. If the signal "dragged along" in Fizeau's experiment has the velocity of light, $v' = c$, what is the result of relativistically adding c to *any* velocity (less than c)?

6.1-21. Sagnac performed his experiment with light of 530 nm wavelength. How much of a fringe shift did he see?

6.1-22. A laser gyroscope contains a coil, 50 mm in diameter, of 100 turns of optical fiber of index 1.49. If the coil spins at 3600 rpm and $\lambda = 633$ nm, by how much will the fringes shift?

6.1-23. Vavilov–Cerenkov radiation can be compared to the shock wave set up by an aircraft flying at supersonic speed. Assume that the particles that produce the radiation travel at velocity v and that the "shock wave" is actually a wavefront. If the medium has an index n and the phase velocity within the medium, therefore, is c/n, what is the sine of the apex half-angle of the shock wave?

6.1-24. In which direction must a dense-to-rare boundary move so that obliquely incident light is not refracted at all?

6.1-25. To what velocity must electrons be accelerated in order to double their mass?

6.1-26. If a proton is accelerated to one-half the speed of light, by what percentage will its mass change?

6.1-27. A cube, carrying the letter F on its front face, is seen from an angle of 45° with the forward direction. If the cube moves at $v = 0.9c$, what does the F look like to a stationary observer?

6.1-28. Imagine looking at a cube of length (width) $L_0 = 1$ moving along at high speed. Plot:
(a) The width of the rear face of the cube as a function of its velocity, increasing from 0 to c.
(b) The width of the side face, for the same velocities.
(c) The total width of the cube.

6.1-29. An experiment like that shown in Figure 6.1-13, left, is set up inside a space laboratory that is accelerated at $10g$. What is the linear deflection of the light at the end of a 100-m-long path? (*Hint:* Find the deflection, h, from the equation that links h with the acceleration, a, and the time, t, by $h = \frac{1}{2}at^2$.)

6.1-30. Which of the optical effects discussed in this chapter are first-order effects and which are second-order, truly *relativistic* effects?

Well, dear reader, we have come to the end of the second edition of my *Introduction to Classical and Modern Optics*. How did you like it? Do you have any comments or criticisms? If you do, I would appreciate it if you would write them down and send them either to the publisher or to me. Many thanks!

Jurgen R. Meyer-Arendt

Answers to Odd-Numbered Problems

Chapter 1.1

1.1-1. 27 m
1.1-3. 8 m
1.1-5. 20 cm; 10 cm
1.1-7. 1.5 cm
1.1-9. 1.6
1.1-11. -6.25 m^{-1}
1.1-13. (a) -80 m^{-1}; (b) -5 m^{-1};
(c) -0.625 m^{-1}
1.1-15. 6.25 Hz
1.1-17. (a) 1697; (b) 5×10^{14} Hz

Chapter 1.2

1.2-1. 52.5°
1.2-3. 48.6°
1.2-5. 22°
1.2-7. (a) 1°; (b) 1.75$^\nabla$; (c) 1.6°
1.2-9. 3.3 mm
1.2-11. Blue inside; red outside
1.2-13. 0.5°
1.2-15. $-2.1°$

Chapter 1.3

1.3-1. Infinity
1.3-3. 1.6
1.3-5. -1.00 diopter
1.3-7. $+2.20$ diopters
1.3-9. (a) 15 cm; (b) 15 cm
1.3-11. -7.5 cm
1.3-15. -15.3 cm
1.3-17. $+20$ cm
1.3-19. (a) -22 mm; (b) $+220$ mm
1.3-21. -5 cm
1.3-23. $+6.9$ cm
1.3-25. (a) 120 cm; (b) 12 mm
1.3-27. 4.675 diopters
1.3-29. 1.578

Chapter 1.4

1.4-3. (a) Excess plus power; (b) the same
1.4-5. $+12.1$ diopters
1.4-7. 12.5 cm
1.4-9. $+7.5$ diopters and $+2.5$ diopters

1.4-11. $+3.2 \text{ m}^{-1}$; 0.12 m^{-1} less; 0.36 m^{-1} more

1.4-13. $33\frac{1}{3}$ cm

1.4-15. $-4.8°$

1.4-17. $x' = 5x_0 + 3y_0$; $y' = 4x_0$

1.4-19. $\begin{bmatrix} 1.02 & 5.2 \\ -0.004 & 0.96 \end{bmatrix}$

1.4-21. Infinity

1.4-23. $+20$ cm

1.4-25. $\begin{bmatrix} 0.5 & 22.5 \\ -0.02 & 1.1 \end{bmatrix}$

1.4-27. $+0.4$

Chapter 1.5

1.5-1. Concave; -12 cm

1.5-3. -6.25 cm

1.5-5. -25 cm; -15 cm

1.5-7. $+2$

1.5-11. $+2$ m

1.5-13. 91 cm

1.5-15. $35°$

1.5-17. $22.5°$

1.5-19. -20 diopters

Chapter 1.6

1.6-1. 1 mm

1.6-3. (a) $+3$; (b) -0.17

1.6-5. (a) -1; (b) $+1.053$; $+3.5$ cm; $+135.8$ cm

1.6-9. $8\frac{1}{3}$ cm

1.6-11. 12 mm

1.6-13. $+2$ diopters

1.6-15. $+11.7$ diopters

1.6-19. -4.3 diopters

1.6-21. (a) 4 cm; (b) 6 cm

Chapter 1.7

1.7-1. $f/2$

1.7-3. Both 0.4%

1.7-5. (a) $53.13°$; (b) $85.65°$

1.7-9. (a) Aperture stop; (b) $f/2.8$; (c) -7.5 cm

Chapter 1.8

1.8-1. (a) Taylor–Cooke triplet; (c) 43 mm

1.8-3. (b) Second image forward and smaller

1.8-5. $+1.2$ diopters; $+6$ diopters

1.8-7. (a) $16\times$; (b) 2 cm

1.8-9. $-10\times$

1.8-11. 1.5 cm away from objective

1.8-13. 6 cm

1.8-15. (a) $1.6\times$; (b) 18.75 mm

1.8-17. $-4\times$

1.8-19. (a) 1×4 versus 3×4; cylinders, axes vertical

1.8-21. $100\times$

1.8-23. (a) $-1000\times$; (b) $-50\times$

1.8-25. (a) 17.5 cm; (b) $\begin{bmatrix} -2.5 & -40 \\ 0.175 & 2.4 \end{bmatrix}$; (c) $-2.5\times$

1.8-27. -16.4 mm

Chapter 2.1

2.1-1. 15 m; 2.2 cm

2.1-3. $2 + 2i$

2.1-5. 1.16 cm

2.1-7. (a) 50 Hz; (b) 8 mm; (c) $0.02 \sin[100\pi(x/0.4 - t)]$

2.1-9. (a) 0.4 m; (b) 0.8 m s^{-1}; 1.6 m s^{-2}

2.1-11. (a) $+2.39$ cm; (b) -4.77 cm; (c) $y = 4.77 \sin(60°t + 30°)$

Chapter 2.2

2.2-1. 1.414 A

2.2-3. (a) 3.576; (b) $36.39°$

2.2-5. $y = \sin x(1 + \cos x)$

2.2-7. (a) 720 nm; (b) red

2.2-9. (a) 10.8 mm; (b) 9 mm

2.2-11. 8.1 mm

2.2-13. 2

2.2-15. The same but more intense

2.2-17. 776 638

2.2-19. 700 nm

2.2-21. 25

2.2-23. 5 fringes

2.2-25. 50

2.2-27. (a) 13.2 μm; (b) 4.4×10^{-14} s

2.2-29. 5 units

Chapter 2.3

2.3-1. 8.6 μm

2.3-3. (a) 0; (b) 1.273; 0.764; 0.347

2.3-5. (a) Refraction; (b) diffraction
2.3-7. 6; 10; 6
2.3-9. 1.8°
2.3-11. 44 arc sec
2.3-13. (a) 26.84×10^{-6} rad;
(b) 0.035 mm
2.3-15. 0.89 mm
2.3-17. (a) 0.067; (b) 1.280
2.3-19. (a) 2.25 m; (b) 75 cm; 45 cm
2.3-21. (a) 3.6 mm; (b) immaterial
2.3-23. Ratio 1:10

Chapter 2.4

2.4-1. 395
2.4-3. (a) 1.2 mm; (b) 2.4 mm
2.4-5. 3.29°
2.4-7. 4 and 5
2.4-9. $\lambda = (da)/m(a^2 + b^2)^{1/2}$
2.4-11. 0.2 nm
2.4-13. 1.735 Å
2.4-15. 29°

Chapter 2.5

2.5-1. 114.13 nm
2.5-3. 570 nm
2.5-5. 0.054 mm
2.5-7. (a) 4 black bands; (b) 4 spectra
2.5-9. 2.85 mm
2.5-11. Narrow and dark on a bright background
2.5-15. 38.2 cm
2.5-17. 4.9 mm
2.5-19. 5.4 nm
2.5-21. (a) Only a thin line is missing
2.5-23. 6.25%
2.5-25. 34%

Chapter 3.1

3.1-1. (a) -7.2×10^{-4} N;
(b) $+9 \times 10^{-5}$ N
3.1-3. 5.9 cm
3.1-5. (a) 1.5×10^{-7} T; (b) 0.12 A m^{-1}
3.1-9. 1 mg
3.1-11. 1:4.16; 1:17.3
3.1-13. (a) 50°; (b) 40°
3.1-15. (a) 0.43%; (b) 1.6%
3.1-17. (a) 5.3% and 0.4%; (b) 89%

3.1-19. (a) 1–2; (b) 1–2, 2–3; (c) 2–3;
(d) none

Chapter 3.2

3.2-3. $4LN\omega$
3.2-5. 529 rev s^{-1}
3.2-7. 2.99×10^8 m s^{-1}

Chapter 3.3

3.3-1. $\dfrac{1}{4\pi\varepsilon_0} \dfrac{p}{\left(R^2 + \dfrac{d^2}{4}\right)^{3/2}}$
3.3-3. 7.5
3.3-5. 2.8×
3.3-7. 10%
3.3-9. 20%

Chapter 3.4

3.4-1. Different ellipticities but same azimuth
3.4-3. Unpolarized; partially linear
3.4-5. (a) 52.2°; (b) 37.8°
3.4-7. (a) 50%; (b) 25%
3.4-9. 26.6°; 39.2°; 50.8°; 63.4°; 90°
3.4-11. 0.15 mm and multiples
3.4-15. (a) Linear horizontal; (b) no transmission
3.4-17. $\begin{bmatrix} 1 & 0 \\ 0 & -1 \end{bmatrix}$
3.4-19. 4.15 mm
3.4-21. 45°

Chapter 3.5

3.5-1. 2.6 mm
3.5-3. 62 cm
3.5-5. (a) 2.8; (b) immaterial
3.5-9. (a) 0.248 . . . ; (b) 14.37°

Chapter 3.6

3.6-1. 26 W
3.6-3. 0.1 sr
3.6-5. 1.38 kW
3.6-7. 24 lx

3.6-9. $25\times$
3.6-11. 22.5 s
3.6-13. 36 cd
3.6-15. No changes
3.6-17. 1.54 lx
3.6-19. (a) 150 cd; (b) 295 lx

Chapter 4.1

4.1-1. (a) 0.25; (b) 0.4
4.1-3. 0.27 Å
4.1-5. (a) 12.6 kHz; (b) yes
4.1-7. 2 Hz
4.1-9. $y = 2A \sin(\omega t) + A \sin(3\omega t)$

Chapter 4.2

4.2-1. 0.2°
4.2-3. (a) B–C; (b) black–white reversed
4.2-5. 40%
4.2-7. (a) Not much; (b) severe degradation; (c) not much

Chapter 4.3

4.3-1. (a) Nothing; (b) clockwise;
(c) spread out
4.3-5. (a) Vertical slit; (b) close to lens
4.3-7. 0.9 mm
4.3-9. (a) Aperture stop; (b) central stop

Chapter 4.4

4.4-1. 0.08 λ
4.4-3. 1.9 μm
4.4-9. Not too critical

Chapter 5.1

5.1-1. 500 nm
5.1-3. (a) 0.62 eV; (b) 4.96 eV
5.1-5. 652 nm
5.1-7. 725 km s^{-1}
5.1-9. 470 km s^{-1}
5.1-11. (b) 6.62×10^{-34} J Hz^{-1};
(c) 2.28 eV
5.1-13. 0.24 Å

5.1-15. 0.6 Å
5.1-17. 3.64 mm

Chapter 5.2

5.2-1. Yes; higher
5.2-3. 3.75 μm; 2.28 μm; 1.27 μm
5.2-5. 9.3 μm
5.2-9. 655 nm; red
5.2-11. (a) 102 nm; (b) 433 nm;
(c) 1.87 μm
5.2-13. 2.18×10^{6} m s^{-1}

Chapter 5.3

5.3-1. (a) 0.42; (b) 0.21
5.3-3. 36%
5.3-5. 45%
5.3-7. 33%
5.3-9. 3.87 m^{-1}
5.3-11. 1 m

Chapter 5.4

5.4-1. 2.87×10^{-19} J
5.4-5. 2×10^{5}
5.4-7. 2.3×10^{-20} J
5.4-9. 20%
5.4-11. (a, b) center; (c) on plane mirror
5.4-13. 77 cm
5.4-15. 1.5 kW mm^{-2}
5.4-17. 6.38 W cm^{-2}
5.4-19. (a) Not at all; (b) to the right;
(c) to the left
5.4-21. 15 W
5.4-23. 31 mJ cm^{-2}

Chapter 6.1

6.1-1. 85.4 km h^{-1}
6.1-3. 15 m
6.1-5. 59°
6.1-7. 16.2°
6.1-9. 30°
6.1-11. 390 nm
6.1-13. 64 km s^{-1}
6.1-15. $\frac{1}{2}c$

6.1-17. (a) 0.17 m s^{-1}; (b) 19.6 min

6.1-19. $[2Lw(1 - 1/n^2)n^2]/c^2$

6.1-21. 1/20

6.1-23. $1/[n(v/c)]$

6.1-25. 2.6×10^8 m s^{-1}

6.1-27. ⊐

6.1-29. 0.05 Å

Index